STRUCTURAL EQUATION GEOMETRY

THE INHERENT PROPERTIES
OF CURVES &
COORDINATE SYSTEMS

To Florence Kaplan

STRUCTURAL EQUATION GEOMETRY

THE INHERENT PROPERTIES OF CURVES & COORDINATE SYSTEMS

J. Lee Kavanau

University of California, Los Angeles

Science Software Systems, Inc.
Los Angeles, California

Copyright 1983 by Science Software Systems, Inc., 11899 West Pico Boulevard, West Los Angeles, California 90064. All rights reserved. No part of this book may be reproduced in any form, by mimeograph or any other means, without permission in writing from the publisher.

ISBN #0-937292-02-8

Library of Congress Catalogue Card Number 83-72244

Published in the United States of America by Science Software Systems, Inc.

INTRODUCTION

Though today's society increasingly requires mathematical, scientific, and technological skills and understanding, the competence of the average student in mathematics and the sciences has declined alarmingly in recent times. And the modern distractions responsible for this decline appear to be on the increase. As remedial measures, educators recently have recommended additional precollege courses in mathematics and the sciences, more homework, longer school years, more exacting graduation requirements, and the provision of better-qualified instructors (including "guest instructors" loaned by industry).

While these measures are highly desirable, they do not address a crucial underlying problem. This is the fact that the average student finds courses in mathematics to be unappealing, boring, or downright distasteful. This situation is not a product of the space age that can be rectified with stopgap measures; it existed long before the advent of modern distractions. Accordingly, it can be suggested that our first line of attack should be directed toward redressing this lack of appeal, and that other remedial measures will not become fully effective until this has been achieved.

Increasing the appeal of mathematics will require extensive revision and revitalization of elementary and intermediate courses. The discipline most in need of attention is Euclidean geometry. This subject not only provides the cornerstones for all fields of mathematics (and physical science) to which the average student subsequently becomes exposed, but impressions gained in elementary geometry courses are the major factor that shapes future attitudes toward other mathematics courses.

A main thrust of my admonitions is that the many fascinating features of elementary mathematics, particularly plane geometry, are neither entirely appreciated nor being used to their full potential in the classroom. In the final analysis, these deficiencies are the products of almost a century of comparative neglect of physical (classroom) geometry by research mathematicians. The little attention this subject has received from time to time has sought to harmonize it with conceptual advances at a level of sophistication and abstraction that is superfluous to the elementary classroom, rather than to make it more meaningful and accessible to the average student.

The Rise to Dominance of Abstract Methods

But neglect is far from being a novel phenomenon in the field of geometry. Of all areas of mathematics, this field has been most susceptible to changing tastes and the vagaries of fashion from age to age. Geometry reached its zenith in classical Greece as a synthetic physical discipline (pure geometry), but fell to its nadir at about the time of the fall of Rome. With the blossoming of the analytical approach, following the innovations of René Descartes (1596-1650) and Pierre de Fermat (1601-1665), geometry in the 17th century stood on the threshold of a new era, only to be all but ignored by research mathematicians for almost two centuries more, as it fell into the shadows of the ever-proliferating branches of analysis.

Geometry's sudden rediscovery and revival, and the ushering in of the "Heroic Age," came chiefly with the dawn of the 19th century, following the efforts of Charles Julien Brianchon (1785-1864) and, particularly, Jean-Victor Poncelet (1788-1867), the effective founder of projective geometry. But the early-to-mid 19th century proved to be a period of often bitterly contesting forces in physical geometry, with Jakob Steiner (1796-1863), generally regarded as the greatest geometer of modern times, effectively championing synthetic methods, and Julius Plücker (1801-1860), undoubtedly the most prolific of all analytic geometers, struggling to advance analytical approaches. In the end, both camps --in fact, the entire discipline of physical geometry--were eclipsed. Compelling new forces had come into play.

To trace the origin of the emergent, and currently prevailing, "Abstract Age," we must turn back to the first third of the 19th century and some of the most brilliant achievements in the field of mathematics. The stage was set, on the one hand, by Evariste Galois (1811-1832), with his theory of groups, and, on the other, by Nicolai Ivanovitch Lobachevsky (1793-1856) and János Bolyai (1802-1860), with their studies (circa 1830) in non-Euclidean geometry.

Building within the discipline of Galois' group theory, Felix Klein (1849-1925) and Sophus Lie (1842-1899) achieved results leading to Klein's remarkable general definition of "a geometry" in 1872, together with his advocacy of the *Erlanger Program*, a new, and ultimately exceedingly fruitful, program of abstract geometrical study.

Because the studies of Lobachevsky and Bolyai compelled geometers to adopt radically different viewpoints, they led to a profound re-examination of the foundations of geometry. The repercussions were so great that the foundations of virtually all fields of mathematics came under critical scrutiny. The most influential figure in the resulting postulational treatment of Euclidean geometry was David Hilbert (1862-1943). His *Grundlagen der Geometrie*, in 1899, broke down the last barriers to wide acceptance of the view of a purely hypothetico-deductive nature of geometry and firmly rooted the same method in virtually every branch of mathematical endeavor.

As a consequence of these achievements, the research interests of academic mathematicians turned increasingly from the concrete to the abstract, and from specifics to the greatest possible generalizations. The major efforts of Euclidean geometers now became channeled into foundation studies, with almost 1,400 articles devoted to this field alone from 1880 to 1910. This turn of events prompted Bertrand Russell's (1872-1970) facetious comment that "mathematics may be defined as the subject in which we never know what we are talking about, nor whether what we are saying is true."

With such concerted efforts along abstract avenues over a span of five generations, analytical research in physical Euclidean geometry inevitably fell into a state of neglect. Most twentieth-century mathematicians are exposed to this field only as students in beginning courses, in which analytic geometry often is covered only as needed to prepare them for the calculus. They usually gain the impression that analytic geometry, like arithmetic and trigonometry, is a "closed book." The few researchers whose efforts continue to touch this discipline employ highly abstract generalized approaches.

Revising and Rejuvenating Treatments of Analytic Geometry

This is the background against which the present treatments of analytic geometry must be viewed. Instead of being reinvigorated periodically, through continued scrutiny, reassessment, and revision by academic researchers, this fountainhead was all but cut off generations ago, leaving the subject virtually in limbo (with consequences that cannot be rectified by more course requirements and homework or better-qualified instructors).

An aim of the present work is to supply raw materials and suggest new approaches for the revision and rejuvenation of analytic geometry at all levels of instruction. The contributions of Chapters I and V-VIII are virtually entirely new. Coverage of the rectangular and polar coordinate systems (Chapters II and III) emphasizes interesting and conceptually-important aspects and relations that often go unmentioned in the classroom or escaped previous notice. Most of Chapter IV, *The Bipolar and Hybrid Polar-Rectangular Coordinate Systems*, also consists of new material. It includes the first in-depth analysis and elucidation of the bipolar system and its equations and curves, and the only explicit recognition and treatment of the hybrid polar-rectangular system, which plays a key role for derivations in most non-Cartesian coordinate systems.

Topics are presented in a form suitable for an advanced undergraduate text, including 155 Problems. Most of these also cover new ground. Specific major recommendations for course revisions, emphasizing applications to beginning courses, are given below and at the end of Chapter IV. Instructors of courses at all levels will find abundant additional pertinent subject matter in the main text.

A strength of the proposed materials and methods is that the same approaches that introduce greater rigor, systematic organization, and more penetrating insights, also provide much-needed substance to which students can better relate. Not surprisingly, these approaches run counter to the mainstream of the last century of mathematical endeavor, namely, in the direction of providing more concrete and specific material, rather than abstractions and generalizations.

The New Subject Matter

Details of how I came to re-examine the subject matter of Euclidean geometry are given in the Historical Epilogue of *Symmetry, An Analytical Treatment* (Kavanau, 1980). I found the field to be in so great a state of neglect that examining it from new viewpoints took on the aura of excitement and anticipation of unearthing and inspecting an ancient treasure, piece-by-piece. By default, physical Euclidean geometry had become the frontier to a vast unknown; almost any step from the beaten track trod on virgin ground. Even for the line and the circle--curves that presumably could hold no secrets--new properties were discovered. The following are some of the highlights of these new findings.

The Equilateral Hyperbola as a Cassinian Oval, and Self-Inversion at 90°

Nineteenth-century mathematicians were amazed that over 100 years elapsed after Giovanni Cassini (1625-1712) studied his famous ovals in 1680 before it was realized that the lemniscate of Bernoulli belonged to the same group. Much more surprising is the fact that an additional 180 years passed before the discovery that the common equilateral hyperbola also is a member of this group (see Fig. 20-1a and text). Yet this remarkable relation could have been guessed 100 years ago. Furthermore, confirming such a guess also would have led to the discovery of a new type of self-inversion, namely, self-inversion at 90°, in which Cassinian monovals inverted about their centers yield curves congruent with, but rotated 90° from, the initial curves.

Eccentricity as a Visible Property

Though the above topics are not commonplace, the next example concerns a subject discussed weekly in thousands of classrooms--the eccentricities of conic sections. The eccentricity of a conic section has been found to be a visible property, in much the same sense that a focus is visible. If the position of a focus is known, the eccentricity also can be illustrated; one simply draws or indicates a tangent-line to the curve at any of its points of intersection with a latus rectum (see illustration below). The magnitude of the slope of this tangent-line is the conic's eccentricity.

Structure Rules--New Analytical Tools

The above examples concern highly specific matters. The next finding has very much broader implications: a major new avenue for the analysis of curves and relations between curves has been discovered. Until a few years ago, virtually all the common equations of analytic geometry were coordinate equations or *construction rules* for curves. These are explicit rules for plotting curves in specified coordinate systems. However, they do not come to grips directly with structural properties of curves.

The recent work has disclosed powerful new tools for studying, characterizing, organizing, and classifying curves; their *structure rules* (the topics of Chapters VII and VIII). Structure rules are derived by employing only single reference elements, rather than the two or more reference elements of a coordinate system (see Figs. 40-42). Since these equations are independent of coordinate systems, they do not give specific information about how to construct curves, or the curves' appearances, but they do give explicit information about other aspects of structure.

The only familiar examples of these largely hidden aspects of structure are the classical symmetries about points and lines, which also are the simplest cases. Because structure rules take their simplest form for the classical symmetries, namely, expressing the equality of two distances, the corresponding underlying structures of the curves are readily perceptible and were the first to be discovered. However, even the most complex of the new structural features can be rendered visible by plotting their structure rules in the rectangular system. Most of the resulting *structural curves* are of unfamiliar appearance and previously were unknown (see cover illustration).

Structure Rules, Inherent Structure,
 and the Classification of Curves

Structure rules cast about single points in the planes of curves are preeminent for analysis. Their great diagnostic power derives from the fact that they describe and define the long sought-after *inherent* structural properties of curves. Classical attempts to probe these properties analytically led up a more or less blind alley to the "intrinsic equations." Like the latter equations, structure rules about points provide unique characterizations. But unlike the relatively sterile intrinsic equations, each *general structure rule* of a curve can be resolved or decomposed into the components or *specific structure rules* that contribute to the unique characterization (see Table 2). None of these general structure rules previously was associated with its curve, not even the 4th-degree structure rule of the line or the 6th-degree structure rule of the circle.

Structure rules about points make it possible to classify curves on the basis of the structural properties that they possess in common, and to construct hierarchical trees of relations, which are essential stepping stones to a deeper understanding of structure. The following is an example of the procedure for constructing a hierarchical tree: many known curves are characterized by a given specific structure rule, A, about a certain point; of these, a number also possess structure rule B about the same point, and some of the latter group also possess structure rule C; of the curves that possess structure rules A, B, and C about the given point, some also have points in their planes for which structure rule D applies, and a fraction of these, in turn, are characterized by structure rule E about the same second point, etc. Thus, large groups of curves whose members share single properties can be subdivided sequentially into smaller groups whose members share greater numbers of properties, leading, ultimately, to unique categorizations in which the identities of the single curves are implicit.

Structure-Rule Analysis and a Coherent
   Theory of Exceptional Points

The points in the planes of curves of greatest interest to pure and applied mathematicians, as well as to physicists, astronomers, and engineers, are the "exceptional" points, such as vertices, foci, double-points, points of infinite curvature, and the poles about which curves self-invert. In the past, points in these different categories were defined in diverse and essentially mutually-exclusive ways.

The incisiveness of structure-rule analysis about points is illustrated by the fact that it provides a common analytical foundation for detecting, defining, and hierarchically ordering points in all the above-mentioned categories (as well as exceptional point-continua, including lines of symmetry and certain asymptotes). Briefly stated, these exceptional points in the planes of curves are the locations for which the curves' structure rules have the lowest exponential degrees (see Tables 10-13). In other words, these are the points about which curves possess the greatest *structural simplicities*. The new coherent theory identifies sixteen such exceptional points and point-continua in the planes of ellipses and hyperbolas.

New Applications and Knowledge of
   Polar Equations

Some of the most productive and elegant analytical procedures of Structural Equation Geometry emerged as by-products of methods for deriving the structure rules referred to above. Among these are intriguing new uses for polar equations (see Chapters III, VII, and VIII). An example is the procedure for defining, and deriving the locations of, the foci of conic sections and Cartesian ovals; until now, this was done primarily through constructions about two points or a point and a line. A lack of knowledge of these procedures frustrated classicists for over 200 years in their attempts to place the study of Cartesian ovals on a secure analytical footing.

Unfortunately, the polar coordinate system and the properties of its curves and equations have been no less neglected than other planar Euclidean systems. The paucity of attention is evident from the absence of a prior explicit treatment and definition of the exponential degrees of polar equations; nor had any prior attention been given to the standardization of equations in the polar system. Instead, polar equations merely were used in their simplest possible forms for the conveniences these forms conferred. In consequence, there has been only a vague appreciation of the significance of the great differences that can exist between the polar equations of differently-positioned congruent curves--a matter that goes to the heart of Structural Equation Geometry. These topics receive detailed consideration in Chapter III.

The Devil's Curve as a Construction from Conics

The Devil's Curve, first studied in the mid-18th century by G. Cramer, is much employed as an example in analytic geometry, for example, in the study of asymptotes. It also finds uses in presenting the theory of Riemann surfaces and Abelian integrals. However, as used in the past, the curve was known only through its equation; close affinities with conic sections were unknown and unsuspected. Structural Equation Geometry revealed the Devil's Curve to be the product of a simple construction from the two best-known central conics, the circle and the equilateral hyperbola. Thus, a useful curve of previously unknown origin now can be derived from common classroom curves and discussed in terms that beginning students can understand.

The most intriguing aspect of the construction of the Devil's Curve is that it is the unique case in which reciprocal derivations yield identical structure rules. Thus, the curve is generated either as the structure rule of an initial circle, overlapping a centered equilateral hyperbola as reference element, or as the structure rule of an initial equilateral hyperbola, overlapping a centered circle as reference element (see Fig. 42e).

Quartic Equivalents of Central Conics

The next example concerns a heretofore unexplored topic of compelling interest: discovering the 4th-degree equivalents of central conics. Based on very close analytical homologies, these have been identified as the *central Cartesians* (see Fig. 22), whose only previously known representatives are Cassinian ovals. The host of engaging known properties of central conics provide but a pale hint of the multi-faceted properties of their 4th-degree relatives. Some of the most compelling of these--all previously unknown--conclude Chapter VIII. In the domain of these properties alone, there probably remains to be discovered more than the totality of the known properties of conic sections.

A striking example concerns the inversion of central Cartesians to curves having true centers, a true center being the point of intersection of two orthogonal lines of symmetry. In all previously known instances of this phenomenon, the poles for such inversions were the true centers themselves, a feature that was assumed to be essential. For example, the equilateral hyperbola and the lemniscate of Bernoulli invert to one another about their true centers. Central Cartesians have proved to be far more versatile, with properties that greatly broaden our perspectives. The biovular members not only invert to three different curves that possess true centers--rather than only one--they do so about five different poles--rather than only about the true center.

Cartesian and Cassinian Ovals Related by Inversion

Despite the fact that Cartesian and Cassinian ovals are two of the most famous groups of 4th-degree curves, it was not realized until recently that Cassinian ovals are inversions of a subgroup of Cartesian ovals (the *parabolic Cartesians*), and vice versa. Accordingly, the immensely interesting central Cartesians mentioned above--the 4th-degree equivalents of central conics--also are the inverses of Cartesian ovals (see Fig. 22). As a concomitant of this finding, limaçons are found to stand in precisely the same relation to central conics as Cartesian ovals stand to Cassinians (and other central Cartesians). It seems evident from these findings that the names "Cartesian" and "Cassinian" are destined to play a much larger role in the future of physical geometry than they did in its past.

Unprecedented New Types of Curves

The new findings of greatest general interest emerged from a systematic study of curves in simple but previously unstudied coordinate systems consisting of two of the three elements--a point, a line, and a circle (the topics of Chapters V and VI). Even when we consider only unconventional manifestations of conic sections, a wealth of unprecedented new curves are commonplace in these other systems. These include: (a) two-arm hyperbolas and parabolas with arms of different sizes; (b) closed curves consisting of segments of dissimilar, as well as similar but different-sized, ellipses (including circles) and hyperbolas; (c) open and closed curves formed from segments of two or more parabolas; and even (d) open and closed curves that include segments of both ellipses and hyperbolas.

Directrices Defined

The same studies revealed that the general analytical significance of directrices extends far beyond their known applications in the construction of conic sections. Many other curves have elegantly simple constructions involving directrices, for example, limaçons and Cartesian ovals. RULE 7 of Chapter V defines a directrix-line of a curve as a line-pole for which there exists a point-and-line (or circle-and-line) construction rule that lacks a constant term. In this connection, many other interesting yet unexploited constructions of conics involve line-poles that are not directrices (see Chapters V and VI).

The New Coordinate System for Deriving Construction Rules

Inasmuch as the polar and rectangular coordinate systems are preeminent for almost all analytical studies, it did not come entirely as a surprise to find that a hybrid system that combines some of the advantages of each achieves preeminence over the parent systems for certain types of analysis. Thus, by using the polar r and the rectangular x, the *hybrid polar-rectangular coordinate system* results. This system provides the key equations for deriving construction rules of curves in coordinate systems that include a point or circle among their reference elements, and for identifying the "point-foci" of curves in these coordinate systems. The latter are the particular points about which the degrees of the construction rules of curves reduce from their values when cast about arbitrarily-located reference elements.

Although, in essence, the hybrid polar-rectangular coordinate system has been in use for hundreds of years, it was unrecognized. It is the implicit coordinate system of classical equations that relate distances from points of curves to arbitrary points in the plane--equations of great practical importance. This system has the potential to become the third-most important coordinate system for studying the properties of curves.

Inversion, Immediate Closure, and a "Natural" Scheme of Classification

A startling finding within an old field concerns a basic property of the inversion transformation (discussed below). It is no coincidence that inversion already has been mentioned several times. In my estimation, it stands alone among geometrical transformations--in beauty, in fascination, in utility, and in its many unique properties.

Inversion long ago proved to be of great use to the mathematician--in the theory of differential equations, in proving difficult geometrical theorems, in deriving "inverse theorems," and in classical geometrical constructions. Carathéodory (1873-1950) regarded inversive geometry as the best avenue of approach to the theory of functions, and H. A. Schwartz's (1843-1921) knowledge of this field provided the foundation for many of his celebrated successes. Inversive geometry also can serve in place of projective geometry, as a common foundation for Euclidean, spherical, and hyperbolic geometries (Wilker, 1981).

Inversion is no less important to the physicist and engineer. Some of its most notable applications occur in potential theory, the theory of elasticity, and the theory of special relativity. These depend upon such important features as the invariance of the n-dimensional Laplacian under inversion, the fact that inversion makes it possible to set up Green's function in closed form for an n-dimensional sphere, and most intriguingly, the fact that inversion brings singularities of functions at infinity into the finite plane, where they can be examined and dealt with.

In view of these many important and long-standing applications, one might have expected that the intensive studies of the inversion transformation by classicists had thoroughly plumbed its depths, and that it could hold no surprises in a fresh examination. Quite the opposite proved to be the case; many of inversion's basic properties, including those that are most beautiful and fascinating, proved to have been overlooked and unsuspected.

Attention here is confined to an astonishing new property of inversion that provides the foundations for the first "natural" paradigm for classifying curves. Consider any initial curve inverted about all points in the extended Euclidean plane (the Euclidean plane plus the ideal line at infinity), to yield a 1st-generation ensemble of inverse curves. The new property--termed *immediate closure*--is that if any one of these 1st-generation curves also is inverted about all points in the plane, no curve not already present in the 1st-generation is obtained. In fact, the ensemble of curves obtained by inverting any 1st-generation curve about all points in the plane possesses the same members as the 1st-generation ensemble itself.

The broad implications of immediate closure for classifying curves are evident. By virtue of this property of the inversion transformation, any given initial curve defines a unique closed group of related curves. No other type of transformation provides this facility. Though limitations of time and space have precluded treating inversion in detail here, this omission is offset by including (Appendix II) a selection of inversion's key mapping properties from *Curves and Symmetry*, vol. 1 (Kavanau, 1982), to which the reader is referred for detailed treatments.

Asymptotes of Inverse Curves

Inverting a curve about a point incident upon it gives rise to an inverse curve that possesses at least one asymptote. Classicists overlooked the elegantly simple and beautiful relationships that determine the number and locations of these asymptotes in the plane of the inverse curve, which provides the last example: *the asymptotes of inverse curves are the inverses of the circles of curvature of the initial curves at their points of inversion.*

Ultimately, these many new findings, new techniques of analysis, and broad new perspectives on the properties of curves and the ways in which curves are related to one another, can be expected to have far-reaching practical applications. At the moment, however, they are largely unknown to the mathematics community. This book presents the new results in a readily-assimilated form and directs attention to their most obvious, immediately-accessible, and urgent applications, namely those for the classroom.

New Approaches, Viewpoints, and Paradigms

Structural Simplicity--The Thread of Continuity

More attractive ways to present standard topics in elementary analytic geometry, and more incisive viewpoints and paradigms, were suggested in the course of studies in Structural Equation Geometry. The guiding theme at this early stage is to try to stimulate and develop the interests of *all* students in geometry. Only the most gifted and imaginative students usually are reached with traditional approaches.

One lays a foundation for achieving this goal by providing a thread of continuity, interest, and anticipation from topic-to-topic--giving students a view of the "forest," also, rather than only of the "trees." To achieve this, the concept of the *structural simplicity* of a curve is introduced. Structural simplicity is merely a logical extension of classical ideas of symmetry to embrace other aspects of structure, with emphasis on the fact that symmetry of position and symmetry of form are equivalent concepts. This equivalence has been placed upon a readily-grasped analytical footing by assessing the simplicities of position and form of curves in terms of the simplicities of corresponding equations.

Two examples drawn from construction rules help to clarify this approach. The first is a special case that will be somewhat familiar, since it involves differences between the equations of circles plotted at different locations in the rectangular coordinate system. Circles with the least structural simplicity (symmetry of position and form) and the most-complicated equations are centered at an arbitrary point in the plane. Circles with somewhat greater structural simplicity and slightly simpler equations are centered on a coordinate bisector. With each increase in the locational symmetry (and structural simplicity) of a circle, its equation simplifies. The simplest and most-symmetrically positioned circles, with the simplest equations, are centered at the origin. Of these, the point-circle (a circle of null radius) has the greatest structural simplicity and the simplest construction rule ($x^2 = -y^2$).

The second example represents the general situation. It concerns lines in the bipolar coordinate system, that is, lines in a system consisting of two point poles (p_u and p_v). The simplest and most-symmetrically-positioned bipolar line is the midline, which has the simplest 1st-degree construction rule, $u = v$. The next-simplest and most-symmetrically-positioned line is the mid-segment, that connects the two poles. It has a more complicated but still very simple 1st-degree equation. Less-simple and less-symmetrically-positioned lines, with slightly less-simple 1st-degree equations, are lateral rays that extend from the poles. These are followed by lines normal to the bipolar axis, with more-complicated equations of 2nd-degree, and the bipolar axis itself, with a still more-complicated equation of 3rd-degree. Lines with the least structural simplicity and symmetry of position are randomly-positioned and have the most-complicated equations, which are 4th-degree.

The concept of structural simplicity is emphasized by highlighting, at every step in the presentation of elementary analytic geometry, the correlations between the simplicities (or symmetries) of the positions and forms of curves and the simplicities of their corresponding construction rules (with respect to given reference elements).

Dimensional Uniformity of Equations

Breaching the barrier between the average elementary student and the use of algebraic equations is a primary objective. One approach to achieving this is to write equations in rigorous fashion. The familiar quadratic equation in the distance variable x, $Ax^2 + Bx + C = 0$, provides a convenient, though very simple, example. In this conventional form, even though each of the three coefficients, A, B, and C, has a different dimension, this multidimensionality is masked. Rewritten in dimensional balance, the equation becomes $Ax^2 + B^2x + C^3 = 0$ (or $A^0x^2 + Bx + C^2 = 0$), where the exponents of the capital letters, like those of the variable x, indicate their distance dimensions; the latter must sum to three (or two) for each term.

Maintaining dimensional uniformity of equations involves only a very modest change in the way they are written. Yet it provides a valuable new pathway for students to gain familiarity with equations, even without understanding fully what the equations mean. Students can immediately begin exercises and problems in balancing equations dimensionally and detecting dimensional imbalance. These are precisely the kinds of down-to-earth contacts needed in early stages to increase the appeal of working with equations. And the students who are likely to find these exercises most rewarding are the ones most in need of stimulation and encouragement.

Maintenance of dimensional uniformity of equations provides a valuable aid for detecting and tracing errors in algebraic derivations, particularly those of Structural Equation Geometry. Its importance was recognized by the eminent mathematical physicist Arnold Sommerfeld, who once took his colleagues to task for the "(bad) habit"--as he put it--of being careless with dimensional balance. The power of "dimensional analysis," an analytical approach that employs the principle of dimensional uniformity for the derivation of specific physical laws from general initial assumptions, is well known to engineers and physicists.

Using More Meaningful Terms than "Equation"

Another effective aid for the average student is to employ more specific and meaningful terms than "equation" wherever possible. When the word "equation" is used in classical analytic geometry, for example, it usually refers to a construction rule. Accordingly, the term "construction rule" should be employed in these cases. It is counterproductive to use a vague general term, to which the average student relates poorly, when a specific term that is meaningful to everyone is available. These examples direct attention to the fact that some of our common approaches to the teaching of mathematics are rooted in an uncritical acceptance of traditional practices.

Introducing and Comparing Coordinate Systems

Tradition also leads us to begin the study of coordinate systems with the rectangular system. Following this route, most students are introduced to three new concepts simultaneously, at a juncture where one would suffice. Thus, in addition to the new concept of a coordinate system, the average student also meets the quite unfamiliar ideas of measuring distances in a negative sense and measuring distances orthogonally from lines. Neither of the latter concepts is needed to introduce a coordinate system. All that is needed for this purpose is the idea of measuring distances from a point, with which even the least-informed students already are familiar.

Nor does any benefit result from introducing coordinate systems with the rectangular system. Not only is this system far from being the simplest conceptually, its properties are highly exceptional. And, though its most interesting and practical features are precisely those in which it is atypical--*degree-restriction*, *singularity of loci*, and *uniqueness of identity*--traditional presentations do not recognize these features as being exceptional; in fact, the level of awareness of them is so low that they were not even named previously. Since the rectangular system is introduced and employed in an entirely matter-of-fact manner, it fails to stimulate the interest and imagination of many students.

The Bipolar Coordinate System--The Introductory System of Choice

Because it combines the merits of being intriguing and the simplest coordinate system, the bipolar system (Chapter IV) is the best introductory system. Geometry instructors are familiar with its rudiments from the constructions of ellipses and hyperbolas about two foci. These constructions point the way to other productive uses.

Though the bipolar system is not a powerful analytical tool, it provides an ideal framework for introducing a coordinate system through the route of constructional puzzles. These are presented in common language for students to solve independently, with chalk and string, pencil, compass, and ruler. One begins with the simplest constructions. For example, what are the shapes and positions of the paths covered by a browsing rabbit when it moves in such a way as to:

(a) always be at equal distances from its two burrows, which are separated by a distance d (the construction rule $u = v$, where u and v are the distances from the two burrows);
(b) keep the sum of its distances from the two burrows equal to the distance between them ($u + v = d$);
(c) keep the difference between its distances from the two burrows equal to the distance between them ($u - v = d$); and
(d) always keep its distance from one burrow equal to twice its distance from the other one ($u = 2v$, the circle of Apollonius)?

Such simple puzzles can be relied on to stimulate interest while preparing the way for more intriguing constructions. These first exercises also introduce the idea of *simplest constructions* and present a line segment and a ray as the products of very simple construction rules, which is not possible in the rectangular system.

Familiarity with these simple constructions provides the foundation for a student to proceed to the next level of complexity--the conventional constructions for central conics. These are presented as logical extensions of the simplest constructions. Eccentricity is introduced in these extensions by modifying the construction rules given above to $u + v = d/e$ and $u - v = d/e$. For the former, letting $e = 1$ yields the connecting line-segment, letting e be greater than 1 yields no locus (introducing the concept of a construction rule that cannot be fulfilled), but letting e be less than 1 generates ellipses, whose eccentricities are defined by the values of e employed.

Similar considerations apply to the construction rule $u - v = d/e$, but now the rays "open" to become the arms of hyperbolas of eccentricity e. The larger the value of e, the closer the locus approximates to the midline; the smaller the value of e--but not less than 1--the closer each arm approximates to a ray; while for values of e less than 1, no construction exists.[The common definition of eccentricity, in terms of focus-directrix construction rules, also is conveniently introduced here, with the intriguing revelation that two entirely different types of constructions take simple forms when expressed in terms of e.]

Some may wish to limit their treatment to the 1st-degree construction rules of conic sections, others will want to show how simple modifications of linear construction rules lead to limaçons, and that the most complex 1st-degree construction rules produce Cartesian ovals. In any event, the simple 2nd-degree construction rule, $u^2 + v^2 = d^2$, should not be overlooked. Since this construction produces a circle, and is equivalent to the Pythagorean theorem, the resulting circle is most appropriately referred to as the *circle of Pythagoras*.

In concluding the treatment of the bipolar system, one draws attention to some key properties of general applicability: the line, ellipses, hyperbolas, etc., have 1st-degree construction rules only when located in the particular highly-symmetrical positions of the above constructions. All departures from these positions lead to more complicated construction rules that are of higher degree. Departures also produce curves with multiple segments, as a result of reflections in the coordinate line(s) of symmetry. These very unfamiliar circumstances are important because they are typical of the vast majority of coordinate systems; *in general, the more symmetrical the position of a curve with respect to coordinate elements, the simpler and/or lower the degree of its construction rule and the fewer the number of segments it possesses.*

The Rectangular and Polar Coordinate Systems

It is against this background that the rectangular system is best introduced --not in the present-day manner, as a "matter-of-fact" practical system, but as a most extraordinary one, with properties that differ remarkably from those of almost all other systems (as treated in detail in Chapter II): practically speaking, it is the only system in which the degrees of construction rules remain unaltered when curves are moved from symmetrical to asymmetrical positions --less-symmetrical positions merely require more complicated construction rules (see Tables 3 and 4). Moreover, the rectangular system makes possible simple expeditious analytical solutions to the previously studied bipolar "puzzles."

Unlike circumstances in the bipolar and most other coordinate systems, linear construction rules in the rectangular system cannot code for a multiplicity of curves; they generate only lines (an aspect of *degree-restriction*). Furthermore, only a single representative of a curve is produced, no matter where its position (*singularity of loci*), and all such representatives are similar to each other (*uniqueness of identity*). Following the usual introductory applications of the rectangular system, it is most instructive to derive the rectangular construction rules that correspond to the "puzzles" of the linear bipolar equations. This will illustrate the simplifications and reductions in degree that occur when the coefficients of u and v are of equal magnitude.

The same comparative approach is most effective for introducing the polar coordinate system--presenting it against the background of the properties of the bipolar and rectangular systems and highlighting its differences and similarities (as treated in detail in Chapter III). Treatments of transformations between rectangular and polar construction rules should emphasize the basis for the frequently differing exponential degrees of equations in the two systems, and for the existence of equations of different degrees for differently-positioned curves, correlating these degrees with the curves' positional symmetries (structural simplicities).

Other Coordinate Systems

The curves of the rectangular and polar coordinate systems are the conventional curves of classicists. The bipolar system adds types that are unconventional only in consisting either of multiple segments, by virtue of reflections, or only of single arms or ovals (as in the conventional bipolar construction of a one-arm hyperbola). When one combines in one coordinate system a line-pole of the rectangular system and a point-pole of the polar system, or either of them together with a circle-pole (Chapters V and VI) new factors come into play; extraordinary curves that are unprecedented in classical experience now become commonplace. Selected examples from these systems are useful at the high-school level to broaden the students' perspectives and provide additional constructions as puzzles.

Graphic Depiction of Eccentricity

After defining the eccentricity of a conic section analytically, it can be shown that this is precisely the magnitude of the slopes of tangents to a conic at the points where the latera recta intersect the curve. [This can be done without appealing to the calculus by translating the origin to one of these points and finding the value of the slope of a line through the same point for which the two intersections with the curve merge.] This sets the stage for the graphic illustration of eccentricity diagrammed below. This leaves a lasting impression on students, whereas the concept based solely on algebraic definitions is easily forgotten.

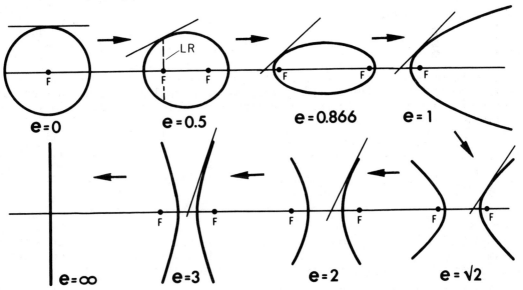

Characterizing Curves by Eccentricity

Eccentricity also provides an ideal, heretofore overlooked basis for identifying curves inverse to conic sections. The inversion transformation has the key property of preserving the magnitudes of angles, such as angles of intersection and tangent angles. Since each conic section can be characterized by an angle related to its eccentricity, for example, the angle between asymptotes, so also can any curve obtained by inverting a conic section. Accordingly, any curve inverse to a conic can be said to have the same eccentricity as the initial conic.

Since a number of well-known curves are the inverses of conic sections, they can be characterized by eccentricities. Some of these curves and their eccentricities are the cardioid ($e=1$), the cissoid of Diocles ($e=1$), the equilateral strophoid ($e=\sqrt{2}$), the equilateral limaçon ($e=\sqrt{2}$), the lemniscate of Bernoulli ($e=\sqrt{2}$), and the trisectrix of Maclaurin ($e=2$). Limaçons parallel conic sections in that their eccentricites also are the magnitudes of the slopes of tangents to the curves at the level of their double-point (see Fig. 19). In fact, it also was unrecognized that the eccentricity of the initial conic enters directly into the conventional polar equation of the inverse limaçon, in the form $r = a - b\cos\theta = a(1-e\cos\theta)$.

An example of the pertinence of identifying such curves in terms of eccentricity comes from comparisons of the structural simplicities of conic sections and limaçons. Employing the various criteria of Structural Equation Geometry, the conics and limaçons with the greatest structural simplicities fall into pairs with reciprocal eccentricities, notably, 0 and ∞, $\sqrt{2}/2$ and $\sqrt{2}$, 1 and 1, and 2 and 1/2.

Eccentricity and the Exceptional Lines of Conics

Once the topic of eccentricity has been introduced, students can be shown very simple relations involving eccentricity that are not now being used to full advantage. One, in particular, makes it easy to remember the positions of all the exceptional normal lines in the planes of central conics, namely, the minor and conjugate axes, the directrices, the vertex tangents, and the latera recta. In terms of e and a, these lie at the easily remembered distances from center of 0, $\pm a/e$, $\pm a$, and $\pm ae$, respectively (or, in units of a, 0, $\pm 1/e$, ± 1, and $\pm e$).

Systematic Nomenclature for Position

Another significant avenue for engaging the interest of students lies in employing systematic nomenclature for the locations and orientations of curves. The need for this was not felt in the past because there never was more than a vague appreciation of relations between a curve's position with respect to reference elements and the degrees and complexities of corresponding construction rules (the structural simplicities of the first example). Since these relations lie at the heart of Structural Equation Geometry, a systematic nomenclature for position becomes desirable. But, again, advantages in the classroom far outweigh practical needs.

Just as adherence to dimensional uniformity provides another avenue for the student to relate to equations, so also does the employment of systematic nomenclature for position provide a rich new avenue for relating to curves. And, just as balancing equations dimensionally provides new opportunities for exercises, independently of the meanings of the equations, so also does describing and deducing the locations and orientations of curves, independently of the significances of the curves.

A few examples of positional nomenclature for the rectangular coordinate system suffice to make its use clear: a *general* ellipse is positioned arbitrarily in the plane; a *general bisector* ellipse has its center incident upon a coordinate bisector; a *general 0°-axial* ellipse has its center incident upon a coordinate axis, with its major axis parallel to the x-axis; a *45°-bisector ellipse including the origin* is a bisector ellipse that passes through the origin and has its major axis oriented at 45° to the x-axis; a *45°-bitangent* ellipse has its major axis oriented at 45° to the x-axis and is tangent to both axes; etc. Students find exercises based on providing descriptive names of this nature for curves in given positions, and positioning curves in accordance with given descriptive names, to be very appealing.

Systematic Nomenclature for Curves

The systematic naming of curves is another topic that received very little attention from classicists, principally because of a paucity of comparative studies. Thus, no adequate scheme of classification existed to render obvious the utility of systematic nomenclature. Yet simple, systematically-applied names provide still another important means for the student to relate to curves. This topic receives detailed treatments in *Symmetry, An Analytical Treatment* (Kavanau, 1980) but some very simple examples drawn from the present work suffice to illustrate one of the approaches.

We saw above how an individual limaçon, the inverse of a conic section about a focus, can be identified in terms of the eccentricity of the initial conic. But what of the names of the three groups of limaçons so obtained? Some possess two loops, some consist of a single closed segment or oval, and one, the cardioid, has a cusp. No classical designation exists for these groups. They are most appropriately named for the parent conic group from which they derive, as *hyperbolic* limaçons, *elliptical* limaçons, and the *parabolic* limaçon, and similarly for all such groups of curves obtained in other non-incident inversions of conic sections. For the inverse curves about the vertices, the corresponding abbreviated names are *hyperbola vertex cubics*, *ellipse vertex cubics*, and the *parabolic vertex cubic* (the cissoid of Diocles).

Algebraic Calculations

A pertinent point should be made for those who employ *Structural Equation Geometry* as a text. Some of the calculations of Structural Equation Geometry are formidable to the uninitiated. The situation is reminiscent of that during the revival period of analytic geometry. Thus, in 1813, Gergonne (1771-1859) observed that "the prolixity often is repulsive" for the calculations involved in some theorems and constructions. Plücker's many improved procedures went a long way toward eliminating such difficulties, at the same time highlighting the elegance of the analytic method, properly employed.

Similar complaints concerning algebraic calculations are voiced today, even by experts, with the consequence that much effort is expended to develop new computer programs and very little effort is expended to find more productive avenues of algebraic analysis. One of the great fascinations of Structural Equation Geometry lies in achieving and understanding the genesis of the ultimate simplifications and degree reductions that frequently materialize in the course of its algebraic procedures. These can be expected whenever relations involving a high degree of structural simplicity are being studied. Some types of calculations that appear to be most formidable, on first encounter, yield readily to appropriate techniques. In fact, the analyses mentioned above that frustrated classicists for over two centuries--the analytic geometry of Cartesian ovals--become elegantly resolved.

Improving the Style of Mathematics Texts

Lastly, the traditional style of elementary mathematics texts leaves something to be desired. Many such books contain page after page of bare equations and plain letter-labelled constructions and line-drawings, more after the style of a research paper than an elementary text. Name labels for curves and figures are used very sparingly, and significant features, such as asymptotes, vertices, foci, directrices, and double-points are not identified with word-labels. Today's student cannot be expected to give high priority to material presented in such austere fashion (in competition with the many more overtly appealing subjects at hand).

Elementary textbooks would much better serve the student and the subject if curves, figures, constructions, and significant features were accompanied by a generous use of name-labels and descriptive word-labels, clearly "tagged" to the identified material by light lines. Every significant equation should be identified by immediately adjacent text, and adjacent step-by-step explanations should be given for all algebraic derivations.

In addition, there should be abundant examples of figures and curves of each type studied. Where appropriate, these should be arranged in sequence, emphasizing changes in form with progressively varying modes of origin or construction, as in the above illustration of the eccentricities of conics. Although the present work is not intended as an elementary text, its almost entirely unfamiliar contents have made similar approaches desirable, leading to many examples of the recommended innovations.

> J. Lee Kavanau
> Los Angeles, California
> December, 1983

FOREWORD

The present work is a much-simplified version of the material of Chapters I-VII, IX, and X of my work, *Symmetry, An Analytical Treatment* (Kavanau, 1980), intended as a sourcebook for instructors of analytical geometry at all levels and as an advanced undergraduate text. For this purpose, treatments of plane Euclidean coordinate systems and the methods and rationale of new approaches to the study of the inherent structure of curves have been greatly elaborated, and 155 Problems have been included. These range from the very simple to the very difficult. Inasmuch as almost all of these Problems are new, detailed outlines of the solutions to most of them are given (Appendix IV).

In the interests of simplicity and greater accessibility, many technical terms have been replaced with more common terminology in the text, although some technical terms remain in figure titles. Thus, *intercept transforms* become *structure rules*, *intercept transform curves* become *structural curves*, *basis curves* become *initial curves*, MAXIMS become RULES, *ordinal and subordinal-rank equations* become XY and X-arc-increment equations, the term *circumpolar* often becomes *about a point*, and the more specific designations, *construction rule* and *structure rule*, frequently replace the general designation, "equation." Although specific, concise new terminology for referring to and classifying curves is available (see Kavanau, 1980, 1982), this also has been employed very sparingly, so as not to overly burden the reader with new designations. Instead of more specific terms for referring to curves and groups of curves, such as *subspecies, species, genus*, etc., the general terms, "member," "subgroup," and "group," are used.

As in the cases of the works, *Symmetry, An Analytical Treatment* (Kavanau, 1980) and *Curves and Symmetry*, vol. 1 (Kavanau, 1982), I have sought to minimize errors by typing the manuscript and preparing the figures myself for direct reproduction. Points of curves first were plotted by pencil, following which the loci were drawn by hand, by inking successive short segments with the aid of French curves and other templates. As an expediency, the cedilla usually has been omitted in typing the name *limaçon*. Abbreviations have been held to a bare minimum--CW, CCW, DP, LR, and eq(s)., for *clockwise, counter-clockwise, double-point, latus rectum*, and *equation(s)*. Use of the square-root sign is conventional, for example, $\sqrt{10}$, $\sqrt{3}/4$, $\sqrt{(x^4+2a^2x^2+a^4)}/2$, and $\sqrt{2}dx^3$, signify the square-root of 10, one-quarter the square-root of 3, one-half the square-root of the bracketed term, and dx^3 times the square-root of 2, respectively.

To avoid confusion, the degree sign, °, always is employed for numerical angular measure. Terms, phrases, and sentences are italicized either for emphasis, in statements of RULES, to introduce new terminology, and as headings or titles. Italics are not employed for existing unusual or little-known terminology or for the unconventional usage of common terms. In the latter cases, quotation marks are used. Footnotes usually are enclosed in brackets, [ ], within the text.

As in the original work, new terms have been made as descriptive as practical and are defined in Appendix III, the *Glossary of Terms*. Existing terminology is adhered to fairly closely, despite not infrequent disadvantages. The term "vertex" is employed in a broader than conventional sense and the term "oval" is used with the broadest classical meaning.

I wish to thank my son Warren for proofreading the final draft of the manuscript. I am greatly indebted to Aaron E. Whitehorn for reading the penultimate draft and for numerous improvements and incisive comments. Technical defects that remain are my own responsibility. Thanks also are due to my sons Christopher and Warren for many helpful suggestions and to Dr. Donald Perry for numerous helpful comments concerning the *Introduction*. I also am indebted to Florence Kaplan for her valuable assistance.

J. Lee Kavanau
Los Angeles, California
December, 1983

THE COVER ILLUSTRATION

The cover illustration is a reproduction of Fig. 52 depicting 11 structural curves of the $e = 0.745$ ellipse for 90°-probes placed at various points in the plane. Each curve illustrates an aspect of the 90°-structural properties of the initial ellipse.

TABLE OF CONTENTS

INTRODUCTION   v

    The Rise to Dominance of Abstract Methods v, Revising and Rejuvenating Treatments of Analytic Geometry vi, The New Subject Matter vii, The Equilateral Hyperbola as a Cassinian Oval, and Self-Inversion at 90° vii, Eccentricity as a Visible Property vii, Structure Rules--New Analytical Tools viii, Structure Rules, Inherent Structure, and the Classification of Curves viii, Structure-Rule Analysis and a Coherent Theory of Exceptional Points ix, New Applications and Knowledge of Polar Equations ix, The Devil's Curve as a Construction from Conics x, Quartic Equivalents of Central Conics x, Cartesian and Cassinian Ovals Related by Inversion x, Unprecedented New Types of Curves xi, Directrices Defined xi, Inversion, Immediate Closure, and a "Natural" Scheme of Classification xi, Asymptotes of Inverse Curves xii.

    New Approaches, Viewpoints, and Paradigms xiii, Structural Simplicity--The Thread of Continuity xiii, Dimensional Uniformity of Equations xiv, Using More Meaningful Terms than "Equation" xiv, Introducing and Comparing Coordinate Systems xiv, The Bipolar Coordinate System--The Introductory System of Choice xv, The Rectangular and Polar Coordinate Systems xvi, Other Coordinate Systems xvii, Graphic Depiction of Eccentricity xvii, Characterizing Curves by Eccentricity xviii, Eccentricity and The Exceptional Lines of Conics xviii, Systematic Nomenclature for Position xviii, Systematic Nomenclature for Curves xix, Algebraic Calculations xix, Improving the Style of Mathematics Texts xx.

FOREWORD   xxi

TABLE OF CONTENTS   xxiii

LIST OF TABLES   xxviii

PART I.   CONSTRUCTION RULES AND STRUCTURAL SIMPLICITY   1

CHAPTER I. THE DISCIPLINE OF STRUCTURAL EQUATION GEOMETRY   1

    Broader Perspectives Provided by Additional Coordinate Systems 1, Inherent Structure and Single Points as Reference Elements 4, The Different Perspectives of Structural Equation Geometry 7.

CHAPTER II. THE RECTANGULAR COORDINATE SYSTEM   9

    Introduction 9, Structural Simplicity and Symmetry of Form 9, Structural Simplicity and Symmetry of Position 11, Analytic Geometry and the Rectangular System 13, Historical 13, The

Rectangular System 14, Degree-Restriction 16, Singularity of Loci 17, Positional Symmetry in the Rectangular System 18, Circles 18, The Overall Simplicity of Structural Equations 22, Ellipses 23, The Rectangular System with Undirected Distances 31, Advantages of the Rectangular System with Directed Distances 34, Singularity of Loci and Uniqueness of Identity 35, Degree-Restriction 35, A Line Has a Maximum of n Real Intersections with a Curve 36, Advantages for the Calculus 36, Advantages for Certain Generalized Symmetry Studies 37, Symmetrical Coverage of the Entire Plane 37.

CHAPTER III. THE POLAR COORDINATE SYSTEM — 41

Introduction 41, Bases for the Widespread Use of Polar Coordinates 42, Curve Plotting 43, The Degree of a Polar Construction Rule 43, Advantages and Shortcomings of Polar-Equation Shortcuts 46, Basic Polar Construction Rules 49, Intersections of Curves at the Polar Pole 50, Structure Rules 51, Polar Curves as Resultants 53, Positional Symmetry in the Polar System 57, Circles 58, Ellipses 62, Elective and Intrinsic Asymmetries of Coordinate Systems 65, Transformations between Polar and Rectangular Coordinates 71, Circles 71, Ellipses 73, The Equilateral Lemniscate 75, The Equilateral Limaçon 78, Biovular Cassinians 80, New Perspectives from Polar Construction Rules 82, Simple Intercept Equations 84, Structural Simplicity and Polynomial Distance Relations 87, Problems 88.

CHAPTER IV. THE BIPOLAR AND HYBRID POLAR-RECTANGULAR COORDINATE SYSTEMS — 90

Introduction 90, Historical 92, The Polarized Versus the Non-Polarized Systems 95, The Most-Symmetrical Bipolar Loci 98, The Rectangular to Bipolar Transformation 101, Bipolar Translations and Rotations 102, Bipolar Foci, Focal Loci, and Point-Foci 105, Bipolar Lines 107, Bipolar Circles 111, Axial Circles 113, Midline Circles 119, Circles in the Plane 120, Lines and Circles and Bipolar Point-Foci 121, Bipolar Parabolas 122, General Parabolas 122, Axial Parabolas 127, Individual and Pole-Pair Bipolar Focal Rank 131, Midline Parabolas 132, Focal Parabolas 134, The Traditional Focus as the Point about Which the Parabola Possesses the Greatest Structural Simplicity 135.

Bipolar Construction Rules and the Hybrid Polar-Rectangular Coordinate System 136, The Degrees of Bipolar Construction Rules and the Foci of Bipolar Curves 139, The Hybrid Polar-Rectangular Coordinate System--Similarities and Differences 142, Bipolar Ellipses 143, General Ellipses 143, Axial Ellipses 146, Subfocal Ranking of the Center and Vertices 150, Midline and Bisector Ellipses 152, Focal Ellipses 153, Focal and Non-Focal Ranking Using Distance Equations 153, Bipolar Limaçons 154, General Construction Rules 155, Point-Foci and Other Axial Poles 158, Axial Construction Rules 159, The Eccentricity of a Limaçon 161, Comparative Bipolar Symmetry of Conics and Limaçons of Different Eccentricities 163.

The Bipolar Equilateral Lemniscate and Dependent Point-Foci 164, Affinities 164, Distance Equations, Bipolar Construction Rules, and Dependent Point-Foci 167, Bipolar Linear Cartesians 170, Introduction 170, Construction Rules of Cartesian Ovals as "Master Construction Rules" for Conic Sections 172, Distance Equations 173, Bipolar Parabolic Cartesians 179, Bipolar Symmetry of Cartesian Ovals 181, Geometrical Correlates 181, Distance Equations 182, Bipolar Construction Rules 184, Comparisons with Linear Cartesians, in General 185, Bipolar Central Cartesians 187, The Bipolar Equilateral Strophoid 193.

Hybrid Distance Equations and the Hybrid Polar-Rectangular System 197, Polar Construction Rules, Polar Symmetry, and Conversions between the Polar and Hybrid Systems 199, Coordinate System Comparisons and Analytic Geometry 201, Identifying Curves 202, Positional Symmetry 204, Absolute and Comparative Structural Simplicity about Points 205, Comparative Utility of the Polar, Rectangular, and Bipolar Systems for Instruction, Enlightenment, and Recreation 207, Problems 210.

CHAPTER V. THE POLAR-LINEAR COORDINATE SYSTEM — 213

Polar-Linear Systems and the Hybrid Polar-Rectangular System 213, Polar-Exchange Symmetry 216, Disparate and Compound Curves 216, Derivation of Polar-Linear Construction Rules 220, Incomplete-Linear Construction Rules 222, Parabolas 223, Lines 229, Central Conics 234, Focus-Directrix Construction Rules 234, Focus-Line-of-Symmetry Construction Rules 236, Vertex-Line-of-Symmetry Construction Rules 237, Focus-Vertex-Tangent Construction Rules 239, The Shapes of Complementary Segments 240, The Incident System 242, Circles 242, Limaçons 246, Focal Axial Limacons 247, Parabolic Cartesians 251, Geometrical Correlates 254, Summary and Comparisons with the Bipolar and Other Coordinate Systems 255, Problems 258.

CHAPTER VI. THE POLAR-CIRCULAR AND LINEAR-CIRCULAR COORDINATE SYSTEMS — 260

THE POLAR-CIRCULAR COORDINATE SYSTEM — 260

Transformation Equations 262, The Construction Rule $u = v$ 264, Polar-Circular and Bipolar Construction Rules 267, The Construction Rule $u = Cv$ 269, Limaçons 269, Cartesian Ovals 271, Parabolic Cartesian Ovals 272, The Construction Rule $u \pm v = \pm Cd$ 275, Summary 280.

THE LINEAR-CIRCULAR COORDINATE SYSTEM — 282

Introduction 282, Transformations between Rectangular and Linear-Circular Coordinates 283, The Construction Rule $u = v$ 284, The Construction Rule $u = Cv$ 287, The Construction Rule $u \pm v = \pm Cd$ 290, The Construction Rule $u^2 = Cdv$ 294, Construc-Rules as Relationships between Curves 301, Summary 301, Other Coordinate Systems 302, Problems 303.

PART II.   STRUCTURE RULES AND STRUCTURAL SIMPLICITY   305

Introduction 305, The Classical Symmetries and Structure Rules 306, Generalizing from the Classical Symmetries 306, Structure Rules about Points and Inherent Structure 310, Structure Rules about Points Versus Construction Rules about Points 311.

CHAPTER VII. STRUCTURE-RULE ANALYSIS, A SURVEY   312

Introduction 312, Structure Rules and Constructions 313, Comparing Structure Rules 314, Deriving Circumpolar Structure Rules 316, The 90°-Structure Rule of Lines 316, Non-Degenerate Structure Rules Can Represent All Combinations of X and Y-Intercepts--Positive, Negative, Null, and Imaginary 318, The 180°-Structure Rule of the Parabola about Its Traditional Focus 319, 90°-Structure Rules of Central Conics about Their Centers 321, The General 90°-Structure Rule of the Circle, 322, Foci, Focal Loci, and Structural Simplicity at the Subfocal Level 323.

The Classification of Curves 326, Inversion Analysis 327, Circumpolar Symmetry Analysis 328, Design and Synthesis of Highly Symmetrical Curves 329, Some Structure Rules of Unusual Interest 331, The "4th-Degree Circle" 332, Central and Focal 90°-Structure Rules of Conics 332, Eclipsing of Foci, and Inverse 4-Leaf Roses as the 90°-Incident Structural Curves of the Equilateral Hyperbola 334, The 90°-Structure Rule of Limaçons about the -b/2 Focus 338, The Devil's Curve--The 0°-90°-Structural Curve of the Circle about a Concentric Equilateral Hyperbola 340.

Overt Symmetry of Structural Curves 340, Circumpolar Point-Foci and Probe-Angles 343, α-Structure Rules and Structure-Rule Formats 343, The Frequency of Occurrence of 1st and 2nd-Degree Structure Rules 345, Arc-Increment Equations and More Specific Characterizations of Inherent Structure 346, The Arc-Increment Equations 349, The Parabola Versus Central Conics 351, Simple Intercept Products--A Fundamental Property of Conic Sections 353, Simple Intercept Products 353, Omnidirectional Circumlinear Self-Inversion of Hyperbolas 355, Types of Circumpolar Point-Foci and Focal Loci 356, Summary 358, Problems 362.

CHAPTER VIII.   STRUCTURE RULES OF CURVES ABOUT POINTS   364

Introduction 364, The Circle 364, Arc-Increment Equations 368, The Parabola 369, Intercept Equations and the General Structure Rule 369, 0° and 180°-Structure Rules and Compound Intercept Formats 371, 90°-Structure Rules 375, Degenerate 90°-Structure Rules and Modular Segments 381, The Structural Curves 382, Central Conics 383, 0° and 180°-Structure Rules 383, Intercept Equations and Formats 389, 90°-Structure Rules 392, Conditions on the Radicand of the 90°-Compound Intercept Format 394, The Latus-Rectum Vertices 398, The a and b-Vertices 398, The Traditional Foci and Centers 400, α-Structure Rules and Focal Conditions for Variable Probe-Angles 402.

Hyperbola Axial-Vertex Inversion Cubics 403, Introduction 403, The Simple Intercept Equation and Format 406, Structure Rules 407, The Double-Point Focus 408, The Loop-Vertex Focus 409, The Variable Focus 409, The Focus at the Asymptote-Point 411, The Loop-Pole at $h = 2b/3$ 412, Limaçons 412, Introduction 412, Simple Intercept Equations and Formats 413, Structure Rules 417, Points on Focal Loci 417, The Double-Point Focus 418, The Focus of Self-Inversion 419, The $Cos^2\theta$-Condition Double-Focus 420, The Axial-Vertex Foci 421.

Parabolic Cartesians 421, Linear Cartesians 423, Central Cartesians 427, Introduction 427, Axial Vertices, Products of Roots, and Self-Inversion 428, Simple Intercept Equation and Format, and the Cartesian Term 429, The Cartesian Condition, Circumpolar Symmetry Analyses, and the Inversion Transformation 430, Coincidences of Cartesian-Condition Poles with Foci 431, Cartesian Ovals and $Cos^2\theta$-Condition Double-Foci 432, Angle-Independent Versus Angle-Dependent Foci of Self-Inversion 433, Vertex Cubics 434, Locating the Foci of Self-Inversion of Cartesian Ovals 435, Ensembles of Inverse Curves of Central Cartesians That Possess at Least One Line of Symmetry 436, The Symmetry Condition of Central Cartesians Inverse to Cassinian Ovals 439, Problems 440.

APPENDIX I. LISTING OF RULES	443
APPENDIX II. INVERSION-MAPPING MAXIMS	448
APPENDIX III. GLOSSARY OF TERMS	455
REFERENCES	462
APPENDIX IV. ANSWERS TO PROBLEMS	463
SUBJECT INDEX	499

LIST OF TABLES

1.	Structural Equation Geometry	2
2.	Circumpolar Structural Equation Geometry	8
3.	Symmetry of Rectangular Circles	19
4.	Symmetry of Rectangular Ellipses	24,25
5.	Symmetry of Polar Circles	61,62
6.	Symmetry of $0°$-Polar Ellipses	64
7.	Elective and Intrinsic Asymmetries of Coordinate Systems	66,67
8.	Hybrid Polar-Rectangular Distance Equations for the Most-Symmetrical Curve-Pole Combinations	198
9.	Overt Symmetry Properties of Complete Structural Curves	342
10.	Inventory of Structure Rules of Low Degrees	347
11.	Degrees of Structure Rules of Conic Sections about Points	390
12.	Degrees of Structure Rules of Cubics about Points	410
13.	Degrees of Structure Rules of Limaçons about Points	410

ART IN MATHEMATICAL DESCRIPTION

When you are at a museum, notice that many visitors are eager to examine the label of an art work. They read the title, author, and date; they refer back to their catalogues and then stand back and examine the work. A description enhances our appreciation of art, as thirst for the intellectual and the visual go hand in hand. When Michelangelo was asked how he drew, he replied, "with my mind." Perhaps the author of this work would say, "my drawings guided my thoughts."

When the art of mathematics is neglected the appreciator is not fully requited. The author's books on geometry are exceptions; in them we have the benefit of descriptions of new geometrical concepts combined with aesthetic graphics.

Frederick I. Sauls

PART I. CONSTRUCTION RULES AND STRUCTURAL SIMPLICITY

CHAPTER I. THE DISCIPLINE OF STRUCTURAL EQUATION GEOMETRY

Classical analytic geometry may be defined loosely as the study of curves and figures in coordinate systems through the use of algebraic representations or eqs. By means of these eqs., the size, shape, and position of a curve can be specified precisely, and its properties can be studied in great depth. The discipline of *Structural Equation Geometry* includes classical analytic geometry but very greatly extends the classical coverage in two principal directions.

Broader Perspectives Provided by Additional Coordinate Systems

First, it greatly broadens our perspectives on curves and their eqs., by employing many more Euclidean coordinate systems than those of classical analytic geometry. The latter, as the reader will be aware, has depended almost entirely on the rectangular and polar systems. Although the concept that any combination of two or more reference elements can make up a coordinate system goes back to Plücker, explorations of other Euclidean systems have been extremely scanty. Likewise, studies of the basic and comparative properties of coordinate systems--one of the main objectives here--have been limited almost entirely to superficial aspects, primarily to satisfy practical needs. Euclidean coordinate systems in their own right have not been studied systematically.

Some of the coordinate systems treated here barely have been touched upon previously, such as one based on a point and a line (*polar-linear*; Chapter V), one based on a point and a circle (*polar-circular*; Chapter VI), and one based on a line and a circle (*linear-circular*; Chapter VI). Certain constructions employing a point and a line, and a point and a circle, are recognized. For example, focus-directrix constructions of conic sections commonly are employed to define conics, their eccentricities, and directrices, and a focus-directrix-circle construction of central conics is known, though obscure (Zwikker, 1950). [Only the least-distance construction of a one-arm hyperbola appears to have been recognized; the complementary greatest-distance construction gives the other (congruent) arm; see Chapter VI and Fig. 33a_1.]

However, these constructions are treated in isolation. There is no explicit recognition of the fact that they are merely simple or simplest examples from coordinate systems whose curves and properties offer much broader perspec-

Table 1.   STRUCTURAL EQUATION GEOMETRY

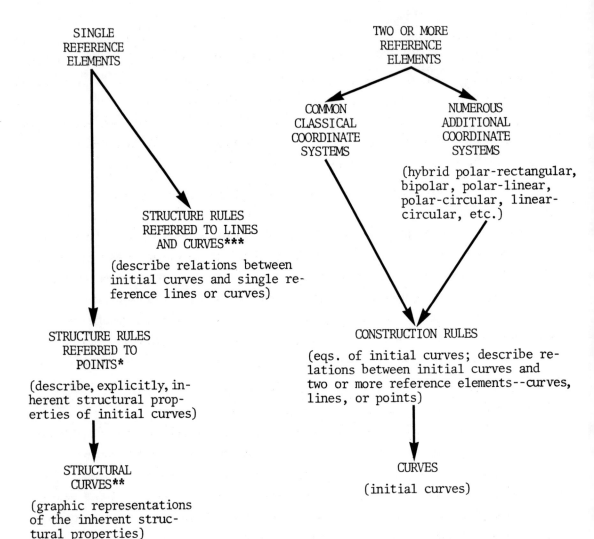

*the circumpolar intercept transform eqs. of Kavanau (1980, 1982)
**the circumpolar intercept transform curves of Kavanau (1980, 1982)
***the circumlinear and circumcurvilinear intercept transform eqs. of Kavanau (1980, 1982)

tives and are far more interesting than those of the rectangular and polar systems.

The bipolar system receives by far the greatest attention here (Chapter IV). This system, consisting of two points as reference elements, achieved limited but very active use in the middle-to-late 19th century. However, despite having been the object of a great deal of attention, the system remained a puzzle; its key characteristics escaped notice and its didactic advantages were unappreciated. No more attention was devoted to elucidating its basic and comparative properties than to those of other Euclidean systems.

One of the newly-employed systems--the *hybrid polar-rectangular system* of Chapter IV--forms the basis for the well-known general distance eqs. These are used prolifically in plane analytic geometry and the calculus to relate distances between an arbitrary point of a curve and an arbitrary point in the plane. But despite this common use, there is no evidence of previous recognition either of the coordinate system itself, the key role it can play for deriving construction rules and defining foci in other coordinate systems, or the fact that distance eqs. are themselves coordinate construction rules for the curves whose distances they specify.

The new, broader studies elucidate the properties of coordinate systems and the fascinating ways in which the various systems interrelate, as well as the close relations between coordinate construction rules and the new concept of *generalized symmetry* or *structural simplicity* (the latter terms are equivalent and are defined in Chapter II and used interchangeably hereafter). Many of the basic and comparative properties of coordinate systems referred to above previously were unrecognized. Others were known but attracted little attention; still others doubtless were taken note of, but with little awareness of their significance in a broad comparative sense.

Only when one studies construction rules and curves in other coordinate systems does one become aware of the fact that the perspectives provided by the polar and rectangular systems not only are extremely limited--they are highly atypical. And while certain properties of the polar system provide hints of more general relations, as they exist in other systems, these properties usually are passed over quickly or ignored completely in the pursuit of practical needs. In consequence, our experience with plane curves has been confined to a most Spartan breadth.

Whereas experience has led us to believe that each of the known curves is unique--such as the circle, the parabola, or an ellipse of a given eccentricity--in fact, these curves are merely the simplest manifestations of their particular types as they occur in other coordinate systems. The most that can be said of the known curves, in respect to the extent to which they give insights into the vast variety of forms of related curves in other coordinate systems, is that they provide the basic templates that are sufficient for drawing these other curves. In other words, if one possesses all sizes of templates of, say, parabolas, one will be capable of drawing the many different varieties of "parabolas" that are represented by non-degenerate eqs. in other coordinate systems. Needless to say, the beauty and fascination of the geometry of curves also are multiplied manyfold in these other coordinate systems.

The inherent properties of coordinate systems referred to in the subtitle of this book include both properties descriptive of the coordinate elements taken together as a system, and properties that coordinate systems impose on curves and eqs. Examples in the former category are *intrinsic* and *elective asymmetries* and *characteristic distances*. Examples in the latter category are *degree-restriction*, *partial degree-restriction*, *polar-exchange symmetry*, possession, or lack, of symmetry of all loci about the coordinate elements, and *compound*, *disparate*, and *segmented* loci (these new designations are defined as they are encountered, and in the *Glossary of Terms*).

Inherent Structure and Single Points as Reference Elements

Second, Structural Equation Geometry very greatly expands our knowledge of inherent properties of curves and relations between curves. Eqs. in two distance variables no longer are regarded only as *construction rules* for specific curves. Every eq. of an initial curve also is a *structure rule*, descriptive of the *inherent* structure of an infinite number of other curves. The term *structural equation* applies generally to eqs. in both the context of coordinate construction rules and the new structure rules (see diagram below).

The new structure rules are not to be confused with the classical "intrinsic" (Cesáro and Whewell) or "natural" eqs. of curves. The latter were arrived at as a result of the classicists' desire to study curves with an analytic method that was independent of an outside framework (coordinate system), that is, a method involving solely relations among elements of curves themselves.

These intrinsic eqs. uniquely describe structural relations that exist within curves. They relate arc length to an arbitrary point on a curve to either the curvature at that point or the angle between the tangent at that point and the tangent at the initial point of the arc.

However, the use of a curve as its own "reference element" and of arc length, curvature, and tangent angles as variables, leads to results that are virtually devoid of utility for comparative purposes. The intrinsic eqs. generated by this extreme approach fail to encompass quite different, and much more useful and interesting, *inherent* properties of curves. These are defined by certain of the *structure rules* referred to above. They may be unique to the curve (like the intrinsic eqs.) or may be shared with infinite numbers of other curves (see Table 2; these properties are referred to as *inherent* to avoid confusion with the properties described by intrinsic eqs.).

[A property of a curve is not any the less *inherent* because it is shared with other curves. For example, curves constructed from circles may share some of the inherent properties of circles, as is true of limacons (see Chapter VII).]

Structure rules, as a general category, remained unknown because, in seeking to achieve independence from outside reference elements, as employed in coordinate systems, classicists overlooked the following essential fact. Insofar as studying the inherent properties of curves is concerned, there is no more reason to regard single non-incident reference points in the plane of a curve as "outside" reference elements than to so regard single incident points (consider the curve to define the area it bounds and the complement of this area in the plane). Any single point in the finite plane of a curve can serve as a reference element for assessing inherent properties.

Thus, to study inherent properties of curves it is unnecessary and, in fact, counterproductive, to eliminate "outside" reference elements and to employ the complicated variables of arc length, curvature, and tangent angles; *mere distances suffice*, whereupon two requirements become crucial. First, positions of points of initial curves must be related to one another in pairs. [So long as the method of expressing such relationships introduces no extraneous structure, the results will be inherent to the curve.] Second, the positions of the paired points must be measured from a single *structureless* reference element. Only single points in the planes of curves satisfy this requirement. Such relations --between the positions of *two* points of a curve assessed from a *single* reference point (structure rules)--express inherent properties explicitly, whereas relations between the positions of *single* points of a curve assessed from *two* reference elements (construction rules) do not.

If the reference element possesses structure, structure rules generally characterize inherent properties of the combination of the element and initial curve. But if the structure of the element is closely related to the structure of the curve, inherent properties also will be revealed. For example, using a line reference element with a curve that possesses a line of symmetry produces structure rules ($X = Y$ and $X = Y + a$) that characterize inherent properties of the curve (see Kavanau, 1980, and examples in Chapter VII).

The fact that the new inherent properties change with the location of a reference point does not mean that these properties are in any sense non-inherent; the changes from point to point also characterize, and are determined solely by, properties of the curve. These changes, in fact, provide one of the bases for the tremendous breadth of the new approaches for characterizing and classifying curves; another powerful contribution resides in the distinctions elucidated by the use of different *probe-angles* (see Fig. 2a).
[A probe-angle is not a variable coordinate reference angle, as in the polar system, but a constant angle between rotating, distance-defining radius vectors (in extreme cases, it is an angle of classical rotational symmetry).]

The diagram below outlines some of the relations referred to above. In subsequent uses, the appropriate one of the three terms--structural eq., structure rule, and construction rule--generally is employed. Otherwise, only the word, "equation," is used (abbreviated as "eq.").

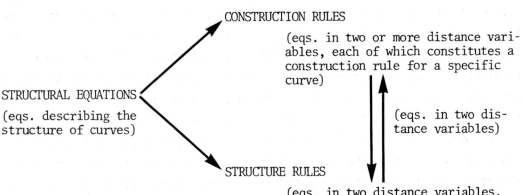

In addition to describing aspects of the inherent structure of an initial curve, each specific structure rule about a point (see Table 2) generally describes aspects of the inherent structure of an infinite number of other curves. Consequently, all these other curves are related to one another and to the initial curve by virtue of the inherent structural property that they share. Every curve, even the line, has an infinite number of both construction rules and structure rules, the former with respect to different positions in infinite numbers of coordinate systems, and the latter with respect to both an infinite number of single reference elements in different positions, and an infinite number of probe-angles. The domains and terminology of Structural Equation Geometry are illustrated in Table 1. Table 2 outlines relations between structure rules obtained using solely point reference elements (the *circumpolar* subdiscipline).

The new studies of structural relations within curves have greatly expanded our knowledge of the properties and interrelations of curves, and have led to the discovery of many new curves, and new ways to describe, organize, and classify curves. In addition, they have revealed ways to construct or formulate open and closed curves with desired structural properties.

The Different Perspectives of Structural Equation Geometry

To "flesh out" the perspectives of Structural Equation Geometry different from those of classical analytic geometry with specific examples, some of the relations conveyed to the structural eq. geometer by the eq., $(y-k)^2 = 4a(x-h)$, are considered. To the classical analytic geometer, this is nothing more than the rectangular construction rule for a parabola with vertex at (h,k). To the structural eq. geometer it is very much more. For example, this eq. also is a construction rule for limacons and *parabolic* Cartesian ovals in the bipolar coordinate system; in the polar-linear system, it is either a construction rule for circles or for new types of conchoids (y, the distance from the line or point, respectively; $h,k = 0$). In the linear-circular system, it is either a construction rule for limacons and Cartesian ovals or for new types of conchoids (y, the distance from the circle or line, respectively). And the same eq. is a construction rule for still other curves in other coordinate systems.

But the eq. has significance in many other connections than as a construction rule for curves. As a *structure rule*, as opposed to a *construction rule*, it defines the structural simplicity (generalized symmetry) of an infinite number of initial curves with respect to single reference elements. Considering only two cases of structural simplicity about a point, this structure rule describes the generalized symmetry of the Norwich spiral about its center ($k=0$) and of the curves, $r = a + b\cos^4\theta$, about their poles ($k=0$). In cases of structural simplicity about single lines, or other single reference elements, the same parabolic structure rule describes the generalized symmetry of still other initial curves.

Table 2.   CIRCUMPOLAR STRUCTURAL EQUATION GEOMETRY

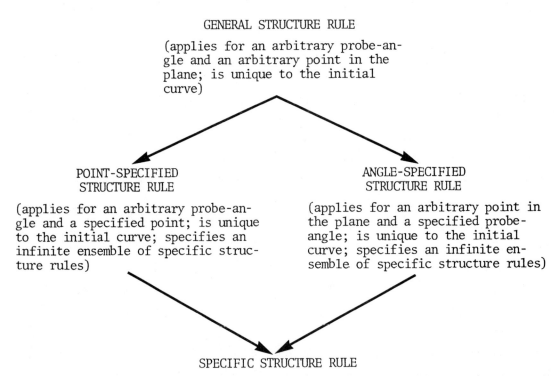

CHAPTER II.   THE RECTANGULAR COORDINATE SYSTEM

Introduction

Structural Simplicity and Symmetry of Form

Symmetry is one of the best-known, most productive, and simplest of all geometrical concepts, with origins tracing back at least to prehistoric cave art. In fact, the aesthetic appeal of simplicity probably is the main reason why geometrical concepts of symmetry remained confined for so long within their strict classical molds. Thus, classical treatments of geometrical symmetry recognized only certain visually obvious aspects of structure, chiefly reflectional and rotational coincidences.

A plane figure that reflects into coincidence with itself along a line has symmetry in the line (Fig. 1a), while one that rotates into coincidence with itself about a point (through an angle less than a complete rotation) has symmetry about that point (Fig. 1b,c,d); for an angle of 180° (half a complete rotation), the figure also has *reflective* symmetry in the point (Fig. 1b). [For detailed treatments of classical geometrical symmetries, the exhaustive treatment by Lockwood and Macmillan (1978) is recommended.]

It is apparent, even from superficial considerations, that classical reflective and rotational symmetries about lines and points are merely extreme manifestations of structural simplicity about these elements. Thus, every case of a classical reflectional or rotational symmetry can be described by the eq. $X = Y$, the simplest and lowest-degree eq. in two unknowns. In the cases of figures with rotational symmetry about a point, X and Y are distances along radius vectors from the point of symmetry to the figure at an angle, α, to each other, where α is the angle of rotational symmetry (Fig. 1c, for 120° or 240°). In the case of reflective symmetry about a line, X and Y are radius vectors to the curve or figure from common points on the line of symmetry, at equal angles (not necessarily 90°) on each side of the line (Fig. 1a,d).

From the realization that the classical symmetries are equivalent to the structure rule, $X = Y$, it is not a very great step to generalizing classical approaches to encompass other aspects of structure, rather than only its simplest manifestations, and to formulating the following RULES.

The Classical Symmetries

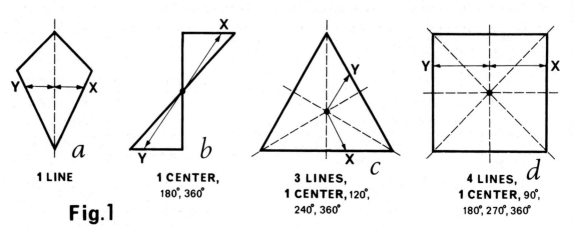

Fig. 1

a — 1 LINE
b — 1 CENTER, 180°, 360°
c — 3 LINES, 1 CENTER, 120°, 240°, 360°
d — 4 LINES, 1 CENTER, 90°, 180°, 270°, 360°

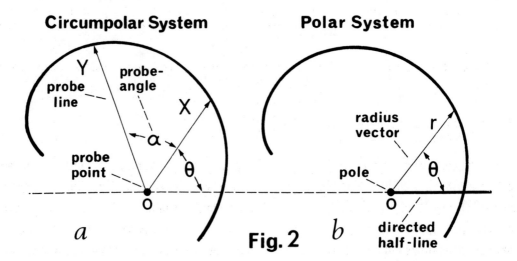

Fig. 2 — *a* Circumpolar System; *b* Polar System

STRUCTURAL SIMPLICITY AND STRUCTURAL EQUATIONS

RULE 1a. *The greater the generalized symmetry (structural simplicity) of a curve or figure with respect to one or more reference elements, the simpler its structural equations (either "structure rules" or "construction rules") and the lower their degrees* (see Chapter I and Tables 1 and 2).

With the understanding that the use of the term "symmetry" from this juncture on refers to *generalized symmetry* or *structural simplicity*, the above relations can be abbreviated to:

RULE 1b. *The greater the symmetry of a curve or figure with respect to given reference elements, the simpler its structural equations and/or the lower their degrees.*

Though these relations long remained unnoticed or unappreciated, they are more or less intuitive. Another aspect of them is expressed by the rules that:

RULE 1c. *For measurements or instructions of comparable simplicity, the fewer the number needed to determine or specify all distances to points of a curve or figure from one or more reference elements, the simpler and/or lower the degree of the structural equation that relates these distances.*

RULE 1d. *For measurements or instructions of comparable simplicity, a 50% decrease in the number needed to determine or specify all distances to points of a curve or figure from one or more reference elements, frequently results in a 50% or 75% decrease in the degree of the structural equation that relates these distances. A decrease in number by less than 50% is insufficient to lead to degree reduction.*

RULE 1e. *For curves specified by measurements or instructions comparable or equal in numbers, the greater the simplicity of the measurements or instructions, the simpler and lower the degree of the equation needed to specify them.*

RULES 1a-e embody the seminal theses of this work. Analytical foundations for them are encountered repeatedly in the following treatments.

Structural Simplicity and Symmetry
    of Position

One gets the impression, at first sight, that symmetry of form and symmetry of position are two different concepts. In fact, however, they are merely two faces of the same coin. The basis for this equivalence is that the evaluation of the positional symmetry of a given curve or figure with respect to any reference element(s) depends ultimately upon the type of symmetry possessed by the curve or figure itself. Thus, a figure that has reflective symmetry in a line (Fig. 1a) will be most-symmetrically positioned with respect to a reference line when its line of symmetry is brought into coincidence with the reference line. A figure that has reflective symmetry in each of two orthogonal lines (Fig. 1d) will be most-symmetrically positioned with respect to two orthogonal reference lines when its lines of symmetry are brought into coincidence with these reference lines. Similarly, a figure with rotational symmetry about a point (Fig. 1b) will be most-symmetrically positioned with respect to a single reference point when its point of rotational symmetry is brought into coincidence with the reference point, etc.

In practice, it is convenient and avoids confusion if the terms "positional symmetry" or "symmetry of position" are used in cases where one is referring specifically to the symmetry of the position (the location and orientation) of a figure with respect to the element(s) of a reference system. However, when curves are arranged in serial order according to their increasing or decreasing positional symmetry about given reference elements, they fall in the same order obtained by ordering them according to their relative symmetries of form (structural simplicities) about the same elements (see Tables 3-6).

In systems with two or more reference elements, it is the simplicities and exponential degrees of *construction rules* of curves that fall in the same serial order as the curves' symmetries of position and form. On the other hand, when only a single reference element is involved, it is the *structure rules* of curves, that is, the eqs. that describe the types of structure that they possess, that fall in this same serial order.

One facet of the analytical approaches developed here is that they enable one to determine the most symmetrical position of a curve possessing any type of classical or generalized symmetry, in relation to any types and combinations of reference elements. Only exceptionally is this determination intuitive. For example, answers are given to such questions as, "when is a figure with two orthogonal lines of symmetry most-symmetrically positioned with respect to a line and a point, or with respect to two points, or a line and a circle, or a point and a circle, etc?"

Answers also are given to such questions as, "what figure has the greatest symmetry of form (structural simplicity) with respect to a point, with respect to two points, with respect to a line, two lines, a point and a line, a line and a circle, etc?" And the figure that has the greatest symmetry of form also is most-symmetrically positioned with respect to the same element(s). The following RULES enable some of these determinations to be made.

RULE 2a. *The curve with the greatest symmetry of form and position with respect to two reference elements is the locus of the construction rule $X = Y$.*

RULE 2b. *A curve is most-symmetrically positioned with respect to one or more reference elements when its structural equation with respect to the same element(s) assumes its simplest and lowest-degree form.*

RULE 2c. *Of a given group of curves with structural equations of the same form, for example, $AX^2 + BY^2 = Cd^2$, the one with the simplest equation, for example, $X^2 + Y^2 = d^2$, has the greatest symmetry of form and position (excluding loci consisting of a finite number of points).*

[The above RULES apply primarily to undirected-distance systems (see Table 7).]

Analytic Geometry and the Rectangular Coordinate System

Historical

There is little agreement among historians of science about the invention of analytic geometry. Even the age in which it occurred is in dispute, partly because of lack of unanimity over the definition of analytic geometry. Those who favor antiquity emphasize the fact that, in ancient surveying, the Egyptians and Romans fixed the position of a point with suitable coordinates, as did also the Greeks in making maps. In fact, Apollonius derived many results for conic sections using the geometrical equivalents of rectangular eqs. of these curves, following an idea that seems to have originated with Menaechmus around 350 B. C. Others see Nicole Oresme (∼1323-1382) as the inventor of analytic geometry. Oresme represented certain relations by graphing the dependent variable against the independent one, as the latter took on small increments. Some credit Oresme with the first explicit introduction of the eq. of a straight line.

To become highly practical, however, an algebraic symbolism was needed. Accordingly, most historians trace the beginnings of the modern spirit of analytic geometry to the decisive 17th-century contributions of the French mathematicians, René Descartes (1596-1650) and Pierre de Fermat (1601-1665). Descartes' claim to the invention of analytic geometry is based on *La géométrie*, one of the three appendices to his philosophical treatise on universal science, published in 1637, in which Cartesian coordinates made a somewhat disguised and informal appearance.

Whereas, to the Greeks, a variable corresponded to a length, the product of two variables to an area, and the product of three variables to a volume, Descartes treated x^2 as the fourth term in the proportion, $1:x = x:x^2$. As such, x^2 can be represented by an appropriate line-segment when x is known. This can easily be constructed using a unit segment (represented herein by the letter "j"). One can, in this way, represent any power of a variable, or the product of any number of variables, by the length of a line, and construct the line of that length with Euclidean tools, when the values of the variables are given. Descartes probably studied his well-known ovals as early as 1629. These "Cartesian ovals" receive detailed treatments here.

On the other hand, Fermat, whose ideas and methods were far in advance of those of Descartes, and much nearer to the modern form, was employing such (Cartesian) coordinates some years earlier in his study of maxima and minima. Fermat's claim to priority rests on his letter to Roberval in 1636, in which he stated that his ideas, even then, were seven years old. However, his own conceptions were published posthumously in 1679 in his *Varia Opera*. Therein, one finds the eqs. of a general straight line and circle, and a discussion of hyperbolas, ellipses and the parabola. In a work on tangents and quadratures, completed before 1637, Fermat defined many new curves.

Where Descartes, to a large extent, began with a locus and then found its eq., Fermat often started with an eq. and then studied its locus. These, of course, are the two inverse aspects of the fundamental principle of analytic geometry. These applications, however, were not made exclusively with the rectangular system. Both Descartes and Fermat, and their successors, down to the middle of the 18th century, also made much use of oblique coordinates (Fig. 6).

The Rectangular System

As the reader will be aware, the rectangular Cartesian coordinate system (hereafter referred to as the "rectangular system;" oblique Cartesian coordinates are referred to as "oblique systems") consists of two reference lines (axes) that intersect at an angle of 90°. Measurements from points of curves to the two axes are made by dropping normals (perpendiculars) from the points to the axes. Since the distances customarily are measured along the parallel-lying axes, rather than along the normals, the (positive or negative) distances to the x-axis are called the "y-coordinates," and those to the y-axis, the "x-coordinates."

The point of intersection of the axes is known as the "origin." Of the several respects in which the origin is a unique point of the system and of the coordinate elements themselves, we take note only of the fact that *it is the sole point that is symmetrically positioned in the system.*

With this relation in mind, any given curve, A, that includes (passes through) the origin, automatically possesses an aspect of greater positional symmetry than a neighboring congruent curve, B, that does not include the origin. This materializes because one of the points of curve A comes to be most-symmetrically positioned in the coordinate system, whereas this is true of no point of curve B. Although this circumstance receives no notice in traditional presentations, it is a crucial one from the point of view of generalized symmetry studies, and serves as our introduction to the analytical bases for assessing structural simplicity.

The special symmetry status of an otherwise unexceptionally-positioned curve, A, that includes the origin, is expressed in a well-known circumstance: the construction rule of curve A possesses one less term than the construction rule of curve B, the unexceptionally-positioned neighboring congruent curve (namely, by virtue of the vanishing of the constant term). In other words, when a curve includes the origin, it requires one less term to specify the positions of all other points of the curve, than is required when the origin is not included (see Table 3; eqs. 1d versus 1f, and 1g versus 1h). This is one aspect of the simplicity of structural eqs. referred to in RULES 1a-d, above. Thus, of two structural eqs. with the same exponential degree, the one with fewer terms generally describes simpler structural relations, or, equivalently, it represents a more-symmetrically-positioned curve (see, for example, Fig. 3 and Table 3).

Although introductions to the rectangular coordinate system in elementary plane analytic geometry are quite matter-of-fact, the rectangular system is a most remarkable one. Its great practical utility depends on several extraordinary inherent properties that have come to be regarded as commonplace. Accordingly, they usually are taken for granted and are mentioned only in passing, rather than with the emphasis that they deserve.

Two of these properties concern us at this juncture, *degree-restriction* and *singularity of loci* (both of which are defined below). Among planar coordinate systems, degree-restriction is unique, and singularity of loci is confined almost exclusively, to directed-distance, intersecting, bilinear systems, that is, to systems whose reference elements consist of two intersecting lines, and in which distances are measured in both a positive and a negative sense.

Barring coincidence of the reference lines of these systems, the angle between them is of no import; nor does it matter whether the distances are least distances from the lines (distances measured along normals to the coordinate lines) or are distances measured parallel to the lines (as in oblique systems; Fig. 6 and Problems 2 and 5). Of these Cartesian systems, all of which have very similar properties, the rectangular system is by far the most useful because of its symmetry and simplicity. For example:

(a) its four segments (quadrants) are of the same shape and size;
(b) certain simple rules apply for reflective symmetry in its axes;
(c) simple eqs., like $x^2+y^2 = R^2$ and $xy = j^2$ (or $x^2-y^2 = a^2$) represent circles and equilateral hyperbolas; and
(d) the normals along which the least distances from the coordinate lines are measured also are parallels to the same lines, obviating the need to employ trigonometric functions (see Problems 2 and 3).

Degree-Restriction

Degree-restriction is the property that the construction rules of a given curve have only a single degree. Thus, all lines have eqs. of 1st-degree, all circles, ellipses, hyperbolas, and parabolas have eqs. of 2nd-degree, all cissoids of Diocles and trisectrices of Maclaurin have eqs. of 3rd-degree, all Cartesian ovals have eqs. of 4th-degree, etc.

The lack of degree-restriction in non-bilinear systems hinges on the fact that the degree of a given curve depends on its position in the system. Accordingly, the property of degree-restriction of intersecting bilinear systems is readily understood in terms of the fact that the degree of a curve does not change with position in such systems. This is evident from the eqs. of transformation for displacements and rotations, which, between them, accommodate all changes in position. The rectangular system is used for illustration.

For translations by amounts, h and k, along the x and y-axes, respectively, say to the right and upward in the 1st quadrant, the new coordinates (x',y') of an original point (x,y) are simply $x' = x+h$ and $y' = y+k$. To obtain the construction rule of a curve in its new location with respect to the original axes, one simply substitutes, in the eq., $f(x,y) = 0$, of the curve in its original location, the value, x'-h, for x and the value y'-k, for y. This yields the new construction rule, $f[(x'-h),(y'-k)] = 0$. For convenience, since the original axes are retained, the prime signs are omitted, and the transformations are referred to simply as x-h and y-k translations. For h and k positive, these move the curve to the right and upward, that is, in the sense of increasing values of the coordinates.

For rotations, the well-known substitutions, $x = x'\cos\theta - y'\sin\theta$ and

$y = x'\sin\theta + y'\cos\theta$, for a CCW (counterclockwise) rotation of the axes convert the eq. of the original curve, $f(x,y) = 0$, to $f[(x'\cos\theta - y'\sin\theta), (x'\sin\theta + y'\cos\theta)] = 0$. In this case, since the eqs. represent the position of the curve in the coordinate system with rotated axes, x' and y', the primes are retained. Now it is evident that, since x' and y', in the cases of all these transformations, are linear functions of x and y, the degrees of the new construction rules will be the same as those of the original ones.

These relations are obvious. What is overlooked is the fact that the consequent property of degree-restriction is not "natural" and is not to be taken for granted. It is extremely unusual and, as mentioned above, unique to intersecting bilinear systems (see Table 7). Because degree-restriction has been

[We shall find in Chapters V and VI that, even the presence of only one reference line, has a restricting influence on degree--*partial degree-restriction*. For example, all polar-linear eqs. of axial curves about a given point are restricted to a degree equal to or twice that of the corresponding hybrid polar-rectangular distance eq.]

taken for granted, and because loci in the polar system almost always are placed in symmetrical locations about the polar pole and initial half-line--with frequent construction rules of degree 1 in r--the absence of degree-restriction in the polar system usually is overlooked. Since the polar and rec-

[The same custom, followed with few exceptions in the bipolar system, also led to the failure to notice the absence of degree-restriction in that system (see Chapter IV).]

tangular systems are employed almost exclusively, even the term, *degree-restriction*, is new. These facts help to explain the lack of prior knowledge of relations between the degrees of construction rules of curves and the curves' symmetry of position and form.

Singularity of Loci

Singularity of loci refers to the property that the non-degenerate rectangular construction rules of curves [rectangular eqs., $F(x,y) = 0$, that are not factorable into the product of several expressions, such as, $f(x,y)g(x,y) = 0$] represent only a single copy of the curve. For example, a rectangular construction rule of a parabola or circle represents only a single parabola or circle. While this property appears to be trivial, it is, in fact, most atypical, yet characteristic of intersecting, directed-distance bilinear systems. Quite different circumstances apply in most other coordinate systems, even the polar one, though the latter fact usually receives little notice.

Positional Symmetry in the Rectangular System

The rectangular system is ideally suited to the purposes for which it customarily has been employed, and with which most readers already are very familiar. It is not ideally suited, however, for studies of the positional symmetries of curves relative to the reference elements of coordinate systems or, equivalently, for comparing the symmetries (structural simplicities) of curves relative to coordinate elements. The three chief reasons for this are the possession of the two atypical properties referred to above and the lack of possession of a *characteristic distance* (see Chapter III, Table 7 and *Elective and Intrinsic Asymmetries of Coordinate Systems*).

In non-bilinear coordinate systems, the following general RULE applies.

RULE 3a. *The greater the positional symmetry of a curve, the lower the exponential degree of, and the simpler, its structural equation.*
(this RULE applies to both construction rules and structure rules).

Because of degree-restriction, the corresponding general RULE for all Cartesian systems is,

RULE 3b. *The greater the positional symmetry of a curve, the simpler its construction rule.*

Circles

This RULE is illustrated first with circles, because orientation of the curve does not play a role, as it does for other curves. The least-symmetrically-positioned circle in the rectangular system is a circle in a general location (Fig. 3a, symmetry index, 5-4), in which h and k are unequal and neither vanishes (h and k are the abscissa and ordinate, respectively, of the center of the circle).

It is evident that an increase in positional symmetry can be achieved by repositioning the circle so that it will be in the same relationship to both coordinate axes, yielding a *general bisector circle*. In reference to construction rule 1a, this occurs when h = k (equality of the abscissa and ordinate of the center), whereupon the center of the circle lies on the coordinate bisector, x = y, yielding eq. 1b (symmetry index, 5-3). At the same time, the number of parameters (h, k, and R) is reduced from three to two (h and R). The symmetry condition of positional equivalence with respect to the coordinate axes (reflective symmetry about the bisector) is designated by the letter A in Fig. 3.

SYMMETRY OF CIRCLES

Table 3 and Figure 3. Symmetry of Rectangular Circles

	no. of terms & parameters*	construction rule	letter symbols	description	
a	5-4 R,h,k,2	$x^2+y^2-2hx-2ky = (R^2-h^2-k^2)$ least symmetrical circle	none	*general circle* (circle in general location)	(1a)
b	5-3 R,h,2	$x^2+y^2-2hx-2hy = (R^2-2h^2)$	A	*general bisector circle* (center of circle on line x = y)	(1b)
c	5-3 R,h,2	$x^2+y^2-2hx-2Ry = -h^2$	B	*general tangent circle* (circle tangent to x-axis)	(1c)
d	5-2 R,2	$x^2+y^2-2Rx-2Ry = -R^2$	AB	*tangent bisector circle* (circle of eq. 1b tangent to axes)	(1d)
e	4-3 h,k,2	$x^2+y^2-2hx-2ky = 0$	C	*general circle, including origin*	(1e)
f	4-2 h,2	$x^2+y^2-2hx-2hy = 0$	AC	*bisector circle, including origin*	(1f)
g	4-3 R,h,2	$x^2+y^2-2hx = (R^2-h^2)$	D	*general axial circle*	(1g)
h	3-2 R,2	$x^2+y^2-2Rx = 0$	BD, BC, or CD	*axial circle, including origin*	(1h)
i	3-1 R	$x^2+y^2 = R^2$	AD	*centered circle*	(1i)
j	2-0	$x^2+y^2 = 0$ most symmetrical circle	ABD, ABC, or ACD	*centered point-circle*	(1j)

Table 3 (continued)

*including non-unit coefficients

[In Tables 3 to 6, for the symmetries of rectangular and polar circles and ellipses, an index of the simplicity and/or degree is given with each eq. For illustrative purposes, the working hypothesis is adopted that: (a) first priority in assessing the overall simplicity of a construction rule goes to its exponential degree; (b) second priority goes to the number of terms; (c) third priority goes to the total number of parameters, non-unit coefficients, and radicals; and (d) within categories in which the sums of parameters, non-unit coefficients, and radicals are the same, fourth priority goes to the eq. with the fewest parameters. Exceptions to category (c) that occur in the cases of general axial rectangular circles (eq. 1g) and general 0°-axial rectangular ellipses (eq. 4s) have their basis in a peculiarity of the rectangular system, as discussed below in the section, *The Overall Simplicity of Structural Equations*.]

Slightly greater simplification of construction rule 1a is achieved for $k = R$, which brings the general circle into tangency with the x-axis (eq. 1c, symmetry index, 5-3)--a *general tangent circle*--designated by the symmetry-condition letter, B. [Although the constants of these eqs. are grouped together and reckoned only as one term, simplifications within this term are taken into account.] Construction rules 1b and 1c simplify further for $h = R$, whereupon the number of parameters is reduced from two to one, yielding eq. 1d (symmetry index, 5-2) and the next-most-symmetrically-positioned circle, a *tangent bisector circle*, that is, a bisector circle that also is tangent to the coordinate axes. The positional symmetry of this circle is designated by the letters, AB.

The next-most-symmetrical location of a circle is not intuitively evident. The working hypothesis is adopted (see Note, Table 3, above) that this is achieved by reducing the number of terms of the eq. from 5 to 4. One way to accomplish this is to have the circle include the origin, that is, by letting $h^2+k^2 = R^2$, whereupon the constant term of eq. 1a vanishes. One point of the circle now is located at the most-symmetrical point in the coordinate system, yielding eq. 1e (symmetry index, 4-3). The condition of a *general circle including the origin* is designated by the letter, C. Accordingly, the working hypothesis amounts to having the condition, C, of including the origin, convey greater positional symmetry than the condition, AB. This assumption is reasonable, since condition C is the first one for which a point of the circle possesses a classical symmetry in the coordinate system.

The next increase in positional symmetry of the circle consists in bringing it into an equivalent relationship to the two coordinate axes, that is, condition, A, for a *bisector circle including the origin*, whereupon the number of parameters of eq. 1e is reduced to one (eq. 1f and symmetry index 4-2). The positional symmetry of this circle is designated as AC.

However, the most-symmetrical rectangular circle with a four-term construction rule is a *general axial circle*, that is, a circle centered in a general position on one of the coordinate axes, for example, the x-axis. This circle is the first to possess a classical symmetry with respect to one of the reference elements. The axial condition is achieved for $k = 0$ (k, the ordinate of the center), whereupon construction rule 1a becomes 1g (symmetry index, 4-3).

Since the circle of eq. 1g is more-symmetrically-positioned than the circle of eq. 1f, it is evident that the vanishing of a linear term in a variable (even in the presence of two parameters, R and h), confers greater symmetry than vanishing of the constant term (and reduction of the number of parameters and non-unit coefficients from three to two; see discussion below). Letting the condition of reflective symmetry in a coordinate axis be denoted by the letter D, this means that, insofar as positional symmetry is concerned, D > AC.

From inspection of eq. 1g, it is evident that the next increase in equational simplicity and positional symmetry of the circle can be achieved by reducing the number of parameters and non-unit coefficients from three to two, and the number of terms from four to three. This is achieved by letting $h = R$, yielding eq. 1h (symmetry index, 3-2), a *tangent axial circle*, tangent to the y-axis, with the symmetry designation, BD (or BC or CD).

An even more symmetrical and simple construction rule than 1h can be obtained by letting both h and k of eq. 1a vanish, giving eq. 1i (symmetry index, 3-1), in which all terms are perfect squares with unit coefficients. Construction rule 1i represents the most-symmetrical non-point circle (Fig. 3i), namely a circle centered at the origin of the coordinate system. This *centered circle* has the symmetry designation, AD.

However, the circle of construction rule 1i is not the most-symmetrically-positioned (nor the most-symmetrical) rectangular circle. It is possible to increase the positional symmetry of the centered circle by bringing the entire circle as close as possible to the most-symmetrical point in the system (the origin), namely by letting R vanish. This reduces the number of terms of eq. 1i from three to two, yielding a *centered point-circle* (eq. 1g and symmetry index 2-0). This is the most-symmetrically-positioned (and most symmetrical) rectangular circle. The symmetry designation of this circle is ABD (or ABC or ACD).

Accordingly, the serial order of positional symmetry (and structural simplicity) of rectangular circles, in terms of the letter designations for positional relationships to the coordinate axes, is as follows.

<u>centered point-circle</u>
ABD (ABC or ACD) > AD > BD > (CD or BC) > D > AC > C > AB > B > A > circle in general location

A = reflective symmetry in the bisector, $x = y$ (same relation to both axes)
B = tangent to an axis
C = including the origin
D = reflective symmetry in an axis

The Overall Simplicity of Structural Equations

Given two structural eqs. of the same degree, the assessment of their comparative simplicities is based upon the number of terms, the number of parameters, the number of non-unit coefficients, the number of radicals, the simplicity of the constant term, the symmetry of the eqs. (in terms of both the presence of, and the coefficients of, the terms in the variables), etc. However, whatever set of rules that one might establish for assessing the comparative simplicities of structural eqs., the resulting assessments would not necessarily correlate with the positional symmetries of the corresponding curves in all coordinate systems, particularly in the cases of coordinate systems with built-in asymmetries (see Table 7).

The rectangular system provides an excellent example of a case in which symmetry of a construction rule in its linear variables does not necessarily confer greater positional symmetry on the curve than asymmetry in these variables (i.e., when one linear variable is absent). Thus, when k vanishes in eq. 1a to yield eq. 1g, the curve is merely translated, rather than transformed to a different curve. In its resulting axial location, it is more-symmetrically positioned than is the case when linear terms in both x and y are present and have identical coefficients, as in eq. 1f. Eq. 1g represents a curve with greater positional symmetry because the curve possesses reflective symmetry in a coordinate axis, whereas the curve of eq. 1f possesses reflective symmetry only in a coordinate bisector.

The view that an axial circle has greater positional symmetry than a bisector circle is supported by considering the (number and) ease of measurements required to specify all points of the curves (see RULE 1c). For the bisector circle of eq. 1f, both the abscissae and the ordinates differ for every pair of points equidistant from the origin, where these pairs can be grouped as follows: $[(a_1,b_1),(b_1,a_1)]$; $[(a_2,b_2),(b_2,a_2)]$; etc. On the other hand, for the axial circle, equidistant pairs of points differ only in the signs of their ordinates, and can be grouped and designated more simply, as follows: $[(a_1,b_1),(a_1,-b_1)]$; $[(a_2,b_2),(a_2,-b_2)]$; etc. As a result, the form of the construction rule required to fit the latter point-pairs (eq. 1g) is regarded as being simpler than the form required to fit the former point-pairs (eq. 1f).

Another way to resolve the matter is to consider the ease of constructing the circles from the given sets of coordinates. In the case of the axial circle, one need merely reflect each point in the x-axis. In the case of the bisector circle, one need merely reflect each point in the coordinate bisector. But first, one must draw the bisector, since it is not a coordinate element. That is the crux of the matter. The axial circle is more symmetrical in the system than the bisector circle because it is reflected in one of the coordinate elements, whereas the bisector circle is reflected in a line that is not a coordinate element.

Ellipses

The least symmetrical rectangular ellipse is one in a general location with a general orientation. The eq. of such an ellipse is obtained from the standard eq. of a centered ellipse (eq. 4a) by rotating the axes of the coordinate system through an angle θ, using the rotation formulae (eqs. 3), yielding eq. 4q (Table 4), and then shifting its center to an arbitrary point in the plane (h,k). After expanding and grouping terms, and relabeling x' and y' as x and y, one obtains the desired general eq. 4a' (symmetry index, 6-10) of Table 4, containing six terms--in x^2, y^2, xy, x, y, and a constant.

(a) $x = x'\cos\theta - y'\sin\theta$,   (b) $y = x'\sin\theta + y'\cos\theta$ rotation eqs.   (3)

$$b^2x^2 + a^2y^2 = a^2b^2 \qquad \text{standard ellipse} \qquad (4a)$$

[One also can translate the curve first and then rotate the axes. In that event, h and k remain the abscissa and ordinate of the center in the unrotated system, so that their values do not give a direct indication of the position of the center in the rotated system. Letting k = 0, for example, does not produce a general axial (x-axis) ellipse (see below), but an ellipse with the ordinate of its center being $h\sin\theta$.]

THE RECTANGULAR SYSTEM

Table 4.   Symmetry of Rectangular Ellipses

terms & parameters*	construction rule	letter symbols	description	
6-10 $a,b,h,$ $k,2,s,$ $c,s^2,$ c^2,sc	$(a^2\sin^2\theta+b^2\cos^2\theta)x^2+(a^2\cos^2\theta+b^2\sin^2\theta)y^2$ $-2x[h(a^2\sin^2\theta+b^2\cos^2\theta)+k(a^2-b^2)\sin\theta\cos\theta]$ $-2y[k(a^2\cos^2\theta+b^2\sin^2\theta)+h(a^2-b^2)\sin\theta\cos\theta]$ $+2(a^2-b^2)xy\sin\theta\cos\theta = [a^2b^2-(b^2h^2+a^2k^2)\cos^2\theta$ $-(b^2k^2+a^2h^2)\sin^2\theta-2(a^2-b^2)hk\sin\theta\cos\theta$	none	*general*	(4a')
6-9 $a,b,h,$ $2,s,c,$ $s^2,c^2,$ sc	$(a^2\sin^2\theta+b^2\cos^2\theta)x^2+(a^2\cos^2\theta+b^2\sin^2\theta)y^2$ $-2hx[(a^2\sin^2\theta+b^2\cos^2\theta)+(a^2-b^2)\sin\theta\cos\theta]$ $-2hy[(a^2\cos^2\theta+b^2\sin^2\theta)+(a^2-b^2)\sin\theta\cos\theta]$ $+2(a^2-b^2)xy\sin\theta\cos\theta = [a^2b^2-(a^2+b^2)h^2$ $-2h^2(a^2-b^2)\sin\theta\cos\theta]$	E	*general bisector* (h = k)	(4b)
6-9 a,b,h $2,s,c,$ $s^2,c^2,$ sc	$(a^2\sin^2\theta+b^2\cos^2\theta)x^2+(a^2\cos^2\theta+b^2\sin^2\theta)y^2$ $+2(a^2-b^2)xy\sin\theta\cos\theta-2hx(a^2\sin^2\theta+b^2\cos^2\theta) -$ $2hy(a^2-b^2)\sin\theta\cos\theta = [a^2b^2+h^2(a^2\sin^2\theta-b^2\cos^2\theta)]$	F	*general axial* (k = 0)	(4c)
6-5 $a,b,h,$ $k,1/2$	$(a^2+b^2)x^2/2 + (a^2+b^2)y^2/2 - (a^2-b^2)xy$ $-x[h(a^2+b^2)-k(a^2-b^2)]-y[k(a^2+b^2)-h(a^2-b^2)] =$ $[a^2b^2-(b^2h^2+a^2k^2)/2-(b^2k^2+a^2h^2/2+hk(a^2-b^2)]$	G	*general 45°* ($\theta = -45°$)	(4d)
6-5 a,b,h $2,1/2$	$(a^2+b^2)x^2/2 + (a^2+b^2)y^2/2 - (a^2-b^2)xy$ $- 2b^2hx-2b^2hy = b^2(a^2-2h^2)$	A (EG)	*general 45°-bisector*	(4e)
6-5 $a,b,1/2,$ $\sqrt{2},\sqrt{}$	$(a^2+b^2)x^2/2 + (a^2+b^2)y^2/2 - (a^2-b^2)xy$ $-\sqrt{2}b^2x\sqrt{(a^2+b^2)} - \sqrt{2}b^2y\sqrt{(a^2+b^2)} = -b^4$	AB (BEG)	*tangent 45°-bisector* $h = \sqrt{[(a^2+b^2)/2]}$	(4f)
6-5 $a,b,1/2,$ $\sqrt{2},\sqrt{}$	$(a^2+b^2)x^2/2 + (a^2+b^2)y^2/2 - (a^2-b^2)xy$ $- \sqrt{2}b^2x\sqrt{(a^2-b^2)} - \sqrt{2}b^2y\sqrt{(a^2-b^2)} = b^4$	A (focus)	*focus-45°* $h = \sqrt{[(a^2-b^2)/2]}$	(4g)
5-10** a,b,h $k,2,s,$ $c,s^2,$ c^2, sc	$(a^2\sin^2\theta+b^2\cos^2\theta)x^2+(a^2\cos^2\theta+b^2\sin^2\theta)y^2$ $-2x[h(a^2\sin^2\theta+b^2\cos^2\theta)+k(a^2-b^2)\sin\theta\cos\theta]$ $-2y[k(a^2\cos^2\theta+b^2\sin^2\theta)+h(a^2-b^2)\sin\theta\cos\theta]$ $+2(a^2-b^2)xy\sin\theta\cos\theta = 0$ (with right side of eq. 4a' = 0)	C	*general, including origin*	(4h)

SYMMETRY OF ELLIPSES

Table 4 (continued)   Symmetry of Rectangular Ellipses

terms & parameters*	construction rule	letter symbols	description	
5-5** a,b,h, k,2	$(a^2+b^2)x^2/2 + (a^2+b^2)y^2/2 - (a^2-b^2)xy -$ $x[h(a^2+b^2)-k(a^2-b^2)]-y[k(a^2+b^2)-h(a^2-b^2)] = 0$ (with right side of eq. 4d = 0)	CG	general $45°$, including origin	(4i)
5-4 a,b, 1/2,√2	$(a^2+b^2)x^2/2 + (a^2+b^2)y^2/2 - (a^2-b^2)xy$ $- \sqrt{2}ab^2x - \sqrt{2}ab^2y = 0$	AC (CEG)	$45°$-bisector, including origin	(4j)
5-5 a,b,h, k,2	$b^2x^2+a^2y^2-2b^2hx-2a^2ky = (a^2b^2-a^2k^2-b^2h^2)$	H	general $0°$	(4k)
5-4 a,b, h,2	$b^2x^2+a^2y^2-2b^2hx-2a^2hy = [a^2b^2-h^2(a^2+b^2)]$	EH	general $0°$-bisector	(4m)
5-4 a,b,h,2	$b^2x^2 + a^2y^2 - 2b^2hx - 2a^2by = -b^2h^2$	BH	general $0°$-tangent	(4n)
5-3 a,b,2	$b^2x^2 + a^2y^2 - 2ab^2x - 2ab^2y = -a^2b^2$	BEH	bitangent $0°$	(4p)
4-6 a,b,2, sc, s²c²	$(a^2\sin^2\theta+b^2\cos^2\theta)x^2+(a^2\cos^2\theta+b^2\sin^2\theta)y^2$ $+ 2(a^2-b^2)xy\sin\theta\cos\theta = a^2b^2$	I	general centered	(4q)
4-3 a,b,1/2	$(a^2+b^2)x^2/2 + (a^2+b^2)y^2/2 - (a^2-b^2)xy = a^2b^2$	GI	$45°$-centered	(4r)
4-4 a,b,h,2	$b^2x^2 + a^2y^2 - 2b^2hx = b^2(a^2-h^2)$	D	general $0°$-axial	(4s)
4-4 a,b,2,√	$b^2x^2 + a^2y^2 - 2b^2x\sqrt{(a^2-b^2)} = b^4$	D (focus)	$0°$-focus	(4t)
3-3 a,b,2	$b^2x^2 + a^2y^2 = 2ab^2x$	BD	tangent $0°$-axial	(4u)
3-2 a,b	$b^2x^2 + a^2y^2 = a^2b^2$	AD	$0°$-centered	(4v)

*includes coefficients; s = sine, c = cosine, in accompanying symmetry index
**less an eliminable parameter

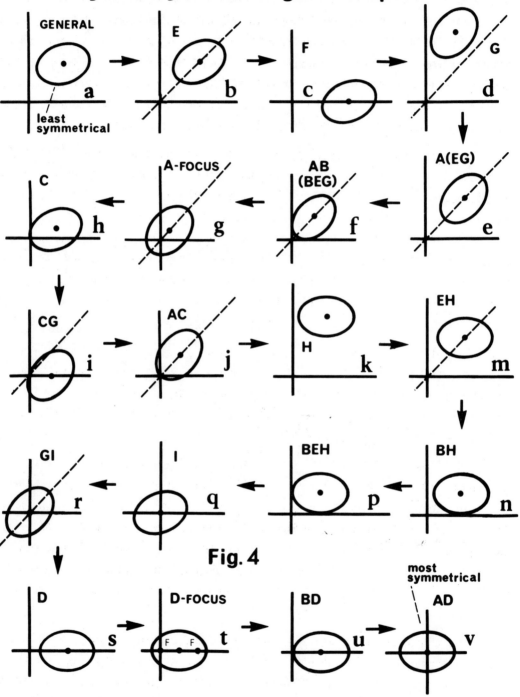

Fig. 4

The smallest increase in the positional symmetry of this ellipse is achieved by placing its center on the coordinate bisector, $x = y$, whereupon $h = k$. The number of parameters and non-unit coefficients thereby is reduced from 10 to 9, and eq. 4a' simplifies slightly, but without decrease in the number of terms, to eq. 4b (symmetry index, 6-9) of a *general bisector ellipse* (Fig. 4b) characterized by the new symmetry condition designated E. Note that the latter differs from the corresponding condition A, for circles, because placing the center of the ellipse on the coordinate bisector does not necessarily bring the ellipse into an equivalent relation to each coordinate axis.

The next-most-symmetrically-positioned ellipse is obtained by letting h or k of eq. 4a' vanish, thereby placing the center of the ellipse on one of the coordinate axes, say the x-axis. This yields the *general axial ellipse* of eq. and Fig. 4c, characterized by the symmetry condition designated F (symmetry index, 6-9). Again, the number of terms does not decrease, but the x, y, and constant terms simplify considerably more than in the change from eq. 4a' to eq. 4b. Similarly, condition F differs from the corresponding condition D, for circles, because the general axial ellipse does not necessarily achieve reflective symmetry in a coordinate axis, whereas the axial circle does.

The next-most-symmetrically-positioned ellipse is attained by making the coefficients of the terms in x^2 and y^2 equal, which is achieved, for example, by letting θ, the initial angle of rotation of the axes, be $-45°$. This yields the *general 45°-ellipse* of eq. and Fig. 4d (symmetry index, 6-5). The corresponding symmetry condition is designated by the letter G.

One obtains the next-most-symmetrically-positioned ellipse by translating the general 45°-ellipse to bring its center onto the coordinate bisector (when $h = k$), yielding the *general 45°-bisector ellipse* of eq. and Fig. 4e (symmetry index, 6-5). This leads to a reduction in the number of parameters from three to two, and a slight simplification of eq. 4d (though the total number of parameters and non-unit coefficients is unchanged). The symmetry condition for this ellipse is A, the same as for the general bisector circle, or, alternatively, EG.

As in the cases of rectangular circles, the next-most-symmetrically-positioned ellipse is established by letting the general 45°-bisector ellipse become tangent to the two axes (Problem 1), that is, the symmetry condition AB or BEG. This provides a better test of the simplification of the construction rule (see RULE 4) than the corresponding case for the circle. In the latter case, h and k simply become equal to R, whereas for the ellipse, the relationship of h and k to the parameters, a and b, is much more complicated, namely, $h = k = \sqrt{[(a^2+b^2)/2]}$. Nevertheless, the constant term simplifies.

> RULE 4. *A curve that includes, or is tangent to, a reference element generally is more-symmetrically positioned in the coordinate system (has a simpler construction rule) than a congruent neighboring curve that does not include, or is not tangent to, the reference element* (see also RULE 40).

[By a "congruent neighboring curve" is meant a congruent curve infinitesimally displaced to an unexceptional position, in this case to a position in which the reference element is not included nor tangent to the curve.]

This is one of only two possible values for h (and k) that leads to a significant simplification of the constant term of eq. 4e and does not introduce new combined functions of a and b (i.e., other than a^2+b^2 and a^2-b^2). The other value of h that accomplishes a comparable simplification is $\sqrt{[(a^2-b^2)/2]}$, which places the origin at the near traditional focus (eq. and Fig. 4g). This [The term "traditional focus" usually is employed to distinguish the classically recognized foci of conics from other circumpolar foci; see Chapters VII and VIII.] offers a hint of the exceptional symmetry status of the traditional focus in coordinate systems that include a point or circle reference element. The corresponding eqs. in these coordinate systems both simplify and reduce in degree. [Letting h be a function of a+b or a-b does not achieve a corresponding simplification.]

Only in the parabola does the chord connecting the points of tangency of the axes with rectangular conic sections with the symmetry condition AB (or BEG) lie at the level of the traditional focus. This follows from RULE 5, since only the parabola has an eccentricity of 1. Similarly, the parabola is the only conic for which the point at the origin (the point of intersection of the two tangents) is at the foot of a directrix.

> RULE 5. *The magnitude of the slope of a tangent to a conic (in standard position) at its latus-rectum vertices (the intersections of the latera recta with the curve) is equal to the conic's eccentricity.*

Following the working hypothesis that, within a given degree category, the number of terms takes priority in determining a rectangular eq.'s simplicity, the next increase in positional symmetry is achieved with a *general ellipse including the origin*, that is, symmetry condition C. This occurs when the constant term of eq. 4a' vanishes, yielding eq. 4h, with one less term than eqs.

4f and 4g but one more parameter (either h or k can be eliminated by using the condition for vanishing of the constant term of eq. 4a'). In considering eq. 4f (or 4g), as compared to 4h, one must bear in mind that the trigonometric functions merely represent numerical coefficients. When these are replaced by specific numerical values, as in the cases of 0° and 45°-ellipses, the eqs. greatly simplify.

[These sequences of positional symmetry are not intended to be complete or definitive. For example, a general ellipse tangent to only one axis is omitted, i.e., condition B, as is a *general tangent bisector ellipse*, i.e., condition BE. Nor are the details of ordering them according to equational simplicity of great importance. The working hypothesis of the note following Table 3 is followed only for illustrative purposes. The general relations between positional symmetry and equational simplicity would be clearly evident even if details of the specific ordering were to be altered.]

The next-most-symmetrically-positioned rectangular ellipse is obtained by having the major (or minor) axis of the preceeding *general ellipse including the origin* make an angle of 45° (or -45° or ±135°) with the x-axis (a -45° rotation of the axes), leading to eq. 4i. The symmetry condition for this *general 45°-ellipse including the origin* is designated by the letters CG (Fig. 4i).

An increase in the positional symmetry of the ellipse of Fig. 4i is achieved by positioning its center on the bisector, $x = y$, yielding eq. and Fig. 4j. The symmetry condition for this *45°-bisector ellipse including the origin* is designated AC or CEG. Since h and k now equal $\sqrt{2}a/2$, only the parameters, a and b of the ellipse itself appear in eq. 4j.

The next-most-symmetrically-positioned rectangular ellipse is the first to have a line of symmetry parallel to a coordinate axis, rather than a coordinate bisector, that is, a *general 0° or 180°-ellipse* (eq. and Fig. 4k). The construction rule of this ellipse not only possesses only 5 terms, it possesses no 2nd-degree term in xy. The coefficients of the x^2 and y^2-terms, of course, must be unequal for an ellipse with this orientation (equality of the coefficients would make the ellipse of Fig. 4k circular), just as they must be equal for 45°-ellipses. The symmetry condition for the general 0°-ellipse is designated by the letter H.

The next-most-symmetrically-positioned rectangular ellipse (eq. and Fig. 4m) is obtained by placing the center of the general 0°-ellipse on the $x = y$ bisector (letting $h = k$), yielding a *general 0°-bisector ellipse*, with symmetry condition designated by the letters EH. Only a modest simplification of eq. 4k is achieved by this change of position, since it represents only a positioning of the center of symmetry of the ellipse on the most-symmetrical locus (the x = y bisector) in the 1st and 3rd-quadrants.

Further increases in the positional symmetry of general 0°-ellipses can be achieved by making them tangent to the x-axis, as represented by eq. and Fig. 4n, or to both axes (eq. and Fig. 4p). The first of these is the *tangent general 0°-ellipse*, with symmetry condition designated BH, and the second is the *bitangent 0°-ellipse*, with symmetry condition designated BEH.

A much greater simplification of eq. 4a' is achieved for a general ellipse centered at the origin (eq. and Fig. 4q), for which h = k = 0. This is the first condition for which the center of symmetry of the ellipse is at the center of symmetry of the coordinate system. This *centered general ellipse* has an eq. of only four terms, with symmetry condition designated I. Again, the relative simplicity of eq. 4q as compared to 4p is increased if the trigonometric functions are represented by some specific values, as in eq. 4r.

Simplification of eq. 4q, giving a much-more-symmetrically-positioned curve, is obtained by letting θ = 45°, leading to equality of the coefficients of the x^2 and y^2-terms and the quite simple four-term eq., 4r. The symmetry condition for this *centered 45°-ellipse* is designated GI. The next-most-symmetrically-positioned ellipse is the first to achieve the classical symmetry condition, D, in which every point of the ellipse is reflected in one of the coordinate axes (in condition C, only one point of the curve is positioned with a classical symmetry). Eq. 4s, for this *general 0°-axial ellipse* possesses four terms but no xy-term. Placing a traditional focus of ellipse 4s at the origin gives eq. and Fig. 4t, but has little influence on the simplicity of eq. 4s.

The positional symmetry of the general 0°-axial ellipse of eq. 4s can be increased markedly by eliminating the parameter, h, by letting h = a, with the vanishing of the constant term. This brings the curve to a position of tangency and, since the origin now is included, the constant term vanishes, yielding the three-term eq. 4u and Fig. 4u, for the *tangent 0°-axial ellipse*, with symmetry condition designated BD.

Finally, the most-symmetrical and most-symmetrically-positioned rectangular ellipse is obtained by centering the general 0°-axial ellipse (h = 0), yielding a *centered 0°-ellipse*, which is in "standard position" and possesses the symmetry condition AD (eq. and Fig. 4v). The only more symmetrical rectangular "ellipses" that can be obtained by simplification of eq. 4v, as noted above, are for $a = b = R$, yielding the centered circle, and $R = 0$, yielding the centered point-circle.

Though there might be some question as to the precise serial ordering of these rectangular ellipses according to equational simplicity and positional symmetry, the major outlines of the ordering are clear. If the ellipses are subdivided into three groups--general $\alpha°$-ellipses (θ an arbitrary angle, α), general 45°-ellipses, and general 0°-ellipses--yielding the three sequences of eqs.: 4a', 4b, 4c, 4h, 4q; 4d, 4e, 4f, 4g, 4i, 4j, 4r; and 4k, 4m, 4n, 4p, 4s, (4t), 4u, 4v; the progressions are more clear-cut.

The sequence of symmetry conditions for rectangular ellipses is summarized below.

centered 0° = AD > BD > D (focus) > D > GI > I > BEH > BH > EH > H > AC > CG > A (focus) > AB > A > G > F > E > general

A = reflective symmetry in the bisector, $x = y$ (same relation to both axes)
B = tangent to an axis
C = including the origin
D = reflective symmetry in an axis
E = centered on a coordinate bisector
F = centered on a coordinate axis
G = with major axis at 45° to x-axis
H = with a line of symmetry parallel to a coordinate axis
I = centered at the origin

The "Rectangular" System with Undirected Distances

It will be noted that RULES 1a-c and 2a do not apply in the rectangular system. Thus, the simplest and lowest-degree eq. in two variables is $x = y$ (Fig. 5a). This does not represent the most-symmetrical curve in the system, though, because it describes only the coordinate bisector in the 1st and 3rd-quadrants. To achieve the most-symmetrical locus, one must include the bisector in the 2nd and 4th-quadrants, $x = -y$ (Fig. 5b). Thus, the "most-symmetrical locus" includes both bisectors and has the eq., $(x-y)(x+y) = 0$, or $x^2 = y^2$ (Fig. 5c), which is not the simplest and lowest-degree rectangular eq.
[References to the "most-symmetrical locus" generally exclude the coordinate elements themselves and their points of intersection with one another.]

The basis for the discrepancy with RULES 1a-c and 2a is that the rectangular system has built into it two arbitrary asymmetries (see Table 7). The one that is of concern at this juncture is the use of directed distances (i.e.,

"Rectangular" Coordinates with Directed & Undirected Distances

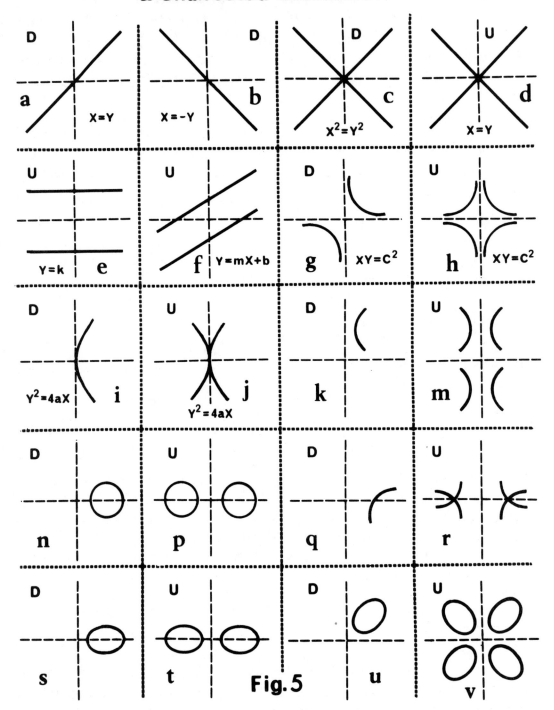

Fig. 5

distances that are measured in both a positive and a negative sense). If this a-symmetry is eliminated by employing undirected distances, the eq. x = y then represents both coordinate bisectors (Fig. 5d), and RULES 1a-c and 2a apply. Other linear eqs. then represent parallel line-pairs and are the next-most-symmetrical loci (Fig. 5e,f). Of these, those that are parallel to one or the other coordinate axis are most symmetrical (known classically as "coordinate curves"), with eqs., x = h and y = k, which are the simplest eqs. of parallel line-pairs (Fig. 5e).

[Strictly speaking, the latter eqs. are of 0-degree and are known as *incomplete-linears*, since both variables do not appear. The loci they represent are the most symmetrical with respect to *single* reference elements.]

If one were to try to compare the relative symmetries of 2nd-degree conic sections in the classical rectangular system, one would have to make a choice between central conics (conics with two lines of symmetry) and the parabola. Central conics would be judged to be more symmetrical, since they can be more-symmetrically positioned with respect to the coordinate axes (Fig. 5g). However, this is only an artifact of the employment of directed distances. If one employs undirected distances, instead, parabolas also become central conics (as they are in some other coordinate systems)--they always have two or four arms and are reflected in both coordinate axes (Fig. 5j,m). Similarly, circles and ellipses have one, two, or four arms, and hyperbolas have two, four, or eight arms (Fig. 5h,p,t,v).

Employing the "rectangular" system with undirected distances, the decision as to the comparative symmetry of 2nd-degree curves can be made upon the basis of their eqs. Excluding from consideration the eqs., $x^2 = y^2$ (which duplicates the locus x = y) and $x^2 = h^2$ and $y^2 = k^2$ (which duplicate the loci x = h and y = k), one inevitably arrives at a comparison of the loci of the three construction rules, 5a-c. Of these, the centered point-circle is indubitably the most symmetrical and most-symmetrically-positioned 2nd-degree locus.

(a) $x^2 + y^2 = 0$ centered point-circle   (5)

(b) $x^2 + y^2 = R^2$ centered non-point-circle, $R \neq 0$

(c) $y^2 = 4ax$ two-arm vertex parabola

Of the other two loci, the parabola would be judged to be most symmetrical on the basis of the number of terms, but the circle would be most symmetrical on the basis of symmetry in the 2nd-degree terms, x^2 and y^2 (both in the fact of the presence of both terms and in the equality of their coefficients). Although priority was given to the number of terms in the above treatments, we did not consider a case of asymmetry in the terms, x^2 and y^2 (i.e., a case of the absence of one or the other quadratic term), but only in the terms in x and y, which did not influence the shape of the curve.

In the present instance, in which different curves are being compared, the centered non-point-circle is considered to be a more symmetrical and more-symmetrically-positioned curve than the two-arm parabola, and its construction rule is considered to be simpler than that of the two-arm parabola. This conclusion is based on RULE 1c. An entire circle can be plotted upon the basis of measurements made on only one octant (or less) of the curve, by employing reflections in the coordinate axes and bisectors. On the other hand, an entire quadrant of measurements are required to plot the two-arm parabola. In other words, although otherwise quite comparable, the circle has more lines of symmetry than the two-arm parabola.

[In applying RULE 1c, the measurements entailed (specified by the construction rules of the curves) are judged to be of roughly equivalent simplicity. Thus, given x, to obtain y for the parabola, one needs to multiply the root of a product by a constant, i.e., $2\sqrt{(ax)}$, while to obtain y for the circle, one needs to take the root of the difference between a constant (R^2) and a square, i.e., $\sqrt{(R^2-x^2)}$. Both measurements involve a root, but that for the parabola also involves two products, and that for the circle, a product and a difference.]

Though the reader may find it strange to deal with two-arm parabolas, four-arm circles, etc., in actuality it is the familiar classical situations, known to us from working exclusively in the rectangular and polar systems, that should be regarded as the relative oddities. This will become clear in following treatments. We also will come to realize that generalized symmetry (structural simplicity of form and position) is a purely relative concept; the locus that is most symmetrical in one coordinate system is not necessarily most symmetrical in another. It will be found, though, that one conic section or another is the most-symmetrical locus in systems that include a point or circle reference element.

Advantages of the Rectangular System with Directed Distances

When pointing out the limitations of the rectangular system engendered by the built-in asymmetry of directed distances, it is not meant to imply that this is a disadvantage in other applications. Quite the contrary situation applies; it is because of its built-in asymmetries that the rectangular system possesses both its unique suitability for classical studies and analyses of curves, and for certain crucial applications in generalized symmetry studies. All are familiar with the great utility and versatility of the system in plane analytic geometry, and many also will appreciate its utility for the calculus. At this juncture, the main features responsible for this utility are considered.

Singularity of Loci and Uniqueness of Identity

The property of *singularity of loci* was considered above, referring to the fact that plots of non-degenerate rectangular construction rules produce only a single copy of a curve. *Uniqueness of identity* of curves is tacit to the above assertion, since it is implied that, say, a curve with two arms, such as a hyperbola, would be regarded as a single curve, rather than two one-arm hyperbolas.

It is an advantageous property of the rectangular coordinate system that each curve plotted in it has the same appearance, regardless of its position in the system. In other words, if one begins with the standard eq. of a parabola, ellipse, hyperbola, or any other curve, and performs translations and rotations (or expansions or contraction) upon it, the resulting loci, no matter where their new locations in the system, will be congruent (or similar) to the initial curve. This is the property of *uniqueness of identity*. Because this property applies in both the rectangular and polar systems, it has monopolized our views of the appearance and identity of curves.

Most other coordinate systems do not possess the practical advantage of uniqueness of identity. As is illustrated in Fig. 5j,m,p,r,t, and v, uniqueness of identity does not apply in the undirected-distance "rectangular" system, where circles may have one, two, or four arms, parabolas may have two or four arms, and hyperbolas may have two, four, or eight arms, depending upon location and orientation. While these examples are sufficient to make the point, they give no hint of the bizarre forms that can be taken on by differently-positioned curves in many other coordinate systems (see Chapters IV-VI).

Degree-Restriction

A second advantage of the rectangular system is possessed by virtue of the fact that it is an intersecting bilinear system. This advantage is shared with all such systems, regardless of the angle between the two lines. *Degree-re-*

striction is the property that the construction rules of a given curve, though they may be infinite in number--with a different construction rule for every position--have the same exponential degree.

Thus, every eq. of a line (in both variables), whether it has two or three terms, is of 1st-degree; every eq. of a circle, whether it has two, three, four, or five terms, is of 2nd-degree; every eq. of an ellipse, whether it has three, four, five, or six terms, is of 2nd-degree, etc. A 2nd-degree eq. never represents a line, a 4th-degree eq. never represents an ellipse, etc., and two eqs. of different degrees never represent the same type of curve (say, ellipses), as typically occurs in other coordinate systems.

A Line Has a Maximum of n Real Intersections
    with a Curve

Uniqueness of identity has numerous practical conceptual consequences, if only because of the great "tidiness" and order it brings to the rectangular system. One aspect of this order is the fact that a line has n algebraic intersections with an nth-degree rectangular curve. By this is meant that, if one solves simultaneously the eqs. of a given line and curve of exponential degree n, there will be n points of intersection (x_1,y_1; x_2,y_2; x_3,y_3; etc.) of the line with the curve. Some or all of these points of algebraic intersection may be imaginary.

On the other hand, given a real (as opposed to imaginary) rectangular curve of degree n, if it is cut geometrically with lines at various angles, one often can find n real points of intersection. Thus, lines will be cut once, circles, ellipses, parabolas, and hyperbolas twice, the Folium of Descartes three times, Cartesian biovals and the lemniscate of Bernoulli four times, etc. In the undirected-distance system, this is not the case; the maximum number of points of intersection reaches 2n.

Advantages for the Calculus

The advantages of the rectangular system for the calculus are evident, because of the correspondence between the derivatives, dy/dx and dx/dy, at given points of curves to the slopes of tangents at these points to the x and y-axes, respectively, and the second derivatives, d^2y/dx^2 and d^2x/dy^2 to the rates of change of these slopes. Similarly, the differential of arc-length, ds, is readily obtained as, $(ds/dx)^2 = 1 + (dy/dx)^2$, as are the curvature, radius of curvature, etc. Inasmuch as few such calculations will be employed, the advantages in question are not pursued further, though they are recognized to be very great. It should be emphasized, though, that, except for the polar system, calculations employing the calculus in other systems typically are much more complex, particularly in non-bilinear systems.

Advantages for Certain Generalized
    Symmetry Studies

The rectangular system is uniquely suited for dealing with the structure rules that characterize inherent structural properties of curves. No other non-bilinear system involving two coordinate distances, that could conceivably be employed for this purpose would give unique characterizations, since no other two-element system is degree-restricted. In other words, if structure rules were to be plotted as structural curves in, say, bipolar coordinates, two structure rules of different degrees, expressing quite different structural properties of curves, could give bipolar loci differing only in their positions or the number of congruent segments possessed (for example, a one-arm, as opposed to a two-arm, hyperbola).

In rectangular coordinates, however, non-degenerate eqs. of different degrees (eqs. that cannot be broken down into a product of factors equal to 0, each of which, taken alone, yields a separate curve) always give dissimilar curves. Furthermore, if two structure rules of the same degree give congruent or similar loci in different locations, the difference in location accurately and meaningfully expresses the underlying structural difference. Thus, the 180°-linear structure rule, $X = Y$, about a point, characterizes a curve with a classical rotational symmetry, whereas the 180°-linear structure rule, $X = Y + a$, does not (though it represents a line parallel to the line, $X = Y$).

Symmetrical Coverage of the Entire Plane

From the perspectives gained in dealing with undirected intersecting bilinear systems, in general, one comes to realize that the "rectangular" system with *undirected distances* is not a single coordinate system, but two congruent systems which, taken together, cover the entire plane. The *45°-undirected bilinear system* (Fig. $6a_4$-d_4), for example, encompasses only the two segments of the plane that lie between the coordinate lines in the 45°-sectors. The remainder of the plane comprises the two sectors of the *135°-undirected bilinear system* (containing the corresponding dashed curves).

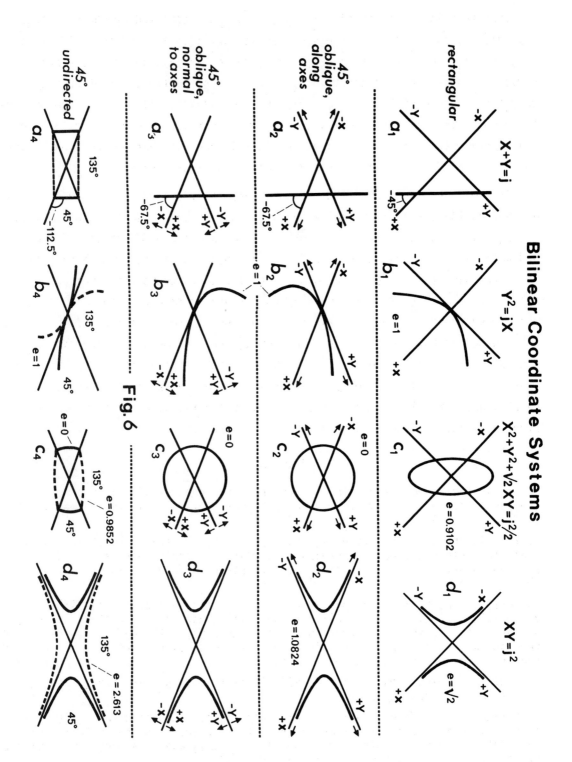

Fig. 6

That this is not an artifical distinction is emphasized by the fact that a single eq. sometimes represents different curves in the 45° and 135°-sectors (Fig. 6c_4,d_4, solid versus dashed curves, respectively), and that the same curves often are represented by different eqs. in the two pairs of sectors. Only for an angle of 90° do these distinctions vanish, allowing the two systems to become joined as one, namely, the system referred to above as the *"rectangular" system with undirected distances* (in which curves that are continuous between the pairs of sectors also have continuous first derivatives).

On the other hand, the distinction between curves in different sectors (say, 45° versus 135°) vanish when directed distances are employed (Fig. 6a_1-d_1, a_2-d_2, a_3-d_3), that is, distances on one side (or along one direction) of a coordinate line are taken to be positive, and those on the other side (or along the other direction) are taken to be negative. But, except for lines and parabolas, the same construction rule represents different curves in non-supplementary systems, say the 90°-system as compared to the 45°-system (compare Fig. 6c_1 and 6c_2 or 6d_1 and 6d_2). Only in rectangular systems do construction rules with symmetry in their coefficients, such as $b^2x^2 \mp a^2y^2 = a^2b^2$, when a = b, also represent curves with symmetry of their axes (i.e., axes of equal length), namely, equilateral hyperbolas (upper sign) and circles, as opposed to other hyperbolas and ellipses.

Problems

1. Derive the values of h and k that bring the general 45°-bisector ellipse (eq. and Fig. 4e) into tangency with both coordinate axes in the 1st-quadrant (eq. and Fig. 4f).

2. What is the transformation, $x_0 = F(x,y)$, $y_0 = G(x,y)$ that carries oblique coordinates into rectangular coordinates? Obtain the rectangular eq. of the oblique curve, $y_0^2 = 4ax_0$. Put this eq. into standard form for an initial 45°-oblique system.

3a. What is the transformation, $x_b = f(x,y)$, $y_b = g(x,y)$ that carries 45°-undirected-distance bilinear coordinates in the 1st-quadrant into rectangular coordinates (i.e., the 45°-sector in the 1st-quadrant, with the x_b-axis being the rectangular x-axis and the y_b-axis being the rectangular coordinate bisector in the 1st-quadrant, as in Fig. 6a_1-d_1, a_4-d_4)?

b. Apply the transformation to the 45°-bilinear eq., $y_b^2 = 4ax_b$, to obtain the corresponding rectangular eq.

c. Rotate the rectangular axes to bring the x-axis into coincidence with the line of symmetry of the curve and express the new eq. in standard form. [Remember that x_b and y_b are distances from the x_b and y_b-axes, respectively, not the y_b and x_b-axes (as in rectangular coordinates).]

d. What are the general eqs. for y_b for the 1st and 2nd-quadrants of θ-bilinear undirected-distance systems?

4. Derive the eccentricity of the ellipse, $x_b^2 + y_b^2 + \sqrt{2}x_by_b = j^2/2$, in the 145°-sector of the 1st and 2nd-quadrants in the 145°-undirected-distance bilinear system (Fig. 6c₄).

5. Derive the eccentricity of the hyperbola, $x_oy_o = j^2$, in the 45°-oblique system (Fig. 6d₂).

6. Prove RULE 5 for ellipses.

CHAPTER III.   THE POLAR COORDINATE SYSTEM

Introduction

The polar coordinate system consists of a directed half-line, as a reference element for angles, and a (terminal) point or pole, as a reference element for distances. It was the first system to be devised after the invention of the Cartesian systems (rectangular and oblique coordinate systems). It made its first appearance with Newton (1642-1727) in his *Methods of Series and Fluxions* (essentially his version of the calculus) in 1671. It was his "Seventh Mode; For Spirals," of nine suggested methods for determining tangents to curves. The Swiss mathematician, Jakob (Jacques) Bernoulli (1645-1705), however, was the first to publish the polar coordinate system and use it in a general manner (*Acta Eruditorum*, 1691 and 1694), rather than only for spirals.

Only rectangular coordinates exceed polar coordinates in their utility and universal use. In fact, these two systems (and oblique coordinates) were used almost exclusively until toward the close of the 18th century. Only then did geometers break away from them in situations where the particular necessities of a problem indicated the greater suitability of some other coordinate approach. In a sense, the polar system is a "stepping stone" to other planar Euclidean coordinate systems; it deviates the least of all non-bilinear systems in its departures from the properties of bilinear systems.

The polar coordinate system is the only system with a single coordinate and single reference element for distance. In consequence, the distance from its point reference element to any point of a curve is expressible directly as the value of a single coordinate. This feature makes the system particularly useful in the study of spirals and rotations, and in the investigation of motions under the action of central forces, such as those of planets and comets. Additionally, the polar system often makes possible the easy recognition and analysis of a curve as the algebraic resultant of two or more other curves, in the sense that the points of the curve in question are obtained by summing, subtracting, multiplying together, or dividing, the distances in the θ-directions to the "component curves," that is, $r = r_1 + r_2$, $r = r_1 - r_2$, $r = jr_1/r_2$, or $r = r_1 r_2/j$ (or combinations thereof).

[The letter "j" represents the unit of linear dimension, which is used to maintain dimensional balance. This greatly aids in the derivation of complicated algebraic eqs., primarily through facilitating error detection. It also helps to prevent carelessness by keeping the student alert to the fact that eqs. that are not in dimensional balance are, strictly speaking, meaningless. In the same vein, the use of symbols, such as A^2 and B^2, as abbreviations for longer expressions, indicates that the expressions have the dimensions of distance squared, i.e., $A^2 = j^2 A$ and $B^2 = j^2 B$, not that $A^2 = A \cdot A$ and $B^2 = B \cdot B$. The capital-letter abbreviations often are used here with this connotation.]

Polar coordinates also are the "natural" coordinate system of the inversion transformation, for which j^2/r is plotted in place of r. The fact that the product of two inversions of an initial curve (i.e., an inversion followed by a second inversion) about the same point restores the initial curve, has far-reaching consequences (in terms of fundamental properties of the transformation).

Already the most important non-trivial geometrical transformation (i.e., excluding rotations and translations), inversion seems destined to assume even greater importance, and is ideally suited for the organization and classification of curves. The polar coordinate system also is uniquely suited for the derivation of *structure rules*, which are the most powerful tools for assessing structural relations within and between curves (see below and Chapters VII and VIII).

Bases for the Widespread Use of Polar Coordinates

The traditional abundant use of polar coordinates was not related to the system being a "stepping stone" to other Euclidean systems, as mentioned above, nor to its unique suitability for deriving structure rules. Instead, the main reasons for its widespread use are:

(1) the fact that the rectangular construction rules of higher-degree curves (3rd-degree and higher), even including curves of infinite degree, often take very simple polar forms;
(2) the ease with which these polar construction rules can be manipulated (solved simultaneously, differentiated, etc.); and
(3) the ease with which polar curves can be plotted (using polar-coordinate paper).

As an example in category (2), in the calculus, in which polar coordinates receive great use, the differential of distance is simply dr, the square of the differential of arc length is simply $ds^2 = r^2 d\theta^2 + dr^2$, the differential of area is simply $dA = r^2(d\theta)/2$, etc. These advantages have so dominated views on the utility of polar coordinates that the significance of the system's ability to provide these facilitations has received little attention. This Chapter is concerned mostly with the properties of the polar system and relations between the positional symmetry of polar curves and their construction rules--topics that consume no more than a few sentences or paragraphs in classical treatments.

Curve Plotting

All readers probably are familiar with the classical method of plotting curves in polar coordinates, so it is reviewed here only very briefly (see also Figs. 2b, 7, and 8, and the section, *Polar Curves as Resultants*). Polar construction rules often can be expressed in the form $r = f(\theta)$ or $r^n = f(\theta)$, as illustrated by eqs. 6 and 7. To obtain the plot of a curve, $r = f(\theta)$, one assigns a value to θ and evaluates $f(\theta)$. One then plots a point at the distance, $r = f(\theta)$, from the pole or origin on a radius vector extending therefrom at an angle θ to the directed half-line (measured CCW for positive θ). The latter also extends from the pole and is known as the polar-axis (the 0°-axis or initial half-line; Figs. 2b and 7). This process is carried out for all values of θ (generally values from 0° to 360° suffice; see below). Sometimes it is easier to assign values of r and compute corresponding values of θ. In still other cases, neither r nor θ is obtained readily, and eqs. of 3rd-degree or higher have to be solved.

The Degree of a Polar Construction Rule

Classicists have not dealt explicitly with the question of the degree in r of a polar construction rule. It is dealt with implicitly, though, in the sense that polar eqs. only very rarely are expressed with θ as a component of a radicand. The construction rule for the nephroid (eq. 6j, below) is one of the rare exceptions. Accordingly, the classical treatments implicitly apply the same principles to polar construction rules that are applied to rectangular ones: the degree of a polar construction rule in r is that of the highest-degree term, after clearing the denominators and radicands of r, and after clearing radicands involving θ.

At first sight, the rationale for the classical approach appears to be substantial. Thus, the taking of roots in attempting to reduce the degrees of polar eqs. often leads to alternate signs before radicals whose radicands involve θ. If the situation were analogous to that in rectangular coordinates, the degree of the eq. would not be reduced by root-taking that leads to alternate signs; this device merely would express the eq. in an alternate form of the same degree, namely, as the degenerate product of the terms that yield each root [thus, $x^2 = y^2$ becomes $x = \pm y$ or $(x+y)(x-y) = 0$]. For example, the degree of the eq., $r^2 = a^2(\cos^2\theta - \sin^2\theta)$, would not appear to be reduced by re-expressing it as $r = \pm a\sqrt{(\cos^2\theta - \sin^2\theta)}$.

However, the polar coordinate situation is not analogous to that in rectangular coordinates. In polar coordinates, the radical with each alternate sign often is for the same locus--the locus for one sign merely is 180° out of phase with the locus for the other sign. Accordingly, one is justified to discard one of the roots, whereupon the degree of the construction rule is halved.

In view of this circumstance, and also considering the matter of the degree of a polar construction rule from the point of view of positional symmetry (see Tables 5 and 6), the classical treatment is seen to be flawed. Such considerations have led to the development of an operational definition of the degree of a polar construction rule for curves of finite degree that is quite different from the definition that is implicit in the classical approach. This is embodied in RULE 6.

> RULE 6. *The degree in r of a polar construction rule of a curve of finite rectangular degree is defined by the lowest-degree, non-degenerate equation that gives a single complete trace of the curve for a single sweep of values of θ ranging between the limits set by two limiting tangent-lines from the polar pole to the curve* (by definition, the limiting tangent-lines intersect the curve only at tangent-points; see Fig. 7).

In essence, RULE 6 demands a strict adherence of the domain of applicability of the polar construction rule to the location of the curve in the system. But before RULE 6 may be applied, the polar construction rules have to be "standardized" as follows:

(1) the only trigonometric functions employed are sines and/or cosines;
(2) The sines and cosines are for the whole angle, θ, only (neither augmented nor diminished by a constant); and
(3) at a maximum, θ varies over a range of 360° (usually 0° to 360°).

In the absence of these standardizations (for curves of finite rectangular degree), the polar system admits of several different ways of expressing construction rules of a given curve in a given location and orientation (see, for example, eqs. 13a,a'), encompassing eqs. with both different numbers of terms and different degrees in r.

It is implicit in the "standard polar system" that the degree of the polar construction rule of any curve of finite rectangular degree, n, which does not include and/or enclose the origin, is n, the same as the rectangular value (see RULE 7a, below). This is because the construction rule must yield a value

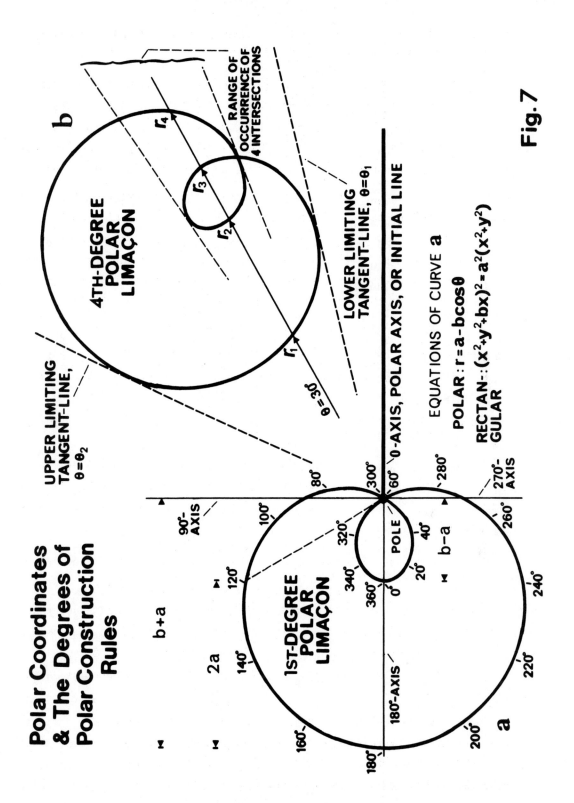

Fig. 7

for r for every point at which a radius vector at a given angle, θ, intersects the curve (for example, the 4 points, r_1-r_4, at which the 30°-radius vector intersects the 4th-degree polar and rectangular curve of Fig. 7b) over a limited range of values (for example, the range between and including the solid limiting tangent-lines in Fig. 7b). There may be as many as n such intersections (i.e., real roots of the polar eq.) at each angle for a curve of polar and rectangular degree n (just as there are 4 points of intersection of radius vectors for every angle between the dashed tangent-lines of Fig. 7b).

There relations often do not apply in the unstandardized system, regardless of whether curves include and/or enclose the pole. Instead, θ may range through 360° or more, as required, so that a radius vector may lie in a given direction for more than one value of θ, and may have different intersections for these different values.

It also is implicit in the standard polar system that the degree of a construction rule for a curve of finite rectangular degree that includes and/or encloses the origin, may be much lower than that of a curve that does not. This becomes possible because multiple intersections may be for two or more values of θ (as in Fig. 7a, where the $\theta = 0°$ intersection at a-b is codirectional with the $\theta = 180°$ intersection at a+b).

Furthermore, the number of codirectional intersections may be much less than n, since some radius vectors from the origin may transect only small portions of the curve. In essence, standardizing polar construction rules removes the ambiguities, redundancies, and the possible multiplicities of eqs. of the unstandardized (classical) system. In the following, the standard polar system often is referred to simply as the *standard* system.

Advantages and Shortcomings of Polar-Equation Shortcuts

Eqs. 6 and 7 illustrate the remarkable ability of the unstandardized polar system to express in concise form the construction rules for curves with high (eqs. 6) and infinite (eqs. 7) values of rectangular degree. This most useful property of the classical polar system and the past tendency to exploit it to the fullest, have several heretofore unappreciated shortcomings. The first of these is a consequence of the use of convenient multiple and fractional angles (eqs. 6a,c,e,g-j), of convenient trigonometric functions (eqs. 6a-c,f,g), of convenient orientations (all eqs. 6) and, above all, of convenient points in

POLAR EQUATIONS

rectangular equation	polar equation	name	
$(a-x)y^2 = x^2(3a+x)$	$r = a\sec(\theta/3)$	trisectrix of Maclaurin	(6a)
$x(x^2+y^2) = 2ay^2$	$r = 2a\sin\theta\tan\theta$	cissoid of Diocles	(6b)
$4(a-x)^3 = 27a(x^2+y^2)$	$r = (a/4)\sec^3(\theta/3)$	Tschirnhausen's cubic	(6c)
$(x^2+y^2+bx)^2 = a^2(x^2+y^2)$	$r = a - b\cos\theta$	limacons	(6d)
$(x^2+y^2)^2 = a^2(x^2-y^2)$	$r^2 = a^2\cos 2\theta$	lemniscate of Bernoulli	(6e)
	$r^2 = a^2(\cos^2\theta - \sin^2\theta)$	eq. 6e in another form	(6e')
$(x-b)^2(x^2+y^2) = a^2x^2$	$r = a + b\sec\theta$	conchoid of Nicomedes	(6f)
$y^2(x^2+y^2) = a^2x^2$	$r = a\cot\theta$	Kappa curve	(6g)
$(x^2+y^2)^2 - d^2(x^2-y^2)/2 = d^4(C^2-1/16)$	$r^4 - (d^2r^2\cos 2\theta)/2 = d^4(C^2-1/16)$	Cassinian ovals (bipolar eq., $uv = Cd^2$)	(6h)
$x^4 = a^2(x^2-y^2)$	$r^2 = a^2\cos 2\theta \cos^4\theta$	lemniscate of Gerone	(6i)
$(x^2+y^2-4a^2)^3 = 108a^4y^2$	$(r/2a)^{2/3} = [\sin(\theta/2)]^{2/3} + [\cos(\theta/2)]^{2/3}$	nephroid	(6j)
$4(x^2+y^2-ax)^3 = 27a^2(x^2+y^2)^2$	$r = 4a\cos^3(\theta/3)$	Cayley's sextic (a = twice small-loop diam.)	(6k)
none	$r = a\theta$	spiral of Archimedes	(7a)
none	$r = a(\sin\theta)/\theta$	cochleoid	(7b)
none	$r = be^{a\theta}$	logarithmic spiral	(7c)
none	$r^2 = a^2/\theta$	lituus	(7d)
none	$r^n = a\theta$	spirals of Fermat	(7e)

the planes of the initial curves as the poles (all eqs. 6; see Problem 1).

Though these shortcuts make it possible to express the eqs. in concise form, and are very useful, they often render the relationships and affinities of the curves, themselves, obscure. For example, expressing the construction rule for the lemniscate of Bernoulli (Fig. 20-2) as a function of 2θ (eq. 6e) obscures the following fact. For given values of θ, the squares of the magnitudes of the radius vectors to the lemniscate are in proportion to the differences between the radius vectors to the leaves of an $r = a\cos^2\theta$ curve (Fig. 8a) and to the overlapping portions of the leaves of an (congruent but orthogonal) $r = a\sin^2\theta$ curve (eq. 6e').

On the other hand, expressing polar construction rules in multiple and fractional angles sometimes facilitates detecting relationships that would tend to be obscured using whole angles. Thus, comparing eqs. 6c and 6k reveals the inverse relationship between Cayley's sextic and Tschirnhausen's cubic. Further, Tschirnhausen's cubic is seen to be an expansion of the trisectrix of Maclaurin about its loop focus by a factor of $\sec^2(\theta/3)$. Cayley's sextic also is seen to be a contraction of the curve inverse to the trisectrix of Maclaurin (inverted about its loop focus) by a factor of $\cos^2(\theta/3)$.

These relationships highlight a manner in which curves inverse to conic sections (in this case, the trisectrix of Maclaurin) are related to curves (in this case, Tschirnhausen's cubic) inverse to the group to which Cayley's sextic belongs. They also reveal one way of defining analytically the otherwise unknown members of the latter group. Thus, by inverting homologues of the trisectrix of Maclaurin (inversions of hyperbolas other than the $e = 2$ hyperbola about their axial vertices) about the homologues of the loop focus (the loop pole at 2b/3) and expanding them appropriately, one obtains the general construction rule of one group of curves to which Cayley's sextic belongs.

In essence, these classical practices have been exploiting the structural simplicity of certain curves about certain points for certain conditions, and the "shorthand" of multiple and fractional-angle expressions, to achieve simplicity of their construction rules (see RULES 1a-c). However, since the exploitation has been applied entirely for practical purposes, and in an unsystematic manner (namely, whatever happens to work is used), the results merely have shown that the shortcuts exist, but usually have taught nothing of their significance.

I do not dwell on polar-eq. shortcomings, which admittedly are two-edged, but only emphasize that, for a systematic rigorous treatment, polar construction rules should:

(a) be cast about an arbitrary point on a line of symmetry of the curve, if such exists;
(b) employ trigonometric functions expressed solely in terms of the sines and cosines of whole angles (inasmuch as these are the expressions of the conversion formulae, $x = r\cos\theta$ and $y = r\sin\theta$); and
(c) be for curves in a single standard orientation.

The casting of construction rules about an arbitrary point in the plane for an

BASIC POLAR CONSTRUCTION RULES

arbitrary orientation is preferable in theory but wholly impractical for eqs. of rectangular degree greater than 2 (because the eqs. become much too complicated). In consequence, regarding conditions (a) and (c), it is preferable to place a line of symmetry of the curve in coincidence with the polar axis.

The second shortcoming is that dealing almost exclusively with construction rules linear in r (eqs. 6a-d,f,g,k; 7a-c), occasionally of 2nd-degree in r (eqs. 6e,i; 7d,e), and rarely of higher degree (eqs. 6h; 7e) or fractional degree (eq. 6j), provided little or no stimulus to develop a theory of the degrees of polar eqs., and, thus, delayed an appreciation of the relationships between a curve's structural and positional simplicity and the degree of its corresponding construction rule (in coordinate systems in general).

Basic Polar Construction Rules

It is clear that the *basic* polar construction rule for any rectangular curve cast about an arbitrary point in the plane, that is, the eq. obtained directly from the rectangular eq. cast about an arbitrary point by substituting $x = r\cos\theta$ and $y = r\sin\theta$, is of the same exponential degree in r as the eq. of the rectangular curve is in x and y. Once RULES 7 are recognized explicitly,

> RULE 7a. *A polar construction rule of a curve derived directly from the curve's rectangular equation about an arbitrary point in the plane (the "basic" polar construction rule) has the same exponential degree in r as the degree of the rectangular eq. in x and y.*
>
> RULE 7b. *The maximum exponential degree in r of the polar construction rule of a curve is the exponential degree of the curve's rectangular equation.*

certain questions naturally arise. Thus, included among eqs. 6 are 6th-degree, 4th-degree, and 3rd-degree rectangular eqs. that yield polar construction rules linear in r, as well as 6th-degree and 4th-degree eqs. that yield quadratic eqs. in r, and even a 4th-degree rectangular eq. that yields a 4th-degree polar eq.

Why do these differences exist? Do polar construction rules quadratic in r exist for rectangular cubics? Do polar construction rules of 3rd-degree in r exist for rectangular quartics, for example, Cassinian ovals? Are there no linear construction rules for the lemniscates of Bernoulli or Gerone? Answering these questions leads to an appreciation of the extent to which the classical formulations of polar eqs. have been arbitrary and unsystematic.

Of course, the answers to some of these questions are apparent from an examination of the construction rules. For example, since the lemniscate of Bernoulli includes the origin of the rectangular system, which is a DP (double-point) of the curve, a term in r^2 factors out of the basic polar construction rule. The same kind of answer applies to some of the other curves of eqs. 6.

We already have concluded that any curve that includes the rectangular origin is more symmetrical in form and position than an unexceptional, similarly-oriented, congruent curve that does not (a *neighboring* curve; because the constant term of the eq. vanishes, and one point of the curve is most-symmetrically positioned in the system). This also is a basis for greater symmetry of curves in the polar system. But when the constant term vanishes from a polar construction rule, the eq. not only has at least one less term (since it is not unusual for more than one term to vanish when the constant term vanishes; see eq. 22, for $h = 0$), it also reduces in degree by at least one unit, since at least one r factors out.

Intersections of Curves at the Polar Pole

A little-appreciated phenomenon is the failure of most polar curves that include the polar pole to yield an analytical solution for an intersection with one another at this point (i.e., when solved simultaneously; see Problem 3). In the rectangular system, on the other hand, when the construction rules of curves that include the origin are solved simultaneously, one solution, $x = y = 0$, is obtained for each intersection at the origin. The explanation for the failure in the polar system is crucial, because it bears directly on the ability of polar construction rules to represent rectangular curves simply, with much lower degrees in r, and is one of the bases for the absence of degree-restriction in polar coordinates (see also note on page 60).

The reason for the divergence between the two systems is that, in the customary employment of polar coordinates, the polar or $0°$-axis is a half-line that does not include the pole. This fact has no influence on intersections at non-zero distances from the pole. Thus, an analytical intersection always occurs when there is a non-zero geometrical intersection. This is not true, however at the polar pole. An analytical intersection of two polar curves at the polar pole occurs only when the values of r for the two curves vanish for the same or equivalent value of θ (see Problem 2).

The situation is illustrated readily by the conventional method of finding the polar construction rule for the rectangular x-axis. In the conversion,

$y = r\sin\theta = 0$, the r is factored out to yield the polar eq. $\sin\theta = 0$. The latter, however, is not the equivalent of the x-axis, and the locus represented by the construction rule, $\sin\theta = 0$, usually will not have an analytical intersection at the pole with polar curves that include the pole. The reason, of course, is that when r is factored out of the eq. $r\sin\theta = 0$, the polar pole (i.e., the solution $r = 0$) is deleted, leaving only the "punctured" x-axis ("punctured" at the pole). Other curves get "punctured" one or more times in the same way (see last paragraph of preceding section and Problem 3).

The factoring out of r's is the principal reason for the reductions of the degrees of polar construction rules from the degrees of the basic eqs., which occurs for all rectangular curves of finite degree that include the origin. As noted above, this reduction of degree in the polar system is the counterpart of the loss of a term of the rectangular construction rule.

In essence, then, one of the bases for the ability of the conventional polar system to give simple representations of rectangular curves of high degree is that one generally casts out the origin as a general point of intersection (i.e., as a point of intersection with any locus that includes the origin). It is clear that this is not an obligatory feature of the polar system. One also could employ an otherwise identical system, but retain the polar pole and the intersections of all curves that include the pole. This would be done, however, at the expense of convenience and simplicity.

Structure Rules

With one's attention drawn to the fact that the degree, as well as the simplicity, of a curve's construction rules can be affected by the curve's positional symmetry with respect to (structural simplicity about) reference elements, other questions arise. By their comparative simplicities and degrees, the various polar construction rules of a given curve merely give an indication of the curve's greater or lesser structural simplicity about the various points taken as poles (or of the positional symmetry of these points in the plane of the curve).

The most compelling question that arises is, can one obtain direct information about the nature of the inherent structure of the curve in the form of eqs. other than construction rules? In other words, do eqs. exist that define the aspects of the inherent structure of curves that are responsible for the differences in the simplicities and degrees of their construction rules? The answer, of course, is "yes." The eqs. in question are the *structure rules* mentioned in the *Introduction* and Chapter I, and their derivation employs a logical extension of the method of obtaining the polar construction rules themselves.

To obtain explicit information about the inherent structure of a curve, one must probe deeper than merely dealing with one point of a curve at a time, as in the method of polar construction rules. In that method, one value of θ yields one value of r (assuming a linear eq. in r), and many values of θ simply lead to a plot of many points of the curve. One must deal, instead, with two points of the curve for each value of θ (dealing with more than two points at a time gives redundant results). That is, for a given value of θ, one must obtain two values, r_1 and r_2, one along each of two radius vectors that are out of phase by a fixed angle, α (Fig. 2a), and this must be done for all possible pairs of values for r that are out of phase by this amount. An eq. that relates all these possible pairs of values to one another is a *structure rule* of the curve, as opposed to a *construction rule*.

To accomplish this objective, the polar construction rules are used to provide information about structural relationships, rather than merely about how to construct the curve. This is done by casting the construction rule in two forms, $r_1 = f(\theta)$ and $r_2 = f(\theta+\alpha)$. Both rules apply to the identical curve (from different starting points), but with plots whose radius vectors are out of phase by the fixed angle, α (Fig. 2a). The latter becomes the fixed reference angle for which the structure rule applies, as opposed to the variable angle, θ, of the construction rule. By giving α its fixed value, and eliminating θ from the expressions for r_1 and r_2 (X and Y), one obtains the sought-after structure rule, $F(r_1,r_2) = f(X,Y) = 0$.

[In the interests of simplicity, the capital letters, X and Y, are used in place of r_1 and r_2 in eqs. derived as structure rules, as opposed to the lower case letters, x and y, in eqs. derived as construction rules (or eqs. of initial curves).]

A structure rule thus is a relation between the lengths of two out-of-phase radius vectors (Fig. 2a) specified by the polar construction rule. Accordingly, it follows that simpler construction rules will tend to yield simpler structure rules (RULE 8a). RULE 8a is not hard-and-fast, though, because the value of the fixed reference angle, α, also exerts an important effect. Assessments of the inherent structure of curves by means of structure rules are the topics of Chapters VII and VIII.

POLAR CURVES AS RESULTANTS

RULE 8a. *Simpler polar construction rules tend to yield simpler structure rules.*

RULE 8b. *In general, the simpler the 90°-structure rule of a curve about a point (the structure rule for a 90°-angle between the probe-lines), the simpler the polar construction rule about the same point.*

Polar Curves as Resultants

We deal here with the unique property that the polar construction rule of an initial curve often can be broken down into components--each of which also represents a curve--of which the initial curve may be considered the resultant. The same treatment amply illustrates principles of plotting, classical symmetries, and orientation of polar curves. Consider, first, the curve represented by the polar construction rule 8a and Fig. 8a, a previously unnamed sextic, which I refer to as a member of the $cos^2\theta$-*group*, for obvious reasons. It

(a) $r = a + b\cos^2\theta$ $[r = (a+b) - b\sin^2\theta]$ curve belonging to the $\cos^2\theta$-group   (8)

(b) $[a(x^2+y^2) + bx^2]^2 = (x^2+y^2)^3$ rectangular equivalent of eq. 8a

is evident that the initial curve (eq. 8a) is the resultant of the addition of a circle, the $r = a$ component, to a *pure $b\cos^2\theta$-group* component (the *symmetry mimic of tangent circles* of Chapter VII). Since the latter also is a sextic (eq. 8b, for a = 0), it is clear that, in this particular case, in which both component curves are symmetrical about both the 0°-180° and 90°-270°-axes, the rectangular degree of the resultant curve does not exceed the rectangular degree of either of its components.

The symmetries of both component curves and the resultant are evident from eq. 8a; the circle, or $r = a$ component, with center at the origin, is symmetrical about the polar axes. The $b\cos^2\theta$-component also is centered and symmetrical about these axes, because the $\cos^2\theta$-function is positive in all four 90°-sectors. [The 0°-180° and 90°-270°-axes, and the 90°-sectors, sometimes are referred to hereafter with their rectangular designations, as the x and y-axes, and as quadrants.]

The segment of a polar curve whose construction rule involves solely $\cos\theta$ always is symmetrical about the x-axis, because the cosine function is positive in the 1st and 4th-quadrants, and $\cos\theta = \cos(-\theta)$, and similarly for symmetry about the same axis in the 2nd and 3rd-quadrants (where the cosine function is negative). Symmetry about the y-axis, also, ensues because the cosine function

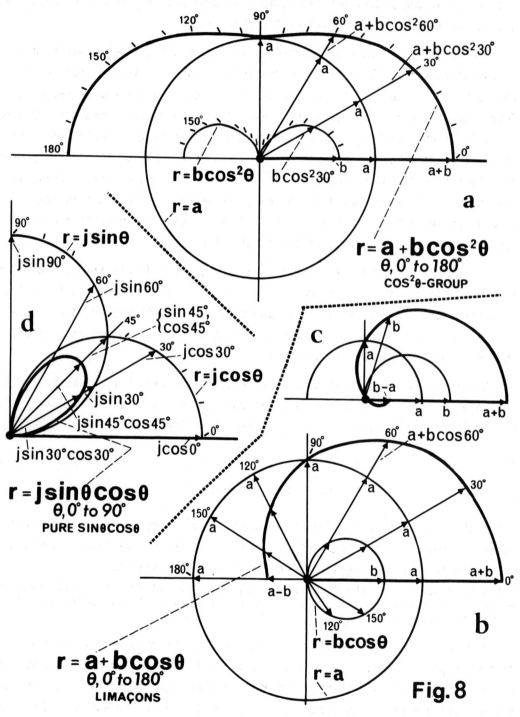

Fig. 8

occurs in eq. 8a as the square, and $\cos^2(90°+\theta) = \cos^2(90°-\theta)$. In other words, just as the magnitudes of the cosine function are equal for equal angles on both sides of the x-axis (0°-180°-axis), so also are they equal for equal angles on both sides of the y-axis (90°-270°-axis).

The manner in which the resultant $\cos^2\theta$-group curve is obtained from its component curves is illustrated in the 1st-quadrant of Fig. 8a. For $\theta = 0°$, the radius vector of the circle, of length a, is added to the radius vector of the "pure" $\cos^2\theta$-group curve, of length $b\cos^2 0° = b$, to yield a resultant of length $a+b$. For larger angles, radius vectors contributed by the circle remain of constant length, but those contributed by the second curve diminish in proportion to the $\cos^2\theta$. At an angle of 30°, the latter has decreased to $3b/4$, so the resultant is of length $a+3b/4$. At 60°, the value has decreased further to $a+b/4$, while by 90° the value of $\cos^2 90°$ is 0 and the length of the radius vector is merely $r=a$, that of the circle alone.

For convenience, the plots of Fig. 8a-c are carried only through 180°. In the present case, it is seen that one obtains only the portion (half) of the resultant curve in the 1st and 2nd-quadrants. On the other hand, the plot for the circle is complete, because the component eq., $r=a$, is independent of θ, and hence must encompass all values thereof.

Consider, now, the related curve of eq. 9a, the well-known limacon, with a rectangular degree of 4. Both component curves now are circles, but the $r = b\cos\theta$ circle is not symmetrical about the origin, though it retains symmetry

(a) $\qquad\qquad r = a + b\cos\theta \qquad\qquad$ limacon $\qquad\qquad\qquad\qquad$ (9)

(b) $\qquad\qquad (x^2+y^2-bx)^2 = a^2(x^2+y^2) \qquad$ eq. 9a in rectangular form

about the x-axis. As a consequence, the rectangular degree of the resultant curve is greater than that of either component, and the resultant curve possesses symmetry about the x-axis but lacks it about the y-axis.
[This is evident from the rectangular eq. by virtue of the fact that replacing y by -y does not alter the eq. (symmetry about the x-axis), whereas replacing x by -x does (lack of symmetry about the y-axis).]

The genesis of the resultant limacon from the component circles now is traced, beginning at 0°. At this angle, the radius vector to the $r = b\cos\theta$ circle is of length b and that to the $r=a$ circle is of length a, its constant value, giving a resultant vector of length $a+b$. As θ increases to 90°, the length of the vector contributed by the $r = b\cos\theta$ circle diminishes and the resultant diminishes by the same amount, until, at 90°, the value of the former is 0 and that of the latter is a. This progression parallels that of Fig. 8a, with the exception that the $r = b\cos^2\theta$ component does not yield a hemi-circle.

With θ increasing beyond 90°, the $r = b\cos\theta$ component becomes negative, with the consequence that it is oppositely directed, into the 4th-quadrant. The resultant now is diminished from the length a by the same amount that it exceeded a in the 1st-quadrant (for equal angles on each side of the 90°-axis), with the consequence that the shape of the curve is altered greatly. At 180°, the resultant reaches its minimum value of a - b. Clearly, when b > a, this resultant becomes negative and is directed back along the x-axis into the 1st-quadrant, as in Figs. 7 and 8c, yielding what I refer to as a *hyperbolic* limacon, as opposed to an *elliptical* limacon (the former inverts to a hyperbola, and the latter to an ellipse).

This example shows that for θ progressing from 0° to 180°, a complete $b\cos\theta$ circle is generated, but only a hemi-limacon. Accordingly, for a complete sweep through 360°, which is required to generate a complete limacon, not only is a second identical $b\cos\theta$ circle generated but, in a strict analytical sense, the limacon must be viewed as the resultant of three circles, not two circles--one "a" circle and two "$b\cos\theta$" circles.

As the reader will have noted, the component $b\cos^2\theta$ curve of Fig. 8a can be regarded as the product of two single 0°-to-180° circles of Fig. 8b (letting b = j and correcting for dimensional balance), that is, $(b\cos\theta)(b\cos\theta)/b = b\cos^2\theta$. The product of two $b\cos\theta$ vectors in the 1st-quadrant of Fig. 8b for 0° to 90° yields the $b\cos^2\theta$ segment in the 1st-quadrant of Fig. 8a. As θ progresses from 90° toward 180°, since the sign of $\cos\theta$ is negative, the product of the two hemispherical segments, $(b\cos\theta)(b\cos\theta)$, each in the 4th-quadrant of Fig. 8b, becomes positive and oppositely directed into the 2nd-quadrant of Fig. 8a, generating a 2nd-quadrant segment of the $r = b\cos^2\theta$ curve. Thus, the hemi-curve generated by $r = b\cos^2\theta$, for θ progressing from 0° to 180°, can be viewed as the product of two complete "$b\cos\theta$" circles.

Lastly, consider a quadrant of the curve (of eq. 10a and Fig. 8d) that is obtained as the product of two non-coincident but congruent hemi-circles, $r = j\sin\theta$ and $r = j\cos\theta$ (neither of which is centered at the origin), for θ progressing from 0° to 90°. The $j\cos\theta$ hemi-circle is familiar from Fig. 8b,c, but the "jsinθ" hemi-circle is 90° out of phase with it, and reflected in the y-axis. In this case, with neither circle centered at the origin, the product curve is of rectangular degree 6, compared to only 4 for the analogous construction of limacons from one centered circle and two coincident non-centered circles.

(a) $\quad r = j(\sin\theta)\cos\theta \quad$ unnamed sextic $\hfill (10)$

(b) $\quad (x^2+y^2)^3 = j^2x^2y^2 \quad$ eq. 10a in rectangular form

Beginning at 0°, the jcosθ vector of Fig. 8d is simply the diameter-vector of the bcosθ hemi-circle, but the resultant j(sinθ)cosθ vector is of length 0, since the vector contributed by the jsinθ hemi-circle is of length 0. At 30°, the jcosθ vector has diminished to length $\sqrt{3}j/2$ and the jsinθ vector has increased to j/2, giving the resultant of length $\sqrt{3}j/4$. At 45°, this is augmented to a maximum length of j/2, at 60° diminished to $\sqrt{3}j/4$, and at 90° diminished again to length 0, since the vector contributed by jcosθ now is of length 0 (see Problem 4).

Positional Symmetry in the Polar System

The polar system represents a marked improvement over the rectangular system for studies of the positional symmetry of curves with respect to coordinate elements, but it still is far from ideally suited for this purpose. An advantage is that lack of degree-restriction adds a third, and highest, priority component to the symmetry index; this is the exponential degree of the construction rule in r. Disadvantages are the facts that (1) with only one distance variable, the number of terms in a polar eq. is only $n+1$, as compared to $[(n+1)(n+2)]/2$ for rectangular eqs. (n, the degree); and (2) like intersecting bilinear systems, the polar system possesses no *characteristic distance*, that is, no distance exists that characterizes either the separation of the coordinate elements or the elements themselves.

A characteristic distance exists for most two-element coordinate systems (see Table 7), in which its value in relation to the parameters of a curve plays an important role in determining the symmetry of the curve's form and position, and its use as the distance parameter of construction rules enables one to recognize and characterize the eqs. of curves (see Chapters IV-VI).

For purposes of simplicity, and comparison with the rectangular situation, circles and ellipses are used as examples for assessing the symmetry of curves in the polar coordinate system. In addition, the corresponding rectangular construction rules of these curves are used as initial eqs. Because of the close relations between the polar and rectangular systems, as exemplified by the simple conversion eqs., $x = r\cos\theta$ and $y = r\sin\theta$, one also expects to find some similarities between symmetries of form and position in the polar and rectangular systems.

For convenience, the rectangular descriptive names and letter symbols for symmetry conditions also are used for the corresponding polar ellipses and circles. Accordingly, the tabulations are not accompanied by additional figures, since the reader usually can refer to Figs. 3 and 4.

Most readers already will be familiar with the simple tests for classical symmetries about the polar pole and axes. A curve is reflected in the 0°-180°-axis if its construction rule is unaltered on replacing θ by $-\theta$; it is reflected in the 90°-270°-axis if its construction rules is unaltered on replacing θ by $(180°-\theta)$, while it is reflected in the pole if its eq. is unaltered on replacing r by $-r$. It should be pointed out, though, since it generally is not appreciated, that the failure to satisfy a test does not necessarily exclude the presence of the corresponding symmetry. For example, the eq. $r = j\sin\theta\cos\theta$ (Fig. 8d) is altered by the test for symmetry about the pole, yet the curve possesses this symmetry. In other words, the application of these tests may merely alter the angular phase of the plot or reflect the curve in a polar axis or the origin, leading to a change in the eq. and a negative test, even though the curve possesses the symmetry in question.

Circles

The least symmetrical polar circle, of course, is the general circle (eq. 11a), a circle arbitrarily positioned in the plane, where h and k are the values of the positive displacements of its center from the polar origin along the 0° and 90°-polar axes, respectively. It is natural to refer displacements to these axes, since the 0°-axis is the reference line (initial half-line) for measurements of θ, and the sine and cosine functions change signs to either side of the 0° and 90°-axes, respectively.

The symmetry index of the general circle (far left column of Table 5) is 2-3-4. This designation signifies that the construction rule (11a) of the general circle is of 2nd-degree, that it possesses three terms, and that it possesses a total of four parameters and non-unit coefficients--in this case, R, h, k, and 2. The eq. (11a) for this circle is derived from rectangular eq. 1a after making the substitutions for converting to polar coordinates. Regardless of the values of h and k (for a circle not enclosing and/or including the origin), a plot of eq. 11a for values of θ between the limiting-tangent values (see Fig. 7) yields a single complete circle.

On the other hand, if θ is allowed to range from 0° to 360°, as in classical treatments, two coincident circles are traced out. This means that eq. 11a can be factored into the product of two 1st-degree eqs. in r, which, taken together, comprise a degenerate 2nd-degree eq. Each of these factors yields a complete circle for θ ranging from 0° to 360°. This 360°-range of θ is not permitted in the standard system, whose more exacting requirements allow θ to range only between the two limiting tangent rays; in the latter case, each 1st-degree eq. yields only a hemi-circle, both of which taken together comprise a complete circle. These matters are treated in detail below in the section on *Transformations between Polar and Rectangular Coordinates*. The following treatments concern primarily positional symmetry.

The first increase in symmetry from the general circle of Fig. 3a is achieved in the general bisector circle (for which h = k) of Fig. 3b and eq. 11b. The construction rule of this circle is slightly simpler by virtue of the possession of one less parameter (symmetry index, 2-3-3). Still simpler, and representing a still more-symmetrically-positioned circle, is eq. 11c, for the general tangent circle of Fig. 3c (for which k = R). But the most-symmetrically-positioned of the circles with symmetry index 2-3-3 is the general axial circle of Fig. 3g and eq. 11d (for which k = 0), which is bisected by the 0°-axis.

Following the progression of symmetry indices, the next-most-symmetrical polar circle is the tangent bisector circle of Fig. 3d and eq. 11e. Note that this order reverses the order of rectangular symmetry for general axial and tangent bisector circles. This results because mirror-image symmetry in the 0°-axis of the polar system is of less significance than mirror-image symmetry in a rectangular coordinate axis (for which a linear term in a variable vanishes from the rectangular eq.).

This lesser significance is a consequence of the fact that the polar 0°-axis is a reference axis for angles, not distances. As a result, the terms of polar construction rules are segregated and reckoned only in powers of r, and no term vanishes from the eq. of a locus upon acquiring mirror-image symmetry in the 0°-axis. The construction rule of the tangent bisector circle is simpler because it possesses one less parameter than the eq. of the general axial circle (since h = k = R). [Although the eq. of the tangent bisector circle has a more complicated linear term than that of the general axial circle, it has a simpler constant term.]

The next-most-symmetrical polar circles have the symmetry indices 1-2-3 and represent the cases for which the degrees of the polar construction rules reduce from 2 to 1. As a first example, consider the general axial circle enclosing the origin (eq. 11f; not represented in Fig. 3), the eq. of which is obtained by completing the square of eq. 11d and taking the expression for the positive root. The symmetry index for this circle is the only index for the circles of Table 5 that includes an accounting for a radical (as shown in the listing below the index).

The next-most-symmetrical circle is the general circle including the origin, of Fig. 3e and eq. 11g (from which the constant term $[R^2 - h^2 - k^2]$ of eq. 11a has vanished and an r has been factored out). The basis for considering
[Once one or more r's have been factored out from a polar eq. in the reduction process, the point $r = 0$, double-points $r^2 = 0$, etc., no longer will appear as an analytical point of intersection of two such loci. Insofar as detecting intersections at the origin analytically is concerned, only the degenerate eq., $r^n f(r,\theta) = 0$, represents the complete locus (see Problem 3 and the section on *Intersections of Curves at the Polar Pole*).]
this circle to represent the weakest case of reduction, by factoring out of an r, lies in experience with higher-degree polar construction rules of other curves. For the latter, the least reduction by any means consists in the loss of the constant term and factoring out of a single r, with a reduction in degree of one unit. The progression of symmetry indices is consistent with this observation.

A more symmetrical circle including the origin is the bisector circle of Fig. 3f and eq. 11h. The symmetry index of this circle is 1-2-2, as compared to 1-2-3 for the general circle including the origin. Although the next-most-symmetrical circle also has the symmetry index 1-2-2, with a single parameter (R) and a single non-unit coefficient (the number 2), its construction rule is much simpler. It is the third-most-symmetrical, and the most-familiar polar circle, namely, the tangent axial circle of Fig. 3h and eq. 11i.
[It will be evident to the reader that, strictly speaking, products of positional parameters and non-unit coefficients, which are being reckoned here both

Table 5. Symmetry of Polar Circles

symmetry index*	construction rule	rectangular symmetry condition	rectangular designation of circle	
2-3-4 R,h,k,2	$r^2 - 2r(h\cos\theta + k\sin\theta) = (R^2 - h^2 - k^2)$ least-symmetrical circle	none	general	(11a)
2-3-3 R,h,2	$r^2 - 2hr(\cos\theta + \sin\theta) = (R^2 - 2h^2)$	A	general bisector ($h = k$)	(11b)
2-3-3 R,h,2	$r^2 - 2r(h\cos\theta + R\sin\theta) = -h^2$	B	general tangent ($k = R$)	(11c)
2-3-3 R,h,2	$r^2 - 2hr\cos\theta = (R^2 - h^2)$	D	general axial ($k = 0$)	(11d)
2-3-2 R,2	$r^2 - 2rR(\sin\theta + \cos\theta) = -R^2$	AB	tangent bisector ($h = k = R$)	(11e)
1-2-3 R,h,√	$r = h\cos\theta + \sqrt{(R^2 - h^2\sin^2\theta)}$, $R > h$	DI***	general axial, enclosing origin	(11f)
1-2-3 h,k,2	$r - 2(h\cos\theta + k\sin\theta) = 0$, $h^2 + k^2 = R^2$	C	general, including origin	(11g)
1-2-2 h,2	$r - 2h(\sin\theta + \cos\theta) = 0$, $2h^2 = R^2$	AC	bisector, including origin	(11h)
1-2-2 h,2	$r = 2h\cos\theta$	BD, BC, or CD	tangent axial ($h = R$)	(11i)
1-2-1** R	$r = R$	AD	centered	(11j)
1-1-0**	$r = 0$ most-symmetrical circle	ABD, ABC, or ACD	centered point-circle	(11k)

* The first number is the degree of the eq., the second is the number of terms, and the third is the sum of the number of parameters, non-unit coefficients, and radicals. These are listed below each symmetry index.
** Strictly speaking, the degrees of these eqs. in r is 0, since r is no longer a variable
*** The letter I denotes that the curve encloses the origin.

as a parameter and as a non-unit coefficient, sometimes should be combined and reckoned as a single parameter (in cases where the positional parameter in question occurs only once or always occurs in the same combination with the non-unit coefficient). For the sake of consistency, this has not been done here. If this were to be done, the symmetry indices (but not their sequence) for the linear eqs. would be altered:

```
              (11f)    (11g)    (11h)    (11i)    (11j)    (11k)
      from;   1-2-3,   1-2-2,   1-2-2,   1-2-2,   1-2-1,   1-1-0
        to;   1-2-2,   1-2-1,   1-2-1,   1-2-1,   1-2-1,   1-1-0
```

Nor would observing the more strict interpretation alter the adopted order of symmetry for rectangular ellipses or circles, though it would reduce by one the sums of the numbers of parameters and non-unit coefficients of eqs. 1b, c, e, f, and h, and 4e (for example, in the latter case, h always occurs in combination with the number 2 as 2h).]

The most-symmetrical non-point polar circle is the centered circle of Fig. 3i and eq. 11j, with symmetry index 1-2-1 (or 0-2-1; see note** to Table 5), in which θ does not appear. As in the rectangular system, the most symmetrical circle is the centered point-circle of Fig. 3j and eq. 11k, obtained by letting $R = 0$. The eqs., $r = 0$ and $\theta = 0$, describe the most-symmetrical polar loci, inasmuch as the former is coincident with the coordinate element from which distances are reckoned (the pole), and the latter is coincident with the coordinate element from which angles are reckoned (the 0°-axis).

Aside from the coordinate elements themselves, the most-symmetrical polar loci are described by: the eqs., $r = R$, the centered non-point circle; and loci with the eq., $\theta = $ constant, which are half-lines extending from the origin at the angle θ to the 0°-axis (the classical "coordinate curves"). Because of the employment of directed distances and angles, the latter is only a single half-line. Of the latter loci, the most symmetrical is the 180°-axis, $\theta = 180°$; since it extends directly opposite the 0°-axis, it is the only half-line for which $|\theta| = |-\theta|$ (i.e., the half-line $\theta = 180°$ coincides with the half-line $\theta = -180°$). The next-most-symmetrical half-lines are the 90° and 270°-axes, since they bisect and make angles of equal magnitude with the combined 0° and 180°-axes (which are the two most-symmetrical half-lines).

[In assessments of the symmetry of loci with respect to the 0°-axis, symmetry of angles, rather than distances, takes priority, since this is a reference axis for angles.]

Ellipses

In the interests of simplicity, only the positional symmetries of 0°-ellipses are considered. The reader easily can extend the treatment to general ellipses and 45°-ellipses. As in the cases of circles and other polar curves, the exponential degrees of construction rules are determined and defined as in the section above (*The Degree of a Polar Construction Rule*; see also the following two sections).

General 0°-ellipses (Fig. 4a') have the basic polar construction rule 12a. This is obtained directly from the corresponding rectangular construction rule, 4k, by the coordinate substitutions. The symmetry index of these ellipses is 2-3-5, indicating a three-term eq. of 2nd-degree, and a total of five parameters and non-unit coefficients. The loss of two terms from eq. 4k (a decrease from five terms to three), of course, is because the r^2-term takes the place of both the x^2 and y^2-terms, and the r-term takes the place of both the x and y-terms. Since the expression $(a^2\sin^2\theta+b^2\cos^2\theta)$ occurs in almost all the construction rules in Table 6, this term is represented in eqs. 12a-f by the abbreviation A^2, for the sake of simplicity (A^2 is used, rather than A, to maintain dimensional balance).

The first increase in positional symmetry from that of general 0°-ellipses is achieved by reduction of the total number of parameters and non-unit coefficients from five to four. There are three ways in which k (or h) can be eliminated from eq. 12a' to yield much simpler eqs. with the symmetry index 2-3-4. One may let k = h, b, or 0, yielding, respectively, bisector 0°-ellipses, tangent 0°-ellipses, and axial 0°-ellipses (Fig. 4m,n,s, respectively). Since this also is the order in which the simplicities of the eqs. fall (as shown by the sequence of eqs., 12b,c,d), it represents the order of increasing positional symmetry.

The next increase in positional symmetry--to eq. 12e and a bitangent 0°-ellipse (Fig. 4p)--is achieved by eliminating both positional parameters (by letting h = a, the length of the semi-major axis, and k = b, the length of the semi-minor axis), This is accompanied by a reduction of the symmetry index from 2-3-4 to 2-3-3. Although the r-term of eq. 12e becomes more complicated because of the addition of a $\sin\theta$-component, this is more than offset by the facts that an additional parameter factors from the r-term, the constant term is simplified, and no positional parameter is present.

As an example of the next-most-symmetrical ellipse, consider a 0°-axial ellipse enclosing the origin (h < a, k = 0), with symmetry index 1-2-4. This is the first instance of a polar ellipse with a 1st-degree construction rule (eq. 12f

Table 6. Symmetry of 0°-Polar Ellipses

symmetry index*	construction rule	rectangular symmetry condition	rectangular designation of ellipse	
2-3-5 a,b,h, k,2	$r^2(a^2\sin^2\theta+b^2\cos^2\theta)-2r(a^2k\sin\theta+b^2h\cos\theta)$ $= (a^2b^2-a^2k^2-b^2h^2)$ letting $A^2 = (a^2\sin^2\theta+b^2\cos^2\theta)$, $r^2A^2-2r(a^2k\sin\theta+b^2h\cos\theta) = (a^2b^2-a^2k^2-b^2h^2)$	H	general	(12a) (12a')
2-3-4 a,b,h,2	$r^2A^2-2hr(a^2\sin\theta+b^2\cos\theta) = [a^2b^2-h^2(a^2+b^2)]$	EH	bisector (k = h)	(12b)
2-3-4 a,b,h,2	$r^2A^2-2br(a^2\sin\theta+bh\cos\theta) = -b^2h^2$	BH	tangent (k = b)	(12c)
2-3-4 a,b,h,2	$r^2A^2-2b^2hr\cos\theta = b^2(a^2-h^2)$	D	axial (k = 0)	(12d)
2-3-3 a,b,2	$r^2A^2-2abr(a\sin\theta+b\cos\theta) = -a^2b^2$	BEH	bi- tangent (h = a, k = b)	(12e)
1-2-4 a,b,h,√	$rA^2 = b^2h\cos^2\theta+ab\sqrt{[b^2\cos^2\theta+(a^2-h^2)\sin^2\theta]}$	DI	axial, en- closing origin (k = 0, h < a)	(12f)
1-2-3 a,b,2	$r = 2ab^2(\cos\theta)/[a^2-(a^2-b^2)\cos^2\theta]$	BD	tangent axial (k = 0, h = a)	(12g)
1-2-3 a,b,√	$r = ab/\sqrt{[a^2-(a^2-b^2)\cos^2\theta]}$ $r = ep/\sqrt{[(1-e^2)(1-e^2\cos^2\theta)]}$	AD	centered	(12h) (12h')
1-2-3 a,b,√	$r = b^2/[a+\sqrt{(a^2-b^2)}\cos\theta]$ $r = ep/(1+e\cos\theta)$	D	focus $(h^2 = a^2-b^2)$	(12i) (12i')

$p = a(1-e^2)/e =$ the focus-directrix distance

* The first number is the degree of the eq., the second is the number of terms, and the third is the total number of parameters plus non-unit coefficients and radicals. The latter are listed directly below the symmetry index.

and Fig. 4t, but without the pole being at a traditional focus). The three most-symmetrical 0°-ellipses all have symmetry indices of 1-2-3 and eqs. that lack a positional parameter (h or k). These curves are considered in the order of the increasing simplicity of their construction rules (to facilitate comparisons between them, the abbreviation A^2 no longer is employed). The first locus to be considered is the tangent axial 0°-ellipse ($h = a$, $k = 0$) of eq. 12g and Fig. 4u. This is obtained from eq. 12d by factoring out an r after the vanishing of the constant term.

The next, and second-most-symmetrical ellipse, is the centered 0°-ellipse ($h = k = 0$) of eq. 12h and Fig. 4v, which is obtained by taking roots of eq. 12d, after letting $h = 0$. But the most-symmetrical polar ellipse is the focus-0°-ellipse (for which $h^2 = a^2-b^2$) of eq. 12i and Fig. 4t, which also is obtained by taking roots of eq. 12d, but after letting $h^2 = a^2-b^2$. Eq. 12i is most often expressed in the form 12i', in terms of e and p, which represent the eccentricity and the focus-directrix distance (see Problem 5): for comparison, the centered 0°-ellipse also is expressed in this form (eq. 12h').

[The expression 12i' (expressed in terms of e and p) must be regarded as involving a radical, since e, itself, involves a square-root, and is present to the 1st-degree in the denominator (the combination ep in the numerator of both eqs. 12h' and 12i' equals $a - ae^2$).]

The change from the rectangular order of symmetry that makes the focus-0°-ellipse the most symmetrical polar ellipse, of course, is a consequence of the influence of the initial half-line reference element of the polar system. Because of its employment, the polar pole is not symmetrically located in the system (as is the rectangular origin). As will be seen in Chapter IV (section on *Bipolar Construction Rules and the Hybrid Polar-Rectangular Coordinate System*), this circumstance has far-reaching implications for the construction rules of central conics (including parabolas) in all coordinate systems that possess point or circle reference elements.

Elective and Intrinsic Asymmetries of Coordinate Systems

The polar coordinate system is merely a variety of incident polar-linear system (see Chapter V)—a half-line and a terminal point—in which one coordinate is a directed distance and the other is a directed angle. Because of the asymmetry introduced by the half-line reference element, a centered ellipse at any angle other than 0° (or 90°, 180°, or 270°), say 45°, is less symmetrical

Table 7.   Elective and Intrinsic* Asymmetries of Coordinate Systems

System	elective a-symmetries, no. & type	intrinsic a-symmetries, no. & type	degree-restriction?	all loci symmetrical about reference elements?	characteristic distance?
bilinear systems					
rectangular	2 distances² polarized¹	0	yes	no	no
oblique	2 distances² polarized¹	1 domain⁵	yes	no	no
oblique, undirected	1 polarized¹	1 domain⁵	yes	no	no
90°, undirected	1 polarized¹	1 domain⁶	yes	no	no
dual-90°, unpolarized	1 distances²	0	yes	no	no
dual-90°, undirected	1 polarized¹	0**	yes	yes	no
dual-90°, undirected, unpolarized	0	0**	yes	yes	no
"polar" systems					
classical	2 distances² polarized³	2 elemental⁷ mensural⁹	no	no	no
undirected angles	1 distances²	2 elemental⁷ mensural⁹	no	yes	no
undirected distances & angles	0	2 elemental⁷ mensural⁹	no	no¹²	no
hybrid polar-rectangular	1 distances¹⁴	3 elemental⁷ distances¹³	no	no	no
bipolar, polarized	1 polarized¹	0	no	no	yes
bipolar, unpolarized	0	0	no	yes	yes
polar-linear, non-incident	1 domain⁴	2 elemental⁷ distances¹⁵	no	no	yes
polar-linear, incident	0	1 elemental⁷	no	yes	no

Table 7 (cont'd.) Elective and Intrinsic Asymmetries of Coordinate Systems

System	elective a-symmetries, no. & type	intrinsic a-symmetries, no. & type	degree restriction?	all loci symmetrical about reference elements?	Characteristic distance?
linear-circular, non-diametric	1 domain[4]	3 elemental[7] distances[8] domain[4]	no	no	yes[10]
linear-circular, diametric	0	3 elemental[7] distances[8] domain[4]	no	yes	yes[11]
bicircular, congruent, non-coincident, polarized	1 polarized[1]	2 domain[4] distances[8]	no	no	yes[10]

* does not include asymmetries in the lines of symmetry (see Chapter IV)

**possesses symmetrical lines of symmetry

[1] reference elements labelled

[2] use of directed distances;   [3] use of directed angles

[4] plane divided asymmetrically by the line and/or the circle(s)

[5] plane divided asymmetrically, as well as an asymmetry between eqs. and the represented loci; for example, $x^2+y^2 = R^2$ is not the eq. of a circle

[6] does not cover the entire plane

[7] asymmetry of reference elements

[8] both greatest and least distances exist

[9] use of distances from one reference element, but angles from the other

[10] the radius of the circle is a secondary characteristic distance

[11] the radius of the circle is the sole characteristic distance

[12] when both distances and angles are undirected, all curves become symmetrically-located in the four 90°-sectors, but this is an asymmetrical angular distribution with respect to the polar axis (Fig. 9d). When only angles are undirected, the curves are not symmetrically positioned in the four 90°-sectors, but they are symmetrically positioned as regards distances from the pole and angular displacements from the 0°-axis (Fig. 9c).

[13] one distance, v, always greater than or equal to the other, x; and one distance, v, always is positive

[14] distance, x, directed in the usual sense; distance, v, always positive but directed in the sense of always being to the same side of the pole as x

[15] distances from point of curve to reference elements along polar axis (i.e., the abscissae) generally disparate

[As the reader easily can verify, for the angles of 90°, 180°, and 270°, the arguments of the trigonometric functions remain whole angles.]
than a 0°-ellipse; its construction rule is more complex than that of a centered 0°-ellipse by virtue of the fact that the expression (θ-45°) replaces θ. These circumstances also apply to the standard polar system (compare eqs. 12h and 13a) and to the classical eqs. (compare eqs. 12h' and 13b,b').

$r = ab/\sqrt{(a^2\sin^2\theta+b^2\cos^2\theta)}$ centered 0°-ellipse in standard form   (12h)

$r^2 = a^2b^2/(a^2\sin^2\theta+b^2\cos^2\theta)$ centered 0°-ellipse in classical form   (12h')

$r = ab/\sqrt{[(a^2+b^2)/2-(a^2-b^2)\sin\theta\cos\theta]}$ centered 45°-ellipse in standard form   (13a)

$r^2 = a^2b^2/[a^2\sin^2(\theta-45°)+b^2\cos^2(\theta-45°)]$ centered 45°-ellipse in one classical form   (13b)

$r^2 = a^2b^2/[(a^2+b^2)/2-(a^2-b^2)\sin(\theta-45°)\cos(\theta-45°)]$ centered 45°-ellipse in another classical form   (13b')

The polar coordinate system possesses no less than four asymmetries. There are two elective asymmetries, in the forms of directed distances and directed angles; and there are two intrinsic asymmetries, in the forms of different types of reference elements (a point and a half-line) and different types of coordinates (a distance and an angle). By comparison, the rectangular system has no intrinsic asymmetries but two elective ones (directed distances and polarized--or labelled--reference elements).

These properties of coordinate systems are brought into perspective in Table 7, where the elective and intrinsic asymmetries of some of the simplest coordinate systems are listed, together with an accounting for the possession or lack of other pertinent properties (the presence of degree-restriction, the existence of a *characteristic distance*, and whether the system is characterized by a symmetrical distribution of all loci around the reference elements).

Referring to this Table, all intersecting bilinear systems are seen to be degree-restricted and to lack a *characteristic distance*, that is, a distance that characterizes the separation of the reference elements (or a property of a reference element itself). Some of these systems divide the plane asymmetrically (Fig. 6a$_2$-d$_2$, a$_3$-d$_3$), with accompanying asymmetries between construction rules and their loci (note 5 of Table 7), others do not cover the entire plane (Fig. 6a$_4$-c$_4$).

Only in the dual-90°-undirected-distance system (i.e., the combination of a 90°-undirected-distance system and the complementary 90°-system) are all curves symmetrically distributed about the reference elements. When bilinear systems are polarized, by labelling their axes, analyses are simplified because of the introduction of useful asymmetries. For example, if the rectangular system were not polarized, all loci would be reflected in one or the other or both coordinate bisectors.

Non-bilinear systems are not degree-restricted (the presence of a single line, however, leads to *partial degree-restriction*; see Chapter V), and most of them have at least one characteristic distance. All loci in the "polar" system lacking directed angles (Fig. 9c) are symmetrical about the reference elements (i.e., they possess symmetry of distances from the pole, and of angles about the 0°-axis). If distances also are undirected (Fig. 9d), this symmetry is lost; all loci now have congruent representatives in each 90°-sector, but these lack angular symmetry about the 0°-axis.

The unpolarized bipolar system, like the unpolarized, undirected-distance dual-90°-bilinear system, is entirely free of asymmetries--elective or intrinsic (omitting a consideration of symmetry in its lines of symmetry); unlike the latter, it is not degree-restricted and it possesses a characteristic distance (the distance between the poles). Both of the listed polar-linear systems have an intrinsic elemental asymmetry but differ markedly in other respects, according to whether or not the point is incident upon the line.

All systems that possess circular reference elements have at least one *secondary characteristic distance*, in the form of the radius of at least one circle, have an intrinsic distance asymmetry, in the form of greatest and least distances, and have an intrinsic domain asymmetry, in that they divide the plane asymmetrically [the interior(s) versus the exterior(s) of the circle(s)]. The *diametric linear-circular system* (the line-pole a diameter) lacks a *primary characteristic distance* (a distance between the reference elements) but possesses a secondary one; when the line is not a diameter, both primary and secondary characteristic distances are possessed. When the line is a tangent, these two distances are equal. Congruent bicircular systems greatly resemble bipolar systems, but they possess a secondary characteristic distance, they divide the plane asymmetrically, and they employ both greatest and least distances.

DIRECTED & UNDIRECTED MEASUREMENTS WITH POLAR REFERENCE ELEMENTS

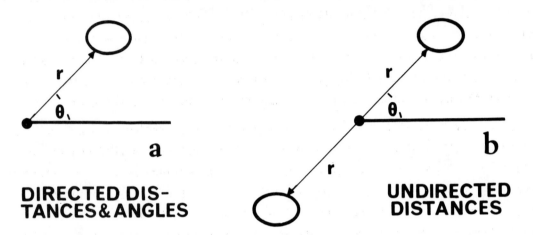

a — DIRECTED DISTANCES & ANGLES

b — UNDIRECTED DISTANCES

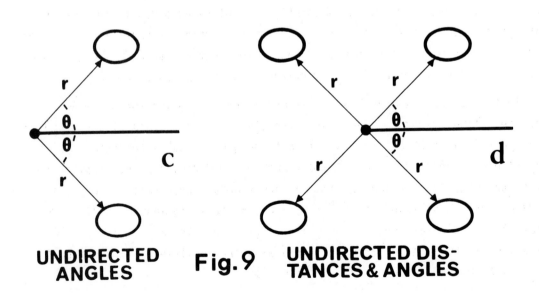

c — UNDIRECTED ANGLES

d — UNDIRECTED DISTANCES & ANGLES

Fig. 9

Transformations between Polar and Rectangular Coordinates

The fact that all rectangular construction rules of a given curve have the same degree, whereas the degrees (in r) of polar construction rules of a curve vary with position, imposes certain properties on transformations between the two systems. As was noted above (RULE 7a), the *basic* polar construction rule for a rectangular curve cast about a point in the plane has the same degree as the rectangular eq. But it also has been noted that if the curve includes the polar pole, the degree must reduce by at least one unit, while if it encloses the pole, reductions may be much greater, even to a value of unity.

Since rectangular curves usually are positioned to enclose and/or include the origin, the degree of a corresponding polar construction rule characteristically is lower than that of the basic eq. and rectangular eq. The reduction generally occurs in one of two ways--by factoring out r's or by taking roots. In the former case, intersections at the origin are lost (as noted above); in the latter case, one root represents a supernumerary or "extra" locus of the curve in question.

In the inverse transformation to the rectangular construction rule from the polar one, the inverse operations are required, that is, one must multiply through by a power of r [or $\pm\sqrt{(x^2+y^2)}$ and/or square the eq. one or more times. In the former case, new points generally are not being added to the rectangular locus (polar lines "punctuated" at the polar pole are exceptions). In the latter case, new loci are not being added; the rectangular construction rule merely is being expressed in a form without radicands involving x or y.

In the following, examples and illustrations are given for a number of representative transformations to illustrate, not the algebra of the process (though this is treated in passing), but the geometrical interpretations of the algebraic procedures and the differences in derivation and plotting between the classical procedures and the new standard procedures. The latter procedures must be followed if one is to achieve the assignment of definitive degrees in r to polar construction rules. Again, circles and ellipses are used as examples, but these are supplemented with the equilateral lemniscate, the $e = 2$ limacon, and the $C = 7/30$ Cassinian biovals.

Circles

Consider a bisector circle centered at point (h,h) in the 1st-quadrant, for which the basic polar construction rule is 14a, obtained directly by substitution in eq. 12b. In the standard system, this eq. represents a circle (Fig. 10a$_1$) with two values of r obtained for each value of θ between the limiting-tangent values (see Problem 6). In the classical treatment, however, it represents two coincident circles, one obtained as in the standard system, and the other as θ progresses between the extensions of the limiting-tangent lines in the 3rd-quadrant. The treatment of a general circle in the plane differs in no essential respect from that of the bisector circle.

polar construction rules of circles

(a) $\quad r^2 - 2hr(\sin\theta + \cos\theta) = (R^2 - 2h^2)\quad$ basic eq. of bisector circle (one circle for θ between limiting-tangent values, two circles for $\theta = 0°-360°$) $\quad(14)$

(b) $\quad r[r - \sqrt{2}R(\sin\theta + \cos\theta)] = 0 \quad$ point at origin plus bisector circle including origin (one circle for θ between limiting-tangent values, two circles for $\theta = 0°-360°$)

(b') $\quad r = \sqrt{2}R(\sin\theta + \cos\theta) \quad$ bisector circle including origin

(c) $\quad [r - h(\sin\theta + \cos\theta) + \sqrt{(R^2 - h^2 + 2h^2\sin\theta\cos\theta)}] \cdot$
$\quad [r - h(\sin\theta + \cos\theta) - \sqrt{(R^2 - h^2 + 2h^2\sin\theta\cos\theta)}] = 0 \quad$ bisector circle enclosing origin (one circle for each bracketed term equated to 0 for a 360° sweep of θ; the two circles are traced 180° out of phase)

(c') $\quad r = h(\sin\theta + \cos\theta) + \sqrt{(R^2 - h^2 + 2h^2\sin\theta\cos\theta)} \quad$ bisector circle enclosing origin (one circle for a 360° sweep of θ)

(d) $\quad r^2 = R^2, \quad r = \pm R, \quad r = R \quad$ centered circle(s)

When a bisector circle includes the origin (for the condition $R^2 = 2h^2$), the constant term of eq. 14a vanishes, allowing an r to be factored to yield the degenerate eq. 14b. This construction rule represents both a point at the origin and the locus or loci obtained by equating the bracketed expression to 0. In the standard system, this construction rule represents a single circle as θ ranges between the limiting-tangents (-45° to 135°), whereas in the classical treatment it represents two coincident circles traced out in succession as θ ranges from 0° to 360°. Discarding the point(s) at the origin (the factored r's)--the usual procedure--leads to the loss of general points of intersection at the origin (eq. 14b').

In the case of a bisector circle enclosing the origin, the basic construction rule is 14a, for $R^2 > 2h^2$. In the classical treatment employing this eq., two circles are represented and traced out simultaneously, but 180° out of phase. In the standard treatment, the square is completed for eq. 14a and roots are taken, yielding the degenerate construction rule 14c. Since this is merely another way of expression eq. 14a, it also represents two coincident circles. This being the case, either term equated to 0 represents the initial bisector circle.

Lastly, for a centered circle, in which case $h = k = 0$, eq. 14a yields 14d-left or middle, for two circles traced out simultaneously and 180° out of phase (assuming a monotonic progression of θ). Classically, this case has been represented by eq. 14d-right, for a single circle, the same result required by the new standard treatment.

Ellipses

Turning now to ellipses, attention again is confined to 0°-ellipses, beginning with the bisector 0°-ellipse of construction rule 15a and Fig. $10b_1$, centered at point (h,h). The circumstances for this ellipse differ in no essential respect from those for a bisector circle. If the major axis of this ellipse is translated into coincidence with the x-axis, to yield an axial 0°-ellipse with center at (h,0), the same circumstances apply, except that the limiting-tangents now are symmetrically positioned about the 0°-axis (this is not the case for the bisector circle; Fig. $10a_1$).

polar construction rules of ellipses

(a) $A^2r^2 - 2hr(a^2\sin\theta + b^2\cos\theta) = a^2b^2 - h^2(a^2+b^2)$ bisector 0°-ellipse (one ellipse for θ between limiting-tangent values, two ellipses for $\theta = 0°-360°$) (15)

$A^2 = (a^2\sin^2\theta + b^2\cos^2\theta)$

(b) $A^2r^2 - 2b^2hr\cos\theta = b^2(a^2-h^2)$ axial 0°-ellipse (loci follow same rules as given above)

(c) $r = (2ab^2\cos\theta)/A^2$ tangent axial 0°-ellipse (one ellipse for $\theta = -90°-90°$, two ellipses for $\theta = 0°-360°$)

(d) $r = [b^2h\cos\theta + ab\sqrt{(A^2-h^2\sin^2\theta)}]/A^2$ axial 0°-ellipse enclosing origin (one ellipse for $\theta = 0°-360°$)

(e) $r^2 = a^2b^2/A^2$, $r = \pm ab/\sqrt{A^2}$, $r = ab/\sqrt{A^2}$ centered 0°-ellipse (2 ellipses for left & center eqs., one ellipse for right eq., $\theta = 0°-360°$)

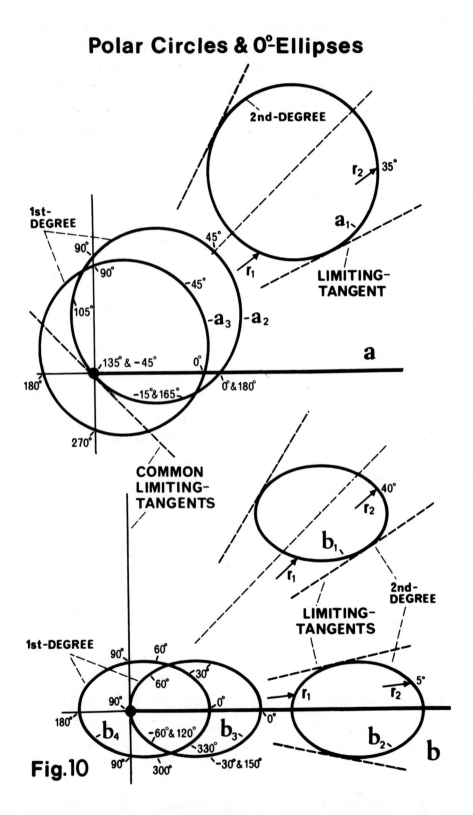

Fig. 10

For the tangent 0°-ellipse (Fig. 10b$_3$), as for the bisector circle including the origin (Fig. 10a$_2$), with the vanishing of the constant term (for a = h), an r factors out of construction rule 15b, yielding 15c. In the classical treatment, as for the tangent axial circle (Fig. 3h), one complete ellipse is traced for θ progressing from 0° to 180°, and a second ellipse for θ progressing from 180° to 360°. In the standard treatment, θ is allowed to progress only between -90° and +90°, the angles of the limiting-tangents, and only the first of these two ellipses is traced.

An axial 0°-ellipse enclosing the origin (Fig. 10b$_4$) is represented by construction rule 15d. Since there are no limiting-tangents, θ progresses from 0° to 360°. Completing the square of eq. 15b, taking roots, and expressing the result as a degenerate eq., yields the product of two factors (see eq. 14c), each of which equated to 0 represents one ellipse. In the standard treatment, one takes the factor with the positive sign for the radical, yielding eq. 15d, for one ellipse. The classical construction rule for this ellipse is 15b, with h < a.

Lastly, the centered 0°-ellipse (with h = k = 0) has the classical construction rule 15e-left or center, both of which represent two ellipses, 180° out of phase. In the standard treatment, only the positive root of eq. 15e-middle is taken, yielding eq. 15e-right, for a single ellipse (not shown in Fig.10b).

The Equilateral Lemniscate

Some simple examples of rectangular 4th-degree curves now are considered, listing their standard polar construction rules for general locations on the 0°-axis and illustrating them in the order of their polar symmetries. The standard practice is followed of orienting the curves with a line of symmetry in coincidence with the 0°-axis and using a center of symmetry or a DP (double-point) as the locational reference point with respect to the polar pole (for an x-h rectangular displacement).

The first example, the equilateral lemniscate (of Bernoulli) of eqs. 6d and Fig. 11-left, is a representative of a group of curves called *central quartics*. This name is derived by analogy with the "central conics" (conics with a true center, specifically, the point of intersection of two orthogonal lines of symmetry), from which central quartics are obtained by inversion about the center.

The standard polar construction rule of 0°-equilateral lemniscates for a general position on the 0°-axis (x-h displacement of the center along the rectangular x-axis), is the 4th-degree eq. 16a, where parameter a is the loop diameter (see Fig. 11a-left). Referring to Fig. 11a-left, it is seen that a radius vector from the pole generally intersects the curve at four points. Since, in the standard polar treatment (see above), θ is allowed to range only between the limiting-tangents, it is clear that a polar construction rule of at least 4th-degree is required to describe a single trace of the curve.

standard polar construction rules of axial 0°-equilateral lemniscates

(a) $r^4 - 4hr^3\cos\theta + [2(2h^2-a^2)\cos^2\theta + (2h^2+a^2)]r^2$ basic eq., center at $r = h$ (16)
    $+ 2h(a^2-2h^2)r\cos\theta + h^2(h^2-a^2) = 0$

(b) $r^4 - 2\sqrt{2}ar^3\cos\theta + 2a^2r^2 - a^4/4 = 0$ pole at loop focus, $h = a\sqrt{2}/2$

(c) $r^3 - 4ar^2\cos\theta + a^2r(2\cos^2\theta+3) - 2a^3\cos\theta = 0$ pole at vertex of left loop, $h = a$

(d) $r^2 = a^2(\cos^2\theta - \sin^2\theta)$, $r = a\sqrt{(\cos^2\theta-\sin^2\theta)}$ pole at center, $h = 0$

If the pole is shifted to the focus of the left loop, as in Fig. 11b-left, θ is permitted to range through the full 360°. Although the linear term in r vanishes, a 4th-degree construction rule (16b) still is required to describe a single trace of the curve. Since eq. 16b is much simpler than eq. 16a, the lemniscate of Fig. 11b is much more-symmetrically positioned than that of Fig. 11a.

Bringing the pole to the vertex of the left loop (with $h = a$) leads to the vanishing of the constant term of eq. 16a and the factoring out of an r (representing the general intersection at the origin), with a reduction to the 3rd-degree four-term eq. 16c. Since a radius vector from the pole now intersects the curve at a maximum of three points at non-zero distances, and θ is limited to a trace between -90° and +90°, a construction rule of only 3rd-degree in r is able to represent a single trace of the curve (allowing θ to range over 360° leads to two traces).

If the pole now is brought to the center of the curve (which also is a DP), for $h = 0$, three terms vanish and an r^2 factors out of the remaining two terms, leaving eq. 16d-left, the classical form in which the construction rule usually

POLAR SYMMETRY & PLOTTING OF RECTANGULAR QUARTICS

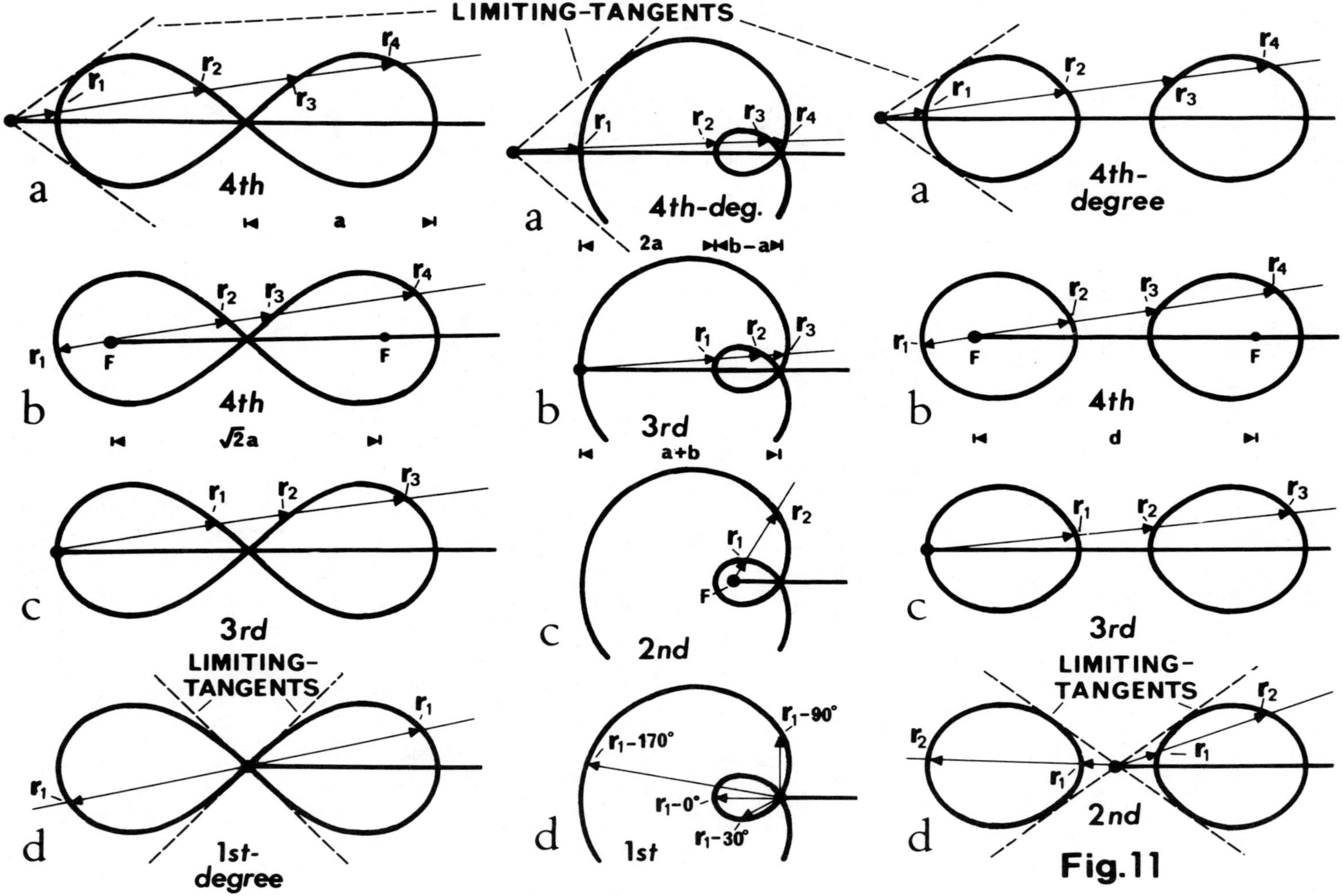

Fig. 11

is given (except that $\cos 2\theta$ customarily is employed in place of $\cos^2\theta-\sin^2\theta$). However, this gives a double trace of the curve, even for θ ranging only between the limiting-tangents (-45° to +45° and 135° to 225°). Accordingly, the positive root is taken, yielding eq. 16d-right for a single trace.

Since eqs. 16a-c and 16d-right are for single traces of the curve, for θ ranging between the limiting-tangents (when they exist), comparisons between them lead to valid assessments of the relative symmetries of position and form of the four loci with respect to the polar reference elements. These fall in the sequence, a to d, both for eqs. 16 and for Fig. 11-left.

The Equilateral Limacon

Turning from a curve with a 4th-degree rectangular eq. and opposed loops, to one with an internalized loop, consider the hyperbolic limacon of Figs. 11a-d-middle, which represents a quite different situation. Lemniscates (central quartics with opposed loops) and hyperbolic limacons are closely related. The former are inversions of hyperbolas about their centers, and the latter are inversions of hyperbolas about their traditional foci (in consideration of which lemniscates and hyperbolic limacons also must invert to each other). [The lemniscate of Fig. 11 is the central inversion of an equilateral hyperbola ($e = \sqrt{2}$), whereas the limacon is the focal inversion of a hyperbola of eccentricity, $e = 2$ (the small loop of the $e = \sqrt{2}$ hyperbolic limacon--the equilateral limacon--is relatively too small for convenient illustration; but see Fig. 8c). The $e = 2$ lemniscate (a lemniscate derived from the $e = 2$ hyperbola) is very similar to the one shown (Fig. 11-left); but its DP-tangents intersect each other at 120°, rather than 90°, and it has relatively broader loops (as e increases, the loops increasingly resemble tangent circles).]

standard polar construction rules of axial 0°-limacons

(a) $r^4+2(b-2h)r^3\cos\theta+[2h(h-b)-a^2+(b-2h)^2\cos^2\theta]r^2$ basic eq. for DP at $r = h$ (17)
$+ 2h[a^2+(h-b)(b-2h)]r\cos\theta+h^2[(h-b)^2-a^2] = 0$

(b) $r^3-2(2a+b)r^2\cos\theta+[a(a+2b)+(2a+b)^2\cos^2\theta]r$ pole at vertex of large loop
$- 2a(a+b)^2\cos\theta = 0$

(c) $r^2 - (b^2-a^2\cos\theta)r/b + (a^2-b^2)^2/4b^2 = 0$ pole at focus of self-inversion

(d) $r = a - b\cos\theta$ pole at DP (h = 0 in eq. 17a)

The standard polar construction rule of a 0°-limacon for a general location on the 0°-axis is the 4th-degree eq. 17a (x-h displacement of its DP along the rectangular x-axis). Referring to Fig. 11a-middle, it is seen that a radius vector from the pole generally intersects the curve at four points, requiring a polar construction rule of at least 4th-degree to describe the curve.

Letting $h = a + b$, places the pole at the vertex of the large loop (Fig. 11b-middle), leading to the vanishing of the constant term of eq. 17a, allowing an r to be factored out to yield eq. 17b. A radius vector ranging between the two limiting-tangents now generally intersects the curve at three points at non-zero distances, and the corresponding construction rule (17b) is of 3rd-degree.

If the pole is placed in the interior of the large loop, for h lying between b-a and b+a (not shown in Fig. 11), a 4th-degree construction rule is required, as is the case for the curve of Fig. 11b-left. This also has five terms (as does eq. 17a) and represents a less-symmetrically-positioned curve than the vertex-limacon of Fig. 11b-middle (with θ varying through the full 360° range).

In the presence of an internalized loop, it becomes possible for the curve to enclose the pole "doubly," that is, for a pole to be within both loops. This is a much more symmetrical position of the pole in the plane of the limacon (and of the limacon about the pole) than for either the general axial 0°-limacon (eq. 17a), the vertex 0°-limacon (eq. 17b), or a 0°-limacon cast about a pole interior to only the large loop. Since θ ranges through the full 360° for this location of the pole, and radii from the pole intersect the curve in at most two points, the curve now can be described with a 2nd-degree construction rule. This is true for a pole in any location within the small loop, not only for axial poles.

Accordingly, eq. 17a reduces to 2nd-degree for any value of h between 0 and b-a (i.e., it may be solved for r^2, representing two values of r that yield the entire curve for a 360° sweep of θ). For convenience, the value selected for h has positioned the pole at the point about which the curve self-inverts, yielding eq. 17c (see Problem 7).

[Hyperbolic limacons self-invert about a pole in the finite plane, whereas lemniscates do not, because the pole of inversion of the parent hyperbola is central for the latter but not for the former. If a central curve does not self-invert about a pole in the finite plane, neither will a central inversion of it do so. On the other hand, a non-central axial inversion of a hyperbola (inversion about a point on a line of symmetry but not at the center) yields a curve with one pole of self-inversion, whereas a non-axial inversion (inversion about a pole that does not lie on either axis) yields a curve with two poles of self-inversion (see Kavanau, 1980, 1982).

In general, a pole of self-inversion in the finite plane of an inverse curve is gained for each line of symmetry lost in the inversion, and the only lines of symmetry that are retained in an inverse locus are those upon which the pole of inversion is incident.]

Lastly, as for Fig. 11d-left, for the pole at the DP of the limacon (Fig. 11d-middle), a radius vector for a given value of θ intersects the curve at only one point, and a 1st-degree construction rule is able to represent the entire curve for θ ranging through 360°. Although some radius vectors have geometrical intersections with the curve at two points, these intersections are for different values of θ and different signs of r; one is for r with one sign at a given value of θ, and the other is for r with the opposite sign for a value of $\theta+180°$.

Biovular Cassinians

For the third example, a curve with two discrete opposed ovals is taken (as opposed to an "annular" curve, in which one discrete oval is wholly interior to another). This is a representative of the famous ovals of Cassini of eqs. 6g, of which the equilateral lemniscate of Fig. 11-left is one *curve of demarcation* (i.e., it is the curve--for a value of $C=1/4$ in eq. 6g--which demarcates, or is the transition locus, between monovular and biovular Cassinians; see Fig. 20-1b). The example of Fig. 11-right employs the ovals obtained for a value of $C = 7/30 = 0.23333$.

The standard polar construction rule of 0°-Cassinians for a general location with center on the 0°-axis (x-h displacement of its center along the rectangular x-axis) is eq. 18a. The treatment of the symmetry and construction rules of axial 0°-Cassinian ovals, (a) at a general position, (b) with the pole at the focus of the left oval, and (c) with the pole at the outer vertex of the left oval (Fig. 11a-c-right and eqs. 18a-c), differs in no essential respect from that for the axial 0°-equilateral lemniscates of Fig. 11a-c-left and eqs. 16a-c.

When the ovals have the pole at their center, however, the treatments diverge, because, unlike the loops of lemniscates, Cassinian ovals do not come into contact. Accordingly, radius vectors lying between the limiting tangents intersect each oval at two points, leading to the requirement for a 2nd-degree construction rule. Letting $h = 0$, and solving eq. 18a for r^2, leads to the root-eq. 18d.

[Note that the centered 0°-equilateral lemniscate (eq. 18e') has greater structural simplicity in the polar coordinate system than biovular Cassinians (eqs. 18d',d"), since its polar construction rule is of only 1st-degree, compared to 2nd-degree for Cassinians.]

standard polar construction rules of axial 0°-Cassinian ovals

(a) $r^4 - 4hr^3\cos\theta + [(4h^2+d^2)/2 + (4h^2-d^2)\cos^2\theta]r^2$ basic eq., center at   (18)
$-h(4h^2-d^2)r\cos\theta + [h^4 - d^2h^2/2 + (1/16-C^2)d^4] = 0$ $r = h$ (d, the distance between the intra-oval foci)

(b) $r^4 - 2dr^3\cos\theta + d^2r^2 - C^2d^4 = 0$ pole at focus of left oval ($h = d/2$)

(c) $r^3 - d\sqrt{(C+1/4)}r^2\cos\theta + Cd^2(1/C^2 + 2 + 4\cos^2\theta)r$ pole at outer vertex of left oval
$- 4Cd^3\sqrt{(C+1/4)}\cos\theta = 0$

(d) $r^2 = (d^2/4)[(\cos^2\theta - \sin^2\theta) \pm m]$ roots of eq. 18a for pole at center
$m = \sqrt{[(\cos^2\theta - \sin^2\theta)^2 - 1 + 16C^2]}$

(d') $[r - (d/2)\sqrt{(\cos^2\theta - \sin^2\theta + m)}] \cdot$ standard eq. of centered biovals in degenerate form
$[r - (d/2)\sqrt{(\cos^2\theta - \sin^2\theta - m)}] = 0$

(d") $r^2 - (dr/2)[\sqrt{(\cos^2\theta - \sin^2\theta + m)} +$ standard eq. of centered biovals
$\sqrt{(\cos^2\theta - \sin^2\theta - m)}] + (d^2/4)\sqrt{(1-16C^2)} = 0$

(e) $r = (d/2)\sqrt{(\cos^2\theta - \sin^2\theta + m)}$ standard eq. of centered monovals

(e') $r = a\sqrt{(\cos^2\theta - \sin^2\theta)} =$ equilateral lemniscate, $C = 1/4$, $m = \sqrt{(\cos^2\theta - \sin^2\theta)}$ and a
$(d\sqrt{2}/2)\sqrt{(\cos^2\theta - \sin^2\theta)}$ (of eq. 16d) $= \sqrt{2}d/2$

In this case, one may not simply take the two roots for the positive sign of the radical (eq. 18d with positive sign) as was done in the case of eq. 16d, as this would give only a double trace (with the traces 180° out of phase) of the portions of the ovals distal to the limiting-tangents. Similarly, taking only the two roots for the negative sign leads to a double trace of the portions proximal to the limiting-tangents. Accordingly, the positive root of r^2 is taken for both cases (both signs of the radical), leading to the degenerate product 18d'. Expressed as a quadratic eq., this becomes 18d".

This complication does not arise when dealing with Cassinian monovals (C > 1/4) because the two roots for the negative sign of the radical are imaginary. Furthermore, since a radius vector from the center only intersects the monovals at one point, a 1st-degree construction rule suffices to describe them, leading to eq. 18e.

[From a comparison of eqs. 18e and 18e', it is seen that the equilateral lemniscate also has greater structural simplicity in the polar system than Cassinian monovals. This is true generally of curves of demarcation, as compared to the loci that they demarcate (see RULE 50).]

New Perspectives from Polar Construction Rules

Having considered the topics of symmetry of position and form in the rectangular and polar coordinate systems, we pause now to inquire into the significance of the findings. In the final analysis their significance is primarily conceptual: they call attention to the properties of the coordinate systems and emphasize the similarities and differences between these properties. In essence, this is true of the studies of symmetries of position and form in any coordinate systems.

Although one system uses two lines as reference elements, and the other a point and a half-line, both are essentially "single-point" systems. In the rectangular system, this is because the intersection of the axes defines a unique point of reference for distances (which transforms to the polar pole). In the polar system, it is because the half-line is a reference element for angles, leaving the pole as the unique point of reference for distances.

These facts, and the correspondences that exist between 0°, 90°, 180°, and 270°-polar axes, on the one hand, **and** the rectangular axes on the other, are the bases for the great similarities between the two systems and their consequent great utility in analysis. The only other non-bilinear coordinate system that possesses a single point as the unique reference element for distances is the hybrid polar-rectangular system (see next Chapter), which combines the use of the polar coordinate r with the rectangular coordinate x.

We now address the question, "How much can one learn about the properties of a curve by inspecting its construction rules for various locations and orientations in the polar and rectangular systems?" In the rectangular system, one can identify only certain very special points and lines, namely, points and lines of reflective symmetry. By making certain comparisons between eqs., one also can identify points of rotational symmetry (i.e., rotational symmetry at angles other than 180°). In addition, one can judge the complexity of the curve from the degree of its construction rules, in the sense of ascertaining the maximum number of intersections of the curve with a line in the finite plane.

All these properties of curves also are readily accessible from inspection of their polar construction rules for various locations and orientations. The polar coordinate system, however, by virtue of two extraordinary properties, adds a critical new perspective that is almost totally obscured in the rectangular system: first, a single coordinate, r, suffices to specify the distances from a unique point (the polar pole) to points of the curve; and, second, in consequence of this fact, and unlike the situations in other coordinate systems, every polar eq. of exponential degree greater than 1 is essentially a degenerate eq. In other words, every polar construction rule of finite degree, n, can, in theory, be expressed in the form 19a. In standard form, this can be re-expressed as eq. 19b.

general polar eqs. expressed in degenerate forms

(a) $[r-f_1(\theta)] \cdot [r-f_2(\theta)] \cdot (r-f_3(\theta)] \cdot [r-f_4(\theta)] \cdots\cdots [r-f_n(\theta)] = 0$ (19)

(b) $[r-f_1(\cos\theta)] \cdot [r-f_2(\cos\theta)] \cdot [r-f_3(\cos\theta)] \cdots\cdots [r-f_n(\cos\theta)] = 0$

The forms taken by the individual factors of eq. 19b are dependent upon the position of the polar pole. For some positions, as we have seen, (a) the $f_i(\cos\theta)$-functions vanish and one or more r's can be factored out; for other positions, (b) at least one function becomes merely the negative of another, whereupon more than one trace is represented, and the supernumerary traces can be factored out (because the traces are for full 360° sweeps of θ); for other positions, (c) the functions simplify, and for still others, (d) more than one of the simplifications and/or reductions (a)-(c) take place.

Whereas reduction (a) occurs for any point of the curve (i.e., for the origin incident upon the curve at any point), reduction (b) and simplification (c) occur only for certain very special positions of the polar pole. The answer to the question of why, precisely, construction rules simplify and/or reduce when cast about these very special positions is deferred to the last section, *Structural Simplicity and Polynomial Distance Relationships*.

Simple Intercept Equations

The topic of extracting the structure rules that describe and define inherent structure of curves is treated in Chapters VII and VIII. At this juncture, only one of the superficial consequences of the above relations is considered. This consequence is that if the polar construction rules of curves are expressed in appropriate forms, they can be used to locate some of the "certain very special positions" of the poles referred to above. In theory, any point in the plane of a curve can serve as the position of reference for casting polar construction rules in these appropriate forms. It greatly simplifies matters, however, if a center or DP is used and if the eqs. are cast about a point at distance h along the line of symmetry to the left (or right) of this reference point, which is the practice followed here. Accordingly, the constrution rules that are employed are the polar eqs. of the curves cast about an arbitrary point on a line of symmetry at distance h to the left or right of the center or DP.

Structure rules, like virtually all construction rules, are expressed in terms of relations between distances (as noted above, the polar coordinate system is highly exceptional in its employment of but a single distance coordinate). It follows that, in the derivation of a structure rule from a polar construction rule, once the probe-angle (see Fig. 2a) has been specified, the variable θ has to be eliminated. This requirement places a very heavy premium on linearity of the eqs. in either the sine or cosine function (but not both together). In fact, the presence of low powers of the cosine or sine function in polar eqs. is to a great extent a more reliable indication of underlying simplicity of a curve's inherent structure and structure rules than the linearity of the eqs. in r.

[It is implicit in this statement (as in RULE 8a) that the structural simplicity of a curve about a point in the polar system is not an absolute guide to the inherent structural simplicity of the curve about the point, i.e. the simplicity of the curve's structure rules about the point.]

These relations are illustrated by formulating some representative polar construction rules in powers of $\cos\theta$, confining attention to eqs. cast about arbitrary points on lines of symmetry that are in coincidence with the polar axis. The following construction rules are *simple intercept equations* of curves already encountered. The use of this name distinguishes these eqs. from polar construction rules expressed in the standard (and classical) forms. Each eq. is arranged in the order of terms in powers of cosθ, followed by terms in powers of r, alone, followed by a "constant term" involving only parameters of the curve and its position. The latter term defines the intercepts of the curve with the 0°-180°-axes. Thus, values of h for which this term vanishes (designated the *constant-condition*) are values for which the curve includes the origin, one or more r's factors out of the eq., and the degree of the eq. reduces by at least one unit.

simple intercept equations

ellipse, referred to center (see eq. 12d)

$$(b^2-a^2)r^2\cos^2\theta + 2b^2hr\cos\theta + a^2r^2 + b^2(h^2-a^2) = 0 \quad (20)$$

cissoid of Diocles, referred to cusp (see eqs. 6b)

$$2(a+h)r^2\cos^2\theta + (r^2+3h^2)r\cos\theta + (h-2a)r^2 + h^3 = 0 \quad (21)$$

equilateral lemniscate, referred to center (see eq. 16a)

$$2(2h^2-a^2)r^2\cos^2\theta + 2h(2r^2+2h^2-a^2)r\cos\theta + r^4 + (a^2+2h^2)r^2 + h^2(h^2-a^2) = 0 \quad (22)$$

limacons, referred to DP (see eq. 17a)

$$(b+2h)^2 r^2\cos^2\theta + 2[(b+2h)r^2 + h(2h^2+3bh+b^2-a^2)]r\cos\theta + r^4 + \quad (23)$$
$$[2h(b+h)-a^2]r^2 + h^2[(b+h)^2-a^2] = 0$$

Cassinian ovals, referred to center (see eq. 18a)

$$(4h^2-d^2)r^2\cos^2\theta + h[(4h^2-d^2)+4r^2]r\cos\theta + r^4 + (4h^2+d^2)r^2/2 + \quad (24)$$
$$[h^4-d^2h^2/2+(1/16-C^2)d^4] = 0$$

In the case of eq. 22, when $h=0$, both the constant term and the coefficient of the cosθ-term vanish (the *cosθ-condition*), that is, the cosθ-term also vanishes. Since this leaves the eq. with only a term in $\cos^2\theta$, it means that the eq. of the curve is cast about a *true center* (the point of intersection of two orthogonal lines of symmetry), since a polar eq. with the $\cos^2\theta$ (or $\sin^2\theta$) as its only function of θ has symmetry about both the 0°-180°-axes and the 90°-270°-axes. Thus, any axial simple intercept eq. that is quadratic in cosθ, for which h is a multiplicative coefficient of the cosθ-term, represents a curve with a true center (actually, this is true if any point exists for which the coefficient of the cosθ-term vanishes, not just if $h=0$).

Note that in every case but eq. 20, values of h exist for which the coefficient of the $\cos^2\theta$-term vanishes. Such a value of h (for example, h = -a, for eq. 21) is said to satisfy the *$\cos^2\theta$-condition* (vanishing of the $\cos^2\theta$-term) and defines the location(s) of one or more high-ranking *circumpolar foci* of the curve--designated as *$\cos^2\theta$-condition foci* (see Chapters VII and VIII). This means that when the construction rule of the curve is cast about the pole h, in question, it takes on a relatively simple form, although its degree in r does not necessarily change.

The $\cos^2\theta$-condition, $\cos\theta$-condition, and constant-condition, however, are by no means the only *focal conditions* that can be ascertained from simple intercept eqs. Another powerful focal condition is known as the *radicand constant-condition*. This is illustrated in the cases of eqs. 20 and 23. If one obtains a solution of these eqs. for $\cos\theta$, employing the quadratic-root formula (or by completing the square), the radicals involved in these solutions are the expressions 20a and 23a. Inspection of the radicands reveals that they become perfect squares for the conditions 20b and 23b. These are the conditions for the traditional foci of ellipses (eq. 20b) and the foci of self-inversion of limacons (eq. 23b)

radicand constant-conditions

$\pm a\sqrt{[r^2(a^2-b^2)+b^2(h^2-a^2+b^2)]}$	radical of quadratic-root solution of eq. 20 for ellipses	(20a)
$h = \pm\sqrt{(a^2-b^2)}$	locations of traditional foci of ellipses	(20b)
$\pm a\sqrt{[b(b+2h)r^2+h^2(a^2-2bh-b^2)]}$	radical of quadratic-root solution of eq. 23 for limacons	(23a)
$h = \pm(a^2-b^2)/2b$	location of foci of self-inversion of limacons	(23b)

[In the employment of eqs. 20 and 23 to derive structure rules, the vanishing of the constant term of the radicand leads to the radicand becoming a perfect square and the vanishing of the radical. This generally leads to a reduction in the degree of the corresponding structure rule by a factor of one-half, which accounts for the power of this focal consistion.

Note that a $\cos^2\theta$-condition also exists for the ellipse of eq. 20. In this case, no positional parameter is involved in the condition, but only the parameters of the curve. This means that the condition in question is not focal for the symmetry of position and form of ellipses, in general, in the coordinate system, but rather it is a comparative symmetry condition among ellipses. Specifically, this condition, a = b, for the vanishing of the $\cos^2\theta$-term, defines the most-symmetrical polar ellipse, the circle.]

STRUCTURAL SIMPLICITY AND POLYNOMIAL DISTANCE RELATIONS

These results, which by no means exhaust the potentials of the simple intercept eq., show that even within the well-known and much-used polar coordinate system, an appropriate analysis of the construction rules of curves in certain forms is capable of revealing both traditional foci and other points about which curves possess exceptional structural simplicity, which previously were defined and detected only by other means (for example, by the constructions of conics about two points or a point and a line).

These considerations lead us to the topics of Chapters VII and VIII, which are concerned with the generalized symmetry (structural simplicity) of curves about single structureless reference elements, namely single points. The aproach differs from that of the polar system in that there is no reference element for angles, in consequence of which there is no directional bias. This approach--*circumpolar symmetry analysis*--achieves the highest level of discrimination of the inherent structure of curves. The eqs. that characterize curves in this unique system are *structure rules* rather than *construction rules*. As will be seen, simple intercept eqs., such as those treated above, provide the raw materials for these studies and, in fact, for all studies of the structure of curves with respect to single reference elements.

Structural Simplicity and Polynomial
   Distance Relationships

As noted above, if the polar construction rules of curves of finite rectangular degree are expressed in "appropriate forms" (powers of $\cos\theta$ or $\sin\theta$), they can be used to locate some of the "certain very special positions" of poles in their planes for which polar construction rules (and circumpolar structure rules) simplify and/or reduce. We inquire lastly as to the basis for simple intercept eqs.--which are the initial eqs. for deriving structure rules--being these "appropriate forms."

The key consideration lies in the conditions for which simple intercept eqs. simplify or reduce, which, for extreme cases of structural simplicity, hinge on whether or not linear solutions of the eqs. in $\cos\theta$ (or $\sin\theta$) involve a radical in the variable r. If not, $\cos\theta$ (or $\sin\theta$) can be expressed as a low-degree polynomial function of r, or the ratio of two such low-degree polynomial functions, whereas if a radical is present, such a low-degree solution is not possible.

Since the polar $\cos\theta$ is simply the ratio of the rectangular x to the polar r, the absence of a radical in r, in solutions to simple intercept eqs., means that the rectangular x (or y) can be solved for explicitly as a simple polynomial function of the polar r. In other words, in the final analysis, the criterion that is being employed to define extreme cases of structural simplicity of a curve about points in its plane is: can the abscissae of the distances to the curve from the point be expressed explicitly as a simple polynomial function (or ratio of simple polynomial functions) of the distances, r, to the curve from the point?

But, since abscissae are simply distances from a line, this criterion amounts to inquiring as to whether or not the distance of every point of the curve from some line can be expressed explicitly as a simple polynomial function of its distance from some particular point (from which it is evident that close ties exist between this criterion and corresponding polar-linear construction rules; see Chapter V). Solutions of simple intercept eqs. for $\cos\theta$ provide rough guides for comparing the structural simplicities of curves about such points. Structure rules--with the added dimension provided by different probe-angles--provide the analytical basis.

Finally, it should be noted, that the *hybrid polar-rectangular coordinate system* of the next Chapter is the system in which precisely the two distances in question--the rectangular x and the polar r--constitute the coordinate distances. This system is the implicit, previously unrecognized, coordinate system of the classical distance eqs. from arbitrary points of curves to arbitrary points in the plane. It is by virtue of its facilitation of the discriminations discussed above that the hybrid coordinate system is the system par excellence for defining the foci of curves in any coordinate system that possesses at least one point or circle as a reference element (RULE 27).

Problems

1. Are the polar and rectangular eqs. of the trisectrix of Maclaurin (eqs. 6a) cast about the same pole?

2. Find the intersections of the polar curves represented by the rectangular eqs. of the equilateral lemniscate (eq. 6e-left) and the circle, $(x - a/2)^2 + (y + a/2)^2 = a^2$.

3. Show that the lemniscate of Problem 2 and the circle $r = a\cos\theta$, both of which include the origin, have no point of intersection at the origin, as determined analytically from their eqs.

4. Where is the segment of the curve of eq. 10a for θ progressing from 90° to 180°, and how does its size and shape compare with those of the segment in the 1st-quadrant?

5. Derive eq. 12i for the focus-0°-ellipse, expressed in terms of e and p, from eq. 12i for the same ellipse in terms of parameters a and b.

6. Find the limiting-tangents to the general bisector circle of eq. 14a.

7. Derive eq. 17c for a limacon cast about the focus of self-inversion at $h = (a^2-b^2)/2b$.

CHAPTER IV. THE BIPOLAR AND HYBRID POLAR-RECTANGULAR COORDINATE SYSTEMS

Introduction

In the non-coincident bipolar system--consisting of two non-coincident point-poles--the construction-rule approach to the symmetry of position and form attains some of its most intriguing manifestations, and certain of its aspects attain their highest level of expression. These singular characteristics trace to several very special (but not necessarily independent) features of a system consisting of two point-poles, as follows.

1. The two elements are the simplest possible.

2. The symmetry of position of the two elements with respect to one another is the simplest possible. [Only a single distance exists between all parts of both elements. If only one element were a point and the other were any curve but a circle (or one or more segments of a circle) centered upon it, there would be an infinite number of different distances from the curve-element to the point.]

3. The relative locations and orientations of the two elements are the most general possible. [Two lines, for example, can have an infinite number of relative orientations, and a line and a circle or a point and a circle can have an infinite number of relative locations.]

4. The amount of directional bias is the least possible in a system possessing multiple point-poles. [In the bipolar system, a point moving in the plane can be either approaching, receding from, or maintaining a constant distance from the bipolar axis (and, in general, from both poles). But in a system of three or more non-collinear points, a moving point in the plane may be approaching one axis while receding from the other two, maintaining a constant distance from one axis while approaching or receding from the other two, etc. (and similarly for the distances from the poles).]

5. The bipolar system is the only "non-redundant" multipolar system. [When three or more poles are present, construction rules involving distances from all of the poles can be recast as construction rules involving distances from only two of the poles (see Problem 18).]

[In connection with the fifth feature, when multipolar construction rules in more than two variables are recast as bipolar construction rules, one or two lines of symmetry may be added to the system. In this sense, many systems of three or more point-poles are not redundant to the bipolar system (see Problem 18).]

Notwithstanding its exceptionally simple features, the bipolar system with unlabelled poles possesses an almost universal intrinsic asymmetry, namely, an *asymmetry in its lines of symmetry*. Thus, the bipolar axis (one line of symmetry) includes the poles, whereas the midline (the other line of symmetry) does not. As will be seen in the following, this circumstance introduces significant asymmetries into transformation eqs. between the bipolar and rectangular systems. These asymmetries derive from the fact that the distances from the poles to an arbitrary point on the bipolar axis can be expressed as a linear function of the displacement from the midpoint, whereas a quadratic eq. is required to relate the corresponding distances of an arbitrary point on the midline (regardless of whether or not the latter is a line of symmetry).

It is easy to understand the power of the bipolar system for construction-rule analyses of symmetry (structural simplicity) from the following point of view. Recall the singular feature of the polar system that carries the analysis of symmetry beyond the limits (reflective and rotational symmetries) of the rectangular system into the realm of generalized symmetry or structural simplicity about a point. We completed our consideration of the polar system with some examples of conditions on simple intercept eqs. that reveal exceptional poles about which the construction rules of 0°-axial curves simplify and/or reduce, and particularly those for which relatively simple structure rules exist.

The same type of situation exists in the bipolar system. The greater the structural simplicity of a curve about two poles, the greater the *bipolar* symmetry and the simpler and/or lower the degree of the construction rule. The bipolar system, however, possesses a greater power for comparing the positional symmetries of poles (or the structural simplicities of curves about poles). This greater power traces to the facts that (a) both reference elements are point-poles (p_u and p_v) and (b) they are equivalent to one another in every respect (except that they may bear different labels in the polarized system).

By reason of the above circumstances, any differences in structural simplicity of a curve about two given poles (residing in the different locations of the poles in the plane of the curve) show up as differences in what is termed the *weighting* of the two poles in the construction rule. This takes the form of differences in the degrees of, coefficients of, and/or number of terms in the two distance variables (u and v) in the construction rule. In other words, the positional symmetries of two poles can be compared directly within a single eq.

Once one learns to interpret differences in weighting, a mere glance at a bipolar construction rule enables one to read out information about (a) classical reflective symmetries, (b) the bipolar symmetries of the curves about the poles--either as an equivalent (reflected in a line of symmetry of the curve) or non-equivalent pole-pair, and, in the latter case, (c) the comparative bipolar symmetry of the curve about the non-equivalent poles.

Bipolar ellipses help to illustrate these properties of bipolar construction rules. The two traditional foci of an ellipse are the only bipolar point-foci of the curve; an ellipse's construction rule cast about these foci is of lower degree than that of any other bipolar ellipse. The construction rule in question is the familiar $u+v=2a$ (more appropriately, $u+v=d/e$, where d is the *characteristic distance* and e is the eccentricity). If, now, one pole is displaced from one of the traditional foci along the line of symmetry, the construction rule becomes of 2nd-degree and the identity of the focal pole is readily discerned from its weighting (see below).

If both poles are displaced along the line of symmetry, the construction rule is of 4th-degree. The most-symmetrically-positioned of the two poles (the pole closest to the ellipse's center) can be ascertained by examining the coefficients of the terms in powers of u and v. If one pole is at a focus and the other is not incident upon the same line of symmetry, the eq. also is of 4th-degree and the focal pole is readily identified.

For both poles lying on a line parallel to a line of symmetry, the construction rule is of 8th-degree and the pole closest to the center is readily identified (by inspecting the coefficients of the terms in powers of u and v). If both poles are positioned arbitrarily, the degree of the construction rule also is 8 but its complexity is greatly increased, including the presence of mixed powers of u and v (products of powers of u and v are absent for 0° and 90°-ellipses).

Since the present treatment of the bipolar system logically draws attention to an important new coordinate system--the *hybrid polar-rectangular system*--a treatment of this system also is included in the present Chapter. In lacking the ambiguities and arbitrariness (particularly with regard to the expression of the angle coordinate) of the polar system as commonly employed, this new system provides a more powerful tool for assessing the structural simplicities of curves about points (and circles) and provides the preeminent construction-rule criterion for the identification of point-foci.

Historical

The idea of using two points as reference elements appears to have originated with Newton. He employed two points in the third of nine Modes he advanced for drawing tangents to curves in his *Methods of Series and Fluxions* (1671). The heyday of bipolar coordinates (then called "vectorial" coordinates) was the mid-19th century (particularly the late 50's through the early 70's), when there was great activity in the posing and solving of problems, mostly by British mathematicians. These are to be found in the pages of the *Educational Times*, *The Proceedings* (and *Transactions*) *of the London Mathematical Society*, *The Messenger of Mathematics*, and the *Quarterly Journal of Mathematics*.

Throughout the 19th century, Cartesian ovals occupied the greatest vectorial-coordinate attention of such figures as Casey, Cayley, Chasles, Crofton, Panton, Quetelet, S. Roberts, Sylvester, Williamson, and Wolstenholme, particularly Cayley, Crofton, and Roberts. Nor are these coordinates entirely forgotten today. For example, Lockwood (1961) devotes five pages to them.

Although analytical procedures frequently were used in the solutions of these problems, the heaviest emphasis in the 19th-century literature pertaining to bipolar curves seems to have been synthetic, employing geometrical relations and constructions. The general character of the problems is given by a few examples (these are not being proposed to the reader for solution).

1. Prove that the arc of a Cartesian oval at any point P is equally inclined to the straight line from P through any one focus and to the circular arc from P through the other two foci (by Crofton).
2. Prove that the sum of the areas of the two ovals of a Cartesian is equal to twice the area of the circle whose center is the triple-focus, and which passes through the points of contact of the double-tangent (by Panton).
3. Prove that if a circle cuts a given circle orthogonally, while its center moves along another given circle, its envelope is a Cartesian oval (by Casey and Quetelet).
4. The vertices of the base angles of a variable triangle move in two fixed circles, while the two sides pass through the centers of the circles, and the base passes through a fixed point on the line joining the centers. Prove that the locus of the vertex is a Cartesian oval (by Williamson).

These problems give only a hint of the rich body of extraordinary properties of Cartesian ovals. In fact, as I show elsewhere (Kavanau, 1982), the inversions of these curves that possess a true center are the 4th-degree homologues of central conics. Cassinian ovals belong to this group and are the most symmetrical of these quartic relatives. The present Chapter touches very [In this connection, constant linear sums of the two bipolar variables yield ellipses and a line-segment, constant differences yield hyperbolas and rays, constant ratios yield circles, and constant products yield Cassinian ovals.] little upon the matters dealt with in the problem examples, but is concerned primarily with the unusual and unique properties of the bipolar system. These are the extraordinary relations that exist between bipolar construction rules and the symmetry of position (and form) of the corresponding curves about the poles.

Although the literature on studies employing bipolar coordinates is rich, no mention is to be found of: (1) the unusual properties of the system listed above; (2) the translation and rotation eqs.; (3) the transformation eqs. between the rectangular and bipolar systems; (4) the fact that exponential degree is not conserved in the system; or (5) the significance of the fact that circles have eqs. of both 1st and 2nd-degree (just as the significance of similar occurrences in the polar system is ignored).

The reason for lacuna (1) probably is the same as that for the lack of prior treatments of the properties of other Euclidean coordinate systems (for example, the properties listed in Table 7), namely, the failure to appreciate the significance of these properties as they impinge on the properties of the loci in the systems. In the case of lacuna (2), the complexity of the transformation eqs. doubtless made (and still makes) such applications unattractive. The explanation for lacuna (3) may be the lack of mention in the research literature, as opposed to the instructional literature of the time.

The bases for lacunae (4) and (5) are several. First, the repertory of bipolar eqs. was extremely limited, including only (with exponential degrees in parentheses); ellipses (1), one-arm hyperbolas (1), limacons (1 and 2), Cartesian ovals (1 and 2), the circle of Apollonius (1), general axial circles (2), lines normal to the bipolar axis (2), and Cassinians (2), including the equilateral lemniscate. Second, and more important, the only positions in which these curves were plotted were 0°-axial, that is, with a line of symmetry coincident with the bipolar axis, and with the curves in some standard relation to the bipolar poles (almost always with the poles at foci).

Classicists sought to avoid having to cope with, rather than to elucidate, the significance of complications introduced by changes of location and orientation. This is the reason why the curves studied had exclusively 1st and 2nd-degree construction rules. Displacements from, or along, the bipolar axis usually lead to equations of much higher degree and complexity, and an inescapable confrontation with the general dependence of degree on position. In the only cases where axial displacements were tolerated--axial circles and normal lines

--the degrees of the eqs. do not change.

Third, the treatments of bipolar curves and their construction rules have been almost cavalier. Nowhere in the classical or modern literature is there mention, nor even a hint of recognition, of the fact that lines, rays, and line-segments have 1st-degree bipolar eqs., and that the simplest of all bipolar eqs., $u = v$, is the eq. of the bipolar midline (in fact, various bipolar lines, rays, and line-segments have eqs. of degrees 1, 2, 3, and 4; see below).

This attitude, and an apparent lack of knowledgeability, is well illustrated by Crofton's (1866) statement that, ".....Cartesian ovals, whose equation being linear, fills the same place in this system which that of the straight line does in rectangular coordinates....." and "The general equation of a straight line in vectorial coordinates is complicated. Hence to find the equation of the tangent to a curve seems to be a problem of little interest or utility in this system." [This is the very purpose for which Newton first employed the system, using Cartesian ovals as his example.] Except for Crofton's first statement, one might assume that the linear bipolar eqs. of a line, rays, and a line-segment were too trivial to mention.

The Polarized Versus the Non-Polarized Systems

The only system recognized and employed by classicists is the polarized system (see Table 7), in which the two poles are assigned fixed labels. Inasmuch as the polarized system, like the rectangular system, possesses an elective asymmetry, it is not ideally suited for the study of the comparative symmetry of form and position of bipolar curves. For example, some of the most symmetrical bipolar curves in the unpolarized system, such as $u = Cv$ (circles of Apollonius), $u - v = d/e$ (two-arm hyperbolas), and $u - v = d$ (opposed rays extending from the poles outward along the bipolar axis) become asymmetrical "half-curves" when plotted in the polarized system--with its artificially-imposed asymmetry.

Accordingly, Fig. 12a-c_2, for the most-symmetrical bipolar loci, represents the situation in the unpolarized system. On the other hand, for correlations between the locations of bipolar poles in the plane and the properties of its construction rules (for example, whether the poles are equivalent, whether a pole is at a focus, nearest the center, etc.), it is best to employ the polarized system. Unless otherwise stated, the following considerations apply to that system.

The Most Symmetrical Bipolar Loci

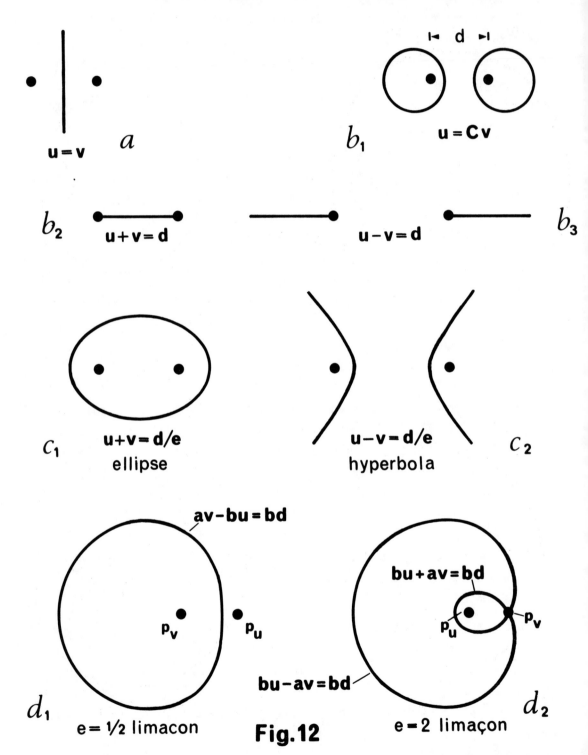

Fig.12

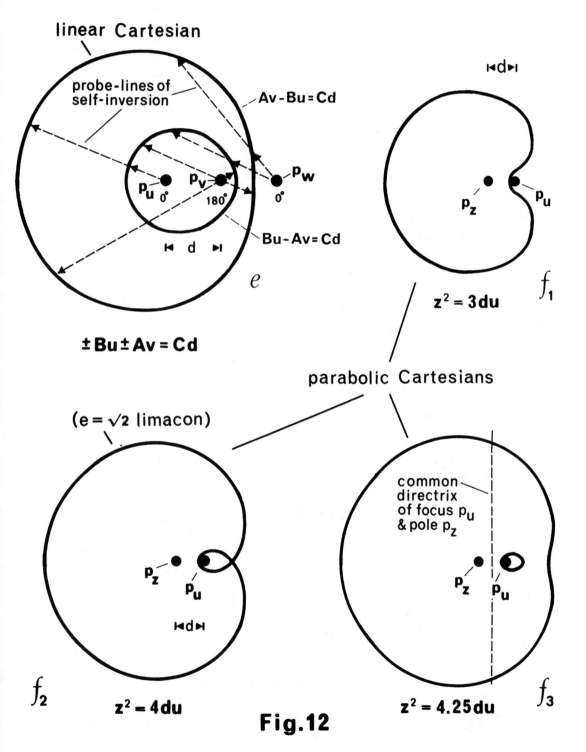

Fig.12 The Most Symmetrical Bipolar Loci

[If the unpolarized system were to be employed in Fig. 12 for all cases, all loci would have to be reflected in the bipolar midline (the line $u = v$). This would then have been necessary, for example, for the limacons and Cartesian ovals of Fig. $12d_1-f_3$.]

The Most-Symmetrical Bipolar Loci

Some of the most-symmetrical bipolar curves are illustrated in Fig. 12 (figures a-c are for the unpolarized system, figures d-f are for the polarized system). As in all two-element undirected-distance coordinate systems, the locus that stands preeminent as regards symmetry of form and position, or, equivalently, of mensural simplicity, is represented by the eq. $u = v$, in this case the bipolar midline. The most-symmetrical point on the midline is the midpoint, as this is the only point on the line for which lines to the poles make equal angles with each other, whether measured CW (clockwise) or CCW (counterclockwise).

Intuition receives a test in deciding on the next-most-symmetrical locus. RULE 1a, however, tells us that it will be the locus of the next-most-simple 1st-degree eq. The three candidates for this distinction are the loci and eqs. of Fig. $12b_1-b_3$. Deciding on the next-most-symmetrical locus then comes down to deciding which of the three eqs. is to be regarded as the simplest. Is it eq. b_1, with only two terms but with an extraneous coefficient, or is it eq. b_2 or b_3, with three terms, of which the third is the *characteristic distance* of the system? The four different lines of attack embodied in RULE 9 lead unequivocally to the conclusion that the connecting line-segment of eq. b_2 (Fig. 12) is the second-most-symmetrical polar locus. First, eq. b_2 is the only eq. that is

RULE 9. *Among loci with comparably simple construction rules in coordinate systems with two congruent elements, the most-symmetrical locus is:*

1. *the one for which the coefficients of terms in u and v are identical; and/or*
2. *the one that is symmetrical about the most symmetrical point in the system; and/or*
3. *the one that includes one or more points of the reference elements; and/or*
4. *the one that retains its symmetry in the identical but polarized system.*

symmetrical in u and v (exchanging u and v does not alter the eq., since their coefficients are identical). Second, the locus of Fig. $12b_2$ is symmetrical about the midpoint. Third, this locus includes poles p_u and p_v. Fourth, locus b_2 is the only one of the three loci that retains its symmetry in the polarized system.

This conclusion and RULE 9 also lead us to yield priority of simplicity (in a system consisting of two congruent reference elements) to the simplest three-term construction rule that includes the characteristic distance, over the simplest two-term construction rule that includes an arbitrary coefficient. It follows from these considerations that locus b_3, for the two opposed axial rays, is the third-most-symmetrical curve, whereupon, by default, circles of Apollonius become the fourth-most-symmetrical locus.

The next-most-symmetrical loci are the central conics of Fig. $12c_1,c_2$, of which, following the criteria of RULE 9, ellipses are most symmetrical. The construction rules of these curves also possess three terms, like those of the connecting line-segment and lateral rays, but in addition possess an extraneous parameter e, the eccentricity. Note that the line-segment and lateral rays are limiting cases of ellipses and hyperbolas for $e = 1$, the value for the parabola. Accordingly, curve b_2 might be regarded as the limiting form of two coaxial parabolas with vertex regions facing away from each other (with joined arms), and curve c_2 might be regarded as the limiting form of two coaxial parabolas with vertex regions facing each other (with opposed arms).

The next-most-symmetrical curves are represented by linear construction rules with three terms, in which one of the variables has a non-unit coefficient and the third term is the characteristic distance, $\pm u \pm (a/b)v = d$, of Fig. $12d_1,d_2$. These represent elliptical limacons or the individual loops of hyperbolic limacons, with one bipolar pole at the pole of self-inversion and the other at the DP. The limacons of Fig. $12d_1,d_2$ are inversions of the $e = 1/2$ ellipse and $e = 2$ hyperbola about a traditional focus.

Note that the possession of a coefficient -1 by one linear variable and +1 by the other (in the absence of other terms in the variables) indicates a non-equivalence of the two poles in the polarized system. One pole of Fig. $12c_2$ is at a traditional focus and the other is at the point where the other traditional focus would be if the hyperbola had two rectangular arms. In the unpolarized system, this distinction is lost, since all curves are reflected in the bipolar midline.

The 1st-degree bipolar curves with the least symmetry of form and position are represented by eqs. in which two of the three terms have non-unit coefficients, $\pm(B/C)u \pm (A/C)v = d$, where A, B, and C are dimensionless coefficients (Fig. 12e). The Cartesian biovals represented by these eqs. are referred to as *linear Cartesians*. Any single valid combination of the alternate signs represents only a single oval, whereas coefficients of the same magnitude with an appropriate sign changed yield an eq. representing a "conjugate" oval. For example, $(B/C)u - (A/C)v = d$ represents the interior oval of Fig. 12e, and $(A/C)v - (B/C)u = d$ represents the exterior oval.

The property of being "conjugate" signifies that the two ovals of Fig. 12e invert to one another and themselves in the three modes illustrated by the arrows that extend from the three foci of self-inversion, p_u, p_v, and p_w. Since the coefficients of u and v are not identical, poles p_u and p_v are not equivalent (not reflected in a line of symmetry).

Lastly, we consider the most-symmetrical bipolar curves with construction rules of 2nd-degree in one variable and 1st-degree in the other (Fig. $12f_1$-f_3). These curves belong to the most-symmetrical subgroup of the Cartesian ovals and have the bipolar construction rule $z^2 = Cdu$ (where pole p_z is not a focus of self-inversion). This single construction rule represents monovals (Fig. $12f_1$; for which the conjugate oval is imaginary), both conjugate biovals (Fig. $12f_3$), and the limacon *curve of demarcation* (Fig. $12f_2$).

Because the construction rule of the curves in this subgroup is that of a standard vertex parabola in the rectangular system, I refer to them as *parabolic Cartesians*. One of the three groups of inversions of these curves that possess a true center (see Fig. 22a,c,f), as mentioned above, is made up of Cassinian ovals (a previously unknown relationship), which are the most symmetrical of the quartic homologues of central conics (the *central Cartesians*).

[It will be noted that d, the characteristic distance of the coordinate system, is used systematically in all of the above construction rules with three terms. This new practice is essential in generalized symmetry studies and bipolar studies in general. The determination of the identity, the comparative symmetry, and the simplicity of the construction rules of curves in the bipolar system (and other coordinate systems that possess characteristic distances) cannot be carried out independently of an accounting for the characteristic distance. This is exemplified in the above treatment of the bipolar symmetry of a line-segment and rays, as compared to central conics. The distance d between the poles was not always ignored in the eqs. of other treatments, but its use was haphazard, without an appreciation of the advantages conferred by employing it systematically.]

The Rectangular to Bipolar Transformation

A convenient method for transforming a rectangular construction rule to its bipolar equivalent is to place the midpoint of the bipolar system at the rectangular origin, with the y-axis coincident with the bipolar midline and the x-axis coincident with the bipolar axis (Fig. 13, upper left). Pole p_u then comes to lie at $x = -d/2$ and pole p_v at $x = +d/2$. Considering a point of a rectangular curve at (x,y), it is readily seen that the coordinates, u and v, the undirected distances of this point from poles p_u and p_v, are given by eqs. 21. Eqs. 22a,a' show that for any point on a rectangular curve (actually any point in the plane), the difference between the squares of the corresponding bipolar coordinates is independent of y. Making use of this relation, y can be expressed as in eq. 22b (see Problem 1).

conversion eqs. between rectangular and bipolar coordinates

(a) $u^2 = (x+d/2)^2 + y^2$ square of distance from pole p_u (21)

(b) $v^2 = (x-d/2)^2 + y^2$ square of distance from pole p_v

(a) $(u^2-v^2) = 2dx, \quad x = (u^2-v^2)/2d$ x, in terms of u and v   (22)

(a') $(v^2-u^2-x_1^2+x_2^2)/2(x_2-x_1) = x$ x, in terms of u and v for p_v at $(x_1,0)$ and p_u at $(x_2,0)$

(b) $y = \pm\sqrt{[v^2-(u^2-v^2-d^2)^2/4d^2]} =$
$\pm\sqrt{[u^2-(u^2-v^2+d^2)^2/4d^2]}$ y, in terms of u and v

(c) $x^2+y^2 = (u^2+v^2-d^2/2)/2$ sum of x^2 and y^2 of eqs. 22a,b

The alternate signs of the radicals of eqs. 22b actually are unnecessary, because the radical must be eliminated in the derivation of bipolar construction rules, and because all bipolar loci are symmetrical about the bipolar axis (in both the polarized and unpolarized systems). Great use is made of eq. 22a', for two axial poles at general positions, x_1 and x_2, in later sections of this Chapter.

A close look at the eqs. for x^2 and y^2 (eqs. 22a,b squared) reveals that the sum, $x^2 + y^2$, is only quadratic in u and v, because the 4th-degree terms cancel (eq. 22c). The same is true of the eqs. for arbitrary axial poles, x_1 and x_2 (see Problems 2 and 3). Since the eq. for x also is quadratic in u and v, a consequence of these relations is readily deduced. The degree of the general bipolar construction rule for points on the line of symmetry of any curve whose rectangular construction rule involves terms consisting only of sums of x^2 and y^2, and of x (i.e., lacking terms in x^2 or y^2 alone, and y or higher powers) will be no greater than the rectangular degree. This, then, is a mark of high bipolar symmetry, and is true, for example, of the rectangular eqs. of centered circles, axial circles, 0°-axial limacons (see eq. 81-left), 0°-axial Cartesian ovals (see eq. 105a) and Cayley's sextic (eq. 6k and Problem 45).

Bipolar Translations and Rotations

Translating a bipolar curve h units to the right (parallel to the bipolar axis) presents a much more complex situation than the corresponding rectangular translation. Only for points incident upon the bipolar axis to the right of both poles are $u' = u+h$ and $v' = v+h$. The general relations are easily shown to be eqs. 23a,b.

eqs. for translations parallel to the bipolar axis

(a) $u'^2 = u^2+h(u^2-v^2+d^2)/d + h^2$ square of distances of points of translated curve from pole p_u (23)

(b) $v'^2 = v^2+h(u^2-v^2-d^2)/d + h^2$ same as eq. 23a for pole p_v

(c) $v'^2-u'^2 = v^2-u^2-2dh$ eq. 23b minus eq. 23a

(d) $v'^2+2dh-u'^2 = 0$ construction rule for a line normal to the bipolar axis at a distance h to the right of the midpoint

It follows from eqs. 23a,b that $v'^2-u'^2$, the difference between the squares of the new coordinates of any point, is independent of the distance of the point from the bipolar axis (eq. 23c). Since the construction rule $u^2 = v^2$ (equivalent to $u = v$, inasmuch as distances are undirected) is the eq. of the bipolar midline, eq. 23c can be used for a direct derivation of the construction rule for a normal line at any point h to the right of the midpoint. Thus, letting $u = v$ in eq. 23c, yields eq. 23d.

Translating a bipolar curve k units normal to the bipolar axis yields eqs. 24a,b. It follows that the difference between the squares of the bipolar coordinates of a point of the translated curve is given by eq. 24c (since the values of the radicals are equal in eq. 22b). Again, the construction rule for

Bipolar Translations, Rotations & 4th-Degree Lines

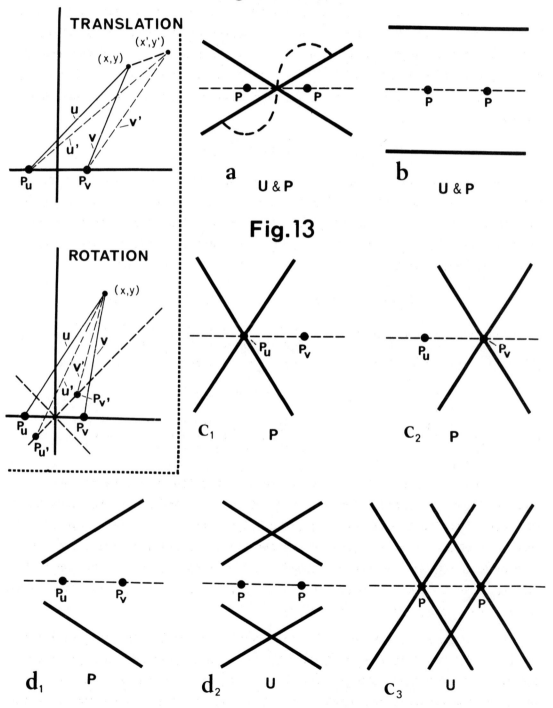

Fig. 13

the bipolar midline--in this case, translated along itself--can be derived by letting $u^2 = v^2$ (eq. 24d).

eqs. for translations normal to the bipolar axis

(a) $u'^2 = u^2+k^2+(k/d)\sqrt{[4d^2u^2-(u^2-v^2+d^2)^2]}$ square of distance of point on translated curve from p_u (24)

(b) $v'^2 = v^2+k^2+(k/d)\sqrt{[4d^2v^2-(u^2-v^2-d^2)^2]}$ same as eq 24a from pole p_v

(c) $u'^2-v'^2 = u^2-v^2$ eq. 24a minus eq. 24b

(d) $u'^2 = v'^2$, $u' = v'$ eq. of midline translated along itself

Eqs. 25a,b represent a general translation--the combination of translations of a bipolar curve to the right and upward (diagonal dashed line of Fig. 13-inset, upper; see Problem 4). Since, from eq. 24c, $v'^2-u'^2 = v^2-u^2$, for translations normal to the bipolar axis, the expression $v'^2-u'^2$, for a general translation (eq. 25c) is the same as eq. 23c for a translation parallel to the bipolar axis.

general translation equations

(a) $u'^2 = u^2+h(u^2-v^2+d^2)/d + h^2+k^2+(k/d)\sqrt{[4d^2u^2-(u^2-v^2+d^2)^2]}$ (25)

(b) $v'^2 = v^2+h(u^2-v^2-d^2)/d + h^2+k^2+(k/d)\sqrt{[4d^2v^2-(u^2-v^2-d^2)^2]}$

(c) $v'^2 - u'^2 = v^2 - u^2 - 2dh$ eq. 25b minus eq. 25a

Lastly, consider eqs. 26 for a CCW rotation of the bipolar axis about the midpoint (Fig. 13-inset, lower). Inspection of eqs. 26a,b reveals the relation 26c between the sums of the squares of the distances to any point (u,v) of the initial curve and the image point (u',v') of the rotated curve. Since the eq. of a centered circle does not change with rotation, eq. 26c can be checked easily using the eq. of a centered circle (including the two poles), $u^2 + v^2 = d^2$. Substitution yields the expected result, $u'^2 + v'^2 = d^2$.

general rotation equations

(a) $u'^2 = u^2+(u^2-v^2)(\cos\theta-1)/2 + \sqrt{[4d^2u^2-(u^2-v^2+d^2)^2]}(\sin\theta)/2$ (26)

(b) $v'^2 = v^2+(u^2-v^2)(1-\cos\theta)/2 - \sqrt{[4d^2v^2-(u^2-v^2-d^2)^2]}(\sin\theta)/2$

(c) $v'^2 + u'^2 = v^2 + u^2$ eq. 26a plus eq. 26b

Although eqs. 23-26 yield interesting information concerning properties of displacements and rotations in the bipolar system, they are not particularly useful for making these transformations. For this purpose it usually is preferable to substitute in the corresponding rectangular construction rules using the conversion eqs. 22. It takes only a glance at eqs. 23-26 to appreciate the advantages of the rectangular system for analytical studies, and to understand why degree is not restricted in the bipolar system (see Problem 5).

[It will be evident from the foregoing that the maximum degree of the construction rule of a conic in the bipolar system is 8, and that, in general, the maximum degree of the bipolar construction rule of any rectangular curve of degree n is 4n (see Problem 12). Thus, any conic can be expressed in the form $Ax^2+Bxy+Cy^2+D^2x+E^2y+F^3 = 0$. Making the substitutions of eqs. 22, clearing of fractions, segregating terms, and squaring, gives the 8th-degree bipolar construction rule

$$\{A(u^2-v^2)^2+C[4d^2v^2-(u^2-v^2-d^2)^2]+2dD^2(u^2-v^2)+4d^2F^3\}^2 =$$

$$[4d^2v^2-(u^2-v^2-d^2)^2][B(u^2-v^2)+2dE^2]^2 . \;]$$

Bipolar Foci, Focal Loci, and Point-Foci

The *general* bipolar construction rule of a curve is the construction rule cast about arbitrary points in the plane. Any pair of points in the plane of a curve for which the general construction rule reduces in degree is a *bipolar focus*. A locus in the plane of a curve, for any two points of which (employed as poles) the general construction rule reduces in degree, is a *bipolar focal locus*. Since general bipolar construction rules simplify and reduce in degree for pole-pairs located on a line of symmetry, lines of symmetry are bipolar focal loci.

Bipolar point-foci are of two types, *independent* and *dependent*. At this juncture, attention is directed solely to the independent type, and, unless otherwise stated, all references to bipolar point-foci refer to this type. An independent point-focus is any identifiable pole in the plane of a curve for which the general construction rule reduces, independently of the identity of the other pole (i.e., for an arbitrary second pole). All bipolar point-foci lie on lines of symmetry.

The following RULES incorporate the known features for reduction of bipolar construction rules, excluding those of lines and circles. The construction rules of the latter loci present certain exceptional aspects (see below and individual treatments).

[This treatment, as well as those following, is meant to apply to the polarized bipolar system. In the unpolarized system all loci have at least two orthogonal lines of symmetry (see Fig. 13).]

RULE 10. *Bipolar point-foci of curves fall on lines of symmetry* (as defined in the rectangular and polar systems).

RULE 11. *Bipolar construction rules typically reduce in degree in steps of 50% or 75%, for example, from 8 to 4 to 2 to 1, and from 16 to 4 to 2 to 1.*

RULE 12. *Bipolar construction rules reduce in degree for each of the following conditions* (illustrated and identified in the diagram below).

1. *One pole an independent point-focus, the other an arbitrary point in the plane.*
2. *Both poles otherwise unexceptional, but incident upon a line of symmetry of the curve* (as defined in the polar and rectangular systems).
3. *One pole an independent point-focus, the second incident upon the same line of symmetry of the first but otherwise unexceptional.*
4. *Both poles independent point-foci incident upon the same line of symmetry.*

RULE 13. *An independent bipolar point focus is an individually-identifiable point on a line of symmetry of a curve which (a) when paired with an otherwise unexceptional pole leads to a construction rule of lower degree than the general bipolar equation, or (b) when paired with an otherwise unexceptional pole on the same line of symmetry leads to a construction rule of lower degree than the general axial bipolar equation.*

ellipses

8th-degree —1→
　　　　　　　→ 4th-degree —3→ 2nd-degree —4→ 1st-degree
8th-degree —2→

limacons

16th-degree —1→ 8th-degree —3→
　　　　　　　　　　　　　　　→ 2nd-degree —4→ 1st-degree
16th-degree —2→ 4th-degree —3→

As will become evident in the following, lines have no identifiable point-foci, that is, there is no point in the plane of a line that can be identified as a bipolar point-focus. In the case of the circle, only the center is a point-focus, as it is in any reference system possessing a point or circle as a reference element. The construction rule of a centered circle in such a system (a *monopolar* circle) can be expressed independently of the number, identity, and location of other reference elements.

In the light of RULES 10-12 and the known degrees of the general bipolar construction rules of conics, the following conclusions apply to the bipolar eqs. of parabolas, ellipses, and hyperbolas:

  a. bipolar eqs. of 4th-degree either are cast about two otherwise unexceptional poles incident upon the same line of symmetry, or about a bipolar point-focus and a second pole not on the same line of symmetry;

  b. bipolar eqs. of 2nd-degree are cast about one bipolar point-focus and a second pole on the same line of symmetry; and

  c. bipolar eqs. of 1st-degree are cast about two bipolar point-foci.

  RULE 14. *A line connecting the poles of a bipolar construction rule of exponential degree 1 or 2 is a line of symmetry of the corresponding rectangular curve and a focal locus, that is, a locus upon which any pole-pair has bipolar focal rank.*

Bipolar Lines

By employing the substitutions of eqs. 22, the general rectangular construction rule of a line, $y = mx + b$ (slope-intercept form), becomes eq. 27a (Fig. 13d$_1$,d$_2$) in the bipolar system. Thus, the "complicated" eq. referred to by Crofton (see above) is of 4th-degree and, in fact, is the simplest *general* bipolar construction rule (i.e., the bipolar eq. of a curve in an arbitrary position).

4th-degree construction rules of bipolar lines

(a) $(1+m^2)(u^2-v^2)^2 + 2d(2bm-d)u^2 - 2d(2bm+d)v^2$ general lines (27)
    $+ 4d^2(b^2+d^2/4) = 0$

(b) $(1+m^2)(u^2-v^2)^2 - 2d^2(u^2+v^2) + d^4 = 0$ lines including bipolar midpoint

(c) $(u^2-v^2)^2 - 2d^2(u^2+v^2) + 4d^2(b^2+d^2/4) = 0$ lines parallel to bipolar axis

(d) $(u^2-v^2)^2 - 2d^2 u^2 - 2d^2[(1-m^2)/(1+m^2)]v^2 + d^4 = 0$ lines including pole p_v

(d') $(u^2-v^2)^2 - 2d^2 v^2 - 2d^2[(1-m^2)/(1+m^2)]u^2 + d^4 = 0$ lines including pole p_u

There are three symmetry conditions on these 4th-degree lines, that is, there are three conditions for which construction rule 27a simplifies but remains of 4th-degree. The most powerful condition is $b = 0$, for a line including the midpoint of the system (Fig. 13a). This line satisfies all but condition (3) of RULE 9 and leads to the simplest 4th-degree construction rule for a line (eq. 27b). In this eq., the coefficients of u^4 and v^4 are equal at unity, and the coefficients of the terms in u^2 and v^2 are equal at $-2d^2$ (in other words, the eq. is symmetrical in the variables u and v). In addition, the locus includes the most-symmetrical point in the system and retains its symmetry in the polarized system (Fig. 13a).

The next-most powerful of these symmetry conditions is $m=0$ (lines of slope 0), leading to the next-most-simple construction rule, 27c. This eq. defines lines parallel to the bipolar axis (Fig. 13b). All such lines have 4th-degree construction rules except the one coincident with the bipolar axis (Fig. 12b$_2$+b$_3$), whose eq. reduces to 3rd-degree but is degenerate (see eqs. 28d,d', below). Construction rule 27c and its general loci also satisfy all but condition 3 of RULE 9 (failing only in not including the reference elements).

The third condition is less evident. Since vanishing of either the u^2 or v^2-term (which occurs by letting 2bm equal either d or -d) simplifies the construction rule but leaves it asymmetrical in the u^2 and v^2-terms (one of which vanishes), neither of these conditions leads to loci that can be considered more symmetrical than general lines. On the other hand, if the lines pass either through pole p_v ($b = -md/2$; Fig. 13c$_1$) or p_u ($b = md/2$; Fig. 13c$_2$), the coefficients of the terms of eq. 27a simplify, and two of them become unity, leading to eqs. 27d,d'. These loci satisfy part of the first condition of RULE 9, since the coefficients of both the u^4 and v^4-terms are unity. They also partially satisfy the third condition, since they include one of the poles. On the other hand, they are not symmetrical about the most-symmetrical point in the system, nor are they symmetrical in the polarized system.

Particular attention should be called to two facts. First, each of the loci referred to above, for eqs. 27a,d,d', consists of two lines in the polarized system and four lines in the unpolarized system. Thus, the former lines are reflected in the bipolar axis (Fig. 13d$_1$,c$_1$,c$_2$) and the latter lines are reflected in both the bipolar axis and the midline (Fig. 13c$_2$,d$_2$). Second, construction rules 27b,c are symmetrical in u and v but 27a,d,d' are not (i.e., exchanging u and v, which does not alter the former two rules, alters the latter three).

Since symmetry in the variables u and v indicates equivalence of poles p_u and p_v (i.e., reflection of the poles in a line of symmetry), the midline is a line of symmetry of the loci of eqs. 27b,c (Figs. 13a,b, respectively), but not of the loci of eqs. 27a,d,d'. Because the bipolar axis is a line of symmetry of all bipolar loci, symmetry in the variable u and v of eqs. 27b,c also means that a true center of the represented loci lies midway between poles p_u and p_v.

The most-symmetrical lines of slope, m = 0, are a line, two rays, and a line-segment, all coincident with the bipolar axis. The construction rules of these lines are obtained either by setting m = 0 in eq. 27b or setting b = 0 in eq. 27c, and rearranging, yielding eq. 27b'. Taking the square roots yields eq. 28a, and rearranging as the product of two factors yields eq. 28b. Factoring eq. 28b (or taking roots) yields eq. 28c, the degenerate product of four terms.

lines coincident with the bipolar axis

$(u^2-v^2)^2+2d^2(u^2-v^2)+d^4 = 4d^2u^2$ degenerate eq. for all loci   (27b')

(a) $(u^2-v^2+d^2) = \pm 2du$ roots of eq. 27b'   (28)

(b) $[(u^2-2du+d^2)-v^2][(u^2+2du+d^2)-v^2] = 0$ eq. 28a expressed as the product of two factors

(c) $(u-d+v)(u-d-v)(u+d-v)(u+d+v) = 0$ eq. 28b expressed as the product of four factors

(d) $(u-d+v)(u-d-v)(u+d-v) = 0$ degenerate eq. of three real parts of the bipolar axis

(d') $(u^3+v^3+d^3)-(uv+d^2)(u+v)-d(u^2+v^2)$ eq. 28d multiplied out
         $+ 2duv = 0$

$(u - v)^2 = d^2$ lateral rays   (29)

$u + v = d$ mid-segment   (30)

In the polarized system, the first term in the product of eq. 28c equated to 0 represents the axial mid-segment (Fig. 12b$_2$), the second term equated to 0 represents the right axial ray extending from p_v, and the third term represents the left axial ray. There is no real solution for the fourth term in an undirected-distance system (since distances cannot be negative, the sum of three distances cannot vanish). Accordingly, three of the terms of eq. 28c suffice to represent the bipolar axis (eq. 28d). When multiplied out, this gives the degenerate 3rd-degree construction rule 28d', an eq. and corresponding locus that satisfy all the conditions of RULE 9.

Note that eq. 28d' is symmetrical in the variables u and v, the first condition of RULE 9, and that this also is true of eq. 28c, from which a term is discarded to obtain eqs. 28d,d'. The reason why eqs. 28d,d' retain symmetry in the variables, even though a term has been discarded, is that the discarded term (right-most term of eq. 28c) also has symmetry in the variables. The same circumstances apply when the terms of eq. 28d are segregated. Thus, removing the first factor, which is symmetrical in u and v, and equating it to 0 gives eq. 30 for the mid-segment (Fig. 12b$_2$). The remaining 2nd-degree product of two factors also is symmetrical in u and v and represents the lateral axial rays (Fig. 12b$_3$).

Lastly, single lines of infinite slope are considered. These cannot all be derived from eq. 27a, since letting $m = \infty$ yields only the most-symmetrical locus in the system, u = v. However, we already have derived the construction rule (eq. 23d) for lines normal to the bipolar axis, repeated below for convenience. This construction rule is not symmetrical in u and v because these lines, in general, are not symmetrically positioned with respect to poles p_u and p_v.

lines normal to the bipolar axis

(d) $\quad v^2 \pm 2dh - u^2 = 0 \quad$ lines normal to the bipolar axis (plus sign to right of center) $\quad\quad$ (23)

(d') $\quad v^2 \pm d^2 - u^2 = 0 \quad$ normal lines including p_u (minus sign) and p_v

(d") $\quad v^2 \pm nd^2 - u^2 = 0, \quad n < 1 \quad$ normal lines between p_u and p_v

The simplicity of eq. 23d can be increased in two obvious ways. The most powerful symmetry condition is $h = 0$, giving the midline, u = v, which is symmetrical in u and v and is the most symmetrical locus in the system. The two lesser conditions are $h = \pm d/2$, which convert the constant term to $\pm d^2$. These conditions lead to normal lines through the poles--through p_v for the positive sign, and p_u for the negative sign.

These considerations do not exhaust the treatment of normal lines. One additional case needs to be considered. Clearly a normal line that lies between the two poles is more-symmetrically positioned than one that does not. The former has one pole on each side of it, whereas the latter has both poles on the same side. The eq. for the former condition is 23d", where n is less than 1.

RULE 15. *For an array of similarly-positioned, congruent bipolar loci, those members that lie between a member of highest symmetry (eq. $u = v$) and a member of second-highest symmetry of the array, possess greater symmetry than the members that do not.*

Bipolar Circles

The circle is the only known closed curve to possess but a single *circumpolar* point-focus (defined in Chapter VII). Another way of expressing this fact, is that the circle is the only curve with a true center, for which any non-central point in the plane is equivalent to an infinite number of other points (namely, the other points at the same distance from center). [In the cases of ellipses, treated below, all axial points occur in equivalent pairs, while all non-axial points occur in equivalent quartets.] As a result of this property, the circle presents the fewest complications of any closed curve for assessing the influences of position on bipolar construction rules (principally because rotations about the center leave the construction rules of circle unaltered).
[The line, of course, possesses no unique point in its plane, but is too singular a curve to be employed for the present purposes; see Problem 6.]

Offhand, one might expect the circle's simplest bipolar construction rules to be those for circles with their centers at the midpoint of the system. However, a linear construction rule for a circle at this location does not exist. This is a consequence of the fact that the lines of symmetry of the non-coincident bipolar system are asymmetrical (the bipolar axis includes the poles, whereas the midline of the unpolarized system does not), whereas any orthogonal pair of lines of symmetry of the circle are symmetrical. Linear bipolar construction rules for ellipses and two-arm hyperbolas (unpolarized system) are possible because their lines of symmetry, like those of the bipolar system, also are asymmetrical. Accordingly, a linear construction rule can be achieved for circles only by locating the curves asymmetrically along the bipolar axis (i.e., with respect to the poles; the circles of Apollonius of Figs. 14-1c, 14-2b, 15a).

In the cases of bipolar circles, the ability to discriminate between the symmetry of position of a curve with respect to one pole as compared to the other (as opposed to symmetry with respect to the pole-pair) is encountered in its simplest form. Ultimately, this traces to the fact that the construction rule of a circle centered on one pole (the *monopolar* construction rule) is the simplest possible, namely $u = R$ or $v = R$. As a concomitant of this circumstance, the coefficient ratios of the v^2 to u^2-terms in 2nd-degree bipolar eqs. (axial circles) become identical with the ratios of the distances of the center of the circle to poles p_u and p_v. [Correspondingly, in 4th-degree eqs., the coefficient ratios of the v^4 to u^4-terms become the squares of these distance ratios.]

Accordingly, in addition to being able to evaluate the relative symmetry of position and form of circles with respect to the bipolar pole-pair on the basis of degree and/or simplicity of construction rules, one also can obtain a precise direct evaluation of the relative positional symmetry of circles with respect to individual poles (on the basis of weighting assessed from coefficient ratios). The closer the center of a circle to a given pole, the greater the positional symmetry of the circle about the pole (and vice versa).

RULE 16. *Bipolar construction rules of circles are weighted heaviest in favor of the pole nearest the center, which is the most-symmetrically positioned of the two poles.*

The general construction rule of a bipolar circle, with parallel displacement, h, of its center to the right of the midpoint and, k, normal to the bipolar axis, is of 4th-degree. This is represented most simply by eq. 31, from which it is evident that no bipolar eq. of the circle can possess terms of both 1st and 2nd-degree in u or v (also true of the line). This is one of the simple aspects that make the circle the best initial example of a closed curve in this system.

construction rules of bipolar circles

$$[d(u^2+v^2)-2h(u^2-v^2)+2d(h^2-k^2-R^2-d^2/4)]^2$$
$$= 4k^2[4d^2R^2-(u^2-v^2-2dh)^2]$$
general circle; p_u at $(-d/2,0)$, p_v at $(d/2,0)$ (31)

(a) $(u^2+v^2)-2h(u^2-v^2)/d+2(h^2-R^2-d^2/4) = 0$ general axial circle   (32)

(b) $(u^2+v^2)-2n(u^2-v^2)+[d^2(2n^2-1/2)-2R^2] = 0$ eq. 32a with $h = nd$

(c) $(u^2+v^2) - 2n(u^2-v^2) + d^2(2n^2-1) = 0$ eq. 32b with $R = d/2$

(c') $(1+2n)v^2 + (1-2n)u^2 = d^2(1-2n^2)$ eq. 32c rearranged

(d) $u = (1+\sqrt{2})v$ circle of Apollonius ($n = \sqrt{2}/2$)

(e) $v = d/2$ p_v-centered circle ($n = 1/2$)

(f) $u^2 + v^2 = d^2$ circle of Pythagoras ($n = 0$)

(g) $3v^2 - u^2 + d^2 = 0$ p_v-including circle ($n = 1$)

(h) $5v^2 - u^2 + d^2/4 = 0$ p_v-enclosing circle ($n = 3/4$)

Centering the general circle on the midline by letting $h = 0$ (eqs. 34) simplifies the general construction rule considerably (only half as many measurements are needed to specify all points of the curve; see RULE 1c), but leaves it of 4th-degree. On the other hand, the general axial circle (eq. 32a) has a construction rule of only 2nd-degree. This reduction of the degree of general axial circles as opposed to general midline circles emphasizes the great positional symmetry of a coordinate line of symmetry that includes or bisects reference elements (the bipolar axis) as opposed to one that merely bisects the plane (the midline). [This distinction is the basis for the bipolar axis being an intrinsic line of symmetry and the midline being only elective.]

Algebraically, the distinction between the construction rules of axial versus midline circles goes back to the fact evident in eqs. 22a,b, that the x-coordinate of a bipolar point is expressible directly as a linear function of the difference between the squares of the bipolar distances, independently of the y-coordinate, whereas the y-coordinate cannot be so expressed (and this is dependent on the fact that the bipolar poles lie on the x-axis).

[Both centering the circle on the bipolar axis and centering it on the midline reduce the number of measurements or instructions needed to specify all points of the curve by a factor of 2, yet only the former change leads to a reduction of degree. The basis for the different results is evident from eq. 31. The entire right side of the eq. vanishes for $k = 0$ (allowing reduction through root-taking on the left), whereas for $h = 0$, the simplifications are of very much lesser significance.]

Axial Circles

Consider, first, axial circles. Expressing h in units of d by letting $h = nd$ in eq. 32a, yields eq. 32b. For $R = d/2$, this becomes eq. 32c, which rearranges to eq. 32d. It is seen from the latter eq. that the ratio of the v^2-term to that of the u^2-term—the v^2/u^2-ratio—is $(1+2n)/(1-2n)$. Beginning with a midpoint circle ($n = 0$), one obtains the circle of eq. 32f (Fig. 14-1a, 14-2c). This is referred to as the *circle of Pythagoras*, since it embodies the Pythagorean theorem, where d is the hypotenuse of the right triangle. A v^2/u^2-ratio of 1 for this curve signifies that poles p_u and p_v are equidistant from the center of the circle.

As n increases from 0, the axial circle travels along the bipolar axis to the right. Concomitantly, the v^2/u^2-ratio increases to infinity (at $n = 1/2$), and thereafter becomes negative and decreases in magnitude monotonically to -1 for n infinite. Since the positional symmetries of p_u and p_v also are equal in the plane of an infinitely-distant axial circle, one sees that a unit v^2/u^2-ratio, whether positive or negative, indicates identical positional symmetries of the two poles.

The circle with an infinite v^2/u^2-ratio ($n = 1/2$) is seen to be centered on p_v (eq. 32e and Fig. 14-1c), which places this pole at the most symmetrical position in the plane of the circle. Thus, the absolute magnitude of the v^2/u^2-ratio--ranging from infinity (circle centered on p_v) to unity (equivalence of p_u and p_v) to 0 (circle centered on p_u)--provides a quantitative index of the relative positional symmetry of pole p_v in the plane of the circle, as compared to that of p_u. As mentioned above, this ratio is the ratio of the distance of the center of the circle from p_u to its distance from p_v (see also Lockwood, 1961). This comparative criterion for individual poles should not, however, be confused with the matter of the positional symmetry of circles with respect to the pole-pair (or vice versa; Fig. 14-2 represents the latter circumstances in the unpolarized system).

Looking at axial circles from the point of view of their positional symmetry with respect to the pole-pair, the circle of Apollonius (eq. 32d and Figs. 14-1c, 14-2b, and 15a) is the most-symmetrical bipolar circle, since its construction rule is simplest--it is the only circle with a 1st-degree bipolar eq. While eq. 32e, for a p_v-centered circle is of 0-degree, this is, strictly speaking, a *monopolar* rather than a bipolar eq., since the variable u does not appear.
[Eq. 32e applies in any coordinate system that possesses a point or circle as a reference element, regardless of the number and identities of the other reference elements.]

The second-most-symmetrical bipolar circle is the one with the simplest and most symmetrical 2nd-degree construction rule (in both variables). This is the circle of Pythagoras (Figs. 14-1a, 14-2c), with a construction rule of only three terms, all with unit coefficients. This circle also satisfies all the conditions of RULE 9. The third-most-symmetrical circles are those of Fig. 14-2d,d', with unit coefficients for only the u^2 and v^2-terms (poles p_u and p_v equidistant from center but not incident). These satisfy three of the four conditions of RULE 9, failing only in not including the poles. Next-most-symmetrical are circles that include one of the poles (Fig. 14-2e,e'), for which two of the three coefficients in the construction rule are equal (or unity, if one divides through by R), but not those of the terms in the variables. These loci

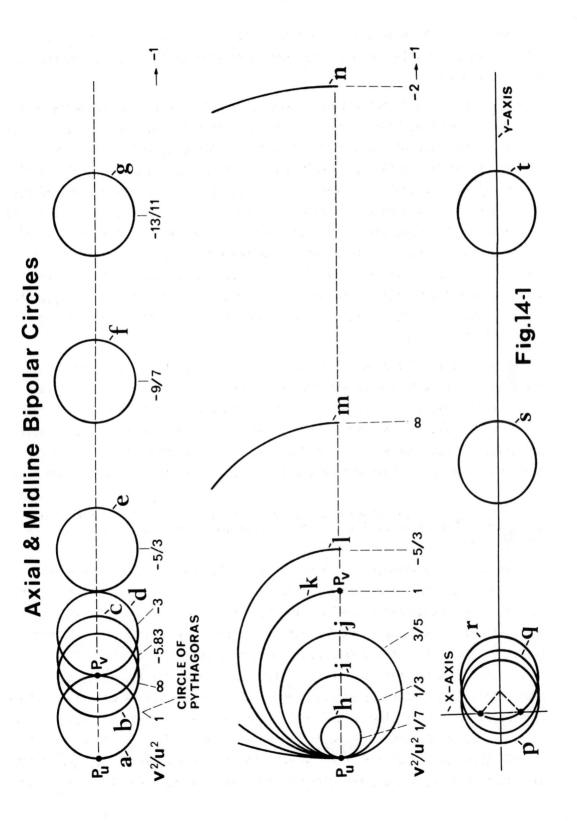

Fig.14-1

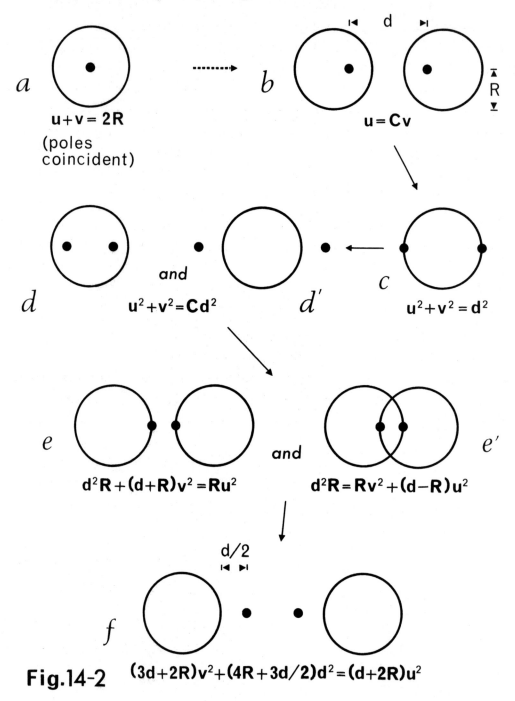

Fig.14-2

AXIAL CIRCLES

satisfy only two of the conditions of RULE 9.

If attention is confined to $R = d/2$ circles, the next-most-symmetrical circle after the circle of Pythagoras is that of Fig. 14-1d, which includes pole p_v, followed by circles that enclose but do not include p_v (excluding the circle of Apollonius, which already has been considered). An example of one of the latter circles is the $n = 3/4$ circle of eq. 32h (not shown in Fig. 14). This circle's construction rule has only one unit coefficient (the minimum possible). Note that though the circle including p_v (Figs. 14-1d, 14-2e) is more symmetrical about the two poles considered as a bipolar pair (i.e., has a simpler eq.), pole p_v considered individually is more-symmetrically positioned than pole p_u in the plane of the circle that encloses it (Fig. 14-1c) than in the plane of the circle that includes it (Fig. 14-1d; based on v^2/u^2-ratios).

Performing similar shifts with an $R = d/8$ circle, which is too small to include both poles, leads to a similar picture, except that p_v can be included in the circle in two ways, without also including p_u (as is the case in Fig. 14-1a). Both p_v-including circles, of course, have construction rules with two unit coefficients (see Fig. 14-2e,e' and eqs. 32g, 33a,b), a characteristic of axial circles that include one pole.

Taking a different approach with an "expanding" ($h = R - d/2$) p_u-including circle (Fig. 14h-n and eqs. 33) gives results consistent with those already obtained but a somewhat different perspective. Of course, the construction rule of every such circle has two unit coefficients (the v^2 and d^2-coefficients). The v^2/u^2-ratios begin at less than unity, signifying greater positional symmetry of pole p_u. They increase to unity for the circle of Pythagoras and exceed unity for all larger circles enclosing p_v, because of the greater positional symmetry (greater proximity to center) of the enclosed pole, p_v, than of the included pole, p_u.

(a) $(d/R - 1)u^2 + v^2 = d^2$ "expanding" p_u-including circles of Fig. 14-1h-n   (33)

(b) $(d/R + 1)v^2 + d^2 = u^2$ "expanding" p_v-including circles like Fig. 14-1d

The view that an enclosed pole is more-symmetrically positioned than an included pole falls in line with the fact that the most-symmetrical bipolar circle--the circle of Apollonius--has an enclosed pole (in the unpolarized system, of course, both poles are inclosed, one in each circle; Fig. 15a).

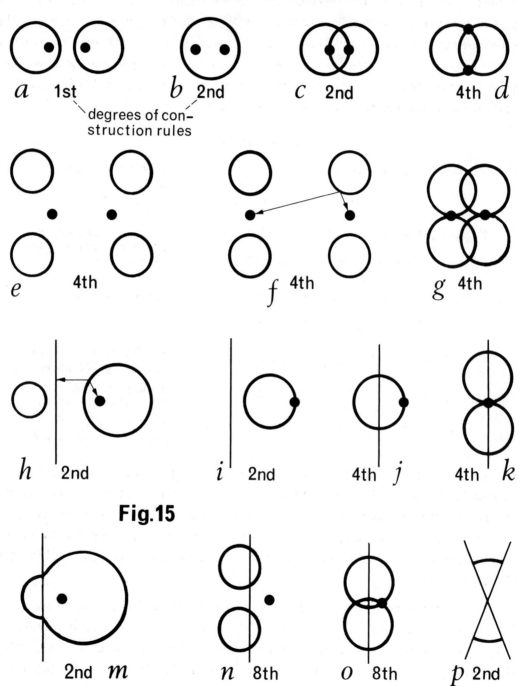

Fig. 15

Midline Circles

Letting $h = 0$ in eq. 31 yields the midline circle of eq. 34a, which, on rearranging and simplifying, becomes 34b. It is evident from this eq. that poles p_u and p_v always have equal weight for midline circles, that is, terms in the same powers of u and v have equal coefficients. [The coefficients of the u^2v^2-terms play no role, since both variables appear in the terms to the same degree.] These circumstances simply derive from the fact that poles p_u and p_v always are equidistant from the centers of midline circles.

midline circles

(a) $[d(u^2+v^2)-d(d^2+4k^2+4R^2)/2]^2$ 　　　general midline circles　　(34)
$= 4k^2[4d^2R^2-(u^2-v^2)^2]$

(b) $d^2(u^2+v^2)^2+4k^2(u^2-v^2)^2-4d^2(k^2+R^2+d^2/4)(u^2+v^2)$ 　　eq. 34a expanded, rearranged, and simplified
$+ 4d^2(k^2+R^2+d^2/4)^2-16d^2k^2R^2 = 0$

(c) $d^2(u^2+v^2)^2+4k^2(u^2-v^2)^2-4d^2(k^2+5d^2/4)(u^2+v^2)$ 　　general $R = d$ circle
$+ 4d^2(k^2+5d^2/4)^2-16d^4k^2 = 0$

(d) $(u^2+v^2)^2 + 3(u^2-v^2)^2 - 8d^2(u^2+v^2) + 4d^4 = 0$ 　　$k = \sqrt{3}d/2$, $R = d$ circle including the poles

(e) 　$(u^4+v^4) - 3d^2(u^2+v^2) + 5d^4/2 = 0$ 　　$k = d/2$, $R = d$ circle enclosing the poles

(f) $4(u^2+v^2)^2+(u^2-v^2)^2-21d^2(u^2+v^2)+377d^4/16 = 0$ 　　$k = d/4$, $R = d$ circle enclosing the poles

There is no noteworthy symmetry condition on eqs. 34a,b, that is, there is no condition for which a simplification occurs, nor for which the 4th-degree or the 2nd-degree terms in u or v vanish. Vanishing of the u^2v^2-terms for a circle centered at $k = d/2$, places the center of the circle and the two poles at the corners of a square (indicated by the dashed lines for the midline circle of Fig. 14-1q).

It is of interest to compare midline circles including the poles with midline circles enclosing the poles. For this purpose, an $R = d$ circle is taken (Fig. 14-1, bottom), yielding construction rule 34c. For the circle including the poles (Fig. 14-1r; see also Fig. 15d), $k = \sqrt{3}d/2$, yielding the relatively simple eq. 34d. For circles enclosing the poles, two cases, $k = d/2$ and $k = d/4$, are considered (eqs. 34e,f). The former (Fig. 14-1q), since it includes the mild symmetry condition, $k = d/2$, mentioned above, is more-symmetrically positioned than the latter (Fig. 14-1p). Accordingly, just as for axial circles (Fig. 14-2), and consistent with RULE 9-3, midline circles that include both poles generally are more-symmetrically positioned with respect to the pole-pair than those that enclose both poles.

Circles in the Plane

As an example of circles in the plane, consider $R = d/2$ circles centered at $h = d/2$ to the right of the midpoint, and let $k = nd$. Substituting in eq. 31 yields construction rule 35a. Expanding and grouping terms yields eq. 35b. It is evident that, for general bipolar circles, the coefficients of the 4th-degree terms take precedence over those of the 2nd-degree terms for assessing the relative positional symmetry of poles p_u and p_v. Thus, the v^2/u^2-ratios may be either greater or less than 1, despite the fact that p_v always is closest to center, whereas the v^4/u^4-ratio always is greater than 1, namely $(n^2+1)/n^2$. The general value of the u^4/v^4-ratio, which is the square of the ratio of the p_u to p_v-distances from the center, is given by expression 36, which also is always greater than 1 (for h positive).

$h = d/2$, $R = d/2$, $k = nd$ general circles

(a) $[v^2 - d^2(n^2+1/4)]^2 = n^2[d^4 - (u^2-v^2-d^2)^2]$ general circles   (35)

(b) $(n^2+1)v^4 + n^2u^4 - 2n^2u^2v^2 - d^2(v^2+4n^2u^2)/2$ eq. 35a expanded and regrouped

  $+ (n^2+1/4)^2 d^4 = 0$

(c) $5v^4 + u^4 - 2u^2v^2 - 2d^2(u^2+v^2) + d^4 = 0$ $h = R = d/2$ circle, including p_v ($n = 1/2$)

(d) $17v^4 + u^4 - 2u^2v^2 - 2d^2(u^2+4v^2) + 5d^4 = 0$ $h = R = d/2$ circle, enclosing p_v ($n = 1/4$)

$[(d+2h)^2 + 4k^2]/[(d-2h)^2 + 4k^2]$ v^4/u^4-ratio for general circles (square of p_u-center to p_v-center distance ratio)   (36)

For an $R = d/2$ circle centered at $h = d/2$ and including p_v ($n = 1/2$), that is, tangent to the bipolar axis at p_v (Fig. 15g, unpolarized system), one obtains the construction rule 35c, while for a circle enclosing p_v ($n = 1/4$), one obtains eq. 35d. The p_v-including circle clearly is more-symmetrically positioned with respect to the pole-pair, since its eq. is simpler, but the p_v-enclosing circle is more-symmetrically positioned with respect to pole p_v, since its v^4/u^4-ratio is greater.

Lines and Circles, and Bipolar Point-Foci

Following RULE 13, the line has no bipolar point-foci. Although the general 4th-degree construction rule reduces to 1st-degree for the midline, the poles are not individually identifiable, since any two points equidistant from the line on a normal to it may serve as poles. Similarly, for lines normal to the bipolar axis, with 2nd-degree construction rules, any two points on the normal to the line may serve as poles.

The situation is similar for points in the plane of the circle, other than the center. Although the circle of Apollonius has a construction rule of only 1st-degree (versus the 4th-degree general eq.), poles p_u and p_v are not individually identifiable. Any point on a diameter at a fractional distance R/C from center may serve as one pole, and a second at the distance CR in the same direction as the other ($C > 1$, and the constant in the eq. $u = Cv$). In the cases of 2nd-degree axial circles, any two points on a diameter may serve as poles.

Just as bipolar construction rules of reduced degree for a line and a circle define points in the planes of the curves that are not individually identifiable as point-foci, they also define lines of symmetry that are not individually identifiable (except in the case of a line coincident with the bipolar axis). Thus, reduction of the construction rules of lines to 1st or 2nd-degree defines the bipolar axis as a rectangular line of symmetry of the curves, and similarly for circles. But in the former case, any line normal to the initial line possesses this property (not just the bipolar axis) and, in the latter case, any diameter possesses it.

We now proceed to a consideration of curves for which bipolar construction rules in the strictest sense (i.e., involving both variables) define individually-identifiable point-foci and lines of symmetry. For these curves, 1st and 2nd-degree bipolar construction rules define at least one bipolar point-focus and one rectangular line of symmetry.

> RULE 17. *First-degree bipolar construction rules of curves other than lines and circles identify two independent bipolar point foci and one individually-identifiable rectangular line of symmetry. Second-degree construction rules define at least one independent point-focus, or two dependent foci, and one individually-identifiable rectangular line of symmetry.*

Bipolar Parabolas

Both the bipolar parabola and the bipolar hyperbola, each with its simplest construction rule, possess only one arm, one bipolar point-focus, and one line of symmetry (in the polarized system). The one-arm parabola has been selected to illustrate the influences of the positions of the bipolar poles on the construction rules of a curve with one line of symmetry. It also serves as a first example for comparing bipolar and circumpolar point-foci. In this connection, the parabola is found to possess only one bipolar point-focus, as opposed to four circumpolar point-foci (the traditional focus, the axial vertex, and the two LR-vertices).

General Parabolas

The construction rule of the general bipolar parabola about points (x_1, y_1) and (x_2, y_2) is derived by taking the rectangular eqs. (37) of a standard vertex parabola, forming the expressions for v^2 and u^2 (eqs. 38) in the usual way, taking the difference (eq. 39a), rearranging and squaring (eq. 39b), solving for x (eq. 40a), and eliminating x between eq. 40a and either of eqs. 38. The resulting construction rule is exceedingly complex, and would require several pages to reproduce in a form suitable for an analysis of special cases. The same objectives can be accomplished more readily by analyzing the construction rules for the special cases, themselves.

deriving the general bipolar parabola

$y^2 = 4ax, \quad y = \pm 2\sqrt{(ax)}$ standard rectangular parabola   (37)

(a) $v^2 = (x_1-x)^2 + (y_1 \pm 2\sqrt{[ax]})^2$ expression for v^2 (38)

(b) $u^2 = (x_2-x)^2 + (y_2 \pm 2\sqrt{[ax]})^2$ expression for u^2

(a) $v^2 - u^2 = x_1^2 + y_1^2 - x_2^2 - y_2^2 - 2x(x_2-x_1) \pm 4\sqrt{(ax)}(y_2-y_1)$ eq. 38a minus eq. 38b   (39)

(b) $[v^2 - u^2 - x_1^2 - y_1^2 + x_2^2 + y_2^2 - 2x(x_2-x_1)]^2 = 16ax(y_2-y_1)^2$ eq. 39a rearranged and squared

(a) $x \cdot (x_2-x_1)^2 = [(x_2-x_1)B^2/2 + 2a(y_2-y_1)^2]$ eq. 39b solved for x,   (40)
           $\pm (y_2-y_1)\sqrt{[2aB^2(x_2-x_1) + 4a^2(y_2-y_1)^2]}$ with B^2 given by eq. 40b

(b) $B^2 = (v^2 - u^2 - x_1^2 - y_1^2 + x_2^2 + y_2^2)$

[General parabolas possess two arms in the polarized system and four arms in the unpolarized system. When both poles are on the rectangular line of symmetry, the numbers of arms are reduced to one and two, respectively. Only one arm is shown in the accompanying illustrations. When the poles are non-axial, the second arm is the reflection of the initial arm in the bipolar axis. If the poles also are unlabelled, the third and fourth arms are reflections of the other two in the midline.]

The first special cases of interest are those for which the general construction rule merely simplifies. As we shall see, these include all special cases except both y_1 and y_2 being 0, and p_u or p_v being the traditional focus. Eqs. 41 list these cases. Cases 41-1 to 41-4 are represented by eqs. 42-45, respectively, where the expressions for B^2, B^4, B^6, and B^8 are obtained by approriate substitutions in eq. 40b.

special cases of bipolar parabolas

1. $y_1 = y_2 \neq 0$ 3. y_1 or $y_2 = 0$ 5. x_1 or $x_2 = 0$ (41)

2. $x_1 = x_2 \neq 0$ 4. $x_1 = x_2 = 0$ 6. $x_2 = -x_1$, or $y_2 = -y_1$

$B^8 + 16a(x_2-x_1)B^6 + 8(x_2-x_1)[(8a^2-x_1x_2+y_1^2)(x_2-x_1)-(x_2v^2-x_1u^2)]B^4$ eliminant of (42)
eqs. 38a and
$+ 64a(x_2-x_1)^2[(x_2x_1+y^2)(x_2-x_1)+(x_2v^2-x_1u^2)]B^2$ 40a for $y_1 = y_2$
$\neq 0$
$+ 16(x_2-x_1)^2[(x_2v^2-x_1u^2)+(x_1x_2-y^2)(x_2-x_1)]^2 = 0$

$[B^2 = v^2-u^2-x_1^2+x_2^2]$

(a) $B^8 + 32a(2a-x_1)(y_2-y_1)^2B^4 - 256a^2(y_2-y_1)^3 \cdot$ eliminant of (43)
eqs. 38a and
$[y_2v^2-y_1u^2+(y_1y_2-x_1^2)(y_2-y_1)] = 0$ 40a for $x_1 = x_2$

$[B^2 = (v^2-u^2-y_1^2+y_2^2)]$

eq. 43a for
(b) $y^4B^8 = 256a^2(y_2^2-4a^2)^3(y_2^2v^2-4a^2u^2)$ $x_1 = x_2 = 2a$,
$y_1y_2 = 4a^2$

(a) $B^8 + 8(2a-x_1)(x_2-x_1)B^6 + 16[(2a-x_1)^2(x_2-x_1)^2+2a(2a-x_1)y_2^2$ eliminant of (44)
38a and 40a
$- (v^2-x_1^2)(x_2-x_1)^2/2]B^4 - 32(v^2-x_1^2)(x_2-x_1) \cdot$ for $y_1 = 0$

$[4ay^2+(2a-x_1)(x_2-x_1)^2]B^2 - 128ay_2^2(v^2-x_1^2) \cdot$

$[2ay^2+(2a-x_1)(x_2-x_1)^2]+16(v^2-x_1^2)^2(x_2-x_1)^4 = 0$

$[B^2 = (v^2-u^2-x_1^2+x_2^2+y_2^2)]$

(b) $B^8 - 8(v^2-4a^2)(x_2-2a)^2B^4 - 128ay_2^2(v^2-4a^2)(x_2-2a)B^2$ eq. 44a for
$x_1 = 2a$
$- 256a^2(v^2-4a^2)y_2^4 + 16(v^2-4a^2)^2(x_2-2a)^4 = 0$

(c) $B^8 = (v^2-u^2+y_2^2)^4 = 256a^2(v^2-4a^2)y_2^4$ eq. 44b for $x_2 = 2a$

Documentation of the assignments of comparative weights to the poles of eqs. 41-44 would require more space than is practical, since expansions of the terms in B^8, B^6, and B^4, grouping of powers of u^8, v^8, u^6, v^6, etc., and comparing the coefficients would be necessary. However, the principles involved are evident from considerations of the terms of lower degree, and discussions based on these terms suffice (see also Problem 35 and eqs. 67c and 138).

Examination of eq. 42 reveals that the general construction rule greatly simplifies but does not reduce for two poles located on a line parallel to the line of symmetry (Fig. 16a), and that there is no condition for which great further simplification will occur. The minor symmetry conditions (conditions for simplification of the eq.), $x_1x_2 = \pm y_1^2$ and $x_1x_2 = y_1^2 + 8a^2$, and their permitted combination, $x_1x_2 = 4a^2$, lead only to simplifications of the coefficients of terms.

The condition $x_1 = x_2$, which leads to reduction of eq. 42 to $u = v$, is the trivial case for the coincident bipolar system (in which the poles are in coincidence; Fig. 14-2a). In this system, the construction rule $u = v$ represents the entire plane. Although the pole closest to the vertex for eq. 42 appears to be weighted heaviest (because of the influence of the terms in $x_2v^2 - x_1u^2$), this apparent differential weighting (in terms of the relative magnitudes of the coefficients) is equalized when the eq. is expanded and terms are combined. It is evident from eqs. 47a,a',b (below) that the comparative weighting of unexceptional poles on the axes of bipolar parabolas is indeterminate (see Problem 35 for additional discussion of the weighting of axial poles).

An examination of construction rule 43a (for $x_1 = x_2$), for both poles on a line normal to the line of symmetry (Fig. 16b), reveals that this condition gives the greatest amount of simplification of all six cases (eqs. 41). Accordingly, these poles have the greatest subfocal rank of the six alternatives (see also eq. 45a, below). Construction rule 43a greatly simplifies for $x_1 = 2a$ (Fig. 16c; Problem 7). For the additional condition $y_1y_2 = 4a^2$, it simplifies still further to eq. 43b (Fig. 16d). The heaviest-weighted pole in eq. 43a is the pole nearest the line of symmetry, since, in the rightmost term, the ordinate of p_u becomes the coefficient of v^2, while the ordinate of p_v becomes that of u^2. In other words, the curve possesses the greatest bipolar symmetry about the pole nearest its line of symmetry.

Turning, now, to eq. 41a ($y_1 = 0$), for one pole on the line of symmetry (Fig.

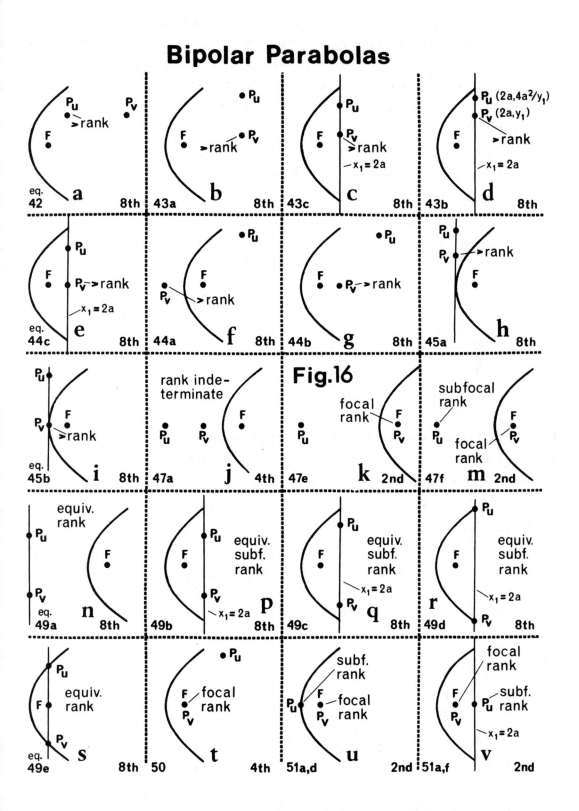

16f), great simplifications from the general construction rule also have occurred. It is clear from this eq. that, if the axial pole is the reflection of the vertex in the focus ($x_1 = 2a$; Fig. 16g), the eq. simplifies still further, yielding construction rule 44b. Letting $x_2 = 2a$, also (both poles on the 2a-chord), yields eq. 44c (Fig. 16e; see also Problem 8), representing the highest-ranking subfocal combination of an axial and non-axial pole (only the traditional focus and a non-axial pole give a simpler eq.--through reduction). This is the same result obtained by letting $x_1 = 2a$ and $y_1 = 0$ in eq. 43a.

Eq. 44a also represents the greatest relative weighting of pole p_v to pole p_u, for an arbitrary axial pole compared to an arbitrary non-axial pole. The variable v occurs in powers of both v^4 and v^2, and these are unmatched by corresponding terms in the variable u. When the axial pole is specified to be at $x_1 = 2a$, the qualitative comparative weighting of p_v to p_u is essentially unchanged, since no term in v is lost (i.e., only the values of the coefficients of the v^2 and v^4 terms change).

On the other hand, when the (lower ranked) non-axial pole is chosen to be one of those with the greatest subfocal ranking, namely, any pole on the 2a-chord, the disparity in the weighting is greatly reduced. It then depends entirely on the presence of a single term in v^2 (eq. 44c). Since the coefficient of this term is $256a^2y_2^4$, the disparity in weighting decreases rapidly as p_u approaches p_v (p_u approaches p_v on the line of symmetry as y_2 approaches 0), and becomes unity when the poles merge (u and v being equally weighted when $B^8 = 0$).

In the case $x_1 = x_2 = 0$, construction rule 43a becomes 45a, for two poles on the vertex-tangent (Fig. 16h). A comparison with eq. 43c reveals that the latter is simpler, confirming the fact that two arbitrary poles on the 2a-chord (Fig.

(c) $B^8 - 256a^2(y_2-y_1)^3[(y_2v^2-y_1u^2)+(y_1y_2-4a^2)(y_2-y_1)] = 0$ eq. 43a for $x_1 = x_2 = 2a$ (43)

(a) $B^8 + 64a^2(y_2-y_1)^2B^4$
$- 256a^2(y_2-y_1)^3[(y_2v^2-y_1u^2)+y_1y_2(y_2-y_1)] = 0$ eq. 43a for $x_1 = x_2 = 0$ (45)

$[B^2 = (v^2-u^2+y_2^2-y_1^2)]$

(b) $(v^2-u^2+y_2^2)^4 + 64a^2y_2^2(v^2-u^2+y_2^2)^2 - 256a^2y_2^4v^2 = 0$ eq. 43a for $x_1 = x_2 = y_1 = 0$

(d) $(v^2-u^2+y_2^2)^4 - 256a^2y_2^4(v^2-a^2) = 0$ eq. 43a for $x_1 = x_2 = 2a$, $y_1 = 0$ (43)

16c) have greater subfocal symmetry rank than two arbitrary poles on the vertex-tangent (and that the curve is more symmetrical about the former pole-pair than about the latter). This also applies when one member of each pole-pair is axial, as can be seen from a comparison of eqs. 45b (Fig. 16i) and 43d. In both cases, of course, the construction rules are weighted heavily in favor of the axial pole p_v, as compared to the non-axial pole p_u.

Cases 5 and 6 of eqs. 41, excluding the case in which both y_1 and y_2 are 0 (both poles on the line of symmetry), are seen to lead to only minor simplifications of eqs. 42-45. The greatest of these simplifications is for either x_1 or $x_2 = 0$. These results enable the formulation of the following general bipolar RULES.

> RULE 18. *General bipolar construction rules of curves with one line of symmetry undergo major simplifications--without reduction--for both poles located on a line either normal or parallel to the line of symmetry, and for one pole on the line of symmetry.*
>
> RULE 19. *The coefficients of the variables of bipolar construction rules cast about poles on a line normal to a line of symmetry are weighted heaviest in favor of the pole nearest the line of symmetry, that is, the pole nearest the line of symmetry has the greater subfocal symmetry rank and the curve is more-symmetrically positioned about it.*

Axial Parabolas

Axial bipolar parabolas possess the lowest-degree construction rules. Taking the general (x-h)-displaced 0°-axial rectangular parabola of eq. 46 and making the substitutions of eq. 22 yields the 4th-degree general axial bipolar parabolas of eq. 47a (Fig. 16g). A glance at this eq. reveals that p_u and p_v can be equivalent poles only for the condition $d = 0$ (coincidence of the poles at the origin) or d infinite (both poles at the axial point at infinity).

Construction rule 47a (and 47a') does not provide a valid criterion of the relative weighting of the poles of axial parabolas because h does not enter into the coefficients of the u^2 and v^2-terms. Analytically (for the eq. of a conic section with a line of symmetry coincident with the x-axis), this is because the rectangular construction rule of the 0°-axial parabola is linear in x, which leaves h dissociated from terms in u and v. Geometrically, this is because the parabola has no line of symmetry normal to the x-axis. As we have already noted, and confirm repeatedly in the following treatments, of two bipolar poles on a line normal to a line of symmetry, the pole closest to the

axial bipolar parabolas

$$y^2 = 4a(x-h) \qquad \text{0°-axial rectangular parabolas} \qquad (46)$$

axial bipolar parabolas (47)

(a) $(u^2-v^2)^2 + 2d(4a-d)u^2 - 2d(4a+d)v^2 + d^2(d^2-16ah) = 0$

(a') $(u^2-v^2)^2 + 2d(4a-d)(u^2-v^2) + d^2(d^2-16ah) = 4d^2v^2$ eq. 47a rearranged

(b) $(u^2-v^2)^2 + 2(x_2-x_1)[(4a-x_2+x_1)v^2 + (4a+x_2-x_1)u^2] + (x_2^2-x_1^2)^2 + 4(x_2-x_1)[(x_2-x_1)x_1^2 + (x_2^2-x_1^2)(2a-x_2)] = 0$ another form of eqs. 47a,a' for the standard rectangular parabola, $y^2 = 4ax$, cast about poles p_v at x_1 and p_u at x_2

(c) $[u^2 - v^2 + d(4a-d)]^2 = 4d^2v^2$ eq. 47a' re-expressed for $h = d/2 - a$ (focus at p_v)

(d) $u^2 - v^2 + d(4a-d) = \pm 2dv$ roots of eq. 47c

(e) $(v+d)^2 = u^2 + 4ad$ eq. 47d rearranged for + sign

(f) $(v+d)^2 = u^2 + d^2$ simplest form of eq. 47e, $a = d/4$

(f') $v^2 + 2dv = u^2$ eq. 47f as three terms

line of symmetry is highest ranking (at the subfocal level).

Essentially the same result is obtained if one allows both poles to be free-ranging along the axis. This is accomplished by casting an eq. about p_v at point x_1 and p_u at point x_2, and comparing the coefficients of the u^2 and v^2-terms for various values of x_1 and x_2. The relevant eq. for a standard vertex parabola is 47b. The coefficients of the u^2 and v^2-terms of this eq. reveal that, in general, the weighting of axial poles of bipolar parabolas is indeterminate. Thus, one coefficient of the quadratic terms is simply 4a plus the characteristic distance and the other is 4a minus this distance (both of which are independent of x_1 and x_2).

On the other hand, eq. 47b reveals a subfocal symmetry condition on all axial bipolar parabolas (see also below), namely, the condition where the characteristic distance is equal to 4a, whereupon a quadratic term vanishes. Whatever the location of one of the poles, parabolas cast about a second pole 4a units distant from the first achieve higher rank at the subfocal level than for other positions.

There is no symmetry condition (i.e., no reduction condition) on eq. 47a for the entire left-hand side considered as a perfect square, since that requires that $(4a-d) = (4a+d)$. Rearranging, however, yields eq. 47a', from which it is readily seen that both sides of the eq. become perfect squares for $h = d/2 - a$. In other words, $h = d/2 - a$ is a bipolar *focal condition*, for which the eq. of an axial parabola reduces to 2nd-degree. Taking the roots of eq. 47c yields eq. 47d, for which only the positive sign produces a solution (the negative sign requires that $ad = 0$), namely, eq. 47e (Fig. 16k).

The bipolar focal condition for axial parabolas, $h = d/2 - a$, requires that the vertex of the parabola be positioned at a distance a to the left of pole p_v (the translation would be to the right only for $d/2 > a$, otherwise to the left). But this puts pole p_v at the location of the traditional focus of the parabola (Fig. 16k). In other words, just as the traditional foci of hyperbolas and ellipses are bipolar point-foci (with linear eqs. $u \mp v = 2a = d/e$, with the minus sign for the hyperbola), so also is the traditional focus of the parabola a bipolar point-focus (RULES 12-3 and 13). The differences between the construction rules are that, in the case of the parabola, any other point on the axis may serve as the second pole, and the construction rule is of 2nd-degree rather than 1st-degree.

When subjected to a "subfocal symmetry analysis," construction rule 47d gives interesting additional information about *focal* axial parabolas. One needs only to inquire as to the condition for which eqs. 47d,e achieve simplest form. Clearly, for the substitution $a = d/4$, a term vanishes from eq. 47d, and all terms of eq. 47e acquire unit coefficients, yielding eq. 47f or f', the sought-after simplest form(s). In other words, eq. 47e achieves the form 47f when pole p_u is 4a units to the left of pole p_v. This places pole p_u at the point of reflection of the focus in the directrix (Fig. 16m). This condition, $a = d/4$, is called a *subfocal condition* on eq. 47e because it only leads to a change to the simplest form, rather than to a reduction of degree. The reflection of the focus in the directrix, correspondingly, has only subfocal rank (see Problem 9). Additional axial parabolas are illustrated in Fig. $17c_1$-c_5.
[It is noteworthy that if one replaces d in eq. 47f by $d/\sqrt{2}$, i.e., if one uses $\sqrt{2}/2$ of the characteristic distance, rather than its full value, the eq. becomes that of the equilateral hyperbola cast about its center (p_u) and right focus (p_v). But only the right arm is represented in the polarized system. This relation illustrates the importance of employing the characteristic dis-

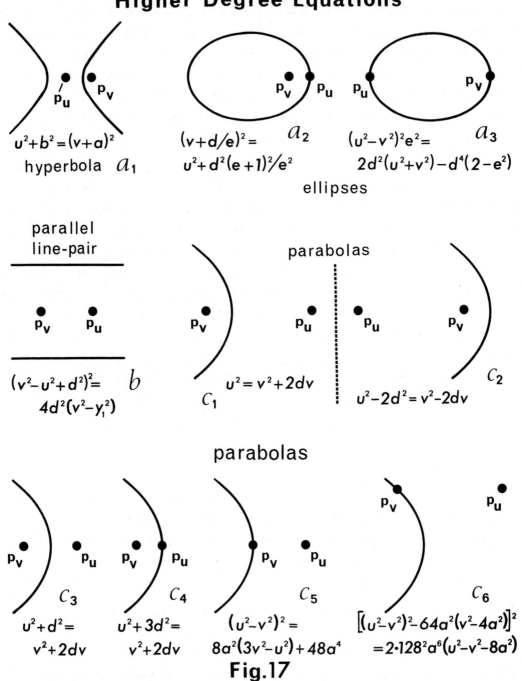

Fig.17

tance for the interpretation of bipolar construction rules.]

Eq. 47e also simplifies if $a = d$, yielding the right-hand term (u^2+4d^2). This condition places pole p_u at the vertex (Fig. 16u), which is a circumpolar point-focus (see Chapter VII). However, this is a very weak subfocal condition to which no great significance can be attached (see further discussion below). In the cases of more complicated construction rules than 47e,f, however, more than one powerful subfocal condition may exist, that is, more than one condition for which the eq. greatly simplifies, but in different ways--such as vanishing of a term, vanishing of two terms, conversion of all coefficients to unity, great simplification of a compound coefficient, etc.

With the realization that only one bipolar point focus exists on the axis of the parabola, we re-examine eqs. 47f,f'. Note that u occurs only as a "pure" square in eq. 47f, whereas v, the distance from the focal pole p_v, occurs in the square of its sum with the characteristic distance. Similarly, in eq. 47f', both u and v occur as pure squares but, in addition, a linear term in v is present. In describing this situation in the most general terms, one says that p_v has greater *weight* in these eqs. than p_u (just as for v^n/u^n-ratios greater than 1 for circles, when all other terms in u and v match up equally). These observations lead to RULE 20.

RULE 20. *In a bipolar construction rule, the focal rank of a pole represented by the square of the resultant of a linear term augmented or diminished by a constant--for example, $(u+d)^2$--generally is greater than the focal rank of a pole for which u or v occurs only as the pure square--for example, u^2.*

Individual and Pole-Pair Bipolar Focal Rank

The bipolar focal rank of a pole-pair is taken to be the inverse of the exponential degree of the construction rule of the curve cast about the pole-pair (assuming that the degree of the eq. is less than that of the general eq.). An *individual* bipolar focal rank can be assigned to the poles only if they are equivalent, or if the degree of the eq. is 1.

In the case of the parabola, no two axial poles are equivalent and no eq. is of 1st-degree, so no pole has individual bipolar focal rank. On the other hand, any two poles reflected in the line of symmetry of the parabola are equivalent, and could be assigned individual bipolar focal rank if the degree of the construction rule cast about them were lower than the degree of the general construction rule. It is shown below, however, that the degree of all eqs. cast about reflected poles is 8, the same as for arbitrary poles (see Note, page 105). Accordingly, no point in the plane of a parabola has individual bipolar focal rank.

Returning to construction rules 47f,f', the bipolar focal rank of the pole-pair (i.e., the two poles taken as a pair) about which these eqs. are cast is 1/2, the reciprocal of the degree of the eqs. RULE 20 allows a comparison to be made between the bipolar focal rank of the two poles, leading to the conclusion that the rank is greater for pole p_v, the traditional focus. In other words, if individual focal ranks could be assigned to the two axial poles in question (as for ellipses, for example), the traditional focus would outrank its reflection in the directrix.

The same conclusion can be reached, even without being able to assign individual bipolar focal rank, using another test. Thus, axial eqs. cast about the focus, p_v, and any other axial pole, p_u, are of 2nd-degree, whereas those cast about the reflection of the focus in the directrix and any other axial pole are of 4th-degree. In other words, the same test that identifies p_v as a point-focus (RULES 12-3 and 13) tells us that if individual bipolar focal rank could be assigned to it, the value would be greater than that for any other axial pole.

These matters become resolved in the consideration of conic sections with two lines of symmetry, but the parabola is the logical curve to consider first, since, aside from the circle, it is the simplest rectangular conic to possess but a single identifiable bipolar point-focus.

RULE 21. *The pole with the greater "weight" in a bipolar construction rule in which the highest degree of both variables is the same, generally is the pole of higher focal rank.*

Midline Parabolas

Having shown that an identifiable point-focus exists on the axis of the parabola, as well as at least one identifiable point with subfocal rank (the reflection of the focus in the directrix), we can now confirm the absence of a second focus. It is clear that all non-axial poles occur in equivalent pairs, since the axis is the line of symmetry. Accordingly, the simplest analysis of non-axial poles is one employing *midline parabolas*, that is, parabolas whose axes coincide with the bipolar midline.

MIDLINE PARABOLAS   133

The initial construction rule (48) is for a rectangular 90°-parabola bisected by the y-axis, with vertex at $y = k$ (employing a y-k translation). The substitutions of eqs. 22 give eq. 49a for the general bipolar midline parabola (Fig. 16n). A few test manipulations are enough to convince one that there is no possibility of achieving a form for which eq. 49a becomes a perfect square (at least not in the real domain), nor can a term be factored out. In other words, there is no focal condition for this eq.--no condition for which its degree will reduce. Consequently, since no pair of bipolar point-foci exists that is reflected in the bipolar midline, no non-axial focus exists.

bipolar midline parabolas

$x^2 = 4a(y-k)$ initial rectangular 90°-parabola   (48)

(a) $128a^2d^4(u^2+v^2) = (u^2-v^2)^4 +$
     $32ad^2(k+2a)(u^2-v^2)^2 + 64a^2d^4(d^2+4k^2)$ general midline parabola   (49)

(b) $128a^2d^4(u^2+v^2) = (u^2-v^2)^4$
     $+ 64a^2d^4(d^2+16a^2)$ eq. 49a for the subfocal condition $k = -2a$ (vertex at $y = -2a$)

(c) $d^6(u^2+v^2) = (u^2-v^2)^4/8 + d^8$ eq. 49b for $d = 4a$

(d) $4d^6(u^2+v^2) = (u^2-v^2)^4 + 3d^8$ eq. 49b for $d = 4\sqrt{2}a$ (incident $k = -2a$ poles)

(e) $8d^6(u^2+v^2) = (u^2-v^2)^4 + 2d^4(u^2-v^2)^2 + 5d^8$ eq. 49a for poles at LR-vertices

It remains only to seek identifiable poles of subfocal rank. These would be defined for any conditions leading to simplifications of eq. 49a--the greater the simplification, the greater the subfocal rank. The most notable subfocal condition on eq. 49a is $k = -2a$, for which the term in $(u^2-v^2)^2$ vanishes, leaving eq. 49b. Therefore, as we already know from our treatment of general parabolas (eq. 43a with $x_1 = x_2 = 2a$), any pair of (reflected) poles on a chord of a parabola (the 2a-chord) at twice the distance of the latus rectum from the vertex has bipolar subfocal rank (Fig. 16p).

The above result allows another test of the comparative positional symmetry of bipolar loci that include the reference poles, as opposed to comparable loci that do not. For lines, we found that a line-segment including both poles (eq. 30) was the second-most symmetrical locus, and a line including both poles was of 3rd-degree (eqs. 28d,d'; in the polarized system), compared to 4th-degree for general lines (eq. 27a). Lines including only one pole were of 4th-degree but more symmetrical than general lines (eqs. 27d,d' versus 27a).

For circles, we found that axial and midline circles that include both poles are generally more-symmetrically positioned with respect to the pole-pair than those that enclose both poles. In the present case of midline parabolas, the most-symmetrically-positioned parabola with poles on the 2a-chord is for the case $a = d/4$, which leads to construction rule 49c, with only one non-unit coefficient. These poles lie on the 2a-chord at the same level as the LR-vertices (Fig. 16q) and, like the reflection of the focus in the directrix, are identifiable poles of bipolar subfocal rank.

For a parabola in the same position but including the poles (eq. 49d and Fig. 16r), two non-unit coefficients (the maximum number for a three-term eq.) are present. To solidify the case against a generally greater symmetry of midline parabolas that include the poles as opposed to those that do not, consider a parabola with poles at the LR-vertices (which also are reflected in the axis and are circumpolar foci). Construction rule 49e for this parabola (Fig. 16s) has four terms and possesses no feature that confers greater simplicity than either eq. 49c or 49d (see Problem 11).

These results emphasize divergences of the bipolar system from the rectangular and polar systems, as well as from the circumpolar system (Chapters VII and VIII). First, a curve that includes one or both bipolar poles (the third condition of RULE 9) is not necessarily more-symmetrically positioned than the same curve in a comparable location but not including the pole(s). [The analytical basis for this is shown below.] Second, circumpolar foci are not necessarily bipolar foci. These observations prepare the ground for RULE 22.

RULE 22. *All bipolar foci are circumpolar foci, but not vice versa.*

Focal Parabolas

Focal parabolas have their traditional focus at one pole (p_v), with the second pole at an arbitrary point in the plane. These parabolas are readily shown to be represented by the 4th-degree construction rule 50 (Fig. 16t; Problem 10), compared to an 8th-degree general eq. A 50% reduction from the degree of the general bipolar construction rule proves to apply to all bipolar eqs. cast about one independent point-focus. This is one of the general relations that provides the bases for RULES 12-1a and 13-1.

FOCAL PARABOLAS

Reduction of the focal construction rule of the parabola to 4th-degree is a consequence of the following simple relation. The bipolar distance v from the focus to any point of the curve can be represented as a linear function of x (eq. 51a), the abscissa of the point (x,y) of the curve as measured from the focus as origin. It is evident that for $y_1 = 0$, eq. 50 becomes the equivalent of eqs. 47d,e for focal axial parabolas.

$$[(u^2-v^2) + 2x_1v - (4ax_1+x_1^2+y_1^2)]^2 = 16ay_1^2(v-a) \qquad \text{focal parabolas} \qquad (50)$$

Note that the construction rule 50 defines the maximum possible weight disparity of terms in u and v for the bipolar construction rules of parabolas, since p_v is a point-focus (the point about which the curve has the greatest structural simplicity) and p_u is an arbitrary point in the plane. Thus, disregarding coefficients, v occurs in powers of 4, 3, 2, and 1, whereas u occurs only in powers of 4 and 2.

The Traditional Focus as the Point about Which the Parabola Possesses the Greatest Structural Simplicity

Just as construction rule 50 highlights the maximum possible differences in weights of terms in u and v, eq. 51a highlights the property of the foci of curves as points about which the curves possess great structural simplicity. In this case, the greatest structural simplicity clearly is possessed about the traditional focus, inasmuch as distances from this point to the curve can be expressed as a linear function of the abscissa of the point (the specifying measurements or instructions of eq. 51a).

distance equations for the parabola

(a) $\quad v = x + 2a$ $\qquad$ distance from focus (p_v) to point on curve with abscissa x $\qquad$ (51)

(b) $\quad u^2 = (x-x_1)^2+(\pm 2\sqrt{[a(x+a)]}-y_1)^2$ $\qquad$ distances from points in the plane (x_1 measured from abscissa of focus)

(c) $\quad u^2 = (x-x_1)^2 + 4a(x+a)$ $\qquad$ distances from axial points

(d) $\quad u^2 = (x+a)(x+5a)$ $\qquad$ distance from the vertex

(e) $\quad u^2 = (x+2a)(x+10a)$ $\qquad$ distance from the reflection of the focus in the directrix

(f) $\quad u^2 = (x+a)^2 + 4a^2$ $\qquad$ distance from $(x_1,y_1) = (a,0)$

Distance eqs. now are employed to assess the structural simplicity of a 0°-axial rectangular parabola with its focus at the origin (in which case, x_1 continues to be measured from the position of the focus). Structural simplicity about axial points (eq. 51c) clearly exceeds that about points in the plane (eq. 51b), inasmuch as the second term of the eq. becomes linear in x (such a simplification is not unexpected, since only half as many distances from the pole in question need be specified).

Using the quadratic-root formula on the right side of eq. 51c allows one to evaluate integral roots other than those of a perfect square (which led to the reduction of eq. 51c to 51a). For the vertex, $(x_1 = -a)$, eq. 51c becomes 51d (Fig. 16u), while for the reflection of the focus in the directrix $(x_1 = -4a)$, it becomes eq. 51e, etc. On the other hand, for a pole on the 2a-chord $(x_1 = a)$, eq. 51c becomes 51f (Fig. 16v).

The former two points $(x_1 = -a, -4a)$ are precisely the points that were found to have highest subfocal rank in combination with the focus (see eq. 47e), and the present development shows the basis for it. These are the points about which the curve has the next-greatest structural simplicity after the focus, in the sense that distances to points on the curve from them are expressed in the simple forms of eqs. 51d,e. They also are the points that lead to relations between a and d of eq. 41e, such that the constant term on the right becomes a perfect square. Thus, the values, $x_1 = -a, -4a$, that lead to the right-hand "integral" products of eqs. 51d,e also lead to values of $(2d)^2$ and d^2 for the constant term on the right of eq. 47e.

Bipolar Construction Rules and the Hybrid Polar-Rectangular Coordinate System

The stage now has been set for an analysis of the factors that are crucial in determining the degrees of bipolar construction rules in general, and in determining the independent point-foci of bipolar curves. It will become evident from this analysis that the same factors are operative in other coordinate systems that include points or circles as reference elements.

It was noted above that the expression in v^2 (or u^2) for the distances from an arbitrary point in the plane to points on a parabola reduces to the linear expression $v = x + 2a$ when the point is the traditional focus, where x is the directed distance from the focus along the x-axis. [If x is measured from the vertex but v is measured from the focus, the eq. becomes $v = x + a$.] It also

DISTANCE EQUATIONS

was noted that, because of this reduction, the degree of the general construction rule of focal parabolas is only 4, compared to 8 for general parabolas. It is natural, then, to examine closely the properties of distance eqs., as they concern the determination of bipolar foci and the properties of bipolar curves.

Consider an algebraic rectangular curve, $y = f(x)$, with a line of symmetry in coincidence with the x-axis, and form the distance eq. 52a from a point in the plane (x_1, y_1) in the 1st-quadrant, to points on the curve (Fig. 18a). The distances x_1 and y_1, of eq. 52a, are measured from the rectangular origin, and if our sole object were to evaluate the magnitude of v, the location of the origin would be of no significance. In the cases of bipolar poles, however, distances are measured from the poles themselves, so it is desirable to bring the point (x_1, y_1), our presumptive bipolar pole, into coincidence with the y-axis, as in Fig. 18b.

[There is no need to bring the presumptive bipolar pole to the origin since y is measured as a function of x, namely f(x), i.e., all distances are measured in terms of the abscissa x.]

distance equations

(a) $v^2 = (x-x_1)^2 + [f(x)-y_1]^2$ conventional distance eq. from point (x_1,y_1) to points on the curve $y = f(x)$ [x measured from the origin at $(0,0)$] (52)

(b) $v^2 = x^2 + [f(x+x_1) - y_1]^2$ *hybrid* distance eq. from point $(0,y_1)$ to points of the curve $y = f(x+x_1)$ [x measured parallel to the x-axis from point $(0,y_1)$]

$(y + y_1) = f(x + x_1)$ eq. of curve of Fig. 18a cast about point (x_1,y_1) as origin; curve generally possesses symmetry about neither the x nor y-axis (Fig. 18c) (53)

(a) $(r\sin\theta + y_1) = f(r\cos\theta + x_1)$ polar eq. of curve of eq. 53; curve generally is symmetrical about neither the 0°-180°-axes nor the 90°-270°- axes (x or y-axis) (54)

(b) $r\sin\theta = f(r\cos\theta + x_1) - y_1$ eq. 54a rearranged

(c) $r^2\sin^2\theta = r^2 - r^2\cos^2\theta =$
$[f(r\cos\theta + x_1) - y_1]^2$ eq. 54b squared; curve now is symmetrical about the 0°-180°-axes (x-axis); Fig. 18c, with curve reflected in the x-axis

(d) $r^2 = x^2 + [f(x+x_1) - y_1]^2$ *hybrid* polar-rectangular eq. (eq. 54c expressed in terms of r and x; curve is symmetrical about the x-axis; Fig. 18c with curve reflected in the x-axis)

Accordingly, an $(x+x_1)$ translation of the curve of eq. 52a is in order.

Thus, x of eq. 52a is replaced by $(x+x_1)$ (i.e., $x' = x-x_1$, whereupon $x = x'+x_1$, or, dropping the prime symbol for convenience, $x = x+x_1$), yielding the *hybrid* distance eq. 52b (see below). The distance x is directed, and is measured in the positive sense to the right of point (x_1,y_1) (which, now being incident upon the y-axis, has the coordinates, $0,y_1$) and in the negative sense to the left of it. The distance v always is positive, and is measured to the same side of the y-axis as the distance x.

Now consider the algebraic curve of Fig. 18a again, but suppose we wish to find its polar construction rule about the point (x_1,y_1), as pole. A combined $(x+x_1)$ and $(y+y_1)$ translation brings the point (x_1,y_1) to the origin, giving construction rule 53, the same eq. employed to derive the basic polar construction rule of a rectangular curve (see Chapter III). The curve represented by eq. 53 generally is not symmetrical about either coordinate axis. The polar transformation substitutions now yield construction rule 54a, the sought-after polar eq. of the curve (still generally not symmetrical about any polar axis).

Rearranging gives eq. 54b, and squaring gives eq. 54c, which consequently is of twice the degree in r. As a result of squaring, eq. 54c possesses only terms in $\sin^2\theta$ and powers of $\cos\theta$. Accordingly, it represents curves that are symmetrical about the 0°-180°-axes (x-axis)--by virtue of being reflected in them (Fig. 18c, with the curve reflected in the x-axis). Replacing $\sin^2\theta$ by $(1-\cos^2\theta)$, and $r\cos\theta$ by x, gives eq. 54d, expressed in terms of the polar r and the rectangular abscissa x.

Comparing eq. 54d with eq. 52b, it is evident that the general distance equation for v^2 from an arbitrary point in the plane, (x_1,y_1) to the curve, $y = f(x)$ [where x is a directed distance measured parallel to the x-axis from the point (x_1,y_1)] is identical with the squared polar construction rule of the same curve cast about the same point but expressed in terms of r and x (where $x = r\cos\theta$ and r replaces v). Accordingly, it becomes evident that distance eqs. formed in the manner of eq. 52b--which are for distances to points of a curve from points in the plane--also are construction rules for curves in a *hybrid polar-rectangular coordinate system*.

In the light of this analysis, it is clear that the bipolar system must share certain properties with the hybrid polar-rectangular system. An obvious common property is that all curves in the two coordinate systems possess at least one line of symmetry. A less obvious property is that all point-foci in the two systems must lie on lines of symmetry of the curves themselves (i.e., the curves considered as rectangular loci), since only under this condition can the eqs. of either system reduce. In the following, the new distance eqs. (52b and 54d) are called *hybrid* distance eqs. to distinguish them from conventional distance eqs. (52a), from which they differ only by an $(x+x_1)$-translation.

The Degrees of Bipolar Construction Rules and the Foci of Bipolar Curves

With the information obtained above at hand, together with the knowledge of how bipolar construction rules are formed from distance eqs. in u and v, a number of conclusions can be drawn concerning the degrees of bipolar construction rules, the conditions for bipolar foci, and the properties of these foci. The following treatment applies generally to bipolar curves with a rectangular line of symmetry, excepting certain properties of circles and lines, which already have been treated.

derivation of bipolar construction rules of parabolas

(a) $v^2 = (x-x_1)^2 + (\pm 2\sqrt{[ax]} - y_1)^2$ conventional distance eq. from point (x_1,y_1) to point (x,y) of curve (55)

(b) $u^2 = (x-x_2)^2 + (\pm 2\sqrt{[ax]} - y_2)^2$ conventional distance eq. from point (x_1,y_1) to point (x,y) of curve

(a) $v^2 - u^2 = 2x(x_2-x_1) \pm 4(y_2-y_1)\sqrt{(ax)}$ eq. 55a minus 55b, $y_1, y_2 \neq 0$ (56)
$\quad + (x_1^2 - x_2^2 + y_1^2 - y_2^2)$

(a') $4x^2(x_2-x_1)^2 - 4x[B^2(x_2-x_1) - 4a(y_2-y_1)^2]$ eq. 56a rearranged and squared
$\quad\quad\quad\quad\quad\quad\quad\quad\quad + B^4 = 0$
$B^2 = (v^2 - u^2 - x_1^2 + x_2^2 - y_1^2 + y_2^2)$

(b) $v^2 - u^2 = 2x(x_2-x_1) \pm 4y_2\sqrt{(ax)} + (x_1^2 - x_2^2 - y_2^2)$ eq. 55a minus 55b for $y_1 = 0$

(c) $\quad v^2 - u^2 = 2x(x_2-x_1) + x_1^2 - x_2^2$ eq. 55a minus 55b for $y_1 = y_2 = 0$

$x = \dfrac{B^2(x_2-x_1)/2 + 2a(y_2-y_1)^2 \pm (y_2-y_1)\sqrt{[2aB^2(x_2-x_1) + 4a^2(y_2-y_1)^2]}}{(x_2-x_1)^2}$ eq. 56a solved for x (57)

$(v^2 - x_1^2 - y_1^2) - (D^3/X^4)[2(2a-x_1)X^2 + D^3 + aY^2] \mp$ eliminant of eqs. 55a and 57 (58)
$\sqrt{(aA^3)}[2(2a-x_1)Y + 2D^3Y/X^2]/X^2 =$ before squaring entire eq.
$\mp (4y_1/X)\sqrt{[aD^3 \pm aY\sqrt{(aA^3)}]}$ $X = (x_2-x_1), \quad Y = (y_2-y_1)$
$A^3 = 2B^2X + 4aY^2, \quad D^3 = B^2X/2 + 2aY^2$

In order to elucidate the following discussion, I illustrate below the derivation of bipolar construction rules using conventional distance eqs. from arbitrary points in the plane to points of a curve, employing as examples the simple cases of parabolas (eqs. 55-58)

The distance eqs. 55a,b, and their differences, eqs. 56a-c, are for three cases: neither pole on the line of symmetry (eq. 56a), one pole on the line of symmetry (eq. 56b), and both poles on the line of symmetry (eq. 56c). To solve eq. 56a for x, one squares and applies the quadratic-root formula, yielding eq. 57. Although squaring introduces a term in B^4, this term enters the radical and is cancelled by the subtraction of a second identical term in B^4 (as is evident from inspection of eq. 56a'). Substituting x from eq. 57 in eq. 55a (or 55b), one obtains eq. 58, which must be squared twice to eliminate both radicals. Reduction of the resultant leads to a general solution of 8th-degree. [This cancellation of B^4 in the radical of the quadratic-root solution is a general occurrence whenever an eq. of the form, $B^2 - 2px = \pm q\sqrt{(sx)}$ is squared and solved for x, since the radicand becomes $(4pB^2+q^2s)^2-16p^2B^4$.]

With this introduction to the derivation of bipolar construction rules from distance eqs., we proceed to a consideration of the implications of the information obtained in the preceding section.

1. The degree in x of the distance eq. 55a or 55b (the *general* distance eq.) from an arbitrary point in the plane to a point of the curve is twice the degree of the basic polar construction rule (because of the need to square the polar eq. to eliminate $\sin\theta$) and the rectangular construction rule (see RULE 7a). This leads to a general bipolar construction rule of the curve of four times the degree of the rectangular eq. (as explained above and already concluded earlier).

2. For a point on a line of symmetry, a general distance eq. reduces in degree by 50%, to the degree of the basic polar construction rule (and rectangular construction rule). This is seen readily from eqs. 55a,b. If either y_1 or y_2 is 0, the radical is eliminated without squaring the entire eq., yielding a 2nd-degree eq. in both v and x. For a curve of rectangular degree 4 (see below), the radicand of the radical of eqs. 55a,b itself contains a radical, so after squaring the right terms, two additional squarings of the entire eq. (rather than only one squaring, as for a 2nd-degree curve) are needed to eliminate all radicals, giving a general distance eq. of degree 8 in v and x. For an axial point, only one additional squaring of the entire eq. is required, yielding an axial eq. of degree 4. These reductions are in accord with RULE 1c, since only half as many measurements are needed from an axial pole to specify all distances to the initial curve (which is symmetrical about the axis), as from a point in the plane.

3. The degree of a hybrid distance eq. (a hybrid polar-rectangular eq.) for a vertex of a curve generally does not reduce from the rectangular degree of the initial curve. An axial distance eq. (a distance eq. from a point on a line of symmetry) reduces only by the method of taking roots. This is a consequence of

the fact that replacing $r\cos\theta$ by x (or $r\sin\theta$ by $\pm\sqrt{[r^2-x^2]}$) eliminates the possibility of reduction by factoring out r's. [The latter can occur only by obtaining r's (to factor out of some terms) from their combination with $\cos\theta$ or $\sin\theta$. If r's occurred in all terms unassociated with $\sin\theta$ or $\cos\theta$, the rectangular construction rule would be degenerate.]

4. Bipolar construction rules do not reduce for the case in which only one of the two poles is an unexceptional point on a rectangular line of symmetry. In such a case, the total number of measurements needed to specify the points of the curve is reduced only by 25%, and this is insufficient (RULE 1d) to lead to reduction. Analytically, we can see how this transpires by referring to construction rule 58. When y_1 of eq. 55a is 0, there is no need to square eq. 58 twice to eliminate all radicals. One squaring is sufficient. However, the 8th-degree construction rule obtained from only one squaring for one axial pole does not reduce (as does the 16th-degree eq. obtained for two squarings for two axial poles), leading to a bipolar construction rule that remains of 8th-degree, despite the need for one less squaring. On the other hand, if both poles are axial, only half as many measurements are needed to specify all points of the curve than if neither is axial. In consequence, both eqs. 55a and 55b become quadratic in the variables, and the eliminant (the eq. obtained by eliminating x) is of only 4th-degree.

5. Any condition for which the degree of an axial distance eq. of a curve reduces to less than the degree of the basic polar construction rule (and rectangular construction rule) defines an independent bipolar point-focus. The construction rule of the curve cast about this pole and an arbitrary point in the plane is only half the degree of the general distance eq., while if the second pole is an unexceptional axial pole, the degree is only 1/4 (or 1/8th for certain quartics) that of the general distance eq. If the second pole also is a bipolar point focus, the reduction in degree is to 1/8th (or to 1/16th for certain quartics) that of the general distance eq. These results are all consequences of (a) the need for fewer squarings of eqs., (b) initial distance eqs. being of lower degree, and/or (c) reduction of one or both distance eqs. from quadratic to linear (or from quartic to quadratic, for certain quartics).

6. The basis for the fact that unexceptional incident points of bipolar curves are not bipolar point-foci is clear from treatments (1) and (5), above. Whereas polar construction rules (and, in consequence, circumpolar structure rules; see Chapters VII and VIII) reduce in degree when cast about any incident point, distance eqs. do so only for those exceptional incident points for which they become perfect squares.

With these results at hand, the following additional RULES are formulated:

RULE 23. *The degree of a "general distance equation"--the equation for the distance from a point in the plane to an arbitrary point of an initial curve (with the abscissa of the arbitrary point as the independent variable)--is twice the rectangular degree of the curve.*

RULE 24. *A general distance equation reduces to the corresponding rectangular degree of the initial curve for an arbitrary point on a line of symmetry, giving the "general axial distance equation."*

RULE 25 *General axial distance equations reduce in degree only through root-taking.*

RULE 26. *Independent bipolar point-foci are defined for any condition for which the degree of a distance equation reduces from the rectangular degree of the initial curve.*

RULE 27. *Independent point-foci are defined by any condition for which the degree of a distance equation reduces from the rectangular degree of the initial curve in coordinate systems employing two points, a point and a line, a point and a circle, a circle and a line, or two circles as reference elements.*

RULE 28. *Vertices of bipolar curves (single intersections of a curve with a line of symmetry) have only subfocal symmetry rank.*

The Hybrid Polar-Rectangular Coordinate System-- Similarities and Differences

We conclude with a summary of the similarities and dissimilarities between the hybrid polar-rectangular coordinate system and the polar and rectangular systems (see also Table 7). With the rectangular system, the hybrid system shares the properties of lacking a characteristic distance and having two distance coordinates. If a rectangular curve includes the origin, the curve in the hybrid system also will include the origin (the point-pole) and all intersections of rectangular curves with each other at the origin are preserved in the hybrid system.

The hybrid system differs from the rectangular system in that only one of its coordinates is directed, in the usual sense, and measured along a line, in that neither coordinate is measured normal to a coordinate element, and in that the coordinate elements are intrinsically asymmetrical. Although any rectangular locus can be represented in the hybrid system, all hybrid loci possess a line of symmetry. This is a consequence of the fact that all hybrid loci are reflected in the coordinate line, with a consequent doubling of the rectangular degree--but without producing degenerate eqs. Lastly, the hybrid system, like most general systems, lacks the property of degree-restriction--that peculiar intrinsic property of bilinear systems.

With the polar system, the hybrid system shares the properties of lacking a characteristic distance and degree-restriction, of possessing intrinsically asymmetrical coordinate elements, of distances being measured from a point, and of one distance being measured along a radius vector to the curve. Unlike the situation in the polar system, one reference element is a line rather than a half-line (or a ray in the basic polar system), only one coordinate is directed (in the usual sense), a distance (the x coordinate) is measured along a coordinate element, all intersections of a curve with the rectangular origin are preserved, all curves are reflected in the coordinate line, and eqs. generally are of twice the degree of the basic polar construction rules. Unlike the situation in either the polar or rectangular system, one coordinate of the hybrid system (v) always is equal to or greater in magnitude than the other coordinate (x) and always is positive.

It seems likely that the hybrid system eventually will become the third-most-important coordinate system for certain calculations and studies of the properties of curves (after the rectangular and polar systems). In a sense, it already is, since all conventional general distance eqs. differ from the corresponding eqs. of the hybrid system only in a linear transformation of the x-variable. Stated otherwise, every hybrid distance eq. is simply the conventional distance eq. of the translated curve (compare eqs. 52a,b). When no translation is needed to bring the point (x_1,y_1) of Fig. 18a onto the y-axis ($x_1 = 0$ in eqs. 52a,b), the conventional and hybrid distance eqs. are identical.

The hybrid system assumes a unique role for the identification of independent foci of curves in coordinate systems possessing point and circle reference elements (see RULES 26 and 27). Furthermore, as we shall see, its eqs. bear very close affinities to construction rules in such systems. In addition, the ability of hybrid construction rules to define independent point-foci provides another approach to obtaining the simplest polar construction rules of curves.

Bipolar Ellipses

In ellipses and hyperbolas, one meets the next-most-symmetrical bipolar curves after the midline, the connecting line-segment, the axial rays, and circles of Apollonius. Since the bipolar hyperbola with a 1st-degree construction rule (in the polarized system) possesses only one arm (one arm of Fig. $12c_2$), ellipses are selected for a detailed examination of a bipolar conic with two "equivalent" foci (foci reflected in a line of symmetry of a curve) and a true center. [All construction rules of ellipses, of course, also represent circles for the parametric condition $a = b$.]

General Ellipses

The need to consider the very complex general construction rule of ellipses can be obviated by considering the special cases of general interest (eqs. 60), for which the general construction rule greatly simplifies but does not reduce.

As for parabolas (eqs. 38-40), one derives the general eq. cast about two arbitrary points, (x_1,y_1) and (x_2,y_2), in the plane by forming the distance eqs. for v^2 and u^2, taking the difference between these eqs., solving for x, and finding the eliminant between this solution for x and either of the initial distance eqs. The only difference from the parabola case is that, for ellipses, the expression $\pm(b/a)\sqrt{(a^2-x^2)}$ replaces $\pm 2\sqrt{(ax)}$.

When $y_1 = y_2$, and both poles are on a line parallel to the major axis (Fig. 18d), one obtains the 8th-degree construction rule 62. The expression $(x_2v^2 - x_1u^2)$ plays the major role in determining the weighting, from which it is evident (as for the parabola) that the pole nearest the major axis has the higher subfocal symmetry rank. The parallel derivation for $x_1 = x_2$ --both poles on a line parallel to the minor axis (Fig. 18e)--yields the 8th-degree eq. 63, which, as expected, has exactly the same form as eq. 62, but the two members of each pair, (a and b), $(x_1$ and $y_1)$, and $(x_2$ and $y_2)$, are exchanged. In this case, the pole closest to the minor axis has the higher subfocal symmetry rank.

bipolar construction rules of ellipses

$b^2x^2+a^2y^2 = a^2b^2$, $y = \pm(b/a)\sqrt{(a^2-x^2)}$ centered 0°-rectangular ellipses   (59)

(a) $y_1 = y_2$, (b) $x_1 = x_2$, (c) $y_1 = 0$ special cases of general ellipses   (60)

(d) $x_1 = y_1 = 0$

$$B^2 = (v^2-u^2-x_1^2+x_2^2-y_1^2+y_2^2)$$ general abbreviation B^2 (61)

$\{a[(x_2v^2-x_1u^2)+(x_2-x_1)(x_1x_2-y_1^2-b^2)] +$ general ellipses cast about two poles on a line parallel to the major axis $(y_1 = y_2 \neq 0)$ (62)

$(b^2-a^2)B^4/4a(x_2-x_1)\}^2 =$

$b^2y_1^2[4a^2(x_2-x_1)^2-B^4]$

$\{b[y_2v^2-y_1u^2 +(y_2-y_1)(y_1y_2-x_1^2-a^2)] +$ general ellipses cast about two poles on a line parallel to the minor axis $(x_1 = x_2 \neq 0)$ (63)

$(a^2-b^2)B^4/4b(y_2-y_1)\}^2 =$

$a^2x_1^2[4b^2(y_2-y_1)^2-B^4]$

(a) $v^2 = (x-x_1)^2 + b^2(1 + x^2/a^2)$ general distance eq. for a pole on the x-axis   (64)

(b) $x = \dfrac{a^2B^2(x_2-x_1) \pm aby_2\sqrt{[4a^2(x_2-x_1)^2+4b^2y_2^2-B^4]}}{2[a^2(x_2-x_1)^2+b^2y_2^2]}$ expression in v^2-u^2 for distance eqs. from a pole on the x-axis, solved for x

Ellipse Constructions

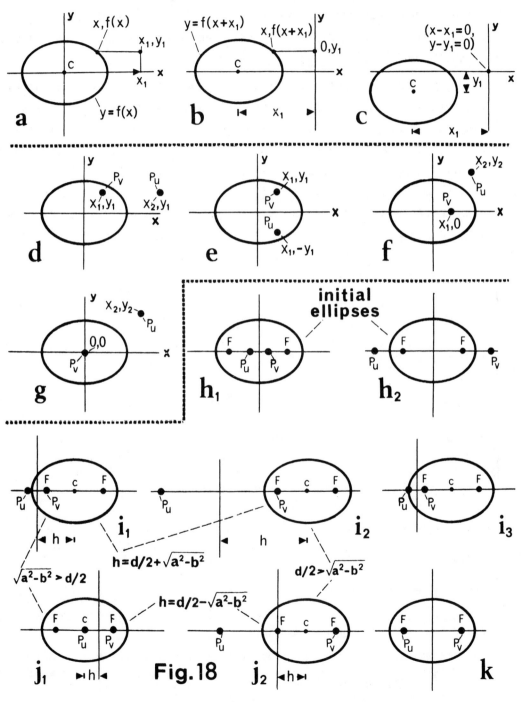

Fig. 18

In the case of an axial pole ($y_1 = 0$; Fig. 18f), the simplifications are not as great (nor were they for the parabola). The results are presented as eqs. 64a and 64b, from which the eliminant must be obtained by substituting for x and x^2 in eq. 64a, squaring the first term on the right (as indicated), segregating terms, and squaring again. It is evident that the eliminant will be heavily weighted in favor of p_v (since the term in v^2 of eq. 64a is unmatched by a term in u^2, whereas u^2 and v^2 of eq. 64b occur together in the powers of B^2). Corresponding results are obtained when one pole is on the minor axis.

For the cases of general *central ellipses* (one pole at the center, the other at an arbitrary point in the plane; eq. 60d and Fig. 18g), one simply lets $x_1 = 0$ in the eliminant of eqs. 64a,b, yielding eq. 65. This construction rule represents the maximum weighting (corresponding to the centered circle of eq. 32e) in favor of a non-focal pole for a central conic other than the circle. These results lead to RULES 29 and 30.

general central ellipses

$$[4(v^2-b^2)(a^2x_2^2+b^2y_2^2)^2 - B^4(a^2-b^2)(a^2x_2^2-b^2y_2^2)$$
$$- 4a^2b^2(a^2x_2^2+b^2y_2^2)(x_2^2+y_2^2)]^2 =$$
$$4a^2b^2(a^2-b^2)^2 x_2^2 y_2^2 B^4 [4a^2x_2^2+4b^2y_2^2-B^4]$$

cast about the center and a point x_2, y_2 (65)

RULE 29. *General bipolar construction rules of curves with a true center undergo major simplifications--without reduction--for both poles located on a line parallel to either line of symmetry & for one pole incident on either line.*

RULE 30. *The coefficients of the variables of bipolar construction rules cast about non-focal poles on lines parallel to a line of symmetry of a curve with a true center are weighted heaviest in favor of the pole nearest the center, that is, the pole nearest the center has the greater subfocal symmetry rank, and the curve is more-symmetrically positioned about it.*

Axial Ellipses

The bipolar construction rules of axial ellipses (eqs. 67a,b) are obtained by making the substitutions of eqs. 22 in eq. 66 for rectangular 0°-axial ellipses (Fig. 18h_1,h_2), for which the centers are translated a distance h from the bipolar midpoint (Fig. 18i_1,i_2). The translation is to the right for positive h. It will be noted that a single term will vanish from eqs. 67 for one of the substitutions of eqs. 68 (for the distance of axial translation, h).

Despite this fact, eqs. 68 do not lead to identifiable points with subfocal rank, because the positions of these points in the plane of an initial ellipse depend not merely upon e, but also upon the value selected for the characteristic distance, d. Accordingly, these conditions are not pursued further.
[The condition 68a, for example, calls for translating the center to the right (upper sign) or to the left a distance obtained by multiplying the distance d/2 by the ratio of the length of the major axis (2a) to the length of the latus rectum ($2b^2/a$), a ratio that exceeds 1 for all ellipses but the circle.]

bipolar axial ellipses

$b^2(x-h)^2 + a^2y^2 = a^2b^2$ initial 0°-axial rectangular ellipses (66)

(a) $(a^2-b^2)(u^2-v^2)^2 + 2d(2b^2h-a^2d)u^2 -$ bipolar axial ellipses (67)

$2d(2b^2h+a^2d)v^2 + d^2(4a^2b^2-4b^2h^2+a^2d^2) = 0$

(a') $(a^2-b^2)(u^2-v^2)^2 + 2d(2b^2h-a^2d)(u^2-v^2) +$ eq. 67a rearranged

$d^2(4a^2b^2-4b^2h^2+a^2d^2) = 4a^2d^2v^2$

(b) $(a^2-b^2)(v^2-u^2-x_1^2+x_2^2)^2/a^2 - 4(x_2-x_1)\cdot$ eq. 67a in another form; that of a centered 0°-axial rectangular ellipse cast about poles at $(x_1,0)$ and $(x_2,0)$

$(x_2v^2-x_1u^2) + 4(x_2-x_1)^2(b^2-x_1x_2) = 0$

(c) $(a^2-b^2)(u^2-v^2)^2 + 2(x_2-x_1)[(a^2+b^2)x_1 -$ eq. 67b expanded and grouped in powers of u and v

$(a^2-b^2)x_2]u^2 - 2(x_2-x_1)[(a^2+b^2)x_2 -$

$(a^2-b^2)x_1]v^2 + (a^2-b^2)(x_2^2-x_1^2)^2 +$

$4a^2(x_2-x_1)^2(b^2-x_1x_2) = 0$

(a) $h = \pm a^2d/2b^2$, (b) $h = \pm a\sqrt{(4b^2+d^2)}/2b$ non-subfocal conditions for simplifying eqs. 67 (68)

(a') $h = \pm d/2(1-e^2)$ eqs. 68a,b in terms of d and e

(b') $h = \pm\sqrt{[4a^2(1-e^2)+d^2]}/2\sqrt{(1-e^2)}$

The weighting of the poles in eq. 67a clearly favors the pole nearest the center. Since terms in u^4 and v^4 have the common coefficient (a^2-b^2), this weighting must be based on the coefficients of the terms in u^2 and v^2. With the displacement of the ellipse to the right (the x-h translation for positive h), pole p_v is closest to center, and the absolute magnitude of the coefficient of v^2 is greater than that of the coefficient of u^2. For h negative, the displacement is to the left and the reverse is the case.

[It is the absolute magnitude that must be used for weighting, since the eq. is unchanged by changing the signs of all terms, but the signs of the u^2 and v^2-terms become exchanged.]

Eq. 67b is for a centered 0°-axial ellipse with general locations assigned to both poles (p_v at x_1 and p_u at x_2). The exchanged coefficients of u^2 and v^2 in the second term from the left confirm the weighting in favor of the pole closest to center, since if p_u is the most distant, its abscissa becomes the coefficient of v^2, etc. In this case, the construction rule is simple enough to expand and group all terms readily in powers of u and v, as in eq. 67c.

Eq. 67c illustrates the general situation for all eqs. cast about two poles that lie on a line normal to a line of symmetry. In all such eqs., the coefficients of terms in equal powers of u and v are either identical (as for the 4th-power terms in eq. 67c) or they become exchanged by exchanging x_1 and x_2 (as for the 2nd-power terms). On the other hand, the constant term is symmetrical in x_1 and x_2 (i.e., it is unchanged by exchanging x_1 and x_2). The comparative weighting in every case must be determined by examining the coefficients that are exchanged. [Recall that in the case of axial poles of the parabola, where no orthogonal line of symmetry is present, the weighting is indeterminate (Fig. 16j).]

There is no condition on the entire left-hand side of eq. 67a that will convert it to a perfect square (just as there was none for the homologous eq. 38 for the parabola). However, if eq. 67a is rearranged to 67a', both sides of the eq. become perfect squares when eq. 69a obtains. Solving eq. 69a for h yields eq. 69b, which specifies focal conditions. Thus, when eq. 69b obtains, eq. 67b reduces to eq. 70a for the positive sign and 70b for the negative sign.

(a) $(2b^2h - a^2d)^2 = (a^2-b^2)(4a^2b^2 - 4b^2h^2 + a^2d^2)$ condition for eq. 67a'-left to be a perfect square   (69)

(b) $h = d/2 \pm \sqrt{(a^2 - b^2)}$ solutions to eq. 69a

(a) $u^2 = v^2 + 2dv/e - 2ad(1-e^2)/e + d^2$ axial ellipses of eq. 67a' for focal condition of eq. 69b (positive sign)   (70)

(b) $u^2 = v^2 - 2dv/e + 2ad(1-e^2)/e + d^2$ axial ellipses of eq. 67a' for focal condition of eq. 69b (negative sign)

$u^2 + d^2(1+e)^2/e^2 = (v + d/e)^2$ eq. 70a for $a = d/(1-e)$, the "focus-vertex" ellipse   (71)

AXIAL ELLIPSES

$$u^2 - d^2(1-e^2)/e^2 = (v - d/e)^2 \qquad \text{eq. 70b for } d = \sqrt{(a^2-b^2)} = ae, \text{ the "focus-center" ellipse} \qquad (72)$$

(a) $\quad u^2 = v^2 \pm 2dv/e + d^2/e^2 \qquad$ eq. 67b for $d = 2ae$, the "bifocal" ellipse $\qquad (73)$

(b) $\quad \pm u = v \pm d/e, \quad u + v = d/e = 2a \qquad$ roots of eq. 73a, the linear bipolar ellipse

Although eqs. 69b, like eqs. 68, also involve the characteristic distance, its presence as a linear increment in the expression for h gives rise to translations to identifiable points in the planes of ellipses. Examples for two initial ellipses are considered below. Ellipse h_1 has its traditional foci distal to the bipolar poles (Fig. $18h_1$); for ellipse h_2, they are proximal (Fig. $18h_2$).

CASE 1: For the positive sign of eq. 69b, the translation of the center to the right of the midpoint is through a distance $d/2 + \sqrt{(a^2-b^2)}$. This carries the left traditional focus to the midpoint [distance $\sqrt{(a^2-b^2)}$] and beyond it a distance $d/2$ to pole p_v (Fig. $18i_1,i_2$). In other words, when p_v is taken to be at the left focus of the ellipse, with p_u somewhere on the major axis to the left of it, eq. 67b reduces to eq. 70a (the proper sign of the linear term in v is ascertained by testing the eq.). This identifies the left traditional focus as a bipolar point-focus. The right traditional focus therefore also is a point-focus (because of symmetry). Fig. $18i_1,i_2$ shows the translated ellipses of CASE 1 for the two initial ellipses, h_1 and h_2, respectively. In CASE 1a, since pole p_u initially is within the ellipse, it will come to lie at the left vertex when $d = a(1-e)$, the focus-vertex distance. By letting $a = d/(1-e)$ in eq. 70a, one obtains eq. 71, the bipolar construction rule of ellipses cast about a traditional focus and a near vertex (Fig. $18i_3$).

CASE $2h_1$: For eq. 69b with the negative sign, and with $\sqrt{(a^2-b^2)} > d/2$, h is negative and the translation is to the left by the amount $\sqrt{(a^2-b^2)} - d/2$. But this is the distance of the right focus from pole p_v (Fig. $18h_1$), whereupon the right focus comes into coincidence with p_v (Fig. $18j_1$), and eq. 67b reduces to eq. 70b. Accordingly, the right traditional focus is identified as a bipolar point-focus (RULE 13-2) and, of course, the left focus also. Since p_u is interior to the ellipse, and to the left of center, and since the center shifts to the left, it is possible for the center to be brought into coincidence with p_u (Fig. $18j_1$). This occurs when $\sqrt{(a^2-b^2)} - d/2 = d/2$, that is, when $d = \sqrt{(a^2-b^2)}$. In that event, one obtains eq. 72, the "focus-center" bipolar construction rule of ellipses.

CASE 2h$_2$: For eq. 69b with the negative sign and with $d/2 > \sqrt{(a^2-b^2)}$, h is positive and the translation is to the right by the amount $d/2 - \sqrt{(a^2-b^2)}$. Referring to Fig. 18h$_2$, this carries the right traditional focus to pole p_v (Fig. 18j$_2$), whereupon eq. 67b reduces to eq. 70b, again identifying the traditional foci as bipolar foci. In this case, since p_u initially is exterior and to the left, and since the shift is to the right, p_u remains exterior. CASE 2h$_2$ thus is for p_v at the right focus and p_u exterior to the curve and to its left (Fig. 18j$_2$).

CASE 3: For eq. 69b and $d/2 = \sqrt{(a^2-b^2)} = ae$, $h = 0$, and p_u and p_v lie initially at the traditional foci (Fig. 18k). In that event, eq. 67b reduces to eq. 73a, and roots can be taken again, yielding eq. 73b-left. Testing to ascertain the proper signs shows that the applicable eq. is 73b-right, the familiar linear construction rule for ellipses about two points. Since eq. 73b-left includes all combinations of signs, it also includes the eq. of hyperbolas as one of the alternatives. These results call attention to the following RULES.

> RULE 31. *Bipolar construction rules that are symmetrical in terms involving the variables u and v represent curves that are either symmetrical with respect to the bipolar midline or have a center of symmetry that lies midway between pole p_u and p_v (see eqs. 27b,c, 29, 30, 32f, 34, 49b-right, 73, and 77b).*

> RULE 32: *If the bipolar construction rule of a curve cast about equivalent poles (poles reflected in a line of symmetry) is of 1st or 2nd-degree, a true center of the curve lies midway between poles p_u and p_v (also frequently true of equations of 4th and 8th-degree; see eqs. 77b,c of 8th-degree, 27b,c of 4th-degree, 29 and 32f of 2nd-degree, and 30 and 73-right of 1st-degree).*

Subfocal Ranking of the Center and Vertices

As we have seen, both the center and the near vertex, when paired with a traditional focus, give rise to relatively simple 2nd-degree construction rules of ellipses (eqs. 71 and 72; also applies to the far vertex). Pole p_v has much greater weight in these eqs. than pole p_u (for the vertex or center), as evidenced by the additional linear term in v (RULE 20). This derives from the fact that the traditional foci are the only bipolar point-foci of ellipses. Of the focus-vertex and focus-center construction rules (eqs. 71 and 72), the focus-center eq. is simplest [since $(1-e^2)$ is simpler than $(1+e)^2 = (1+2e+e^2)$], in consequence of which the center has higher subfocal rank than a vertex. [When construction rules 71 and 72 are expanded, one obtains eqs. 71' and 72', of which the focus-vertex eq. would appear to be slightly simpler since it involves only $2d^2/e$, as opposed to $2d^2/e^2$. However, when allowance is made for the fact that e represents a term containing a radical, namely, $e = \sqrt{(a^2-b^2)}$, the greater simplicity of eqs. 72 and 72' holds up (since squaring e eliminates the radical); Figs. $17a_2$ and $18j_2$.]

2nd and 4th-degree construction rules of bipolar ellipses

$v^2 + 2dv/e = u^2 + d^2 + 2d^2/e$	focus (p_v)-vertex (p_u)	(71')
$v^2 - 2dv/e = u^2 + d^2 - 2d^2/e^2$	focus (p_v)-center (p_u)	(72')
$(u^2-v^2)^2 - 2d^2u^2 - 2d^2(2-e^2)v^2/e^2 = d^4(3-4/e^2)$	vertex (p_u)-center (p_v)	(74)
$e^2(u^2-v^2)^2 - 2d^2(u^2+v^2) = d^4(e^2-2)$	vertex-vertex	(75)

The higher subfocal rank of the center as compared to a vertex is confirmed by an examination of the "vertex-center" eq., 74. Since the ratio $(2-e^2)/e^2$ always is greater than 1 for ellipses, pole p_v (the center) has greater coefficient weight, giving the center higher subfocal rank (see also RULE 30). Even though these comparisons favor the center, they are biased against it, since one is comparing a singular pole with a pole for which an equivalent pole exists, as treated in RULE 33.

RULE 33. *Comparative subfocal ranking of the poles of a bipolar construction rule is not necessarily valid unless the initial curve possesses an equivalent pole for either both or neither of the poles being compared. When the latter conditions are not fulfilled, the ranking is biased against the singular pole.*

Judging from these results, the subfocal ranks of the center and vertices would not be expected to differ greatly, in which case one expects the vertex-vertex construction rule to be simpler than that for the vertex-center (since the measurements are of comparable simplicity--see eqs. 80b,c--but only half as many are needed for the vertex-vertex construction; see RULE 1c). This is confirmed by eq. 75 (see also Problem 24) which, though also 4th-degree, is simpler and more symmetrical than eq. 74 (see Fig. $17a_2$).

In the treatments that follow, we shall see that the distance eqs. of the hybrid polar-rectangular system enable essentially definitive bipolar subfocal ranking to be achieved (even between different point-foci). These subfocal rankings also are valid in other coordinate systems that possess one or more points or circles as reference elements (and employ solely distances as coordinates).

Midline and Bisector Ellipses

Some construction rules of ellipses of both general interest and as they bear on RULES for bipolar eqs., now are considered. The midline 0°-ellipses of eq. 76 are 8th-degree but they are symmetrical in terms in the variables u and v since the curve is symmetrical about the midline. [The term $\{4d^2v^2 - (u^2-v^2-d^2)^2\}$ is symmetrical in u and v since it can be re-expressed as $(u^2-v^2)^2 + d^4 + 2d^2(u^2+v^2)$.] These ellipses (eq. 76) are an example of a case for which a true center does not lie between equivalent bipolar poles (see RULE 32). If the ellipse is rotated 45°, and its center is translated to the bisector (eq. 77a), then symmetry is lost in the terms for u and v (by virtue of the two $[u^2-v^2]$-terms), since the bipolar poles (at $+d/2$ and $-d/2$ on the bipolar axis) no longer are symmetrically located about a line of symmetry of the curve (nor about the center of the curve).

midline and bisector ellipses

$$[(b^2-a^2)(u^2-v^2)^2 + 2a^2d^2(u^2+v^2) + 4a^2d^2(k^2-b^2-d^2/4)]^2 \quad \text{midline 0°-ellipses} \quad (76)$$
$$= 16a^4d^2k^2[4d^2v^2-(u^2-v^2-d^2)^2]$$

(a) $d^2[2d(a^2+b^2)(u^2+v^2) - 8b^2h(u^2-v^2) - (a^2+b^2)d^3 + 8b^2d(a^2-2h^2)]^2 = 4[(a^2-b^2)(u^2-v^2) - 4b^2dh]^2 \cdot [4d^2v^2-(u^2-v^2-d^2)^2]$ general 45°-bisector ellipses   (77)

(b) $(a^2+b^2)^2d^4\{(u^2+v^2) - [d^2/2 + 4a^2b^2/(a^2+b^2)]\}^2 = (a^2-b^2)^2(u^2-v^2)^2[4d^2v^2-(u^2-v^2-d^2)^2]$ centered 45°-bisector ellipse

(c) $64a^4b^4[(u^2+v^2) - 8a^2b^2/(a^2+b^2)]^2 = (a^2-b^2)^2(u^2-v^2)^2\{32a^2b^2v^2/(a^2+b^2) - [u^2-v^2-8a^2b^2/(a^2+b^2)]^2\}$ centered 45°-bisector ellipse including poles; $d = 2\sqrt{2}ab/\sqrt{(a^2+b^2)}$

On the other hand, if the ellipse is translated along the bisector to bring its center to the bipolar midpoint (eq. 77b), then symmetry in the u and v-terms is recovered (since the poles are symmetrically located about a center of the curve, even though not incident upon a line of symmetry). Therefore, this is an example of an 8th-degree construction rule that satisfies RULES 31 and 32. If the curve of eq. 77b is made to include the poles (eq. 77c), the constant term in the brackets on the left simplifies, but the overall simplicity of the eq. is not greatly altered.

Focal Ellipses

It remains to consider the general construction rule for bipolar *focal ellipses*. Eq. 78a is cast about the left traditional focus, p_v, and about a point in the plane, $(x_1,y_1) = p_u$ (see also Problem 13). This construction rule is only 4th-degree (eq. 78a; RULE 12-1)--the same value as for general axial ellipses (RULE 12-2). Like eq. 50 for focal parabolas, eq. 78a defines the maximum possible difference in the weights of terms in u and v for the bipolar construction rules of ellipses. Again, v occurs in powers of 4, 3, 2, and 1, whereas u occurs only in powers of 4 and 2. For $y_1 = 0$, the condition for *axial focal ellipses*, eq. 78a reduces and simplifies to the 2nd-degree eq. 78b. Since the construction rules of chief interest for specific axial focal ellipses already have been derived (eqs. 71, 72, and 73b-right), we shall not pursue a further analysis of eq. 78b.

focal ellipses

(a) $[v^2-u^2+x_1^2+y_1^2-2x_1(av-b^2)/\sqrt{(a^2-b^2)}]^2 =$ cast about left traditional fo- (78)
$(4b^2y_1^2/a^2)[2av-b^2-(av-b^2)^2/(a^2-b^2)]$ cus, p_v, as origin, and a point in the plane, p_u, at (x_1,y_1)

(b) $v^2-u^2+x_1^2-2x_1(av-b^2)/\sqrt{(a^2-b^2)} = 0$ axial focal ellipses; eq. 78a with $y_1 = 0$

Focal and Non-Focal Ranking Using Distance Equations

The hybrid polar-rectangular distance eqs. enable us to do more than merely identify point-foci (RULES 26 and 27) in coordinate systems that employ one or more points or circles as reference elements. They also enable us to compare the ranks of point-foci (at the subfocal level) as well as non-focal poles.

Applying the distance eq. 52b to the standard rectangular ellipse of eqs. 59, and expanding and grouping terms, leads to the general axial distance eq. 79. The condition for the right-hand side to become a perfect square identifies the two bipolar foci at $x_1 = \pm\sqrt{(a^2-b^2)} = \pm ae$ and gives the linear focal distance eq. 80a for the left focus (see Problem 15; eq. 80a', for a one-arm hyperbola, is given for comparison). Letting $x_1 = 0$ in eq. 79 gives the center distance eq. 80b, while letting $x_1 = a$ for the right a-vertex gives eq. 80c. A similar derivation for the upper b-vertex leads to eq. 80d (where x is the distance from the b-vertex along the minor axis--positive values above the vertex and negative values below it). The corresponding eq. for the left a-vertex and lower b-vertex employ a plus sign for the rightmost term instead of a minus sign.

distance equations of ellipses

$v^2a^2 = (a^2-b^2)x^2 - 2b^2xx_1 + b^2(a^2-x^2)$ general major-axis distance eq. (79)

(a) $v = ex + a(1-e^2)$ distance eq. for left focus (80)

(a') $v = a(e^2-1) - ex$ eq. 80a for one-arm hyperbolas

(b) $v^2 = e^2x^2 + a^2(1-e^2)$ center distance eq.

(c) $v^2 = e^2x^2 - 2ax(1-e^2)$ right a-vertex distance eq.

(d) $v^2 = e^2x^2/(e^2-1) - 2ax/\sqrt{(1-e^2)}$ upper b-vertex distance eq.

It is clear from these results that the two traditional foci are the only real bipolar point-foci of ellipses. Among the axial points of interest that possess only subfocal rank (i.e., points for which eq. 79 or the corresponding minor-axis eq. only simplify rather than reduce; see Problem 14), the center is highest ranking (simplest 2nd-degree distance eq., 80b), the a-vertex is next-highest ranking (eq. 80c), and the b-vertex is lowest ranking (eq. 80d). These results confirm the comparative subfocal symmetry rankings of the center and a-vertex derived from comparisons of the several pertinent bipolar construction rules of the curves (see above).

Bipolar Limacons

It was noted above that limacons (eqs. 81) possess linear bipolar construction rules when cast about the DP (the abbreviation "DP" applies to the double-points of both hyperbolic and elliptical limacons, which are the two solutions of eqs. 81 for $[x,y] = [0,0]$) and the pole of self-inversion. This identifies both of these poles as bipolar point-foci (RULES 13 and 17). As the first quartics treated, limacons are the first examples of curves whose axial bipolar construction rules reduce to 1/4 of the degree of the general construction rule about two arbitrary points in the plane. A more or less general derivation of limacon bipolar construction rules is pursued to the extent necessary to demonstrate the basis for this four-fold reduction, followed by some derivations that illustrate the uses of distance eqs. as initial eqs.

General Construction Rules

Standard rectangular limacons (Figs. 7a, 12d$_1$,d$_2$, and 19) are represented by eqs. 81; the two general distance eqs. about points (x_1,y_1) and (x_2,y_2) are eqs. 82a,b. The basis for the four-fold reduction of degree for the axial eq. is seen by letting $y_1 = y_2 = 0$. Before any cancellations, this yields eqs. 83a,b and the difference eq. 84. From these eqs. it is evident that since the terms in x^2 of eqs. 83 cancel one another, the eq. (85a) obtained after segregating terms and squaring is 4th-degree in v (or u) but only 2nd-degree in x. Since the difference eq. 84 is quadratic in u and v but linear in x, the eliminant of eqs. 84 and 85--the general axial construction rule 92--is only 4th-degree in u and v. In the corresponding situation for conics (for example, the squares of eqs. 38a,b), the homologue of eq. 85a is 4th-degree in x, and includes cubic, quadratic, and linear terms.

limacon equations

$(x^2+y^2+bx)^2 = a^2(x^2+y^2)$ standard eqs. of rectangular limacons (81)

$y^2 = f^2(x) = a^2/2 - x^2 - bx \pm (a/2)\sqrt{(a^2-4bx)}$

(a) $v^2 = (x-x_1)^2 + (\pm\sqrt{[f^2(x)]} - y_1)^2$ distance eq. for pole (x_1,y_1) (82)

(b) $u^2 = (x-x_2)^2 + (\pm\sqrt{[f^2(x)]} - y_2)^2$ distance eq. for pole (x_2,y_2)

(a) $v^2 = (x-x_1)^2 + f^2(x) = x^2 - 2xx_1 + x_1^2 +$ eq. 82a for pole $(x_1,0)$ (83)

$\quad a^2/2 - x^2 - bx \pm (a/2)\sqrt{(a^2-4bx)}$

(b) $u^2 = (x-x_2)^2 + f^2(x) = x^2 - 2xx_2 + x_2^2 +$ eq. 82b for pole $(x_2,0)$

$\quad a^2/2 - x^2 - bx \pm (a/2)\sqrt{(a^2-4bx)}$

$x = (v^2 - u^2 - x_1^2 + x_2^2)/2(x_2-x_1)$ eq. 83a minus eq. 83b (84)

(a) $[v^2 + x(2x_1+b) - x_1^2 - a^2/2]^2 = (a^2/4)(a^2-4bx)$ general axial distance eq. 82a rearranged and squared (85)

(b) $x = \{-[(2x_1+b)(v^2-x_1^2-a^2/2) + a^2b/2] \pm$ general axial distance eq. 85a solved for x

$\quad a\sqrt{[v^2b(b+2x_1) - x_1^2(2bx_1+b^2-a^2)]}\}/(2x_1+b)^2$

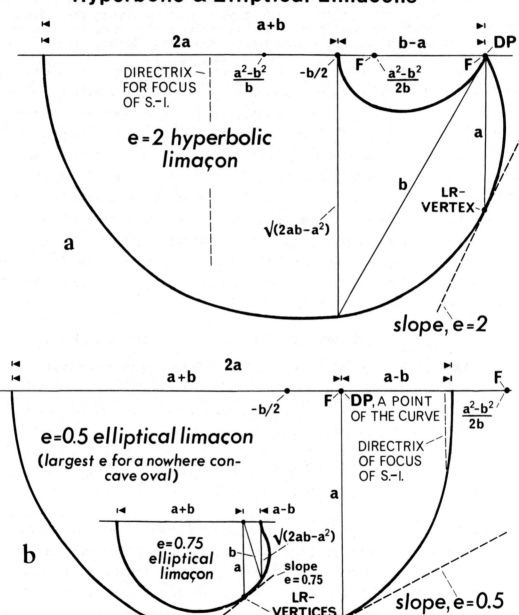

Fig. 19

[The basis for the cancellation of the x^2-terms in eqs. 83a,b is the presence of the term $(x^2+y^2+bx)^2$ in eq. 81-upper. Upon squaring this term and segregating terms, one obtains $y^4+2y^2(x^2+bx-a^2/2)+(x^2+bx)^2-a^2x^2 = 0$. Since the coefficient of the y^4-term is 1, application of the quadratic-root formula leads to a term in x^2 in the solution for y^2. This occurs whenever a term in $(x^2+y^2+bx)^2$ occurs in a quartic (whether $b=0$ or not) unaccompanied by other 3rd or 4th-degree terms in the variables. The relatively great circumpolar symmetry of curves possessing such terms already has been noted (Kavanau, 1980).]

Having illustrated the basis for the four-fold reduction of degree for the axial construction rules, it now is shown that despite the relatively greater bipolar symmetry of limacons about axial poles (the corresponding reduction for conics is only two-fold), their general bipolar construction rules that include only a single axial pole remain 16th-degree (see RULE 1c) and their general distance eqs. remain 8th-degree (similarly, the general focal eqs. remain 8th-degree; see Problem 16). To this end, eq. 82 is expanded and squared twice to eliminate radicals. The first squaring gives eq. 86a (for which $y_2 = 0$ leads to eq. 85a, the corresponding eq. in v and x_2) and the second squaring gives eq. 86b. Inspection of eq. 86b reveals that it is 8th-degree in u and 4th-degree in x, including cubic, quadratic, and linear terms in x, confirming that the general distance eq. is 8th-degree (RULE 23).

general distance eq. for limacons for pole (x_2,y_2)

$[D^2\pm(a/2)\sqrt{(a^2-4bx)}]^2 = 4y_2^2[a^2/2-x^2-bx\pm(a/2)\sqrt{(a^2-4bx)}]$ eq. 82b expanded and squared   (86a)

$D^8+2D^4[a^2bx-a^4/4-2y_2^2(a^2+2x^2+2bx)]+4a^2y_2^2(a^2-4bx)D^2 +$ eq. 86a squared   (86b)

$a^4(a^2-4bx)^2/16 - 2a^2y_2^2(a^2-4bx)(a^2/2-x^2-bx) +$

$16x^2y_2^4(x^2+2bx+b^2-a^2) = 0$

$D^2 = [u^2-x_2^2-y_2^2-a^2/2 + x(2x_2+b)]$

To obtain the bipolar construction rule about an axial point $(x_1,0)$ and a point in the plane $(x_2,y_2 \neq 0)$, one need only substitute the value of x from eq. 85b in eq. 86b. However, since eq. 85b involves a radical, one additional squaring is needed, leading to a limacon bipolar construction rule of 16th-degree for an unexceptional axial pole and an arbitrary point in the plane. Thus, even in the case where a 50% reduction in the number of measurements needed to specify the points of a curve leads to a four-fold reduction in degree of the general eq., a 25% reduction in the number of measurements is insufficient to lead to degree reduction (RULES 1c,d).

Point-Foci and Other Axial Poles

Eqs. 85a,b now can be employed to identify the bipolar point foci of limacons. These are defined by any condition for which the degree of these eqs. in v reduces from 4. These are found to be the conditions for which the radicand of eq. 85b becomes a perfect square. This occurs for the two conditions $x_1 = 0$ and $x_1 = (a^2-b^2)/2b$, leading to eqs. 87 and 88, respectively.

conventional distance eqs. for axial poles of
limacons and their domains of applicability
(axial pole)

(a) $bx + v^2 = av$ large loop, and oval of elliptical limacons   distance eqs. for DP (both eqs. apply at the DP itself)   (87)

(b) $bx + v^2 = -av$ small loop

$bv^2 \pm b^2v + a^2x = (a^4-b^4)/4b$ minus sign for all cases but elliptical DP   pole of s.-i.   (88)

$-a^2bx = (v^2-b^2/4)(v^2-b^2/4-a^2)$ all cases,   pole at $-b/2$ (89)

$[v^2+(2a-b)x-(a-b)^2-a^2/2]^2 = (a^2/4)(a^2-4bx)$ all cases,   (a-b)-vertex   (90)

$[v^2-(2a+b)x-(a+b)^2-a^2/2]^2 = (a^2/4)(a^2-4bx)$ all cases,   -(a+b)-vertex   (91)

It is evident from eqs. 85a,b that the condition $x_1 = -b/2$ leads to great simplifications, yielding eq. 89. But since the latter is 4th-degree, the pole at $x_1 = -b/2$ is not a bipolar focus. From comparisons with eqs. 85a,b, though, it is clear that it is a pole of very high subfocal rank. In fact, this pole is a very high-ranking circumpolar focus, about which limacons have structure rules of compelling interest (see Chapters VII and VIII). In subsequent treatments, this "pole at $-b/2$" is referred to repeatedly when dealing with its homologues in other curves.

Other axial poles of interest are the vertices of the small (a-b) and large (-a-b) loops of hyperbolic limacons (or the right and left vertices of elliptical limacons, respectively). However, eq. 85a does not simplify notably for either vertex (eqs. 90 and 91). Since all vertices are circumpolar point-foci, (Chapters VII and VIII), these circumstances lend additional emphasis to RULE 22.

[The failure of the pole at $-b/2$ to possess bipolar focal rank correlates with the fact that it is a *variable circumpolar focus*. In some hyperbolic limacons ($e = b/a > \sqrt{2}$) it is within the small loop, in others ($e < \sqrt{2}$) it is outside the small loop but within the large loop; in the equilateral limacon ($e = \sqrt{2}$) it is at the small-loop vertex, while it is within the oval of elliptical limacons ($e < 1$) and the cardioid ($e = 1$). Since ordinary vertices have only subfocal symmetry rank (RULE 28), this pole would not be expected to be a bipolar focus.]

Axial Construction Rules

The general axial construction rule 92, about any two poles, $(x_1,0)$ and $(x_2,0)$, can be derived as the eliminant (i.e., by eliminating x) between eqs. 84 and 85a (or 85b). Although the comparative weighting of poles in eq. 92 is not indeterminate, as for the parabola, it involves complicated calculations. In the following, we do not treat the topic of the comparative weighting of axial poles based on the coefficients of terms for curves that lack an orthogonal line of symmetry, that is, for curves that lack a true center.
[It can be shown by expansion (Problem 21) that all terms of eq. 92 in powers of u and v either have identical coefficients or have coefficients in which the abscissae of the two poles are exchanged symmetrically.]

general axial bipolar construction rule of limacons

$$[2(x_2v^2-x_1u^2)+2x_1x_2(x_2-x_1)-a^2(x_2-x_1)+b(v^2-u^2-x_1^2-x_2^2)]^2 =$$
$$a^4(x_2-x_1)^2-2a^2b(x_2-x_1)(v^2-u^2-x_1^2-x_2^2) \qquad \begin{matrix} p_v \text{ at } (x_1,0), \\ p_u \text{ at } (x_2,0) \end{matrix} \quad (92)$$

hybrid distance eqs. of limacons and their domains
of applicability   (axial pole)

(a) $bx + v^2 = av$ large loop and oval   distance eqs. for
                          of elliptical limacons   DP (both eqs. apply   (87')
(b) $bx + v^2 = -av$ small loop   at the DP itself)

$bv^2 \pm b^2v + a^2x + (a^2-b^2)^2/4b = 0$ minus sign for all cases   pole of   (88')
                                              but elliptical DP   s.-i.

$-a^2bx = (v^2-b^2/4)^2 - a^2(v^2+b^2/4)$ all cases,   pole at $-b/2$ (89')

$[v^2-x(b-2a)-a(b-a/2)]^2 = (a^2/4)[(2b-a)^2-4bx]$, all cases, $(a-b)$-vertex (90')

$[v^2-x(b+2a)+a(b+a/2)]^2 = (a^2/4)[(2b-a)^2-4bx]$, all cases, $-(a+b)$-vertex (91')

(a) $x_{s-i} = x_{DP}+d = x_{DP}+(b^2-a^2)/2b$, (b) $x_{-b/2} = x_{DP}+d = x_{DP}+b/2$, relations be- (93)
(c) $x_{-b/2} = x_{s-i}+d = x_{s-i}+a^2/2b$, (d) $x_{slv} = x_{DP}+d = x_{DP}+(b-a)$, tween x's of
                                                                                    eqs. 87'-90'

s-i = self-inversion,   slv = small-loop vertex

All axial construction rules can be derived from eq. 92 by making the appropriate substitutions. Alternatively, the bipolar construction rule for pairwise combinations of the specific poles of eqs. 87-90 can be derived by a simultaneous solution of the corresponding two conventional distance eqs., since x, the variable to be eliminated, is measured from the DP in all eqs. (see

Problem 19).

Although the conventional distance eqs. are most conveniently used for deriving bipolar construction rules, the hybrid distance eqs. are of greatest interest for the comparative ranking of bipolar poles and point-foci. Accordingly, these are given for the corresponding poles of eqs. 87-91 as eqs. 87'-91'. To obtain the bipolar construction rules for pairwise combinations of the poles of these eqs., the substitutions of eqs. 93 must be made (see also Problems 25, 26, 28, and 29) before simultaneous solution of the corresponding eqs. can be undertaken (of course v must be replaced by u in one of the eqs.).

For example, to obtain the construction rule about the DP and the focus of self-inversion, one first substitutes $x+d$ of eq. 93a in eq. 88' and replaces v of eq. 88' by u. One then eliminates x between eqs. 87' and 88', yielding eqs. 94. For the focus of self-inversion and the pole at $-b/2$, one first substitutes $x+d$ from eq. 93c into eq. 89', replaces v of eq. 89' by u, and then eliminates x between eqs. 88' and 89', yielding eq. 95, etc. Eqs. 95 and 96 are the first parabolic bipolar construction rules we have encountered, that is, they are eqs. in which one variable is squared and the other occurs only in linear form, calling attention to RULES 34 and 35.

bipolar construction rules of limacons about two
axial poles (parameters a & b)   (axial poles)

(a)	$bu \pm av = bd$	hyperbolic limacons,	minus sign for large loop,	DP (p_v) and pole of self-inversion	(94)
(b)	$av - bu = \pm bd$	elliptical limacons,	minus sign for the DP itself,		
	$u^2 + b^2/4 = b(d \pm v)$		minus sign for elliptical DP; plus sign for all other cases	pole of s.-i. (p_v) and pole at $-b/2$ (p_u)	(95)
	$u^2 - b^2/4 = \pm av$		plus sign for large loop, and oval and DP of ellipticals, minus sign for small loop	DP (p_v) and pole at $-b/2$ (p_u)	(96)

RULE 34. *If a bipolar construction rule is linear in v but not in u, pole p_v is a point-focus and outranks pole p_u*

RULE 35. *The presence of a linear variable in a bipolar construction rule indicates point-focal rank of the corresponding pole.*

The Eccentricity of a Limacon

Inasmuch as limacons are inversions of conic sections about their traditional foci, they can be named and identified in terms of the eccentricity of the parent conic, for example, the $e = \sqrt{2}$ limacon, or equilateral limacon, is the inverse of the equilateral hyperbola; the $e = 1$ limacon, or cardioid, is the inverse of the parabola, etc.

[This convention is more than just an arbitrary convenience; it actually characterizes properties of the limacon itself. For example, since the inversion transformation is conformal (preserves the magnitude of angles) the angle between the asymptotes of hyperbolas (which invert to circles tangent to the DP-tangents at the DP) becomes the angle between the DP-tangents. In the cases of both ellipses and hyperbolas, the magnitude of the slope of the angle between the tangent at an LR-vertex and the major or transverse axis is equal to the eccentricity (RULE 5). This angle also is preserved in the limacon as the slope of the tangent at a limacon LR-vertex (the points of intersection with the curve of a normal to the line of symmetry at the DP; see Fig. 19 and Problems 42 and 43).]

Limacons also can be named after the parent conic. For example, *hyperbolic* limacons comprise all representatives with two loops ($e > 1$), and are the inverses of hyperbolas. *Elliptical* limacons comprise all non-cusped ovular representatives ($e < 1$), and are the inverses of ellipses, etc. Moreover, in the standard polar construction rule of limacons ($r = a - b\cos\theta$) the eccentricity, e, is equal to b/a. Within this framework, the limacon axial construction rules (94-96) now are recast in terms of d, the characteristic distance (the distance between the poles) and e, the absolute magnitude of the slope of a tangent at one of the LR-vertices at the level of the DP (see Problems 42 and 43).

Construction rules 94a'-c' (about the DP and focus of self-inversion) resemble greatly the linear bipolar construction rules of central conics (eq. 94d'). In fact, the eqs. are homologous, since the focus of self-inversion is the inverse of the contralateral traditional focus, and the DP is the reciprocal inversion pole (i.e., it occupies the position of the pole of inversion--the traditional focus--in the plane of the limacon). The only differences between the two construction rules are that the coefficients of v and d are exchanged, and changes of signs of coefficients occur in some cases (since limacons lack a true center, the coefficients of corresponding terms in the variables cannot be identical, as, for example, in eq. 94d' for ellipses).

In the cases of the construction rules about the DP and the pole at -b/2 (eqs. 96'), the eq. for elliptical limacons lacks being identical with the homologous eq. for ellipses (cast about a traditional focus and its reflection in

construction rules of limacons and central conics
about two axial poles

(axial poles)

(a')	$eu + v = ed$	small loop	(b') $eu - v = ed$	large loop	limacon DP (p_v) & focus of s.-i. (94)
(c')	$v - eu = ed$	elliptical ovals			
(d')		$ev \pm eu = d$			homologous eq. for central conics (plus sign for ellipses)
(a')	$u^2/e^2 = 2dv + d^2(2-e^2)$				limacon focus of s.-i. (p_v) & pole at $-b/2$ (p_u) (95)
(b')	$u^2 = v^2 - 2dv + d^2(2-e^2)$				homologous eq. for central conics for rt. focus (p_v) & reflection of left focus in left directrix
(a')	$u^2 - d^2 = \pm 2dv/e$	plus sign for all cases except small loop			limacon DP (p_v) and pole at $-b/2$ (p_u) (96)
(b')	$u^2 - v^2 = 2dv/e$				homologous eq. for central conics for lt. focus (p_v) & its reflection in the directrix
(a)	$d^2 - u^2 = \pm 2dv/e + (1-2/e^2)v^2$				limacon DP (p_v) and pole at $(a^2-b^2)/b$ (homologue of central conic center); minus sign for large loop & ellipticals, both signs apply to elliptical DP (97)
(b)	$v^2 - u^2 = \pm 2dv/e \pm (1-2/e^2)d^2$				homologous eq. of ellipses (plus signs) & hyperbolas, for the focus (p_v) and center (p_u)
(a)	$[(e+2)v^2 - (e-2)u^2]^2 - 2d^2[(3+2e)v^2+(3-2e)u^2] = d^4(e^2-2)$				both limacon vertices, all cases (p_v on small loop & right side of oval) (98)
(a')	$(v^2+u^2-d^2)(v^2+u^2-d^2/2) = 0$				eq. 98a for $e = 0$, the circle of Pythagoras (left factor) & DP (right factor)
(b)	$e^2(v^2-u^2)^2 - 2d^2(v^2+u^2) = d^4(e^2-2)$				homologous eq. for central conics

the near directrix; see Problem 23) by the replacement of the term in v^2 in the ellipse eq. by the term in d^2 in the limacon eq.

The homologue of the center of central conics, in the plane of limacons, is the pole at $(a^2-b^2)/b$. This lies at twice the distance of the focus of self-inversion from the DP (since the center of the initial ellipse lies at half the distance to the contralateral focus, which inverts to the focus of self-inversion). The construction rule about the pole at $(a^2-b^2)/b$ and the DP is eq. 97a. The corresponding eq. for central conics is 97b. The two homologous eqs. differ only in a change of sign and interchanging of the terms in v^2 and d^2, just as

for eqs. 96a' and 96c'. However, these close correspondences are found only when the limacon DP (the reciprocal inversion pole) participates in the homologous eqs. (note the greater differences between the homologous eqs. 95a' versus 95b' and 98a versus 98b, in which the DP does not participate).
[This is not unexpected, since it is the location of the pole of inversion (in this case the DP) that determines the identity and properties of the inverse curve.]

Comparative Bipolar Symmetry of Conics and Limacons of Different Eccentricities

The comparative bipolar symmetry of the various limacons and conics of eqs. 94'-96', 97, and 98 can be evaluated by examining the conditions for which their construction rules achieve simpler and simplest forms. The question of the comparative symmetry of bipolar curves with linear eqs. (Fig. 12 and *The Most-Symmetrical Bipolar Loci*) already has been treated. In the cases of eqs. 95a' and 95b', it is evident that the $e=2$ limacon (Fig. 19a) and $e=2$ hyperbola are most symmetrical, since the constant term vanishes from their construction rules. The same can be said for the limacons and ellipses of eqs. 96a' and 96b', for which the coefficients of the linear terms in v become unity for $e=2$.

For the limacons and ellipses of eqs. 97, the situation is somewhat different. Now, the equilateral limacon is most symmetrical, because a term vanishes when $e=\sqrt{2}$. The $e=2$ limacon is next-most symmetrical because, for it, the coefficient of the linear term in v becomes unity, that is, $2dv/e$ becomes dv (as in construction rules dealt with earlier, the characteristic distance possesses the same status as a variable in assessing the simplicity of coefficients).

The curves represented by eqs. 98 present very interesting cases. Again, it is evident that the equilateral limacon and equilateral hyperbola ($e=\sqrt{2}$) are highly-symmetrical curves, since the constant term vanishes for them. On the other hand, for the "equilateral" ellipse, or circle ($e=0$), eq. 98b reduces to 2nd-degree, and represents the circle of Pythagoras. The same condition on eq. 98a, for the "equilateral elliptical limacon," or circle, leads to the much simpler degenerate 4th-degree eq. 98a', which represents a circle of Pythagoras with a bipolar point-circle at its center--the homologue of the limacon DP. On the basis of the symmetry and simplicity of its construction rule, the circle of Pythagoras has greater bipolar symmetry than any curve with the construction rule $u^2-v^2 = Adv + Bd^2$ (see below).

In summary, of limacons cast about the pole at -b/2, in combination with either the DP or the focus of self-inversion, the e = 2 limacon has the greatest bipolar symmetry. Of the homologous central conics cast about a traditional focus and a reflection of either focus in the near directrix, the e = 2 hyperbola has the greatest symmetry. On the other hand, of curves cast about the DP and the homologue of the central conic center, it is the equilateral limacon that has the greatest bipolar symmetry, and similarly for the equilateral hyperbola cast about the homologous poles.

For poles at the axial vertices (or, more precisely, at the axial points of intersection with the curve other than the DP), the equilateral limacon and the equilateral hyperbola have the greatest bipolar symmetry among 4th-degree loci. But both construction rules reduce to 2nd-degree for e = 0, representing the circle of Pythagoras, which is the most symmetrical of central conic and limacon loci cast about the axial "vertices."

The Bipolar Equilateral Lemniscate and Dependent Point-Foci

Affinities

The $e = \sqrt{2}$, or equilateral, lemniscate (Fig. 20-1b, 20-2) now is considered briefly for the additional light it sheds on bipolar curves and properties of the bipolar system (it is our first example of *dependent point-foci*), and on relationships between rectangular conics and quartics. This curve has long been known as the "lemniscate of Bernoulli," as a member of the Cassinian ovals (since the end of the 18th century; Fig. 20-1,2), and as the curve obtained by inverting an equilateral hyperbola about its center.

Its membership in the Cassinian ovals emphasizes the fact that, excluding curves inverse to conics, Cassinian ovals are the closest quartic relatives of central conics (see Kavanau, 1982). [These curves were studied by G. D. Cassini (1625-1712) in 1680 in work concerning the relative motions of the earth and sun.] In fact, we now know that the equilateral hyperbola itself is a member (the only open member) of the Cassinian ovals (Kavanau, 1982). This can be understood in terms of the fact that Cassinian monovals *congruent-invert* (invert to congruent loci) about their centers (Fig. 20-1a,b).

[The conventional view of hyperbolas is quoted from Zwikker (1950). "The projective geometer assigns to the hyperbola two normal points at infinity, the inversion geometer one double point; both consider the hyperbola as consisting of one single track and both have a simpler idea of the hyperbola than the lay-

The Equilateral Lemniscate & The Equilateral Hyperbola As Curves of Demarcation of Monovular & Biovular Cassinians & Their Central Inversions

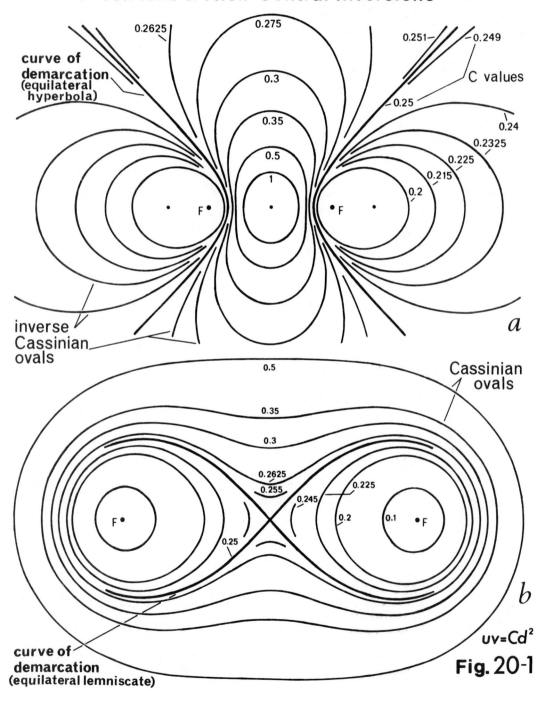

Fig. 20-1

Bipolar Equilateral Lemniscates And Cassinian Ovals

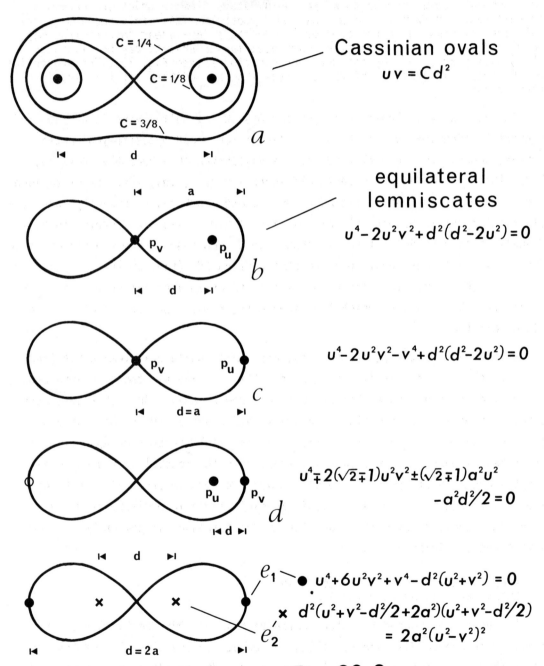

Fig. 20-2

man, who assigns four infinitely remote points to it and who considers it as consisting of two separate branches."

In point of fact, the viewpoints of projective geometers, inversive geometers, and laymen are all narrowly conceived--solely within the framework of rectangular and polar hyperbolas. The properties of hyperbolas (using the term in a general sense) depend upon the coordinate system in which the curves are represented. For example, linear bipolar "hyperbolas" have only one arm in the polarized system, while the arms of 1st-degree polar-linear "hyperbolas" may be of disparate sizes, partial or complete, may intersect, may be closed in the finite plane, and may possess four asymptotes (see Fig. 30). The properties of the points at infinity of these curves must be modified accordingly, in all three viewpoints.]

It has long been known that biovular Cassinians self-invert at 0° and 180° (invert to identical curves along coincident-radii and oppositely-directed chord-segments, respectively) but it was discovered only recently (Kavanau, 1980) that the monovular members self-invert at 90° (i.e., they invert to identical curves along radii directed at 90° to one another). Accordingly, when the classical ensemble of monovals and biovals of Fig. 20-1b are inverted conventionally (along coincident-radii) about the common center, the biovals maintain their positions, but the monovals become rotated 90° (Fig. 20-1a). At the same time the equilateral lemniscate "opens out" to an equilateral hyperbola that demarcates the monovals (which lie between the arms) from the biovals (which lie within the arms).

Such curves as the equilateral hyperbola and equilateral lemniscate (and, in fact, this is true of all hyperbolas, ellipses, and all their inverse curves in relation to other ensembles of quartics; see Kavanau, 1982) that demarcate between ensembles of monovals and biovals, different ensembles of monovals, etc., are called *curves of demarcation*. From a still broader perspective they are known as *curves of intersection*. Thus, both the ensemble of hyperbolas and the ensemble of *hyperbolic central quartics* (the inverses of hyperbolas about their centers) *intersect* Cassinian ovals in their equilateral members (i.e., the equilateral members are the curves possessed in common by each ensemble and Cassinian ovals).

Distance Equations, Bipolar Construction Rules, and Dependent Point-Foci

Eq. 6e is reproduced here for convenience as eq. 99. The first step in forming the hybrid distance eq. 100a, for points on the line of symmetry, reveals that the terms in x^2 cancel, just as they do for limacons. Accordingly, the general axial bipolar construction rule (102a), like that of limacons, is

4th-degree. From the specific examples of axial eqs. in Fig. 20-2b-e_2, it is evident that the comparative subfocal ranks of the center and vertices agree with those for ellipses (see also Problem 14).

Eq. 102a provides a good test of RULE 19, that the pole nearest a line of symmetry is weighted heaviest in bipolar construction rules, since both the quartic and the quadratic terms in u and v have coefficients that involve the abscissae of the poles. Moreover, since coefficients consisting solely of the abscissae are simply exchanged for the v^4 and u^4-terms (right side after squaring), these terms give an unequivocal comparative weighting, in which the pole closest to the center is weighted heaviest. For internal consistency, the weighting based on the quadratic terms must lead to the same result. The two quadratic terms are 102b,c. Factoring $4(x_2-x_1)$ from each term leaves the terms $x_2(2x_1^2-a^2)u^2$ and $-x_1(2x_2^2-a^2)v^2$. It is readily confirmed that for all poles exterior to the curve, that is, for both $|x_1|, |x_2| > a$, these terms lead to the same comparative weighting as the quartic terms; the exterior pole closest to the orthogonal line of symmetry is highest ranking.

equilateral lemniscate distance eqs. and bipolar construction rules

$(x^2+y^2)^2 = a^2(x^2-y^2)$,	standard rectangular equilateral lemniscate	(99)
$y^2 = -x^2-a^2/2 \pm (a/2)\sqrt{(a^2+8x^2)}$		
(a) $v^2 = x^2 + f^2(x+x_1) =$ $-2xx_1-x_1^2-a^2/2 \pm (a/2)\sqrt{[a^2+8(x+x_1)^2]}$	first step in forming the hybrid axial distance eq. from point $(x_1,0)$	(100)
(b) $(v^2+2xx_1+x_1^2+a^2/2)^2 = (a^2/4)[a^2+8(x+x_1)^2]$	hybrid axial distance eq. from point $(x_1,0)$	
(a) $\quad (v^2+a^2)^2 + 2\sqrt{2}axv^2 = 5a^4/4$	hybrid distance eq. from point $(\sqrt{2}a/2,0)$	(101)
(b) $\quad (u^2+a^2)^2 - 2\sqrt{2}axu^2 = 5a^4/4$	hybrid distance eq. from point $(-\sqrt{2}a/2,0)$	
(a) $4[x_2v^2-x_1u^2+(x_2-x_1)(x_1x_2+a^2/2)]^2 =$ $a^4(x_2-x_1)^2 + 2a^2(v^2-u^2-x_1^2+x_2^2)^2$	general axial bipolar construction rule	(102)
(b) $4x_2(x_1-x_2)(2x_1^2-a^2)u^2$ (c) $4x_1(x_2-x_1)(2x_2^2-a^2)v^2$	quadratic terms of eq. 102a	
(d) $\quad uv = a^2/2 = d^2/4$	45°-equilateral hyperbolic bifocal bipolar construction rule	

The hybrid axial distance eq. 100b also is of 4th-degree but will not reduce (see Problem 26). However, for the condition $2x_1^2 = a^2$, of the expanded eq. 100b, the terms in x^2 vanish and the 4th-degree eq. in v becomes linear in x (eqs. 101a,b). This is the condition for the traditional loop foci of the equilateral lemniscate, that is, the poles at $x = \pm\sqrt{2}a/2$ are point-foci.

These point-foci within the loops of equilateral lemniscates are of a different type from those encountered to this juncture, since they become focal (i.e., construction rules cast about them reduce) only when they are paired with each other. The other bipolar point-foci with which we have dealt are focal when paired with any other point in the plane. For this reason, the former poles are referred to as *dependent* point-foci and the latter poles as *independent* point-foci.

The above observation calls to mind the case of the limacon pole at $-b/2$. Its distance eq. also is quartic in v and linear in x (eq. 89), but it is not a bipolar point-focus, neither dependent nor independent. The circumstances differ in the fact that the present focus can be paired with an equivalent pole, whereas the limacon pole at $-b/2$ cannot be so paired. These poles, the $-b/2$ pole of limacons and the $\pm\sqrt{2}a/2$ foci of the equilateral lemniscate, are homologues, with properties in common (for example, both are $cos^2\theta$-*condition* circumpolar foci; see Chapter VIII). In fact, they are also the reciprocal inversion poles of one another in inversions between the equilateral lemniscate and the equilateral limacon (but this does not apply for limacons and lemniscates of other eccentricities).

Knowing already that the $-b/2$ pole is not a bipolar point-focus of limacons, that the poles at $\pm\sqrt{2}a/2$ of the equilateral lemniscate are focal only by virtue of being paired with one another, and that the axial bipolar construction rules of both curves are 4th-degree (eqs. 92 and 102a; i.e., 1/4 the degree of the general bipolar construction rules, rather than 1/2, as in conics), it is evident that no bipolar eq. that includes only one of the loop point-foci can be of lower degree than the corresponding general construction rule. The general distance eq. for a point in the plane (x_2, y_2), for example, is eq. 103, which is 8th-degree in u and 4th-degree in x; substituting for x from the distance eq. 101a for the right focus (after converting to a conventional distance eq.) leads to a 16th-degree eq., the same value as for the general eq. about two unexceptional points in the plane.

general conventional distance eq. of the equilateral
lemniscate about a point in the plane (x_2,y_2)

$$[(u^2-x_2^2-y_2^2+2xx_2+a^2/2)^2+(a^2/4)(a^2+8x^2)+4y_2^2(x^2+a^2/2)]^2 = \quad (103)$$

$$a^2(2y_2^2-u^2+x_2^2+y_2^2-2xx_2-a^2/2)^2(a^2+8x^2)$$

The bifocal construction rule of the equilateral lemniscate is readily derived from the hybrid distance eqs. for the two dependent foci (eqs. 101a,b). After translating, so that the x-coordinate of both eqs. is measured from a common point (see eqs. 93), x is eliminated and the resulting eq. is reduced by factoring and taking roots (Problem 27). This yields the quadratic "$45°$-*equilateral hyperbolic*" construction rule 102d (so-called because the same eq. represents a rectangular 45°-equilateral hyperbola). The most-symmetrical bipolar curve with a construction rule of this form, $uv = d^2$, is the central oval of Fig. 20-1a.

In summary, the only bipolar construction rules of the equilateral lemniscate that reduce from 16th-degree are the axial eqs. All of these are 4th-degree (Fig. 20-2b-e_2) except the one cast about the two dependent loop-foci, which is 2nd-degree. The equilateral lemniscate is typical of the other curves studied, in that axial poles possess focal rank only when paired with one another. It is atypical in that its axial point-foci also must be paired with one another to possess point-focal rank. For all bipolar point-foci that participate in 1st-degree construction rules (and the traditional focus of the parabola), their use, alone, as one pole of a construction rule ensures degree reduction, as compared to the eq. for two unexceptional poles or an axial pole and an unexceptional pole.

Bipolar Linear Cartesians

Introduction

We now turn to several groups of curves of unparalleled fascination. Just as we reach another level of complexity in passing from conics to quartics, so also do we reach another level of fascination. Corresponding to each interesting property of a conic--of which scores to hundreds were known even centuries ago--there are several or many interesting properties of Cartesian ovals (Figs. 12e,f_1,f_2; 21; 22e,e'; 23a_2) and *central* Cartesians (curves with two orthogonal lines of symmetry obtained by inverting Cartesian ovals; Figs. 20-1; 20-2; 22a,c,f; 23a_1,a_3). These properties are barely touched upon here (but see Chapters VII and VIII and Kavanau, 1980, 1982), where our primary purpose is to illustrate the properties of bipolar loci and the bipolar system.

Classical general analytical treatments of Cartesians by any method are unknown. Crofton (1866) summed the situation up as follows:

"The Cartesian ovals seem at all times to have attracted considerable attention, though no great success has attended the attempts made to investigate their properties. Of these, the most remarkable which has been discovered, is no doubt that of the third focus, given by Chasles without proof, which evidently teaches us, that as there is another focus possessing the same properties and the same importance as the two comprised in the definition, we should extend our attention to all three if we wish for a general and complete view of the curve.

The unsuitable nature of the coordinates, polar or rectangular, which have been employed, is doubtless the reason of the limited progress which has been made in the investigation of these curves. Vectorial coordinates (i.e., bipolar coordinates).....seem to be naturally indicated by their very definition. The method has difficulties of its own, and indeed has yet to be created; but every step made in the study of these coordinates will necessarily be a step in the theory of the Cartesian ovals....."

Though the algebraic difficulties encountered by the early workers were formidable, these curves present less than average difficulty when analyzed by the methods of Structural Equation Geometry. In fact, it is an axiom of Structural Equation Geometry that no matter how formidable a calculation of a specific structural property of a curve may appear to be at first sight, if the curve is one of high symmetry (i.e., of great structural simplicity), the final results will be elegantly simple. Cartesian ovals and central Cartesians are just such curves of high symmetry.

Some of the methods referred to above are used here to analyze *linear* Cartesians. The approach leans heavily on conventional distance eqs. and many of the derivations are treated in the Problems. In these analyses of Cartesian ovals, it is not the analytical difficulties that impress one, but the most remarkable and intriguing properties possessed by the curves. All the more interest attaches to Cartesian ovals by virtue of their relatedness to central Cartesians (by inversion), and their close equational relationships to conic sections (as pointed out below, in their broadest aspects the construction rules of Cartesian ovals include those of conic sections as special cases). In fact, Cartesian ovals are related to central Cartesians in the same way that limacons are related to central conics, with precise analytical homologies and very close parallels.

Following the convention of referring to construction rules in other coordinate systems by type, after the names of the corresponding rectangular curves, Cartesian ovals represented by linear eqs. (which always are cast about two foci) are referred to as *linear* Cartesians. Cartesian ovals represented by 2nd-degree construction rules (cast about one focus and a second non-focal axial pole) are referred to, in general, as *quadratic* Cartesians, etc.

Since all linear construction rules of Cartesian ovals have exactly the same form, $\pm Bu \pm Av = Cd$, and all the parameters must be unequal (see below) and non-zero, these eqs. do not provide a basis for equational comparisons of the bipolar symmetry (structural simplicity) of the curves. This objective can be achieved, however, using the construction rules of *quadratic* Cartesians. The simplest of these is the eq. $z^2 = Cdu$, whose form corresponds to the eq. of a standard rectangular vertex parabola. In consequence, the members of the sub-group of the Cartesian ovals that are represented by this eq.--which are the most symmetrical in the group--are referred to as *parabolic* Cartesians. So as not to be misled by the employment of this designation, the reader must keep in mind the fact that all Cartesian ovals have parabolic construction rules (as well as linear eqs. [for biovals], hyperbolic eqs. [but not equilateral hyperbolic eqs.], elliptical eqs., etc.; see Problem 41).

Construction Rules of Cartesian Ovals as
 "Master Construction Rules" for Conic
 Sections

The treatment of Cartesian ovals using the general linear construction rule (and the three parameters A, B, and C) is particularly intriguing because, algebraically, it encompasses all the bipolar curves treated until this juncture except Cassinian ovals. Thus, taking the general linear construction rule 104, if equalities between, and vanishing of, the dimensionless coefficients, A, B, and C, are allowed, one also obtains a line (C = 0, A = B), a line-segment (A = B = C, paired opposite signs), "monopolar" circles (A or B = 0) and circles of Apollonius (C = 0, paired opposite signs), central conics (A = B), and all limacons except the cardioid (A = C or B = C).

These same relationships apply for all construction rules of Cartesian ovals and all their transforms (for example, inversions), when the eqs. employ the parameters A, B, and C. By substituting the desired relation between the parameters, one obtains the corresponding eqs. of the conic or limacon (or of the corresponding transform curves). Thus, these construction rules and transform

LINEAR CARTESIANS

eqs. of Cartesian ovals also are "master construction rules" for conic sections and all of their transforms.

bipolar linear Cartesians

$\pm Bu \pm Av = Cd$ general eq., $A \neq B \neq C$, real, dimensionless, > 0 (104)

(a) $[(x^2+y^2)(A^2-B^2) - 2A^2dx + d^2(A^2-C^2)]^2 = 4B^2C^2d^2(x^2+y^2)$ rectangular eq., p_u at $(0,0)$, $d = d_{uv}$ (105)

(b) $y^2 = 2B^2C^2d^2/(A^2-B^2)^2 - x^2 + 2A^2dx/(A^2-B^2)$ eq. 105a solved for y^2

$\quad - d^2(A^2-C^2)/(A^2-B^2) \pm [2ABCd/(A^2-B^2)^2] \cdot$

$\quad \sqrt{[2d(A^2-B^2)x + (C^2+B^2-A^2)d^2]}$

[As we shall see, for studies in Structural Equation Geometry, and for didactic purposes in general, it is highly desirable to retain individual coefficients for each variable, and the characteristic distance, in eq. 104 (in the case of general treatments this is essential to allow complete flexibility for the vanishing of terms). The customary practice of dividing through by one coefficient merely obscures analytical relationships, particularly when one deals with highly complex eqs. (such as those dealt with here).

We see how simple it is, in the above treatment, to express the coefficient relationships for the different groups of curves when all three coefficients are retained (note that all three coefficients are taken to be positive, and that two alternate signs allow for all combinations of signs.]

If pole p_u is taken to be the origin of the rectangular coordinate system, and pole p_v is taken at $(d,0)$ (See Problem 14), with the line of symmetry coincident with the x-axis, the rectangular construction rule of linear Cartesians is eq. 105a. It is apparent from a superficial examination of this eq. that if $A = B$, it reduces to 2nd-degree and represents central conics (but not circles), whereas if $C = 0$, it represents circles, and if $A = C$, it becomes the eq. of limacons (eq. 81-upper). Solving for y^2 yields eq. 105b. Since, like the corresponding construction rules for limacons and the equilateral lemniscate (eqs. 81-lower and 99-lower), eq. 105b possesses a term in x^2, the general axial bipolar eqs. of Cartesian ovals are 4th-degree.

Distance Equations

Instead of pursuing a general analysis of linear Cartesians, their properties are explored using distance eqs. The conventional distance eq. 106a is formed for pole p_u at $(x_1,0)$ and y^2 is eliminated using eq. 105b. Performing the necessary operations--combining terms, segregating and cancelling terms, squaring, etc.--leads ultimately to eq. 106b, expressed in powers of x. Several alternative approaches are available in the further analysis of eq. 106b. The aim, of course, is to find the conditions for its reduction, since these define the curves' bipolar foci (RULE 26), which lie on the line of symmetry (RULE 10).

locating the bipolar foci of linear Cartesians

(a) $\quad u^2 = (x-x_1)^2 + y^2 \quad$ conventional distance eq. in u (106)

(b) $(u^2-x_1^2)^2(A^2-B^2)^4 + 2(u^2-x_1^2)(A^2-B^2)^2 \cdot$
$[d^2(A^2-B^2)(A^2-C^2) - 2B^2C^2d^2] + d^4(A^2-B^2)^2(A^2-C^2)^2$
$= x^2\{4(A^2-B^2)^2[x_1(A^2-B^2) - A^2d]^2\} +$
$4(A^2-B^2)^2x\{(u^2-x^2)(A^2-B^2)[x_1(A^2-B^2) - A^2d] +$
$x_1[d^2(A^2-B^2)(A^2-C^2) - 2B^2C^2d^2] - A^2d^3(A^2-C^2)\}$

eliminant of eqs. 105b and 106a, squared and rearranged in powers of x

One approach is to complete the squares of both sides of eq. 106b and find the conditions for which the remainders vanish (and the eqs. reduce). These conditions either will specify the locations of the foci of Cartesian ovals, in general, or specify particular members of the Cartesian ovals whose eqs. reduce by this method. Combining the remainders and equating to 0 yields eq. 107a. Since this remainder eq. contains the variable u, it cannot vanish unless the coefficient of the u-term vanishes. One condition for the vanishing of this coefficient is $A^2 = B^2$, the condition for central conics. In other words, the distance eqs. of central conics reduce by this method (for example, eqs. 79 and 80a). However, if $A^2 = B^2$, the entire left term vanishes, whereupon C^2 would also have to vanish. The combination of these conditions specifies the bipolar midline, u = v.

A second condition (eq. 108b) is $D^3 = 0$ or $x_1 = A^2d/(A^2-B^2)$. This specifies the location of the pole p_z, the homologue of the limacon -b/2 pole. However, completing the squares for this condition is not allowed because then the x^2-term on the right of eq. 106b vanishes. Accordingly, the first approach to reduction does not apply for Cartesian ovals.

Next, the square is completed only on the right side of eq. 106b and the term needed to balance the completed square (the negative of the square of half

POINT-FOCI OF LINEAR CARTESIANS   175

locating the bipolar foci of linear Cartesians

$(A^2-B^2)^2\{(u^2-x_1^2)(A^2-B^2)D^3+d^2x_1[(A^2-B^2)(A^2-C^2)-2B^2C^2]$ sum of remainders (107)
$-A^2d^3(A^2-C^2)\}^2-4B^2C^2d^4D^6[(A^2-B^2)(A^2-C^2)-B^2C^2] = 0$ of eq. 106b after completing the squares on both
$D^3 = [A^2d-x_1(A^2-B^2)]$ sides

(a) $A^2 = B^2$ and $C^2 = 0$, or   conditions of eq. (108)
                                              107 for eq. 106b
(b) $D^3 = 0$; $x_1 = A^2d/(A^2-B^2)$ to become perfect squares on both sides

$4A^2B^2C^2E^3d^3(u^2-x_1^2)+(A^2-C^2)^2d^4E^6-A^4d^6(A^2-C^2)^2$ condition for the (109)
                                                                   right-hand side of
$-x_1^2d^4[(A^2-B^2)(A^2-C^2)-2B^2C^2]^2 + 2A^2d^5x_1(A^2-C^2)\cdot$ eq. 106b becoming
                                                                      a perfect square
$[(A^2-B^2)(A^2-C^2)-2B^2C^2] = 0$

$E^3 = [x_1(A^2-B^2)-A^2d] = -D^3$

$4x_1B^2C^2d^3\{A^2(A^2-B^2)x_1^2-dx_1[2A^4-A^2(B^2+C^2)] +$ condition for the (110)
                                                                vanishing of all
$A^2d^2(A^2-C^2)\} = 0$ but the leftmost
                                                                term of eq. 109

(a) $x_1 = 0$, for p_u,   (b) $x_1 = d$, for p_v roots of eq. 110; (111)
                                                                the three bipolar
(c) $x_1 = (A^2-C^2)d/(A^2-B^2)$, for p_w foci

the bracketed coefficient of x) is equated to the left-hand side of the eq..
When this is done the 4th-degree terms in u cancel, leaving (after several additional operations) eq. 109. For this to become a term in u^2 alone (together with its coefficient), which would make possible the taking of roots on both sides of the modified eq. 106b, all but the leftmost term of eq. 109 must vanish. The application of this condition, together with a number of additional algebraic manipulations (simplifications by collecting terms, cancellations, etc.), leads to eq. 110, which is cubic in x_1 (the abscissa of the reference pole for the distance eq.). Eqs. 111 give the three roots of this eq., which specify the locations of the three bipolar foci of linear Cartesians (RULE 26; Fig.21-1b,c). [These foci are defined by another method in Chapter VIII; see eq. 331b.]

Returning to eq. 106b, with the square of the right side completed and all terms but the one in u^2 having vanished, leaves eq. 112. Making the three substitutions for x_1 of eq. 111 leads to the three conventional distance eqs. 113 for the three foci of linear Cartesians, where x_u is the directed distance from

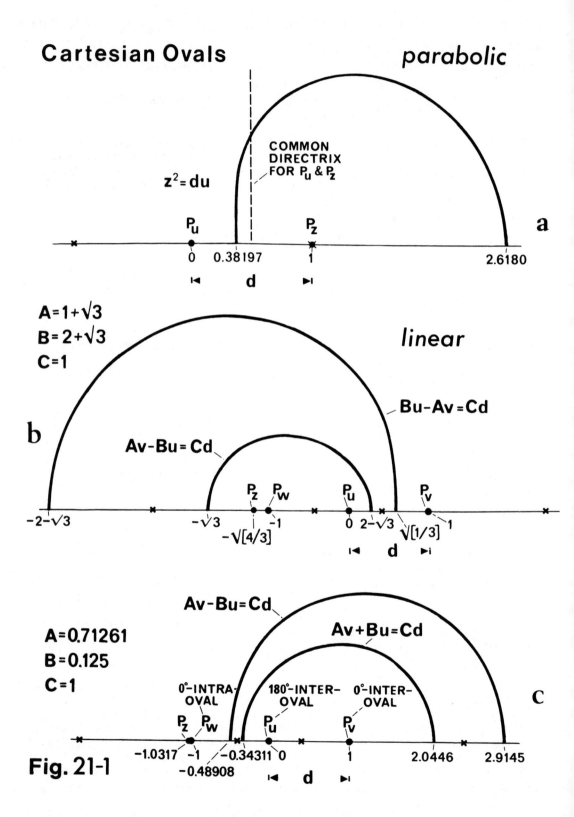

Fig. 21-1

pole p_u (at the origin of the rectangular system). With x_u, x_v, and x_w related as in eqs. 114, eqs. 113a,b,c yield the hybrid distance eqs. 113a,b',c' (since pole p_u is at the origin, eq. 113a is both the conventional and hybrid eqs.)

derivation of the distance eqs. for the foci of
linear Cartesians

$4v^2A^2B^2C^2d^3[A^2d-x_1(A^2-B^2)] = \{2x[x_1(A^2-B^2)-A^2d]^2$ eq. 106b modified to   (112)
$+ (u^2-x_1^2)(A^2-B^2)[x_1(A^2-B^2)-A^2d] +$ yield the distance eqs. applicable for the three
$x_1d^2[(A^2-B^2)(A^2-C^2)-2B^2C^2]-A^2d^3(A^2-C^2)\}^2$ bipolar foci of eqs. 111

distance eqs. $(d = d_{uv})$

(a) $A^2x - (A^2-B^2)u^2/2d - d(A^2-C^2)/2 = \pm BCu$ distance eq. for focus p_u (113)
at $x_1 = 0$ (plus sign for large oval)

(b) $B^2x - (A^2-B^2)v^2/2d - d(B^2+C^2)/2 = \pm ACv$ distance eq. for focus p_v at $x_1 = d$ (plus sign for large oval)

(b') $B^2x - (A^2-B^2)v^2/2d + d(B^2-C^2)/2 = \pm ACv$ hybrid distance eq. for focus p_v

(c) $C^2x-(A^2-B^2)w^2/2d-(A^2-C^2)(B^2+C^2)d/(A^2-B^2)$ distance eq. for focus p_w
$= \pm ABw$ at $x_1 = (A^2-C^2)d/(A^2-B^2)$ (plus sign for small oval)

(c') $C^2x-(A^2-B^2)w^2/2d+(A^2-C^2)(C^2-B^2)d/(A^2-B^2)$ hybrid distance eq. for focus p_w
$= \pm ABw$

(a) $x_u = x_v + d$, (b) $x_u = x_w + (A^2-C^2)d/(A^2-B^2)$ relations between x_u, x_v, and x_w of eqs. 113   (114)

An examination of the hybrid distance eqs., with the object of ranking the three foci, reveals that the eqs. are symmetrical and of equal simplicity. Accordingly, from this point of view, the foci have the same focal and subfocal rank, which means that the curve has equal structural simplicity about each of them. Of course, this already was known from the fact that linear eqs. of equal

[We shall see below that, although this view applies when the eqs. are expressed in terms of one or the other of the distances between the three foci, it does not apply for parabolic Cartesians when distances are expressed in terms of the p_u-p_z-distance d_{uz}, where p_z is the -b/2 limacon homologue. Focus p_u emerges as by far the highest ranking focus at the subfocal level, with a distance eq. that applies to both ovals. This is another example of the significance of the distance units employed--in the case in point, the characteristic distance is that of the parabolic Cartesians, $z^2 = Cdu$.]

The Equilateral Limacon As The Curve of Demarcation of Bipolar Parabolic Cartesians

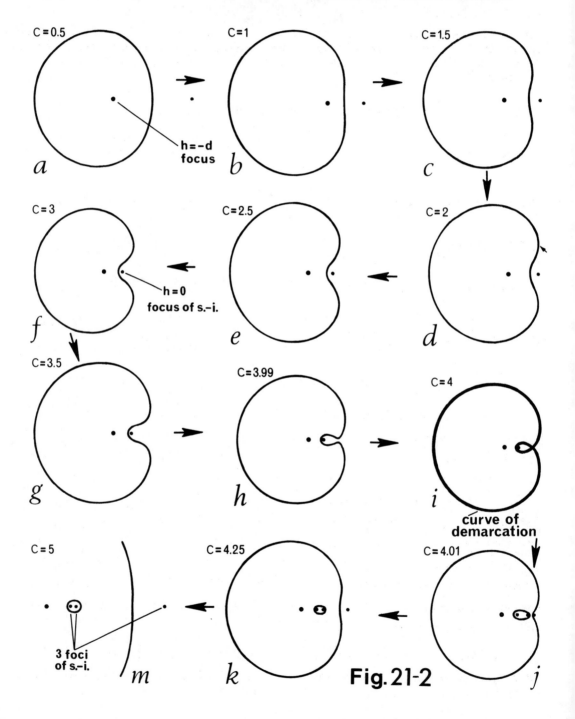

Fig. 21-2

PARABOLIC CARTESIANS 179

simplicity (eqs. 115) accommodate the three foci pairwise (Problems 28-30). The apparent greater complexity of eqs. 113c,c' for p_w is illusory, dependent upon the fact that "d" of eqs. 113 is the p_u-p_v-distance d_{uv}. If the symbol "d" is assigned to the p_u-p_w-distance d_{uw}, the eq. for p_v has the more complex term, etc. (illustrated in eqs. 115; see also Problems 28-30).

[In the following derivations and Problems for Cartesian ovals, the proper values of the alternate signs of roots, and for the large and small ovals, have to be ascertained by testing, using the specific values of the examples of Figs. 21 and 28.]

1st-degree eqs. of linear Cartesians

(a) $Av \pm Bu = Cd_{uv}$ $+ = $ large oval   bipolar eq. equivalent to eqs. 105 for p_v and p_u (115)

(b) $Aw \pm Cu = Bd_{uv}(C^2-A^2)/(A^2-B^2)$ $+ = $ small oval   eq. 115a for p_u and p_w, using d_{uv}

(b') $Aw \pm Cu = Bd_{uw}$ $+ = $ small oval   eq. 115b using d_{uw}

(c) $Bw \pm Cv = Ad_{uv}(C^2-B^2)/(A^2-B^2)$ $+ = $ small oval   eq. 115a for p_v and p_w, using d_{uv}

(c') $Bw \pm Cv = \pm Ad_{vw}$ $+ = $ small oval   eq. 115c using d_{vw} (sign of constant term depends on specific values of A,B,C)

Bipolar Parabolic Cartesians

Linear and parabolic Cartesians belong to groups of rectangular and polar curves that possess both monovals and biovals. First-degree bipolar construction rules (eqs. 104 and 115), however, can represent only one of the two ovals of the biovular rectangular and polar curves, whereas 2nd-degree bipolar construction rules can represent both of the biovals and the monovals (see Problem 41). The special properties of the curves designated *parabolic* Cartesians are that they have the simplest 2nd-degree construction rule ($z^2 = Cdu$, where p_z is the homologue of the -b/2 pole of limacons) and that they are the most symmetrical Cartesian ovals (by all the measures of structural simplicity employed in this work). These curves are referred to hereafter simply as "parabolic Cartesians."

The failure of linear construction rules to represent the groups of Cartesian ovals whose members possess only a single (real) oval hinges on the fact that bipolar construction rules become linear only when cast about two point-foci (excluding the eqs. of circles and lines). Accordingly, since the monovular members possess only a single (real) bipolar point-focus, they, like the parabola, cannot be represented by a linear construction rule.

In any group of biovular Cartesians (for example, the parabolic group, $z^2 = Cdu$), as the two ovals approach one another more and more closely (Fig. 22e,e', for C = 4.004), with progressive changes in the parameter values (as C decreases toward 4), two of the three foci also approach one another (the two open circles on each side of "pole to C" of Fig. 22e'). Ultimately (for C = 4) they come into coincidence at the DP of the resulting limacon curve of demarcation (Fig. $12f_2$). This is the limiting position at which a linear construction rule (see eqs. 94) can represent the curves. When the loops of the limacon open (for example, for C infinitesimally less than 4) to yield a monoval (Figs. $12f_1$ and 21-2h), the two coincident foci separate again but now are imaginary, and a linear construction rule (with real coefficients) no longer can represent the curve.

On the other hand, eqs. of rectangular parabolic form can be cast about the combination of p_u, the common focus of self-inversion of monovals and biovals (referred to as the *monoval-bioval focus*), and the -b/2 limacon homologue p_z (but these have the simplest form, $z^2 = Cdu$, only in the group of highest symmetry, designated "parabolic Cartesians"). The former pole, p_u, always is within the small loop or oval (Figs. $12f_2,f_3$; 21-1b,c; 22e,e'; and $23a_2$) and exterior to monovals (Figs. $12f_1$ and 21a). The latter pole, p_z, always is within the confines of the large loop or oval (but either interior or exterior to the small loop or oval; Figs. $12f_2,f_3$ and 21-1b,c). Since these poles always are real, the parabolic eqs. cast about them are able to represent both monovals and biovals.

Attention is called again to the fact that construction rules of Cartesian ovals also represent limacons as special cases. A particular limacon is the curve of demarcation between the biovals and monovals of every group of Cartesiand ovals (Fig. $12f_1$-f_3). But limacons invert to central conics about their double-points, and Cartesian ovals invert to central Cartesians about a homologous pole (see Fig. 22e', "pole to C;" this "pole to C" is *modular* to the monoval-bioval focus of self-inversion, p_u, and comes to lie at the DP of limacons). Accordingly, every central conic also is the curve of demarcation of a (certain) group of central Cartesians.

[The word "certain" is employed above because it now is known (see Kavanau, 1982) that every biovular Cartesian inverts to three different central curves (in the case of parabolic Cartesians, to curves a, c, and f of Fig. 22, for $C = 4.005$). Central conics are the curves of demarcation for only one group of the central curves (in the case of parabolic Cartesians, the central curves are Cassinian ovals).]

In the case of parabolic Cartesians ($z^2 = Cdu$), the limacon curve of demarcation is achieved for $C = 4$, whereas in general cases of linear Cartesians (eqs. 104 and 105a) it is achieved for either $A^2 = C^2$ or $B^2 = C^2$ (but not both). As either condition is approached, the ovals approach one another and come into contact in the limit.

Bipolar Symmetry of Cartesian Ovals

Judged solely from the linear construction rules of the biovals, all biovular Cartesians have essentially the same bipolar symmetry, since no two of the coefficients, A, B, or C are allowed to be equal (if any two are equal, the eqs. represent central conics, limacons, etc). However, as mentioned above, the comparative bipolar symmetry of all Cartesian ovals can be assessed from their parabolic construction rules, cast about the always-real poles, p_u and p_z. The curves with the construction rule, $z^2 = Cdu$, are most symmetrical because they are represented by the simplest parabolic eq. Within this group, the most-symmetrical curve is the representative for $C = 1$ (Fig.21-1a), that is, the curve represented by the eq., $z^2 = du$, just as for their inverses the eq. $uv = d^2$ represents the Cassinian oval with the greatest bipolar symmetry (central oval of Fig. 20-1a). [In the circumpolar approach of Chapters VII and VIII, it is the curve of demarcation that possesses the greatest symmetry.]

Geometrical Correlates

Geometrical correlates generally are readily apparent for the curve of a group that possesses the greatest symmetry of any type. In the case of the circumpolar symmetry of parabolic Cartesians, the most-symmetrical curve--the equilateral limacon--is the curve that demarcates between monovals and biovals. In the cases of central conics, the most symmetrical curves are the ones (the circle and the equilateral hyperbola) for which the two orthogonal axes (major and minor axes for ellipses, transverse and conjugate axes for hyperbolas) are equal in length.

In the case of the bipolar symmetry of parabolic Cartesians, the most symmetrical representative ($C = 1$) is the transition curve between the members that are convex at the vertex nearest the focus of self-inversion ($C < 1$) and the ones that are concave at the same vertex ($C > 1$). For Cartesian ovals, in general, the most symmetrical representatives (parabolic Cartesians) are the ones for which pole p_z comes to lie at modular distance from the monoval-bioval focus p_u (as explained below; see also Figs.21-1a & 23a$_2$), the distance eq. for focus p_u applies to both ovals (eq. 119, without a need for alternate signs), and poles p_z and p_u have a common polar-linear directrix (see Fig. 23 and Chapter V).

Distance Equations

The standard rectangular construction rules of parabolic Cartesians ($z^2 = Cdu$) cast about the focus of self-inversion, p_u, at the origin, with "d" the distance between this focus and the -b/2 limacon homologue, p_z (at $x = d$; Fig. 21-1a), are eqs. 116. One notes immediately from eq. 116b that all axial bipolar construction rules of parabolic Cartesians will possess a degree no higher than 4, since a term in $-x^2$ occurs in the expression for y^2 (see discussion on page 157). Forming the conventional distance eq. about an axial pole x_1 gives eqs. 117a,b. From these eqs., it is clear that great simplifications occur for $x_1 = d$ (defining the location of p_z, the limacon -b/2 homologue), whereupon the linear term in x on the left of eq. 117b vanishes.

derivation of the bipolar foci of parabolic Cartesians

(a) $(x^2+y^2-2dx+d^2)^2 = C^2d^2(x^2+y^2)$ standard rectangular eq. (116)

(b) $y^2 = C^2d^2/2+2dx-d^2-x^2 \pm Cd\sqrt{(2dx-d^2+C^2d^2/4)}$ eq. 116 solved for y^2

(a) $v^2 = (x-x_1)^2+y^2 = -2xx_1+x_1^2+2dx-d^2 +$ conventional distance eq. (117)
about pole $(x_1,0)$
$\quad C^2d^2/2 \pm Cd\sqrt{(2dx-d^2+C^2d^2/4)}$

(b) $[(v^2-x_1^2-C^2d^2/2+d^2)+2x(x_1-d)]^2 =$ eq. 117a rearranged and squared
$\quad C^2d^2(2dx-d^2+C^2d^2/4)$

(a) $4x_1^3 + x_1^2d(C^2-8) + x_1d^2 = 0$ reduction conditions for eqs. 117 (118)

(b) $x_1 = 0$ for p_u (c) $x_1 = (8+CE-C^2)d/8$ for p_v roots of eq. 118a; the three foci of self-inversion

(d) $x_1 = (8-CE-C^2)d/8$ for p_w, $E = \sqrt{(C^2-16)}$

Squaring eq. 117b and rearranging in powers of x on the left, with the remaining terms on the right, completing the square on the left (it already is known from the analysis of linear Cartesians that completing the squares on both sides is inappropriate; see page 174), and letting the combination of the remaining terms equal 0, yields an eq. in v^2 (Problem 35). To be able to extract roots from both sides of the eq., all terms on the right but the one in v^2 must vanish. The condition for this is eq. 118a, of which the three roots are eqs. 118b,c,d, representing the three bipolar foci of self-inversion (RULE 26). The $x_1 = 0$ focus, p_u is the common focus of both monovals and biovals. The other two foci, p_v and p_w, are real only in biovals ($C > 4$).

Returning to eq. 117b, substitution of the three roots of eqs. 118b,c,d leads to the conventional distance eqs. 119a-c, for the three foci of biovular parabolic Cartesians (see also Problem 37). The corresponding eq. (120) for pole p_z at $x_1 = d$ is obtained directly from eq. 117b. Though the three foci have equivalent subfocal rank in distance eqs. expressed in terms of their distances from p_u, it is evident from these eqs. that focus p_u has by far the highest subfocal rank in distance eqs. expressed in terms of distances from p_z.

conventional distance eqs. for the foci of parabolic Cartesians ($d = d_{uz}$)

(a) $u^2 - Cdu + d^2 = 2dx$ eq. for focus p_u at $x_1 = 0$ (both ovals)   (119)

(b) $v^2/d \pm \sqrt{(8C)}v/\sqrt{(C-E)} + C^2d(8+CE-C^2)/32 = C(C-E)x/4$

$E = \sqrt{(C^2-16)}$

eq. for focus p_v (eq. 118c) (plus sign for small oval)

(c) $w^2/d \pm \sqrt{(8C)}w/\sqrt{(C+E)} + C^2d[4(3C+E) - C^2(C+E)]/16(C+E) = -C(C+E)x/4$

eq. for focus p_w (eq. 118d) (plus sign for small oval)

$z^4 - C^2d^2z^2 + C^2d^4 = 2C^2d^3x$ eq. for pole p_z ($x_1 = d$)

Not only is eq. 119a much the simplest of the three eqs., it also represents distances from both ovals, whereas eqs. 119b,c represent the small ovals for the plus sign and the large ovals for the minus sign.
[Eqs. 119b,c cannot be simplified materially by substituting d_{vz} or d_{wz} for d_{uz} (the distance between poles p_z and p_u) and/or by converting to the hybrid eqs.]

Accordingly, the bipolar equivalence of foci p_u, p_v, and p_w at the subfocal level (as assessed from linear eqs.) is of a strictly limited nature. This is not a surprising finding when one takes into account the facts that the biovals self-invert by a different mode about each focus (see Figs. 12e and 23b) and that two of the three foci are imaginary in monovals, whereas the third focus is real in all members.

Bipolar Construction Rules

All the bipolar construction rules cast about the four poles taken pairwise now can be derived by eliminating x between pairs of the distance eqs. or by using any one of the distance eqs. in conjunction with the general axial distance eq. $(v^2-u^2-x_1^2+x_2^2)/2(x_2-x_1) = x$ (see Problems 31-34). Those cast about the paired foci, p_u and p_v, and the paired foci, p_v and p_w (the two imaginary foci in monovals) are eqs. 121a,b, respectively. Since eq. 121b' is manifestly simpler and more symmetrical than eq. 121a', the structural simplicity of the biovals is greater when cast about the paired lower-ranking foci, p_v and p_w, than when cast about the higher-ranking focus, p_u, paired with either of the lower-ranking ones.

[although foci p_v and p_w are not equivalent in the sense of being reflected in a line of symmetry, similar factors are at play here to those that form the basis for RULE 33.]

bifocal construction rules of parabolic Cartesians

(a) $\sqrt{[C(C-E)]}u \pm 2\sqrt{2}v = C\sqrt{[C(C-E)]}d_{uz}/2$ eq. for foci p_u and p_v (plus   (121)
$-\sqrt{[16C/(C-E)]}d_{uz}$ sign for small oval); d_{uz} and C as in the eq. $z^2 = Cd_{uz}u$

$E = \sqrt{(C^2-16)}$

(a') $\sqrt{[C(C-E)]}u \pm 2\sqrt{2}v = 4\sqrt{[C/(C-E)]}d_{uv}$ same as eq. 121a but employing d_{uv}, the p_u-p_v-distance

(b) $\sqrt{(C+E)}v - \sqrt{(C-E)}w = \pm E\sqrt{(C/2)}d_{uz}$ same as eq. 121a but for foci p_v and p_w

(b') $\sqrt{(C+E)}v - \sqrt{(C-E)}w = \pm 4\sqrt{(1/2C)}d_{vw}$ same as eq. 121b but employing d_{vw}, the p_v-p_w-distance

(a) $z^2 \pm \sqrt{(8C)}d_{uz}v/(C-E) = Cd_{uz}^2(C-E)/4$ eq. for focus p_v and pole p_z (122)
(plus sign for small ovals)

(b) $z^2 \pm 8\sqrt{(8C)}d_{vz}v/C(C-E)^{3/2} = 16d_{vz}^2/C(C-E)$ eq. 122a in terms of d_{vz}, the p_v-p_z-distance

For comparison with the simple parabolic eq., $z^2 = Cd_{uz}u$, and to confirm the much higher subfocal rank of focus p_u than p_v or p_w, when paired with p_z, the construction rule cast about the pole-pair p_v and p_z also is given (eq. 122a). Although this eq. is parabolic, it is much more complicated than the simple eq. $z^2 = Cd_{uz}u$, cast about focus p_u and pole p_z, even when expressed in terms of the p_v-p_z-distance, d_{vz} (eq. 122b).

Comparisons with Linear Cartesians, in General

To confirm that members of the parabolic subgroup of Cartesian ovals have greater bipolar symmetry than other Cartesian ovals, it is necessary to compare the parabolic construction rules of both groups (though it is a foregone conclusion that no parabolic construction rule can be more simple than $z^2 = Cdu$). In other words, what is the general eq. of Cartesian ovals cast about focus p_u and pole p_z, as compared to the simple eq. $z^2 = Cdu$ of the parabolic subgroup?

The construction rule in question is obtained readily, employing the conventional distance eq. 113a for focus p_u of linear Cartesians and the general axial distance eq. $(z^2-u^2-x_1^2+x_2^2)/2(x_2-x_1) = x$, where $x_2 = 0$ for p_u, and $x_1 = A^2 d_{uv}/(A^2-B^2)$ for p_z (this pole is defined by the condition for vanishing of the x^2-term of eq. 112 after the squaring of the right side). Eliminating x from these eqs. and simplifying leads to eq. 123a. Re-expressing this eq. in terms of d_{uz}, the p_u-p_z-distance (eq. 123b), rather than d_{uv}, the p_u-p_v-distance (p_v is the pole at $x = d$), yields eq. 123c.

parabolic construction rules of Cartesian ovals

(a) $z^2 = (A^2C^2+A^2B^2-B^2C^2)d_{uv}^2/(A^2-B^2)^2$ general eq. of linear Cartesians (123)
$\pm 2BCd_{uv}u/(A^2-B^2)$ about poles p_u and p_z (plus sign for small ovals)

(b) $d_{uz} = \pm A^2 d_{uv}/(A^2-B^2)$ p_u-p_z-distance, d_{uz}

(c) $z^2 = (A^2C^2+A^2B^2-B^2C^2)d_{uz}^2/A^4$ eq. 123a in terms of d_{uz}
$\pm 2BCd_{uz}u/A^2$

(d) $A^2C^2+A^2B^2=B^2C^2$, (e) $2BC_L/A^2 = C_p$ conditions defining the parabolic subgroup of Cartesian ovals

Since construction rule 123c is very much more complicated than the eq. $z^2 = Cdu$, for parabolic Cartesians, the latter subgroup has very much greater bipolar symmetry than Cartesian ovals, in general. Since eq. 123c applies to all Cartesian ovals, that is, both to the inclusive group named *linear* and to the subgroup named *parabolic*, it also specifies the special symmetry requirement for the latter group, namely, the conditions for which eq. 123c simplifies to $z^2 = Cdu$. These conditions are given by eqs. 123d,e (see Problem 36).

[It will be evident to the reader that, since linear construction rules of Cartesian ovals exist for all three bipolar foci taken pairwise, a linear construction rule for all three foci taken together as a trio also exists, $\pm Bu \pm Av = Cw$ (see Problem 48). This carries us into the domain of tripolar coordinates (see Kavanau, 1980).]

Lastly, eq. 116a, the standard rectangular construction rule of parabolic Cartesians (cast about focus p_u, with pole p_z at $x = d$) is re-expressed in the form of eq. 105a, the standard rectangular construction rule of linear Cartesians (cast about focus p_u, with focus p_v at $x = d$). One set of values of A, B, and C (eqs. 124) of the linear bipolar eqs. of parabolic Cartesians already are known from eq. 121a', cast about p_u and p_v. Substituting these values in eq. 105a yields eq. 125a.

rectangular eqs. of parabolic Cartesians

(a) $A_L = 2\sqrt{2}$, (b) $B_L = \sqrt{[C(C-E)]}$ coefficients of linear eqs. of (124)
(c) $C_L = 4\sqrt{[C/(C-E)]}$, $E = \sqrt{(C^2-16)}$ parabolic Cartesians expressed in terms of C (of eq. $z^2 = Cdu$)

(a) $\{(8-C^2+CE)(x^2+y^2)-16d_{uv}x +$ rectangular eq. 105a of linear (125)
$[8-16C/(C-E)]d_{uv}^2\}^2 = 64C^2 d_{uv}^2(x^2+y^2)$ Cartesians recast in terms of C for parabolic Cartesians, (of eq. $z^2 = Cd_{uz}u$)

(b) $[(x^2+y^2)+8d_{uv}x+16d_{uv}^2]^2 = 400d_{uv}^2(x^2+y^2)$ eq. 125a for $C = 5$, $d_{uv} = j/4$

(c) $[(x^2+y^2)-2d_{uz}x+d_{uz}^2]^2 = 25d_{uz}^2(x^2+y^2)$ eq. 116a for $C = 5$, $d_{uz} = j$

For $C = 5$ (Figs. 21-2m and 23a$_2$), eq. 125a becomes 125b. This can be compared to eq. 116a, with $C = 5$, given as eq. 125c. These two eqs. differ in two respects. First, the curve of eq. 125b has reversed right-left orientation to that of eq. 125c (evident from the opposite signs of the linear terms in x, which are indicative of a reflection in the y-axis). Second, d_{uv} of eq. 125b, the p_u-p_v-distance, is only 1/4 of d_{uz}, the p_u-p_z-distance of eq. 125c (for $C = 5$ parabolic Cartesians). Thus, for a common value of d, the curve of eq. 125b, though similar, has four times the linear dimensions and reversed orientation to the curve of eq. 125c.

Bipolar Central Cartesians

The extraordinary group of curves known as *central* Cartesians (eqs. 126; Kavanau, 1980, 1982) now are considered briefly (a more extended treatment is given in Chapter VIII). These curves were known previously only through their most symmetrical representative, the Cassinian ovals (eqs. 6h and 127). *Central* Cartesians, in general, are obtained by inverting Cartesian ovals about certain poles (see "pole to c," "poles to a," and "poles to f" of Fig. 22e'). In the case of a Cartesian monoval, there is only one inverse curve with a true center, whereas for Cartesian biovals there are three such curves, only one of which was known previously. This finding, of more than one inverse central curve, was quite unexpected (even more unexpected was the finding that central curves can self-invert about other poles than their centers; see Kavanau, 1982).

central Cartesians

(a) $(x^2+y^2)^2 + B^2x^2 + C^2y^2 + D^4 = 0$ general eq., centers at the origin   (126)

(a') $B^2 \ne C^2$; B^2, C^2, D^4, real, $\ne 0$

$B^4 \ne 4D^4$, $C^4 \ne 4D^4$

parameter restrictions, parameters unrelated to those used previously

(b) $y^2 = -x^2 - C^2/2 \pm \sqrt{[x^2(C^2-B^2)+C^4/4 - D^4]}$ eq. 126a solved for y^2

$(x^2+y^2)^2 - d^2x^2/2 + d^2y^2/2 + d^4(1/16 - C^2) = 0$ eq. 126a for Cassinians (d and C have the values of the bipolar eq. $uv = Cd^2$)   (127)

Representatives of all three types of central Cartesian are shown in Fig. 22a,c,f, for the case where the C = 4.0048 parabolic Cartesian (Fig. 22e,e') can be considered to be the initial curve. Cassinian ovals are obtained in inversions about the two axial poles *modular* to the interior 0°-focus of self-inversion (i.e., the axial poles that lie at a distance on each side of this focus at the unit of linear dimension for self-inversion; Fig. 21-1b,c, inter-oval poles marked with x's; Fig. 22e', "pole to c"). Annular (one oval within the other) *central parabolic Cartesians* are obtained by inverting parabolic Cartesians about (1) the two points of intersection ("poles to f" of Fig. 22e') of the modular circles for the two 0°-foci of self-inversion (circles about these foci with radii equal to the respective modular distances; also called "circles of inversion") and (2) the two axial poles modular to the exterior 0°-focus of

Relations Between Highly Symmetrical Cubics & Quartics That Self-Invert In Multiple Modes

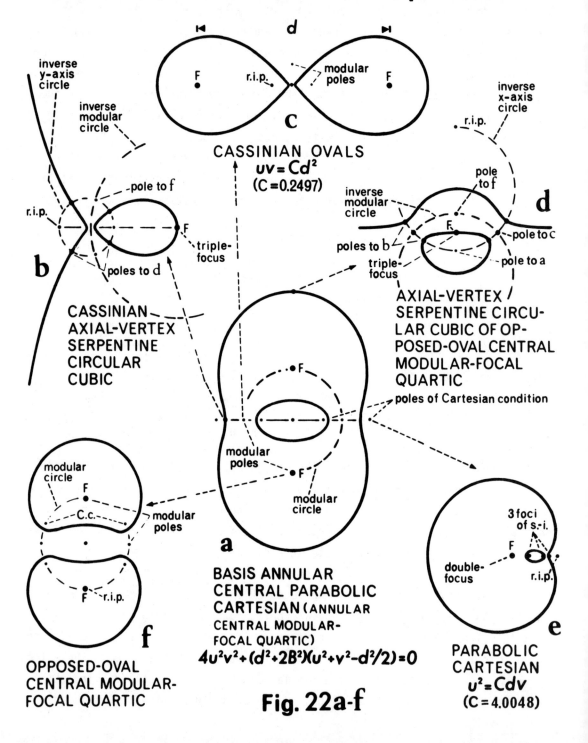

Fig. 22a-f

Relations Between Highly Symmetrical Cubics & Quartics That Self-Invert In Multiple Modes

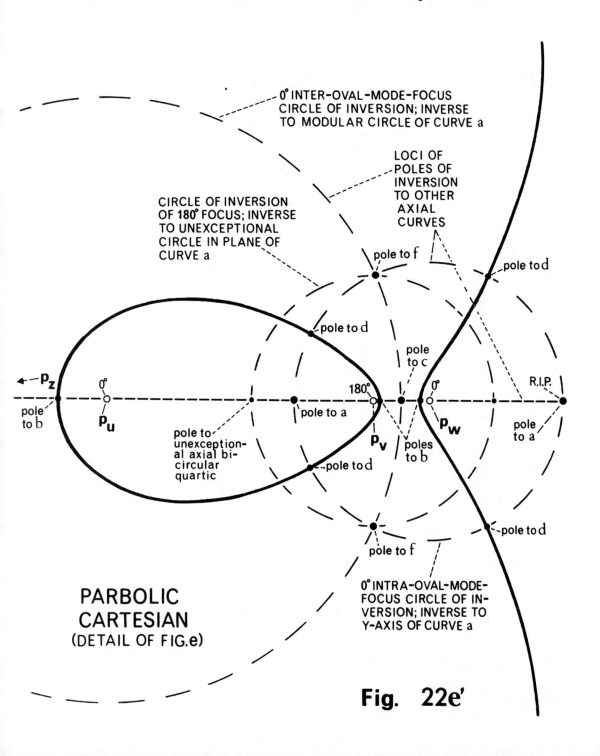

Fig. 22e'

self-inversion.

[As mentioned above, another aspect of the exceptional symmetry of members of the parabolic Cartesian subgroup is that one of the poles modular to focus p_u is coincident with pole p_z, the -b/2 limacon homologue (Fig. 21-1a and 23a$_2$). In consequence of this fact, pole p_z is one of the points about which inversion of parabolic Cartesians produces a central curve.]

Central Cartesians, in general, like their Cassinian subgroup, possess both monovular and biovular representatives. The conditions for annular biovals are D^4 positive, and greater than $B^4/4$ and $C^4/4$, and B^2 and C^2 negative; if the other conditions apply but B^2 (or C^2) is positive, the ovals are opposed. The biovals self-invert about their centers in both the 0° and 180°-modes (sometimes referred to as "positive" and "negative" self-inversion, respectively), but self-inversion of the monovals occurs only exceptionally--as in Cassinian monovals--and, then, is known only at 90° (see Fig. 20-1a,b).

Central Cartesians are of particular interest here as examples of a group of curves whose members, in general, possess no bipolar point-foci, but possess four non-central poles of high bipolar subfocal rank. As will be seen below, when these poles come into coincidence, pairwise, as in the Cassinian representatives, they acquire *dependent* bipolar point-focal rank.

It is evident from eq. 126b that the general axial bipolar construction rule 128 is only 4th-degree, even though x^2 occurs in the radical (because the external terms in x^2 vanish; see page 157). It follows from eq. 128 that the exterior pole closest to center is weighted heaviest (see RULE 19 and Problem 44). The conventional distance eq. is set up in the usual manner (eq. 129a).

$$[v^2x_2 - u^2x_1 + x_1x_2(x_2-x_1) + C^2(x_2-x_1)/2]^2 =$$
$$(C^2-B^2)(v^2-u^2-x_1^2+x_2^2)^2/4 + (C^4/4-D^4)(x_2-x_1)^2$$

general axial bipolar eq. of central Cartesians (128)

(a) $v^2 = (x-x_1)^2 + y^2 = x_1^2 - 2xx_1 - C^2/2$
$\pm\sqrt{[(C^2-B^2)x^2 + C^4/4 - D^4]}$

1st step in forming the conventional distance eq. about pole $(x_1, 0)$ (129)

(b) $[(v^2 - x_1^2 + C^2/2) + 2xx_1]^2 =$
$(C^2-B^2)x^2 + C^4/4 - D^4$

eq. 129a rearranged and squared

(c) $x^2(4x_1^2 + B^2 - C^2) + 4(v^2 - x_1^2 + C^2/2)xx_1 +$
$(v^2 - x_1^2 + C^2/2)^2 + D^4 - C^4/4 = 0$

eq. 129b squared and grouped in powers of x

(a) $x_1^2 = C^2/2$, (b) $x_1^2 = (C^2-B^2)/4$

conditions for simplification of eq. 129c (130)

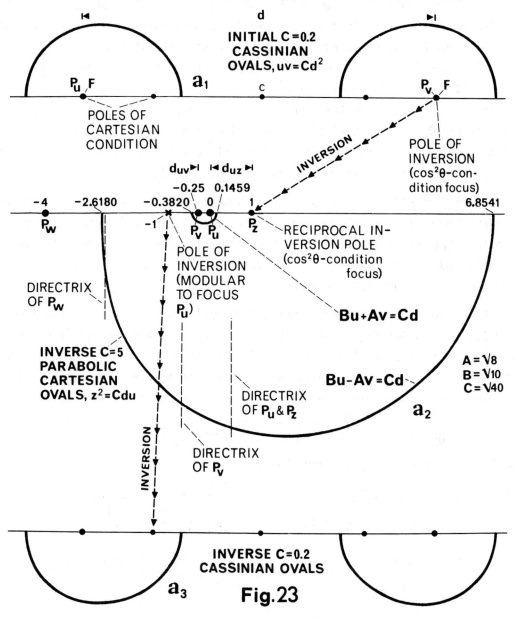

Fig. 23

Rearranging and squaring yields eq. 129b, from which it is evident that for $x_1^2 = C^2/2$ the eq. greatly simplifies. Squaring the term in brackets and rearranging again yields eq. 129c, from which a second condition for simplification is seen to be $4x_1^2 = (C^2-B^2)$. The latter eq. specifies the location of the $\cos^2\theta$-*condition* circumpolar foci, which are homologues of the limacon pole at $-b/2$ and lie interior to one (in the annular members) or both ovals. The reader can readily confirm that there is no general condition for which eq. 129c will reduce, from which it can be concluded (RULE 26) that central Cartesians, in general, possess no *independent* bipolar point-foci.

On the other hand, for condition 130b, eq. 129c, which is 4th-degree in v, becomes linear in x. Recall that this also is the case for the $-b/2$ pole of limacons (eq. 89) and the homologous poles of the equilateral lemniscate (eqs. 101) and Cartesian ovals (pole p_z; eq. 120). For limacons and Cartesian ovals, the pole is not a bipolar focus and has no equivalent pole, whereas an equivalent pole exists in the equilateral lemniscate. When the bipolar construction rule of the equilateral lemniscate about these two poles is formed, the poles are found to be *dependent* bipolar point-foci.

Accordingly, the possibility exists that the poles of eqs. 130a,b also are dependent point-foci, that is, that the bipolar construction rule cast about one or the other pair (or each pair), eq. 130a or 130b, reduces from 4th-degree. The two construction rules in question are eqs. 131 and 132. However, there is no general condition for which either of these eqs. reduces (no condition that applies to all central Cartesians). In fact, the sole condition for which the eqs. reduce is $B^2 = -C^2$, which is the condition for Cassinians. Thus, these two pairs of poles, the members of which possess only subfocal bipolar rank in all other central Cartesians, come into coincidence as a single pair in Cassinians, and thereby acquire dependent point-focal rank.

$$2C^2(u^2+v^2)^2 - (C^2-B^2)(u^2-v^2)^2 = 2C^2(C^4-4D^4) \quad \text{construction rule cast about subfocal poles of eq. 130a} \quad (131)$$

$$[v^2+(B^2+C^2)/4][u^2+(B^2+C^2)/4] = (C^4/4 - D^4) \quad \text{construction rule cast about subfocal poles of eq. 130b, the limacon } -b/2 \text{ homologues} \quad (132)$$

$$uv = Cd^2 \quad \text{common solution of eqs. 131 and 132 for } B^2 = -C^2 \quad (133)$$

This is representative of a common phenomenon. Whenever two exceptional poles, whether at the focal or subfocal level, come into coincidence in a specific member of a group of curves, the symmetry (structural simplicity) of the curve about the pole(s) of coincidence is increased or, equivalently, the structural eq. (construction rule or structure rule) of the curve about the pole(s) is simpler and/or lower degree (compare eq. 133 with eqs. 131 and 132).

To take an example from the realm of structure rules about points (which define and describe *inherent* structure), the 0° and 180°-structure rules of limacons about the pole at -b/2 (a circumpolar focus, though not a bipolar one) generally are 4th-degree, while those about the small and large-loop vertices generally are 6th-degree (see Chapter VIII). In the $e=2$ ($b=2a$) limacon, the -b/2 pole comes into coincidence with the small-loop vertex. In consequence, the degrees of the individual 4th and 6th-degree structure rules reduce to 2 for the coincident poles (the latter structure rule has "elliptical" form, i.e., it consists of the eq. of an ellipse).

The Bipolar Equilateral Strophoid

The most-symmetrical cubics by most criteria are the cissoid of Diocles (eqs. 6b and 134a) and the equilateral strophoid (eq. 136a). The former is a vertex inversion of the parabola, the latter of the equilateral hyperbola. The homologous inversions of the circle and the line--the other two most symmetrical conics--are not cubics, but lines. Because the equilateral strophoid is a well-known curve and possesses a loop, it is used to illustrate the bipolar construction rules of cubics.[For the bipolar construction rules of selected sextics, see Problems 45-47.]

When central conics are inverted about their axial vertices, the inverse curves are the *central conic axial-vertex inversion cubics* of eq. 135a (see Problem 40), which retain one line of symmetry. The parameters a and b of this construction rule are identical with those of the rectangular construction rules of conics (the plus sign is for initial hyperbolas). Expressed in terms of eccentricity (e) and the parameter a, eq. 135a becomes 135b.

It is evident from eq. 135b (as mentioned above) that all the axial-vertex inversion cubics (hereafter referred to simply as "vertex cubics") can be identified in terms of the eccentricities of the initial curves. In fact, this is true of all curves inverse to parabolas, ellipses, and hyperbolas. Geometrically, the two expressions, $a/2(e^2-1)$ and $a/2$, are the distances of the asymptote and the loop vertex from the DP, respectively. For convenience, these two distances are assigned new designations, a and b, respectively in the vertex cubics (eq. 135c). Like the corresponding conics, the vertex cubics can be identified by measuring either two lengths or an angle (for hyperbola vertex cubics, this is the angle between the DP-tangents).

central conic axial-vertex inversion cubics and
their hybrid distance eqs.

(a) $x(x^2+y^2) = 2ay^2$ standard rectangular cissoid of Diocles   (134)

(b) $v^2 = 2ax^2/(2a-x)$ hybrid distance eq. of the cissoid of Diocles

(a) $y^2(x \pm a^3/2b^2) =$ central conic axial-vertex inversion cubics   (135)
                                  (conic parameters); plus sign for hyperbolic
       $x^2(a/2 - x)$ members (i.e., inversions of hyperbolas)

(b) $y^2[x+a/2(e^2-1)] =$ eq. 135a in terms of the eccentricity e and
                                  the central conic parameter a
       $x^2(a/2 - x)$

(c) $y^2(x+a) = x^2(b-x)$ standard rectangular hyperbola axial-vertex inversion cubics (vertex cubic parameters; DP at $x = 0$, asymptote at $x = -a$, vertex at $x = b$)

(a) $y^2(x+a) = x^2(a-x)$ equilateral strophoid (DP at $x = 0$, asymptote   (136)
                                  at $x = -a$, loop vertex at $x = a$)

(b) $u^2 = 2ax^2/(a+x)$ hybrid distance eq. of the equilateral strophoid about the DP

(c) $x = -2av^2/(a^2+v^2)$ hybrid distance eq. of the equilateral strophoid about the loop vertex

(c') $v^2 = -a^2x/(2a+x)$ eq. 136c solved for v^2

(d) $x_u = x_v + a$ relation between x's of eqs. 136b,c,c'

The most symmetrical of the hyperbola vertex cubics is the equilateral strophoid (Fig. 24a), the inverse of the equilateral hyperbola, for both of which curves $a = b$ (eqs. $b^2a^2 - a^2y^2 = a^2b^2$ and 135c). For the equilateral strophoid, this equivalence signifies that the asymptote and the loop vertex are equidistant from the DP on each side. Another geometrical correlate of the high symmetry of the equilateral strophoid is that the homologue of the -b/2 pole of limacons is coincident with the loop vertex, which also is the pole about which the curve self-inverts.

[In the equilateral limacon, the corresponding geometrical correlate is that the -b/2 pole is modular to the pole of self-inversion, which places it at the same distance on one side of this focus as the DP is on the other side.]

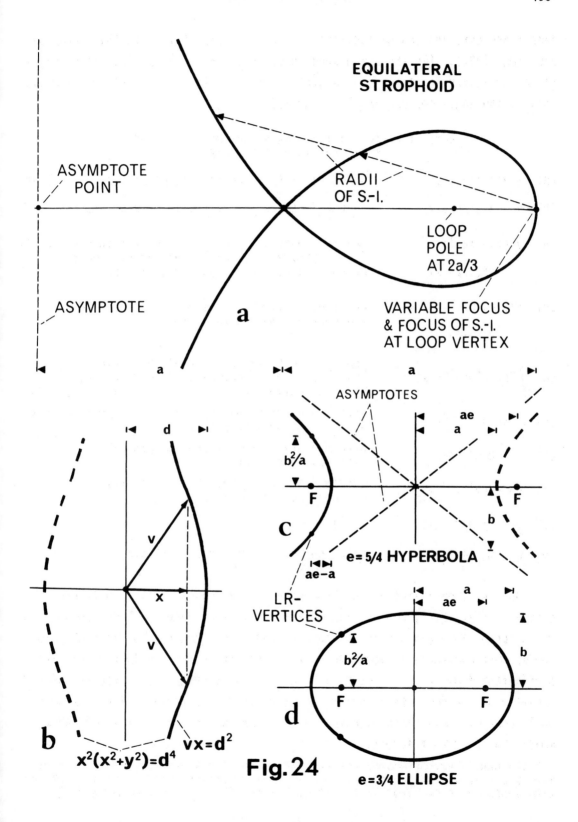

Fig. 24

Inasmuch as the axial distance eqs. of cubics with a line of symmetry also are cubics (RULE 24), and distance eqs. are not known to reduce by methods other than root-taking, it is evident that cubics can possess no independent bipolar point-foci (RULE 26). The possibility exists, however, that cubics possess dependent bipolar point-foci. By showing that neither of the two highest-ranking bipolar poles (which are circumpolar foci) is a dependent point-focus in the equilateral strophoid, the presumption becomes very strong that neither dependent nor independent bipolar point-foci exist in any cubic (RULE 36), and that dependent bipolar point-foci exist only as equivalent pairs (RULE 37), as in Cassinian ovals.

RULE 36. *Bipolar point-foci do not exist in cubics (neither dependent nor independent).*

RULE 37. *Dependent bipolar point-foci exist only as equivalent pairs.*

The DP is the highest-ranking angle-independent circumpolar focus of curves inverse to a conic, so this point is taken as one of the poles for the presumptive simplest and lowest-degree bipolar construction rule of the equilateral strophoid. The loop vertex is taken as the other pole. This is a circumpolar *triple-focus*, being the point of coincidence of the vertex, the focus of self-inversion, and the $\cos^2\theta$-condition (variable) focus (the $-b/2$ limacon homologue).

Eqs. 136b,c, respectively, are the hybrid distance eqs. for the DP and loop vertex. These are, by far, the simplest distance eqs. of the curve. From a comparison of these two eqs., it is evident that the confluence of three exceptional poles confers the highest bipolar subfocal rank on the loop vertex, since x occurs only to the 1st-degree in its distance eq. (see also eq. 136c'), as opposed to 2nd-degree for the DP. This linearity in x correlates with that of the distance eq. for the $-b/2$ pole of limacons, and of all the distance eqs. thus far encountered for its homologues in other curves.

The bipolar construction rule about the DP and loop vertex can be derived either from the two hybrid distance eqs., employing the relation 136d, or by using either of these distance eqs. together with the general axial bipolar eq. in terms of x, $(v^2-u^2-x_1^2+x_2^2)/2(x_2-x_1) = x$, leading to the sought-after eq. 137, which is 4th-degree in u and v (see Problem 38). Pole p_v possesses higher subfocal rank than pole p_u, since v is weighted greater than u in eq. 137 (terms in v^4 and $-2a^2v^2$, as compared to only a^2u^2). This confirms the subfocal ranking based upon the distance eqs. (136b,c).

bipolar construction rules of the equilateral
strophoid

$$u^2(a^2+v^2) = (a^2-v^2)^2 \qquad \text{DP } (p_u) \text{ and loop vertex } (p_v) \qquad (137)$$

general axial eq.; $p_v = (x_1,0)$, $p_u = (x_2,0)$

$$v^4(x_2-a)+u^4(x_1-a)+(2a-x_1-x_2)u^2v^2 + (x_2-x_1)[2x_1(x_2-a)+x_2^2]v^2 + \qquad (138)$$
$$(x_1-x_2)[2x_2(x_1-a)+x_1^2]u^2 = (x_2-x_1)^2[a(x_2^2+x_1^2)-x_1x_2(x_2-x_1)]$$

It remains to show that the degree of construction rule 137 is not less than the degree of the general axial construction rule, because if it were, p_u and p_v, of eq. 137, would be dependent point-foci (it is evident from their distance eqs. that neither is an independent point-focus). The general axial construction rule 138 is derived by eliminating x and y from the general axial eq. in x (see text above and Problem 39), the general axial distance eq., $v^2 = (x-x_1)^2+y^2$, and eq. 136a. Since eq. 138 also is 4th-degree, neither of the poles of eq. 137 has bipolar point-focal rank.

Hybrid Distance Equations and the Hybrid Polar-Rectangular System

For convenience of reference, the hybrid distance eqs. of various curve-pole combinations now are listed in Table 8, in essentially the order of the structural simplicity that they represent. In other words, this list orders the hybrid eqs. according to the symmetry of the represented curves about specific poles (excepting eq. 145); the same order applies in any two-element coordinate system employing point or circle reference elements.

[In the interests of simplicity, applications and implications of the hybrid distance eqs. are limited to two-element coordinate systems, but equally facile applications obviously can be made to colinear tripolar systems, etc.]

A great utility of hybrid distance eqs. is that they provide the starting points (initial eqs.) for deriving construction rules in all two-element coordinate systems possessing one or more point-poles or circle-poles. Since, aside from the rectangular system, these are the coordinate systems of greatest interest, and the only ones treated here (except in Problem 48), these eqs. are used repeatedly in Chapters V and VI to derive polar-linear, polar-circular, and linear-circular construction rules. Accordingly, it should come as no surprise to find that the construction rules in these other systems usually bear

198   THE HYBRID SYSTEM

Table 8.   Hybrid Polar-Rectangular Distance Equations for the
            Most-Symmetrical Curve-Pole Combinations

equation	degree in v & x	degree in v	degree in x	No. of terms	No. of parameters	focus	curve-pole	
$v = R$	–	0	–	2	1	yes	circle, center	(139)
$v = x$	–	1	1	1	0	no*	ray, terminal point	(140)
$v = x + 2a$	–	1	1	3	1	yes	parabola, focus	(141)
$v = \pm ex + a(1-e^2)$	–	1	1	3	2	yes	ellipse, focus (plus sign for left focus)	(142)
$v = a(e^2-1) \pm ex$	–	1	1	3	2	yes	arm of hyperbola, focus (plus sign for rt. arm)	(143a)
$v = ex - a(e^2-1)$	–	1	1	3	2	yes	right arm of hyperbola, left focus	(143b)
$v^2 = 2xR$	–	2	1	2	1	no	circle, incident point	(144)
$v^2 = (R^2 - x_1^2) - 2xx_1$	–	2	1	3	2	no	circle, axial point	(145)
$v^2 + aex = av$	–	2	1	3	2	yes	elliptical limacons & lge. loop, DP (-av for small loop)	(146)
$v^2 - Cdv + d^2 = 2dx$	–	2	1	3	2	yes	parabolic Cartesians, monoval-bioval focus	(147)
$v^2 - aev + ax/e = -a^2(1-e^2)^2/4e^2$	–	2	1	4	2	yes	limacons, pole of s.-i. (+aev for ellipticals)	(148)
$(A^2-B^2)v^2/2d \pm BCv + d(A^2-C^2) = A^2x$	–	2	1	4	4	yes	linear Cartesians, focus (+BCv for large oval)	(149)
$v^2 = e^2x^2 + a^2(1-e^2)$	–	2	2	3	2	no	ellipses and hyperbolas, center	(150)
$v^2 = -a^2x/(x+2a)$	3	2	1	3	1	no	equilateral strophoid, loop vertex	(151)
$v^2 = 2ax^2/(x+a)$	3	2	2	3	1	no	equilateral strophoid, DP	(152)
$v^4 - C^2d^2v^2 + C^2d^4 = 2C^2d^3x$	–	4	1	4	2	no	parabolic Cartesians, p_z	(153)
$(v^2 - a^2e^2/4)^2 - a^2(v^2 + a^2e^2/4) = -a^3ex$	–	4	1	4	2	no	limacons, pole at -b/2, all cases	(154)

* There are compelling reasons also to regard the terminal point of a ray as a focus, whereupon linear terms in the variables of all hybrid eqs. would represent focal poles. However, adopting this view here would require lengthy additional discussions and qualifications of RULES.

close similarities to the corresponding hybrid eqs., and, in some cases, are identical (except for the substitution of u or v for x).

The utility of these eqs., however, is limited to coordinate systems and *construction rules* (as opposed to *structure rules*). Comparable listings of curve-pole combinations according to the simplicities of their structure rules (see Table 10) entail the specification of a probe-angle, and often differ for different angles (for example, 90°-structure rules as opposed to 180°-structure rules). Moreover, an infinite number of curves possess, say, the 180°-structure rule $X = Y$, whereas the hybrid distance eq. (construction rule) $v = x$ is unique to a ray about its terminal point. Nor are the foci of construction rules in one-to-one correspondence with those of structure rules, though all foci defined by the hybrid distance eqs. (RULES 26 and 27) also are circumpolar foci (RULE 22).

Polar Construction Rules, Polar Symmetry, and Conversions between the Polar and Hybrid Systems

The reader will recall that the hybrid distance eqs. differ from polar construction rules because the use of an angle as a polar coordinate allows the x of the hybrid eqs. to be re-expressed as $v\cos\theta$ ($r\cos\theta$). This often either allows v's (or r's) to be factored from the polar eq., with consequent reductions in exponential degree, or allows the number of terms to be reduced. For example, $v^2 = ax$ becomes $r = a\cos\theta$, and $v = x+2a$ becomes $r(1-\cos\theta) = 2a$.

A corresponding listing of curve-pole combinations according to their polar symmetry would find entries in the hybrid list moved up one or more degree categories in every case in which a hybrid distance eq. lacks a constant term. For example, the second entry, $v = x$, would become $\cos\theta = 1$, equivalent to $\theta = 0°$ or, more generally, to $\theta =$ constant. This eq. falls in the 0°-category with the eq. $v = R$ ($r = R$) for the centered circle. The degree categories and relative orders of listing of the parabola and central conics remain unchanged in the polar system, but the number of terms decreases from 3 to 2.

The eq. of a circle about an incident point changes from the 2nd-degree, $v^2 = 2xR$, to the 1st-degree eq. $r = 2R\cos\theta$, which, having only two terms, now advances to the level of the parabola in order of simplicity. Similarly, eq. 146 for an elliptical limacon about the DP, $v^2 + aex = av$ loses a term and moves up one degree category to $r = a(1-e)\cos\theta$. Other curve-pole combinations not appearin the list move up into the 1st-degree category.

Note that the only hybrid distance eqs. with terms that possess products of powers of v and x are those of the equilateral strophoid. This can occur for any rectangular curve of higher degree than 2. For example, the simple hybrid eq., $vx = d^2$, for a curve (the right arm of Fig. 24b) not previously considered, becomes $r^2\cos\theta = d^2$ in the polar system, and the 4th-degree rectangular construction rule $x^2(x^2+y^2) = d^4$ (curve of Fig. 24b). If the eq. $vx = d^2$ (representing only the right arm) is ranked on the basis of degree and number of terms and parameters, it ranks with the circle about an incident point, in 6th-position in the Table [The origin for this curve is not a focus (RULES 26 and 27), since the 4th-degree hybrid eq., $v^2x^2 = d^4$, is required for both arms.]

Note, also, that when the conversion is made from the hybrid system to the polar system, curves with one arm, one loop, and one oval acquire the second arm, loop, or oval of the corresponding polar and rectangular curves. Furthermore, each of the eqs. 142 and 143a, of ellipses and hyperbolas, converts to a polar eq. that represents both ellipses and hyperbolas. To clarify these circumstances and to elucidate further the workings of the hybrid system, the ellipse and hyperbola of Fig. 24c,d are taken as examples.

Consider, first, eq. 143a with the negative sign, for one-arm hyperbolas (left arm of Fig. 24c). The variable v represents distances from the hybrid pole (the traditional focus of the left arm) to points on the curve. It is the hypotenuse of the triangle formed by the x-vector and the normal from the reference line to the curve (as shown in Fig. 24b). Accordingly, v is always positive but is *directed* in the sense that it is measured to the same side of the pole as x, which itself can be positive or negative.

It is clear from eq. 143a, for the left arm, that the maximum value x can assume and still satisfy the requirement, $v \geq x$, is ae-a, in which case v also equals ae-a (the focus-vertex distance; Fig. 24c, left arm). But if one lets x equal ae-a, which is the distance to the vertex of the non-existent right arm, v becomes equal to -(ae+a), which has no locus. Similarly, for x = ae, v equals -a. On the other hand, x can take on negative values without limit.

Converting eqs. 142 and 143a to the polar construction rules yields eqs. 142' and 143a'. Now if one lets $\theta = 0$ in eq. 143a', r becomes equal to ae-a, specifying the vertex of the left arm. Correspondingly, all positive and negative values of θ, short of the radius vector becoming parallel to an asymptote, yield positive values of r and lead to a tracing of the entire left arm. For other values of θ, however, the situation is quite different. Thus, if one lets θ be 180°, the value of r becomes $-(ae+a)$, which directs r back along the 0°-axis to the vertex of the right arm, which was not included in the hybrid locus. As θ takes on values progressively greater and less than 180°, negative values of r are obtained and the right arm is traced again, until the radii parallel the asymptotes.

$$r = a(1-e^2)/(1-e\cos\theta) \quad \text{polar eqs. of ellipses about their left foci} \quad (142')$$

$$r = a(e^2-1)/(1+e\cos\theta) \quad \text{polar eqs. of hyperbolas about their left foci} \quad (143a')$$

If we return to construction rule 143a', for $\theta = 0°$, and reconsider the value $ae-a$ for r, this expression is valid for all values of e, not just those for hyperbolas ($e > 1$). For an ellipse (Fig. 24d), this leads to a negative value for r, which places the point of the plot at its left vertex. As θ progresses toward 90°, r assumes the value $a(e^2-1)$, the negative of the length of the hemi-LR, which places the point of the plot at the left lower LR-vertex. A similar progression occurs for θ ranging between 0° and -90°, terminating at the left upper LR-vertex.

For $\theta = 180°$, r assumes the value $-(ae+a)$, placing the point of the plot at the right vertex. As θ ranges from 180° to 270°, and from 180° to 90°, negative values of r also are obtained, tracing the entire portion of the ellipse to the right of the left LR. In essence, then, where the hybrid eq. 143a represents only the left arm of the hyperbola, the polar eq. 143a' traces an entire ellipse and right arm of a hyperbola for negative values of r, and the left arm of the same hyperbola for positive values of r. The situation for the polar construction rule 142', derived from the hybrid eq. 142, parallels that described above, except that now the ellipse and the right arm of the hyperbola are traced for positive values of r and the left arm is traced for negative values. In both cases, the curves are traced only once.

Coordinate System Comparisons and Analytic Geometry

This Chapter concludes with an overview of the comparative utility of the bipolar, polar, and rectangular coordinate systems, and additional suggestions and comments (see also the *Introduction*, pages xiii to xx) for revisions of courses in analytic geometry. The first phase of the treatment deals with three topics. What are the strengths and weaknesses of construction rules of

curves in these three systems for:
1. identifying the curves;
2. assessing the positional symmetries of the curves with respect to the coordinate elements; and
3. assessing the absolute and comparative simplicities of the structures of the curves about specific points?

The hybrid polar-rectangular coordinate system (also referred to as the *hybrid* system) also enters the discussions of the third topic. The question of the practical utility of the systems for conventional studies is not treated; the polar and rectangular systems are indisputably preeminent for this purpose. Discussions of didactic utility close the Chapter.

Concerning the identification of curves, the bipolar and rectangular systems possess major advantages in different respects, with the polar system being very much less utile. All three systems are useful in roughly equal measure for assessing positional symmetries. In regard to assessing structural simplicities about points, the polar system is preeminent for absolute simplicity in regard to *structure rules*, the hybrid system for assessing absolute simplicity in regard to *construction rules*, and the bipolar system for assessing comparative simplicity in regard to *construction rules*. The latter system also is very useful for assessing absolute simplicity. The rectangular system possesses utility for assessing structural simplicity about points only by virtue of the fact, and to the extent, that one can visualize the polar conversions and simplifications merely from inspecting the rectangular eqs.

Identifying Curves

In considering the topic of identifying curves, it is necessary to divide construction rules into two categories: those for curves in random positions (locations and orientations), and those for curves in standard positions. Concerning the first category, the rectangular system (and other bilinear systems) stands preeminent by virtue of its property of degree-restriction. Thus, any 1st-degree construction rule represents a line, and any 2nd-degree construction rule a circle, parabola, ellipse, or hyperbola. Of these quadratics, only circles and equilateral hyperbolas can be identified readily from their eqs., regardless of orientation (circles are the only quadratics whose eqs. are invariant under central rotations, and equilateral hyperbolas are the only quadratics for which the coefficients of the x^2 and y^2-terms either are equal and of opposite signs or both 0).

One cannot readily distinguish between randomly-positioned members of the other three types of quadratics upon the basis of their construction rules. For identifying curves of higher degree, the rectangular system is practically useless in any connection except ascertaining degree categories, for which it possesses the same utility as the polar system.

Virtually the only utility of the polar system for identifying curves in random positions is that the rectangular degree is immediately evident, inasmuch as degree values for such curves are the same in both systems; the degree of the basic polar eq. in r is the same as the rectangular degree in x and y (RULE 7a). The bipolar system is essentially without value for identifying randomly-positioned curves from their construction rules. One can conclude only that the maximum degree of the corresponding rectangular construction rule does not exceed 1/4 of the value of the bipolar degree.

The situation is quite different for non-randomly-positioned curves. In the rectangular system, one readily recognizes the identity of all the conic sections--for example, 0° and 90°-ellipses, hyperbolas, and parabolas. But there are few instances in which higher-degree rectangular curves in standard positions possess distinctive construction rules. In the polar system, only a very few common and uncommon curves in any positions are identifiable from their construction rules.

The bipolar system possesses some of its most impressive advantages in the realm of non-randomly-positioned curves. From linear construction rules with two and three terms, alone--and through the most simple distinctions in signs and absolute and comparative values of coefficients--one can identify the eqs. of a line, rays, a line-segment, circles, ellipses, hyperbolas, limacons, and Cartesian ovals. The 2nd-degree category adds easily-identified eqs. of Cassinian ovals ($uv = Cd^2$), parabolic Cartesians ($u^2 = Cdv$), centered circles ($u^2 + v^2 = Cd^2$), the circle of Pythagoras ($u^2+v^2 = d^2$), normal lines ($v^2-u^2 = Cd^2$), and normal lines that include a pole ($v^2-u^2 = \pm d^2$).

The curves represented by displaced parabolic, circular, elliptical, and hyperbolic construction rules are far less readily identified, but include only curves already represented by the linear, centered quadratic, and vertex parabolic eqs. (see Problem 41). The identification of curves from their higher-degree construction rules is even more difficult in the bipolar system than in the rectangular system.

Positional Symmetry

Positional symmetry with respect to coordinate elements can be determined by two methods. The first is direct and the second indirect. The direct method is the only one encountered in classical treatments, and the only one for which the rectangular system has utility. This method gives the same amount of information in both the polar and rectangular systems. Every reader already will be familiar with the criteria for axial symmetries and symmetries about a center in the rectangular and polar systems (see Chapters II and III).

Since all bipolar curves have symmetry about the bipolar axis, no specific test for this property exists. Nevertheless, one still can ask the question, "Is this symmetry possessed solely by virtue of the reflection of two separate congruent segments in the bipolar axis?" If not, each oval, loop, or segment is bisected by the bipolar axis and possesses a line of symmetry at the position of the axis when plotted in either the rectangular or polar system. Since the answer to this question depends upon the indirect approach, it is temporarily deferred.

To test the construction rules of bipolar curves for symmetry about the coordinate midline is simple and straightforward: there must be complete equivalence of pole p_u and p_v, that is, u and v must participate in the bipolar construction rule in a precisely symmetrical manner in every respect--degree, occurrence, coefficients, etc. There is no specific test for a true center of curves whose eqs. fulfill these requirements, since all bipolar curves that are symmetrical about the bipolar midline possess true centers (the points of intersection of two orthogonal lines of symmetry). Nevertheless, the presence of a true center of the corresponding rectangular locus can be ascertained in special cases by the indirect method. Rectangular curves that possess a center of symmetry that is not a true center (for example, the dashed s-shaped curve of Fig. 13a) acquire a true center by virtue of reflection in the bipolar axis.

The rectangular system has no utility, whatsoever, in the indirect method of determining positional symmetry, because this method depends solely on the exponential degrees of construction rules. Since the rectangular system is degree-restricted, the degree of a rectangular eq.--as the sole consideration--can give no indication of positional symmetry. The polar and bipolar systems are quite different in this respect. In them--with few exceptions--information about positional symmetry (structural simplicity) with respect to coordinate elements is conveyed by the degree of a construction rule, alone.

A great variety of transcendental polar construction rules of 1st-degree in r represent rectangular curves of high degree that are symmetrical with respect to either the polar center or the 0°-180° or 90°-270° polar axes, or all three. The capability for such representations is one of the major advantages of the polar system in its common uses. Almost without exception, polar construction rules are cast about the points in the planes of curves (placed in appropriate orientations) that lead to the lowest-degree construction rules. But notwithstanding the universality of this practice, classicists have not inquired into the significance of the reductions (as opposed to the mechanisms).

Structural Equation Geometry recognizes the points or loci in question to be polar foci of the curves, that is, points or loci about which the curves possess the greatest symmetry or structural simplicity. Any incident point of a polar curve possesses focal, but not necessarily point-focal, rank. Point-focal rank is reserved for poles about which degree reduction exceeds one unit.

The situation in the bipolar system bears many similarities to that in the polar system. A major difference is that the presence of a second reference pole eliminates the possibility of a pole acquiring focal rank solely by virtue of its being included in a curve. This circumstance generally confers only subfocal rank--even for the circle of Pythagoras, which includes both poles.

On the other hand, any 1st or 2nd-degree bipolar construction rule represents a curve whose symmetry in the bipolar axis traces to the possession of a classical line of symmetry in the rectangular or polar system (i.e., the curve does not merely acquire a line of symmetry by virtue of being reflected in the bipolar axis). For any 1st or 2nd-degree construction rule in which the poles are equivalent, one also can assert that a true center of the curve lies midway between the poles.

Absolute and Comparative Structural Simplicity about Points

The absolute and comparative structural simplicities of curves about points is another category in which the rectangular system provides, at most, subtle hints (see eq. 4g and accompanying text). On the other hand, the polar system is preeminent among coordinate systems in the implications of its construction rules for the absolute simplicities of curves about points. It is the key system for the derivation of all structure rules (see Chapters VII and VIII, and

Kavanau, 1980), whether with respect to points, lines, or other curves. The overall simplicity of a curve's standard polar eq. (expressed solely in terms of the sine and/or cosine functions)--including its degree-is a direct, though crude, indication of the comparative simplicity of the curve's *structure rules* about the polar pole (as well as of its structure rules about lines and other curves).

Polar construction rules remain a good index of structural simplicities of curves in coordinate systems that include one or more points or circles among their reference elements, but preeminence passes over to the hybrid system, which also provides the key distance equations for deriving construction rules in these systems.

In the same application, that is, as an index of the absolute simplicities of the structures of curves about points considered as coordinate elements, the bipolar system also provides excellent indices. Thus, any poles about which bipolar construction rules are linear in the variables, or merely possess a linear term in one variable (regardless of whether terms of higher degree occur), are points about which the curves possess a high degree of structural simplicity. If attention is confined to linear eqs. or eqs. with a linear term in a variable, the bipolar system is on a par with the hybrid system. But for cases in which the lowest power of a variable is greater than 1, the bipolar system generally loses its utility in this respect.

On the other hand, in regard to the *comparative* simplicity of the structure of curves about points as assessed from single construction rules, the bipolar system assumes the preeminent role. Individual construction rules in the polar and hybrid systems are of much less use for this purpose. RULES 20, 21, and 34 deal with the utility of bipolar construction rules in this regard, primarily in terms of the comparative focal and subfocal ranks of the two poles.

Regarding the broader question of preeminence for judging comparative structural simplicity about points, without limiting consideration to the utility of *single* construction rules, the hybrid system rises to the level of the bipolar system or exceeds it. A direct comparison of two hybrid eqs. (construction rules), like the direct comparison of the two poles of a single bipolar construction rule, often gives precisely the same information and with much the same facility. In cases where one of the poles being compared possesses an equivalent pole but the other does not, the hybrid system assumes preeminence (see RULE 33).

Polar construction rules are next-most utile in the same application, but are not entirely reliable when one deals with eqs. from which r's have been factored. For example, the pole of the simple linear construction rule of the cissoid of Diocles (eq. 6b) or the trisectrix of Maclaurin (eq. 6a), from which two r's have been factored, are not point-foci of construction rules in other coordinate systems--only in the polar system. On the other hand, the *subfocal* ranks of such poles usually are high in other coordinate systems (see Problem 30).

Comparative Utility of the Polar, Rectangular, and Bipolar Systems for Instruction, Enlightenment, and Recreation

The first factor to be taken into account in considering the topic of comparative utility of the polar, rectangular, and bipolar systems for didactic purposes is the fact that valid comparisons cannot be made within the framework of present approaches to the teaching of analytic geometry for two principal reasons: (a) typical presentations of the rectangular and polar systems are solely utilitarian, and (b) these presentations have not been based upon a solid foundation of knowledge of their properties nor benefited from the perspectives provided by knowledge of other planar systems.

To entice today's average student into more than a superficial interest in analytic geometry, one needs to do much more than impress him with its practical applications--these are too remote from his everyday needs. What is needed to attract today's students is an exposure to the most stimulating and intriguing aspects of geometry. Only these can rival the appeal of other subjects and attractions. The most intriguing aspects of the two coordinate systems in question are their basic and comparative properties, most of which can be approached from the viewpoint of the presence or absence of symmetry, or of one sort or another of equality. The appeal that follows from understanding and comparing the properties of the two systems and their construction rules and curves, becomes compounded and reinforced with each new system encountered, each of which possesses its own peculiar characteristics.

Accordingly, assuming a much more effective approach to the teaching of analytic geometry through emphasizing the compellingly interesting aspects of curves, construction rules, and coordinate systems, rather than only their practical or applied aspects, where does the bipolar system fit into the picture? As the reader can guess from the amount of space devoted to this system (see also the *Introduction*), I recommend that the bipolar system be the system of choice for introducing coordinate systems.

As reference elements from which to measure distances, not only are two points conceptually simpler and easier for students to relate to than two lines, but undirected distances also are coneptually much simpler to relate to than directed distances. Points of reference (for example, towns on a roadmap) and undirected distances are commonplace in our everyday lives; reference lines and directed distances are not.

But the use of the bipolar system should not extend beyond simple practical geometrical constructions and demonstrations, using curves represented by linear construction rules and the simplest 2nd-degree eqs. One needs only a pencil, ruler, compass, and calculator or slide rule to confirm the essentials of the relations embodied in these construction rules. Paper and blackboard drawings of hyperbolas, ellipses, limacons, and Cartesian ovals could be produced readily using templates. This approach would introduce the student to a host of fascinating curves and emphasize the close relationships that exist between them at the very simple level of linear construction rules in the simplest of coordinate systems.

[The approaches to the properties and relatedness of curves through structure rules, whenever simple and practical, would complement those through construction rules. An appreciation of the merits of the structure-rule approach, however, will have to await Chapters VII and VIII. Needless to say, I would employ the more specific terms, *construction rules* and *structure rules* rather than "equations" in analytic geometry, and use the general term *structural equation*, to encompass both concepts.]

Paralleling these treatments would be an introduction to the concept of the *structural simplicity* of curves about points, using linear bipolar eqs. and the corresponding curves as examples (the curves of Fig. 12). An impressive illustration is the traditional ellipse chalk-and-string construction about two points, with conversion of the constructed locus to a circle as the poles come together. Thus, as the two foci, the points about which ellipses possess the greatest structural simplicity (in terms of constructing the curve), approach one another, the curve becomes ever more symmetrical about them. The symmetry (structural simplicity) of the curve (the circle) reaches the maximum possible about the two merged foci, with the construction rule reducing to 0-degree and taking on the simplest possible form (u or v = R).

Still another impressive illustration associates the use of poles about which ellipses possess the greatest structural simplicity with the presence of the simplest possible terms (1st-degree) in the corresponding construction rules. If any other axial point than the second focus is employed in conjunction with one focus, the construction rule becomes more complicated than the eq. cast about both foci and its degree increases to 2, but it continues to possess a linear term in the variable for the focal pole (eq. 78b). If neither pole is a focus but both lie on the same line of symmetry, the degree of the much-more-complicated construction rule increases to 4 and no linear term is present (eqs. 67a,a'). On the other hand, if a focus is paired with an arbitrary point in the plane, the degree of the still-very-complicated construction rule also is 4, but now a linear term in the variable of the focal pole again is present (eq. 78a).

Thus, if a point about which the curve possesses the greatest structural simplicity is employed as a bipolar pole, a simplest (1st-degree) term in its variable always is present. If two such points are employed, 1st-degree terms in both variables are present, whereas if no such point is employed, no 1st-degree term is present.

Though these proposals would introduce greater complexity into the initial treatments of analytic geometry, the complexity is well within the conceptual grasp of high school students. Moreover, they carry with them the crucial elements of intrigue and fascination that the traditional approaches lack. The vital factor for the beginning student of analytic geometry is not to achieve efficiency and simplicity, but to present a much more appealing format and content to stimulate his interest and imagination. Inasmuch as the bipolar system is one of the most intriguing and easiest to relate to of all planar systems, it provides the logical starting point.

After the introductory treatment of the bipolar system, I would present the rectangular system, emphasizing as much the fundamental properties of the system as dealt with here, as its curves and construction rules. Comparisons between the different systems always would be emphasized, keeping foremost the main theme of how changes in the positional symmetry of a curve influence its construction rules. This would be followed, as is customary, by an introduction to the polar system, featuring the topics of Chapter III. Depending upon progress and course length, I would follow the treatments of the polar system with the essentials of the polar-linear system. Aspects of the latter system rival or exceed those of the bipolar system in their fascination. Treatments of the

Problems

1. Derive the conversion eqs. from the rectangular to the bipolar system for p_u at the origin and p_v at $(d,0)$.

2. What is the sum of the rectangular coordinates, x^2+y^2, in terms of u and v for p_u at $(-d/2,0)$ and p_v at $(d/2,0)$? From the answer, derive the bipolar eq. of the circle of Pythagoras.

3. What is the sum of the rectangular coordinates, x^2+y^2, in terms of u and v for p_v at $(x_1,0)$ and p_u at $(x_2,0)$?

4. Derive eq. 25a.

5. Which of the relations of eqs. 23-26 applies in the rectangular system?

6. What information can be obtained from the bipolar eqs. of lines concerning the relative distances of the lines from the poles?

7. Show that eq. 43a applies for two specific poles on the line $x=2a$.

8. Show that eq. 44c applies at an LR-vertex.

9. Derive the bipolar eq., $(v-d)^2 = (u^2-d^2)$, for a parabola cast about the traditional focus and a pole 4a units to the right of the focus (the reflection of the focus in the directrix--used in the text--is 4a units to the left of the focus). Is this eq. of comparable simplicity to eq. 47f (which would give the pole 4a units to the right of the focus equal subfocal rank to the pole 4a units to the left of the focus)?

10. Derive eq. 50 for focal parabolas.

11. Derive eq. 49e for a midline parabola with poles at the LR-vertices.

12. Show that the maximum bipolar degree of a rectangular curve of degree n is $4n$.

13. Derive the bipolar eq., $[v^2-a^2(1+e^2)-a^4e^2(v^2-u^2+b^4/a^2)^2/4b^6]^2 = a^4e^2[a^2(v^2-u^2+b^4/a^2)^2/b^6-4]$, of a hyperbola cast about the left traditional focus and the left upper LR-vertex.

14. Show that centered minor-axis bipolar ellipses have the 4th-degree eqs., $(b^2-a^2)(u^2-v^2)^2-2b^2d^2(u^2+v^2) + 4a^2b^2d^2 + b^2d^4 = 0$. What is the eq. for the centered minor-axis bipolar ellipse with the b-vertices as poles, expressed in terms of d and e?

15. Derive the distance eq. for the right focus of ellipses.

16. Derive the general axial expression homologous to eq. 64b for ellipses, for the abscissa x of an incident point of a limacon (eqs. 81) in terms of v--the distance from the DP--and u--the distance from a point (x_1,y_1) in the plane. Use this to derive the bipolar eq. of a limacon cast about the DP and the pole (x_1,y_1).

17. Derive eq. 92. Prove the validity of this eq. for $x_1=0$, $x_2=-b/a$, $b=2a$.

18. Derive the linear eq. 94a for limacons using the distance eqs., $v^2 = \pm av - bx$ and $u^2 + a^2 x/b = bu - (a^2-b^2)^2/4b^2$, for the DP and another axial point, and the focus of self-inversion and another axial point, respectively.

19. Derive the bipolar eq. cast about the focus of self-inversion and the $-b/2$ pole of limacons from the distance eqs. 88 and 89.

20. Show that $u^2 = 4dv$ (Fig. $12f_2$) is the bipolar eq. of the equilateral limacon ($e = \sqrt{2} = b/a$) about the focus of self-inversion and the pole at $-b/2$.

21. Show that all the terms of eq. 92 in powers of u and v either have identical coefficients or coefficients in which the abscissae of the two poles are exchanged symmetrically.

22. Derive the bipolar eq. of limacons cast about the DP and the $(a-b)$-vertex.

23. Derive eq. 96b'.

24. Derive the bipolar eq. of hyperbolas cast about the two axial vertices. Express the eq. in terms of d and e.

25. By means of hybrid distance eqs. show that the center of the equilateral lemniscate has higher bipolar subfocal rank than the right vertex ($x_1 = a$).

26. Show that the hybrid distance eq. for a point on the transverse axis of the equilateral lemniscate will not reduce from 4th-degree.

27. Derive the bipolar bifocal eq. of the equilateral lemniscate, $uv = a^2/2 = d^2/4$, from eqs. 101a,b.

28. Derive the linear bipolar eq. of Cartesian ovals about the two foci p_u and p_v from the hybrid distance eqs. 113a,b'.

29. Derive the linear bipolar eq. of Cartesian ovals cast about foci p_u and p_w, for the small oval, from the hybrid eqs. 113a,c'.

30. Derive the linear bipolar eq. of Cartesian ovals cast about pole p_v and p_w, for the small oval, using the conventional distance eqs. 113b,c.

31. Derive the bipolar eq., $z^2 \pm \sqrt{(8C)}d_{uz}v/\sqrt{[C-\sqrt{(C^2-16)}]} = Cd_{uz}^2[C - \sqrt{(C^2-16)}]/4$ for parabolic Cartesians, where p_z is the pole at $x = d_{uz}$ (the $-b/2$ homologue) and p_v is the focus at $x = [8d + Cd_{uz}\sqrt{(C^2-16)} - C^2 d_{uz}]/8$.

32. Derive the conventional distance eq. of parabolic Cartesians for the pole at $x = d$ (the $-b/2$ limacon homologue) and use it to derive the eq. $z^2 = Cdu$ for parabolic Cartesians.

33. Derive the linear bipolar eq. for parabolic Cartesians cast about p_v and p_w, the foci of self-inversion at $x_1 = d(8+CE-C^2)/8$ and $x_2 = d(8-CE-C^2)/8$, respectively, and express it in terms of the distances between the two foci.

34. Derive the linear bipolar eq. for parabolic Cartesians cast about the foci of self-inversion, p_u and p_v, at $x_1 = 0$ and $x_2 = d(8+CE-C^2)/8$, respectively, and express it in terms of the distances between the foci. From this eq. derive the corresponding eq. for the equilateral limacon about the focus of self-inversion at $x_1 = (a^2-b^2)/2b$ and the DP at $x_2 = 0$.

35. Derive eq. 118a for the reduction conditions for the axial distance eq. for parabolic Cartesians.

36. Show that the symmetry condition of eq. 123d applies for the parabolic Cartesians of eq. 121a'.

37. Using the relations of eqs. 124a-e, show that for parabolic Cartesians the conventional distance eqs. 113a,b for foci p_u and p_v, which are expressed in terms of A_L, B_L, C_L, and d_{uv} become the conventional distance eqs. 119a,b, which are expressed in terms of C_P (of $z^2 = Cdu$).

38. Derive eq. 137 either directly from the hybrid distance eqs. or from eq. 136c and the general axial distance eq., $(v^2-u^2-x_1^2+x_2^2)/2(x_2-x_1) = x$.

39. Derive eq. 138.

40. Derive eq. 135 by inverting central conics about the left vertices (let j, the unit of linear dimension, equal a, the length of the semi-major or semi-transverse axis).

41. Show that the general axial focal bipolar eqs. of Cartesian ovals (eqs. cast about one focus and a non-focal pole) are quadratic, and may be circular, elliptical, hyperbolic, or parabolic.

42. Why are the slopes of limacons at the two LR-vertices at the level of the DP equal in magnitude to their eccentricity?

43. How are the angles of the slopes of inverse curves at inverse points related?

44. On the basis of the terms in u^4 and v^4, show that the exterior pole of central Cartesians closest to center is weighted heaviest.

45. Find the general axial bipolar eq. of Cayley's sextic (eq. 6k) cast about poles p_v at x_1 and p_u at x_2. By examining this eq. select the two poles of highest subfocal rank. Find the eq. of the curve about these poles. Which pole is highest ranking? Is there a focal condition?

46. Find the bipolar focus of the sextic of Fig. 8a, $(x^2+y^2)^3 = [a(x^2+y^2)+bx^2]^2$. What is the corresponding focal axial eq?

47. Derive the general axial bipolar eq. of the sextic of Problem 46 cast about (a) two equivalent axial poles, and (b) two arbitrary axial poles, p_v at $(x_1,0)$ and p_u at $(x_2,0)$. What is the general axial distance eq. from point $(x_1,0)$?

48. Find the linear tripolar eqs. in u, v, and w for the C = 5 parabolic Cartesian of Fig. 23. Show the redundancy of the colinear tripolar system by reexpressing these eqs. in terms of linear eqs. in u and w alone.

CHAPTER V.   THE POLAR-LINEAR COORDINATE SYSTEM

Though I have dealt at length with the unique and most-simple properties and relationships between the reference elements of the bipolar system, and the fascinating properties of its construction rules, there are other coordinate systems that rival the bipolar system in the interest and fascination they hold. In fact, some other systems are much more intriguing in certain respects. One of the most interesting of these is the non-incident polar-linear system, consisting of a line and a non-incident point as reference elements. This system possesses no less than three intriguing new properties, of which there is no hint in any of the systems treated in earlier chapters or in the classical literature. In addition, it is the first system in which intuitive concepts of positional symmetry receive severe tests, and one has to rely heavily on the criterion of the comparative simplicities of construction rules.

Polar-Linear Systems and the Hybrid Polar-Rectangular System

The coordinate systems dealt with in this Chapter and Chapter VI employ distances alone as coordinates. These are measured from points on curves to: (a) the reference element (in the case of a point-pole); (b) the nearest point of the reference element (in the case of a line-pole); or (c) either the nearest or the most distant point of the reference element (in the case of a circle-pole). However, these are not the only methods for employing reference elements in coordinate systems. For example, a linear element could be an initial reference line, ray, or half-line (as in the polar system) for measuring angles, or it might be employed as an element along which, rather than from which, distances are measured (as in the hybrid system). Another possibility is to employ a point and a rotating line or ray, as in conchoid-type constructions (see Kavanau, 1982; Fig. 3-7).

On the other hand, the measurement of coordinate distances *along* a fixed line, as opposed to *from* (normal to) a fixed line are essentially identical procedures. Recall that the rectangular coordinates, x and y, which are distances from lines (the y and x-axes, respectively), are measured along lines (the x and y-axes, respectively). Thus, the magnitude of a coordinate distance measured *along* a first reference line (say, the coordinate line of the hybrid system) is the same as that of a coordinate distance measured *from* a second reference line, if the latter line is normal to the first line at the point of origin for the measurements (say, a line through the hybrid pole normal to the hybrid axis). In the event that one measurement is directed and the other is undirected, the two distances can differ in sign.

Comparisons of the degrees of construction rules obtained by employing one or the other of the above methods of measurement reveal that the degrees of eqs. often are twice as great in undirected-distance systems as in directed-distance systems. Thus, a squaring of the directed coordinates often is needed to make all values positive for the undirected-distance system. A parallel displacement of the second reference line from the point of reference on the first line may or may not lead to a change in degree, depending upon the location of the locus. The implication of these relationships for the present considerations is clear. If one considers only polar-linear line-poles that are normal to the hybrid axis, very close parallels often exist between construction rules in the two systems.

Thus, if the point-pole of the hybrid system is incident upon a normal line-pole (the incident polar-linear system), and one considers only the portion of the hybrid locus to one side or other of this line-pole, the construction rules in both systems have the same degree. Moreover, if the symbol x, for the distance from the hybrid point-pole, is replaced by u, the distance from the polar-linear line-pole (p_u), the construction rules become identical, except for possible differences in the signs of odd-degree terms in u. In other words, *the hybrid polar-rectangular coordinate system is an incident polar-linear system with directed distances.*

If one compares the construction rules for given complete loci in the incident polar-linear and hyrid systems, they will either be identical, differ only in the signs of the odd-degree terms in u, or the incident polar-linear construction rule will have twice the degree of the hybrid eq. (and be degenerate, consisting of two terms, one each for the portions of the locus in the right and left half-planes). The two construction rules will be identical for cases in which the locus lies entirely to the right of p_u, and for cases in which the hybrid eqs. possess only even-degree terms in x; they will differ only in the signs of odd-degree terms in u for cases in which the locus lies entirely to the left of p_u; and the degree of the polar-linear eq. will be twice that of the hybrid eq. when the locus lies on both sides of p_u and the hybrid eq. possesses odd-degree terms in x. Thus, for all curves with hybrid eqs. that contain only even-degree terms in x (curves with a true center at the hybrid pole;

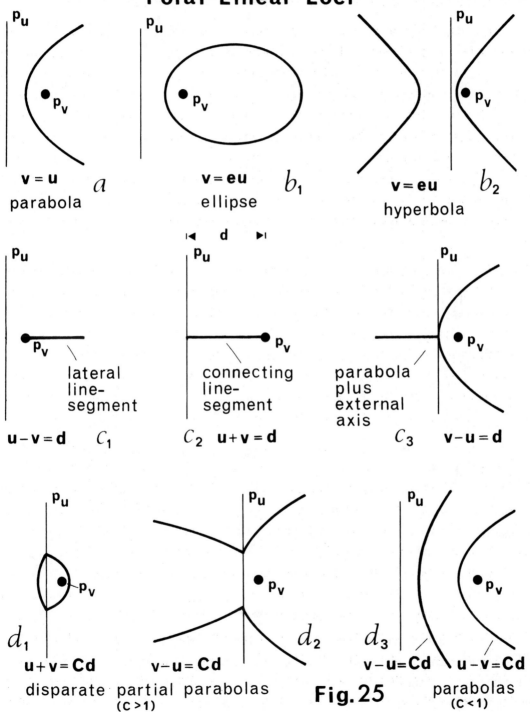

Fig. 25

eqs. 150), the construction rules will be identical.

Similar considerations with regard to degree apply if the hybrid pole is not incident upon p_u. But the situation is complicated by the fact that the plane is divided into non-equivalent portions. The construction rules then differ in the respect that terms in $f_1(\pm u)$, $f_2(\pm u)$, $f_3(\pm u)$, etc., of the polar-linear construction rule become $f_1(\pm u \pm d)$, $f_2(\pm u \pm d)$, $f_3(\pm u \pm d)$, etc. These relations are specified by the transformation eqs. 157-159 and clarified and illustrated with examples.

Polar-Exchange Symmetry

Polar-linear systems are unique in combining in the most simple fashion (i.e., with only one representative of each) the two simplest of all reference elements--a point-pole and a line-pole. The intrinsic asymmetry of the coordinate elements is responsible for one of the newly-encountered properties, *polar-exchange symmetry*. This property characterizes all coordinate systems employing solely distance coordinates and possessing dissimilar, or merely incongruent, reference elements. For example, given an eq. in u and v in such a two-element coordinate system, corresponding to every locus represented by the given eq., there exists a *polar-exchange-symmetrical* locus represented by the same eq. with the variables exchanged.

The most-symmetrical locus (the curve represented by the eq. u = v) in any two-element coordinate system with dissimilar or incongruent reference elements, in this case a parabola (Fig. 25a), always is polar-exchange symmetrical with itself, since exchanging the poles leaves the eq. and the locus unchanged. On the other hand, letting p_v be the point-pole and p_u the line-pole, the ellipse, v = eu (e < 1), is polar-exchange symmetrical with the hyperbola of reciprocal eccentricity, since exchanging the poles leads to the construction rule for a hyperbola, ve = u (e < 1), or v = u/e (Fig. 25b_1,b_2). Similarly, the lateral line-segment u - v = d (Fig. 25c_1) is polar-exchange symmetrical with a parabola and its exterior axis, v - u = d (Figs. 25c_3 and 26e).

Disparate and Compound Curves

The fascinating properties of the bipolar system reside largely in relations between the variables of its construction rules as they correlate with the positions of the represented curves. Differences between curves in the polarized bipolar system, as compared to their corresponding polar and rectangular

Parabola Constructions

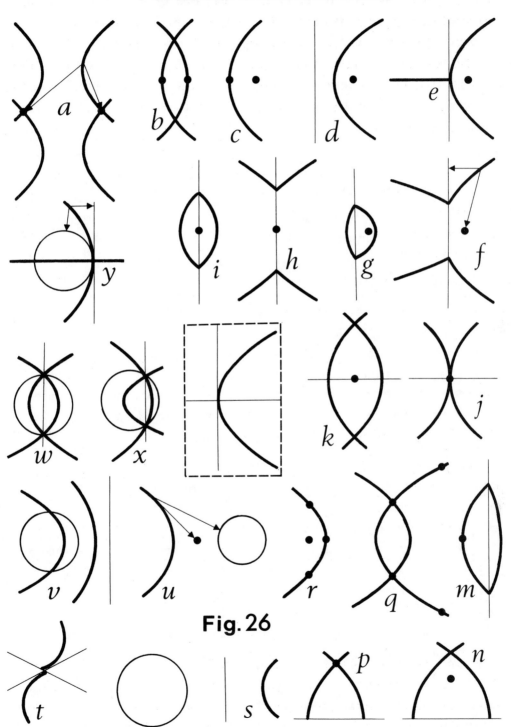

Fig. 26

equivalents, often consist in the possession of only one arm or oval, as opposed to two. This occurs, for example, for hyperbolas and Cartesian ovals that are represented by linear bipolar eqs. On the other hand, bipolar curves sometimes consist of multiple representatives of the curves in the other systems (Figs. 15a,c-g; 26a,b--the unpolarized system). But the forms of the individual segments of the bipolar curves are the same as those in the polar and rectangular systems, and the multiple representatives of these bipolar segments are congruent.

Highly interesting properties of the polar-linear system reside both in its construction rules and its curves. Polar-exchange symmetry is essentially a combination of the properties of construction rules and curves, since the relation between polar-exchange-symmetrical curves has interest only by virtue of the exchange relationship between the variables of their construction rules. The other two newly-encountered properties possessed by the polar-linear system, however, reside primarily in its curves.

In this system one meets types of curves not previously encountered, whose existence is unrecognized in the classical literature. I call these new types of curves *disparate curves* (consisting of one or more pairs of similar but incongruent complete arms, ovals, or segments) and *compound curves* (consisting of joined incomplete arms or segments of common curves). Both types include open and closed representatives. Examples of the former are given in Figs. $25c_3$ and 27j,k,q, and of the latter in Figs. $25d_1,d_2$ and 27s,t.

The occurrence of *compound* and *disparate curves* owes its origin to a property of the non-incident polar-linear system not encountered previously; this is the fact that the line-pole divides the plane into non-equivalent half-planes. Thus, the point-pole occurs only in the one half-plane or the other. Assuming a curve that lies entirely to the right of the point-pole (see Fig. 27a), a consequence of this non-equivalence is that the distance ("abscissa") of a point of the curve from the point-pole along the polar-linear axis (the line of symmetry of the coordinate system) always is greater than its distance from the line-pole in the left half-plane and less than this distance in the right half-plane (reminiscent of the hybrid system, in which v always is greater than x).

As will be seen, a consequence of this non-equivalence of the two half-planes is that the locus of a given construction rule on one side of the line-

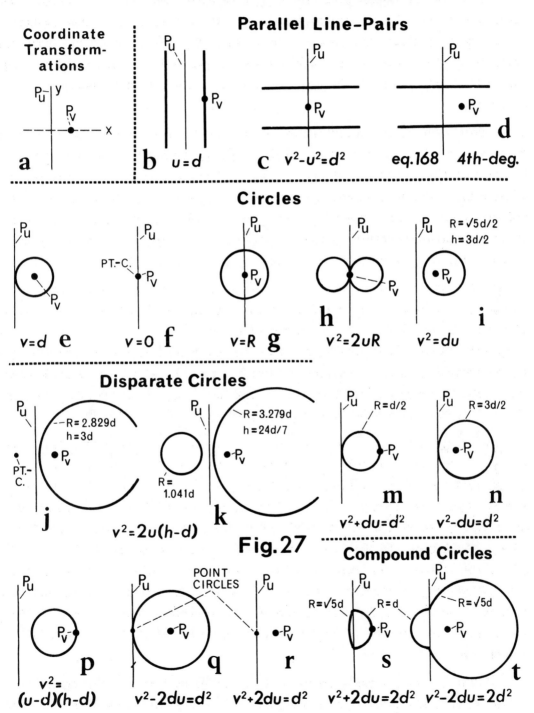
Fig. 27

pole commonly is either of finitely different size from that of a similar locus on the other side (for example, Fig. $25d_1,d_2$), of infinitely different size (for example, Fig. $30b_1-b_3$), real as opposed to non-existent or imaginary (for example, Fig. $25a,b_1$), or dissimilar (see Problems 11 and 15). A degenerate eq. having twice the degree of the eq. for the portion of an initial locus in one half-plane often accommodates the complete initial locus in both half-planes, often accompanied by a complete disparate locus. The construction rule of the hyperbola of Fig. $25b_2$ is the only linear eq. (in both variables) in the non-incident system that represents congruent loci on both sides of the line-pole.

Perhaps the best approach to understanding the unusual properties of this coordinate system (and other systems with one line-pole) is to regard the two half-planes as comprising two separate coordinate systems that meet at the line-pole; all loci in the one system that intersect the line-pole are continuous at the points of intersection with the corresponding loci in the other system (Fig. 25s,t; a point of tangency is not regarded as a point of intersection).

Derivation of Polar-Linear Construction Rules

As concomitants of the other newly-encountered differences characteristic of the polar-linear system, one commonly meets with supernumerary (disparate or dissimilar) loci and degenerate eqs. For example, the polar-linear eq. derived from an initial rectangular eq. may be degenerate, since more than one non-degenerate polar-linear construction rule may be required to represent the rectangular locus in both half-planes. Similarly, when an initial polar-linear eq. of a given locus is converted to a rectangular eq., the locus represented by the rectangular eq. may possess segments not represented by the initial eq.

It also frequently happens that a non-degenerate polar-linear eq. derived to represent a given rectangular curve also represents a similar but incongruent rectangular curve (for example, Figs. $25d_1,d_2$; $27k,s,t$), a dissimilar rectangular curve (see Problems 11 and 15), or what is generally regarded as a different rectangular curve (for example, Figs. $25c_3$; $27j,q$). In the latter case, however, the locus is more appropriately regarded as a similar curve of infinitely different size.

If one takes the y-axis as the line-pole and an axial point (d,0) as the point-pole (Fig. 27a) of the standard system for conversions between rectangular and polar-linear coordinates, it follows that the general expression for

TRANSFORMATION EQUATIONS

general transformation eqs. between the polar-
linear and rectangular coordinate systems
(p_u, the y-axis; p_v, the axial point
at $x = d$)

(a)	$u = \pm x$	general polar-linear u (distance from y-axis, p_u)	(155)
(a')	$u = x$	polar-linear u for curve in 1st and 4th-quadrants	
(a")	$u = -x$	polar-linear u for curve in 2nd and 3rd-quadrants	
(b)	$v^2 = (x-d)^2 + y^2$	polar-linear v^2	
(b')	$v = \sqrt{[(x-d)^2 + y^2]}$	polar-linear v (distance from point-pole, p_v)	
(a)	$x = \pm u$	rectangular x	(156)
(b)	$y^2 = v^2 - (\pm u - d)^2$	rectangular y^2	
(b')	$y = \pm\sqrt{[v^2 - (\pm u - d)^2]}$	rectangular y	

transformation eqs. between the polar-linear and
hybrid coordinate systems

line-pole d units to left of point-pole

(a)	$u = \pm(x+d)$	if the curve intersects the line-pole	(157)
(b)	$u = (x+d)$	if the curve is to the right of the line-pole	
(c)	$u = -(x+d)$	if the curve is to the left of the line-pole	

line-pole d units to right of point-pole

(a)	$u = \pm(x-d)$	if the curve intersects the line-pole	(158)
(b)	$u = (x-d)$	if the curve is to the right of the line-pole	
(c)	$u = (d-x)$	if the curve is to the left of the line-pole	

incident polar-linear system

(a)	$u = \pm x$	if the curve intersects the line-pole	(159)
(b)	$u = x$	if the curve is to the right of the line-pole	
(c)	$u = -x$	if the curve is to the left of the line-pole	

u is simply $\pm x$. Since u is an undirected distance, it always is positive, which means that the upper sign of $\pm x$ represents the initial curve in the 1st and 4th-quadrants, and the lower sign represents it in the 2nd and 3rd-quadrants. The distance from the point-pole is given by the distance eqs. 155b,b'. The sign of the radical is positive in eq. 155b' because v is an undirected distance. The eqs. for the inverse transformations are 156a,b,b'.

When the eqs. obtained after eliminating radicals contain expressions with alternate signs, an additional squaring is needed to obtain the final construction rule. The latter always is degenerate, and expressible as the product of several factors, each of which, equated to 0, represents a segment of the initial curve. Despite this degeneracy, the degree of the polar-linear construction rule must be taken to be that of the degenerate eq., since all of the factors are needed to represent the entire initial curve. Of course, if the curve lies wholly to the right or left of the line-pole (eqs. 155a',a"), squaring will be unnecessary, and the polar-linear construction rule and corresponding hybrid distance eq. will have the same degree.

Although the transformation eqs. 155 and 156 are very useful, a great deal of our attention will be directed to the transformation eqs. 157-159, for use with hybrid distance eqs., since the latter already have been derived and/or inventoried in Table 8 (eqs. 139-154). Moreover, there is no need to derive general construction rules to ascertain polar-linear degree categories; RULES 23 and 24 define the degrees of general distance eqs., and it is evident from the transformation eqs. 157-159 that the maximum degree of the polar-linear construction rule of a rectangular curve is twice the degree of its hybrid distance eq., whence RULES 38 and 39.

RULE 38. *The degree of the general polar-linear construction rule of a rectangular curve is four times the rectangular degree.*

RULE 39. *The degree of the general axial polar-linear construction rule of a rectangular curve is twice the rectangular degree.*

Incomplete-Linear Construction Rules

Incomplete-linear construction rules, or *incomplete-linears*, in any coordinate system, are linear eqs. in which not all of the coordinate elements are represented by variables. In non-coincident two-element coordinate systems, two such construction rules exist. In the present system, these eqs. are v = constant and u = constant, representing a circle centered on the point-pole (the "monopolar" circle) and a parallel line-pair with the two members equidistant from and parallel to the line-pole (known classically as "coordinate curves"). The most symmetrical such loci are the circle, v = d, tangent to the line-pole

(Fig. 27e), and the parallel line-pair, u = d, one member of which includes the point-pole (Fig. 27b), since all coefficients in both construction rules are unity and no extraneous parameters occur.

RULE 40. *Within given degree categories of given curves in given orientations and general locations in two-element coordinate systems with undirected distances, the most-symmetrical representatives (i.e., the representatives with the simplest construction rules) frequently are those that include and/or are tangent to the coordinate elements. In the cases of two loci with indistinguishably simple construction rules, the locus that includes and/or is tangent to both coordinate elements is the more symmetrical.*

Parabolas

Among construction rules in which both variables participate, the most symmetrical locus, of course, is the parabola, u = v (Fig. 25a). The derivation of this construction rule from the corresponding hybrid distance eq. (141) is illustrated here. The pertinent transformation eq., when the line-pole is situated to the left of the point-pole and does not intersect the curve, is 157b. Eliminating x between these two eqs. yields the construction rule 160a. If the line-pole is 2a units to the left of the focus, this simplifies to the sought-after construction rule 160b (Fig. 25a). There can be no locus to the left of the line-pole (directrix), since v always is greater than u in this half-plane.

construction rules of point-focal axial
parabolas

(a) v = u + 2a - d   axial parabola cast about its focus and a line-   (160)
pole d units to the left of the focus (d > a)

(b) v = u   most-symmetrical polar-linear parabola

(c) v - u = d   point-focal axial parabola cast about the
vertex-tangent as the line-pole

It is evident from eq. 141 and the transformation eqs. 157-159, that all point-focal axial parabolas with a non-intersecting line-pole (parabolas for which the point-pole is the focus, the line of symmetry is coincident with the polar-linear axis, and the line-pole is normal to the axis and does not intersect the curve) have linear construction rules, of which eq. 160b merely is the simplest case.

Accordingly, all lines normal to the polar-linear axis and not intersecting the curve are to be regarded as line-pole foci of parabolas, since the degree of the point-focal construction rule is greater than 1 for intersecting line-poles and for non-orthogonal line-poles, regardless of whether they intersect the curve. It also is apparent that parabolas will be among the most-symmetrical curves in other coordinate systems that possess a single point or circle as a coordinate element. This is a consequence of the fact that hybrid distance eq. 141, which provides the initial eq. for deriving the construction rules of focal parabolas in such systems, is the second-most simple hybrid distance eq.

If, instead of selecting the classical directrix as line-pole, one employs the vertex-tangent, by letting $a = d$ in eq. 160a, one obtains the construction rule 160c (Fig. 25c$_3$). Since this rule allows values of v (the distance from the point-pole) greater than u by precisely the amount d (the characteristic distance), a locus also exists in the left half-plane. This is a ray consisting of the parabola's exterior axis. The latter is the limiting form of the left half-plane locus of Fig. 25d$_2$ as C decreases to 1, that is, it is a parabola for which the parameter a, of the eq. $y^2 = -4ax$, is 0.

[It is exceptional for a locus tangent to p_u in one half-plane, as in the present instance, to have a complementary locus in the other half-plane.]

If the line-pole p_u intersects the curve, say for $d = a/2$, in which case p_u is a chord midway between the vertex-tangent and the LR, relation 157a becomes the pertinent transformation eq. This rearranges to eq. 157a'. Eliminating x between eqs. 157a' and 141 yields construction rule 161a-left. Letting $a = 2d$ yields eq. 161a-right. Rearranging and squaring yields the 2nd-degree construc-

construction rules of point-focal axial parabolas

$x = \pm u - d$	transformation eq. 157a rearranged	(157a')
(a) $v = \pm u + 2a - d = \pm u + 3d$	eliminant of eqs. 157a' and 141	(161)
(b) $u^2 = (v - 3d)^2$	point-focal axial parabolas reflected in line-pole midway between vertex-tangent and LR	
(c) $(v + u - 3d)(v - u - 3d) = 0$ lt. half- rt. half- plane plane	eq. 161b as the product of two factors; the half-planes are identified for segments of the initial locus	
(a) $u^2 = (v - 2a)^2$	point-focal axial parabolas reflected in LR	(162)
(b) $(v+u-2a)(v-u-2a) = 0$	eq. 162a as the product of two factors; each factor equated to 0 represents a reflective partial parabola (a compound parabola)	

tion rule 161b, or the product of two factors, 161c.

The factor on the left of eq. 161c represents the segment of the initial parabola in the left half-plane plus a segment of a confocal disparate parabola, $y^2 = -2a(x-a)$, in the right half-plane (p_u, the y-axis; Fig. 25d$_1$; see Problem 1), while the factor on the right represents the complementary segments in the right and left half-planes (Fig. 25d$_2$). Accordingly, the construction rule that represents the entire curve is degenerate for line-poles that intersect the curve. When expressed as the product of two factors, one factor represents one compound parabola and the other represents another compound parabola, each consisting of the complementary segments of the other (Fig. 25d).

[In the following, identification of bracketed or parenthetical terms as representing portions of a locus in one or the other half-plane refers to the portions of the initial locus; if the locus intersects p_u, portions of a disparate or dissimilar locus are represented in the opposite half-plane by the same construction rule.]

The open locus (Fig. 25d$_2$) v-u = Cd (C > 1) is polar-exchange symmetrical with a conventional point-focal axial parabola u-v = Cd (same value of C) in the right half-plane (Fig. 25d$_3$-right). There is no real locus for u+v < d or u-v > d. The construction rule v-u = Cd (C < 1) also represents a conventional parabola in the right half-plane (Fig. 25d$_3$-left).

If, now, the LR is taken to be the line-pole of focal parabolas, one passes over to the incident polar-linear system, in which all loci are symmetrical about both the line-pole and the polar-linear axis, and the characteristic distance is 0. The pertinent transformation eq. now is 159a. Substituting for x in eq. 141 yields the 2nd-degree construction rule 162a, representing two complete parabolas, each the reflection of the other in the LR (composite of Fig. 28c$_1$,c$_2$). Just as construction rule 161b could be factored, eq. 162a factors to 162b. Because the line-pole now is a line of symmetry of the coordinate system, each linear factor equated to 0 represents a partial parabola in each half-plane, reflected in, and becoming continuous with one another at, the line-pole (Fig. 28c$_1$,c$_2$). Accordingly, they are called *reflective* (or *reflected*) *partial parabolas*.

If the axial pole p_v is taken to be non-focal, the degree of the axial distance eq. is 2 (RULE 24), whence it is evident from eqs. 157-159 that parabolas cast about non-focal axial poles have eqs. of either 2nd or 4th-degree, depending upon whether the line-pole intersects the curve. Examples of these for

The Most Symmetrical Incident Polar-Linear Loci

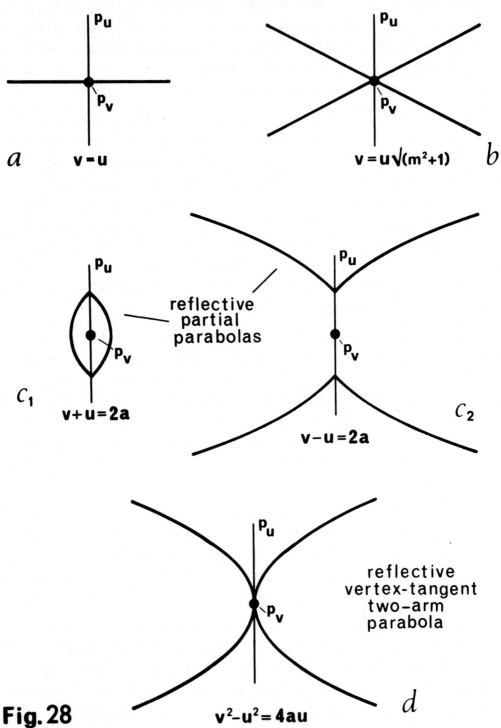

Fig. 28

cases in which the point-pole is at the vertex in the non-incident system, are given in Fig. 29b_1,b_2 for conventional parabolas, and Fig. 29c_1,c_2 for disparate partial parabolas (compound parabolas), in which the line-poles intersect the curves. If entire conventional parabolas are to be represented, 4th-degree construction rules are required in the latter cases, but as these are degenerate, each factors into terms that represent either closed (Fig. 29c_1) or open (Fig. 29c_2) disparate partial parabolas.

For line-poles coincident with the line of symmetry and non-axial point-poles, the hybrid distance eqs. are 4th-degree (RULE 23). Since the line-pole now intersects the curve, a degenerate 8th-degree construction rule is needed to represent an entire initial parabola, in which case a second complete parabola is reflected in the polar-linear axis, giving *reflective dual parabolas* (plus complementary loci; see Problems 2 and 9-11). To represent only the initial ensemble on one side of the line-pole or the other, however, only a 4th-degree construction rule is needed, since the 8th-degree rule is degenerate. An example of such a *reflective dual hemi-parabola* is Fig. 29d_1, for which the point-pole is at a distance x_1 from the vertex (along the axis).

An examination of the construction rule for this curve (Fig. 29d_1) reveals two conditions for which it simplifies materially. From the point of view of coefficient simplicity, the most-symmetrical curve finds the point-pole at an LR-vertex (Fig. 28d_2), which is a circumpolar focus, whereas from the point of view of the number of terms (which takes priority; see RULE 42), this role falls to loci for which the point-pole is incident upon the line at $x = 2a$ (the 2a-chord; see Problem 9). Recall that this line is a locus of high bipolar subfocal rank.

Accordingly, both conditions for notable increases in the structural simplicity of polar-linear reflective dual hemi-parabolas correspond to conditions for increased structural simplicity in other systems—focal in the case of *structure rules* about a point, subfocal in the case of *construction rules* about two points.

Returning to the incident system and considering examples for which the line-pole is coincident with the line of symmetry of the curve, consider first the case for which the point-pole is focal. This condition leads to a non-degenerate 2nd-degree construction rule for a reflective two-arm parabola (Fig. 29a_2). For the point-pole at the vertex, the reflected arms become tangent to

Quadratic And Higher Degree Polar-Linear Parabolas

	incident system	
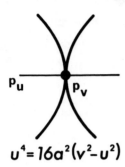 a_1 $u^4=16a^2(v^2-u^2)$	**reflective two-arm**	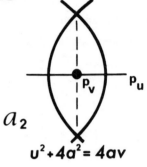 a_2 $u^2+4a^2=4av$
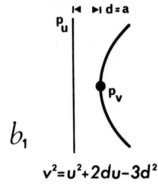 b_1 $v^2=u^2+2du-3d^2$	**non-incident system** **one-arm**	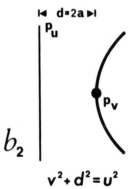 b_2 $v^2+d^2=u^2$
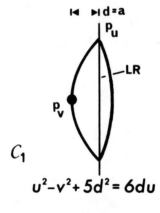 c_1 $u^2-v^2+5d^2=6du$	**disparate partial**	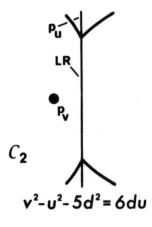 c_2 $v^2-u^2-5d^2=6du$
d_1 $16a^2v^2=u^4+8au^2(2a-x_1)$ $-32a^2du+16a^2(d^2+x_1^2)$	**reflective dual hemi-** Fig. 29	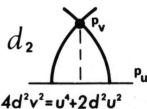 d_2 $4d^2v^2=u^4+2d^2u^2$ $-8d^3u+5d^4$

one another at their vertices, with a 4th-degree construction rule (Fig. 29a$_1$), as for other such non-focal axial point-poles (see Problem 2). This is the most-symmetrical non-focal parabola having its axis coincident with the line-pole, inasmuch as its construction rule is simplest (the numerical coefficient of the right-hand term is arbitrary; see also Problem 2).

The next-most-symmetrical parabola in this group has the point-pole at the reflection of the vertex in the focus, with the construction rule $u^4 = 16a^2(v^2-4a^2)$. The corresponding case of vertex-tangent parabolas (with the line-pole tangent at the vertices; Fig. 28d) has only a 2nd-degree construction rule, since the line-pole also is a focus (as is any non-intersecting line-pole normal to the line of symmetry).

Lines

For the construction rules of lines, line-segments, and rays in coincidence with the polar-linear axis, one employs the hybrid distance eq. 140, $v = x$, for a ray about its terminal point, in conjunction with the transformation eqs. 158 and 159. Considering first a ray that does not intersect the line-pole, and lies to its right, eq. 157b yields the linear construction rule $u-v = d$ (Fig. 25c$_1$). As noted above, this ray is polar-exchange symmetrical with the focal vertex-tangent parabola and its exterior axis, $v-u = d$ (Fig. 25c$_3$). If $d = 0$, the point-pole becomes incident upon the line-pole, yielding the incident system. The ray now is reflected in the line-pole to yield the line $u = v$ (Fig. 28a), the most-symmetrical curve in the incident polar-linear system.

For the line-pole intersecting the ray at a distance d to the right of the point-pole (eqs. 158), the ray to the right of the line-pole has the construction rule $v-u = d$ (the same as for the exterior axis--of the parabola--in the left half-plane of Fig. 25c$_3$). Employing eq. 158c, for the portion (line-segment) of the ray to the left of the line-pole, yields the construction rule $u+v = d$, representing a line-segment that is polar-exchange symmetrical with itself (Fig. 25c$_2$). The construction rule representing a line coincident with the polar-linear axis combines the line-segment and two peripheral rays in a degenerate 3rd-degree eq. (163d) which is identical with the one for a bipolar axial line (eq. 28d).

The general polar-linear construction rule for lines now is derived by substitution from the transformation eqs. 156a,b' in the rectangular slope-inter-

230   THE POLAR-LINEAR SYSTEM

the axial line and its components

(a) $u - v = d$ non-intersecting axial ray extending   (163)
                          to the right of p_u from p_v

(b) $v - u = d$ intersecting axial ray extending to the left
                          of p_u, plus an axial parabola

(c) $u + v = d$ axial line-segment between p_u and p_v

(d) $(u+v-d)(u-v-d)(v-u-d) = 0$ degenerate eq. of the axial line,
                                    including an axial parabola

polar-linear lines

$[v^2-2u(mb-d)-u^2(m^2+1)-(b^2+d^2)]\cdot$ general lines corresponding to   (164)
   right half-plane   the rectangular line $y = mx+b$

$[v^2+2u(mb-d)-u^2(m^2+1)-(b^2+d^2)]=0$
   left half-plane

(a) $[v^2-(m^2+1)(u-d)^2]\cdot$ general lines, including point-   (165)
                                                  pole p_v, before taking roots
      $[v^2-(m^2+1)(u+d)^2]=0$

(b) $[\sqrt{(m^2+1)}u + v - \sqrt{(m^2+1)}d]\cdot$ lines of slopes +m and -m that
         between p_v and p_u include point-pole p_v, plus a
                                                       hyperbola of eccentricity $e = \sqrt{(m^2+1)}$, with point-pole p_v
      $[\sqrt{(m^2+1)}u - v - \sqrt{(m^2+1)}d]\cdot$ as its right focus
         rt. half-plane beyond p_v

      $[v - \sqrt{(m^2+1)}u - \sqrt{(m^2+1)}d]=0$
         left half-plane

      $v = \sqrt{(m^2+1)}u$ lines including point-pole p_v (166)
                                                  in the incident system

$[x+(m^2+2)d/m^2]^2 - y^2/m^2 =$ rectangular hyperbola comple-   (167)
                                                  menting the lines of eq. 165b
      $4(m^2+1)d^2/m^4$

cept eq. for a line, $y = mx+b$, yielding the degenerate 4th-degree construction rule 164. The reduction condition for this eq., that is, the condition for which roots can be taken for both factors (equating each to 0), is $m = -b/d$ ($d \neq 0$), namely, the condition for lines that include the point-pole. In other words, for $d \neq 0$, general lines that include the point-pole are more symmetrical than those that do not (RULE 40). Substituting for b in eq. 164 yields eq. 165a, and taking roots yields the degenerate 3rd-degree construction rule 165b.

Each factor of eq. 165b, equated to 0, represents a portion of the lines of the complete eq. (the discarded linear factor has no real locus). The left-most factor represents the line-segments between p_u and p_v, the middle factor represents the rays beyond p_v, and the right-most factor represents the rays in the left half-plane. Letting $m=0$, of course, leads to the degenerate 3rd-degree construction rule for the axial line, which is the most-symmetrical non-incident polar-linear line. Each axial portion of the line (i.e., for $m=0$) is more symmetrical than the corresponding non-axial portion, both intuitively and because its construction rule is simpler (all coefficients are unity).

Having noted that an axial parabola in the right half-plane complements the ray in the left half-plane of the 3rd-degree axial line (eq. 163d), it is natural to inquire as to whether a right half-plane locus also complements the rays in the left half-plane (eq. 165b, right-most term) of the 3rd-degree lines that include pole p_v (eq. 165b). Taking the right-most term of eq. 165b and making the substitutions of eqs. 155a,b', yields the construction rule for a rectangular hyperbola (eq. 167) of eccentricity $e = \sqrt{(m^2+1)}$. The right-most polar-linear term represents only a portion of the right arm of this hyperbola (in the right half-plane), of which the right focus is pole p_v (Fig. 30b_1). In fact, the line-segments between p_v and p_u (left-most term of eq. 165b) and the intersecting rays extending from p_v (middle term of eq. 165b) in the right half-plane, also have complementary segments that belong to the same rectangular hyperbola (eq. 167) and become continuous with it at the line-pole (Fig. 30b_3,b_2, respectively).

The line-segments are complemented in the left half-plane by the segment missing from the right arm mentioned above (Fig. 30b_3), while the rays are accompanied in the left half-plane by the entire left arm (Fig. 30b_2). Accordingly, the 3rd-degree construction rule for lines that intersect at point-pole p_v also represent a complete rectangular hyperbola of eccentricity $e = \sqrt{(m^2+1)}$, with p_v as its right focus. In fact, the intersecting lines in question are parallel to the asymptotes of this hyperbola (they represent the limiting form of a hyperbola of the same eccentricity, as the parameters a and b of the rectangular eq. approach 0).

In the case of intersecting lines with $d=0$, whereupon p_v is incident upon p_u, a parallel derivation leads to the linear construction rule 166. This represents intersecting lines that include the point-pole in the incident polar-linear system (Fig. 28b). The most-symmetrical of these lines is the line of

The Most Symmetrical Non-Incident Polar-Linear Loci

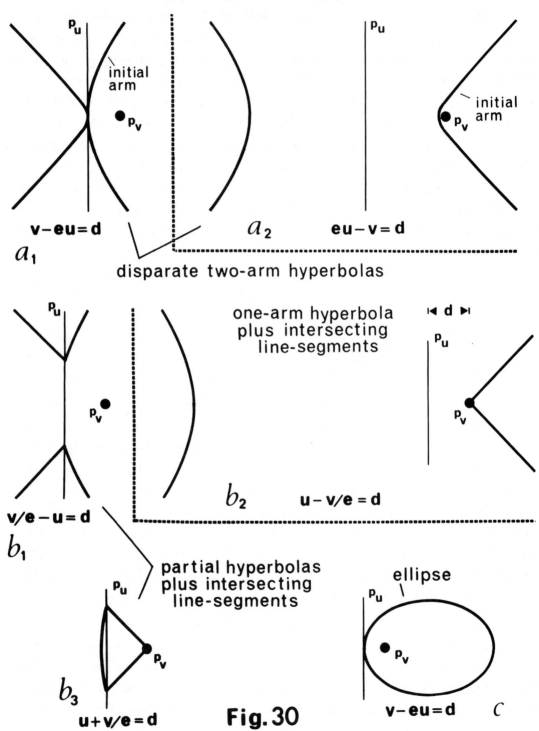

Fig. 30

slope 0 (Fig. 28a), with the construction rule v = u, which also is the most symmetrical locus in the incident polar-linear system.

In determining the second-most-symmetrical locus in the bipolar system, the fact that both poles were points made a decision clear-cut, since this demanded that priority be given to equivalence of the coefficients of terms in u and v (RULE 9). Since the two elements of the present system are dissimilar, this factor does not have priority. By the same token, the comparative weighting of the poles is not a criterion for assessing relative structural simplicity about the coordinate elements. In discriminating between polar-linear construction rules with two, as opposed to three, terms, priority goes to the eq. with two terms. Accordingly, two lines intersecting at the point-pole (of non-zero slope; Fig. 28b) are the second-most-symmetrical loci in the incident polar-linear system, while ellipses and hyperbolas (Fig. 25a,b) are the second-most-symmetrical loci in the non-incident system.

RULE 41. *The second-most-symmetrical loci in two-element coordinate systems with dissimilar elements are those represented by the construction rule $u = Cv$.*

RULE 42. *In discriminating between simplicities of construction rules (in given degree-categories) in both variables of two-element coordinate systems with dissimilar elements, priority goes to the equations with fewest terms.*

Our intuition still stands us in good stead in confirming the greater symmetry of the parabola than of central conics in the non-incident system. This depends, in essence, on the fact that both the parabola and the coordinate system have only one line of symmetry. [In fact, these same circumstances account for the greater symmetry of the parabola (and ray), as compared to central conics, in the hybrid polar-rectangular coordinate system.] For example, if one aligns an ellipse in the polar-linear system such that its two b-vertices are symmetrically positioned in the system (Fig. 25b$_1$), its two a-vertices become asymmetrically positioned, etc. This is not the case for a parabola, whose symmetry matches that of the system.

Lastly, consider the cases of line-pairs parallel to the polar-linear axis (Fig. 27d). Using the rectangular construction rule $y^2 = b^2$ as the initial eq. ($y = b$ gives the same result, since all polar-linear loci are reflected in the polar-linear axis), in conjunction with eq. 156b (or 156b'), yields the degenerate 4th-degree construction rule 168, expressed as the product of two factors. For b = 0, of course, this becomes the product of four factors, one of which can be discarded (since it represents no real locus) giving the 3rd-degree eq. 163 for an axial line (plus a complementary parabola).

(a) $[v^2-(u+d)^2-b^2][v^2-(u-d)^2-b^2] = 0$ parallel line-pair parallel to   (168)
      rays in left   rays in right   the polar-linear axis of the
       half-plane   half-plane   non-incident system

(b) $v^2 - u^2 = b^2$ eq. 168a for the incident system

In the incident system (Fig. 27c), the non-degenerate 2nd-degree construction rule 168a becomes 168b, for which there is no polar-exchange-symmetrical locus. For the minus sign changed to a plus sign (yielding a "circular" eq.), eq. 168b becomes a unique locus in the incident system, namely, an $e = \sqrt{2}/2$ ellipse with its minor axis coincident with the polar-linear axis (Fig. 31d$_2$). In the non-incident system, circular eqs. also represent $e = \sqrt{2}/2$ ellipses (Fig. 31c), the most symmetrical of which, $u^2+v^2 = d^2$ (the analogue of the circle of Pythagoras), is tangent to the line-pole (see RULE 40). The ellipses of these circular eqs. are polar-exchange symmetrical with themselves. [One sees that the criterion for equivalence of poles in the bipolar system becomes that for "self" polar-exchange symmetry in the polar-linear system (and other coordinate systems with dissimilar elements).]

Central Conics

Focus-Directrix Construction Rules

The hybrid distance eqs. of central conics cast about their left foci are employed to derive the standard focus-directrix construction rules, as well as construction rules cast about a focus and a contralateral directrix, and a directrix and a contralateral focus. Taking the left directrices as line-poles p_u, and the left foci as point-poles p_v, $u = d-x$ (eq. 158c) for hyperbolas and $u = d+x$ (eq. 157b) for ellipses. Substituting for x in the hybrid distance eqs. 143 and 142, respectively, yields eqs. 169a and 170a.

The normal lines having the highest subfocal rank (the line-poles about which the curves have the simplest construction rules) are the ones for which the constant terms of eqs. 169a and 170a vanish. These conditions (eqs. 169b and 170b) define the traditional directrices, at distances a/e from center for ellipses and hyperbolas. Accordingly, making the substitutions of eqs. 169b and 170b in eqs. 169a and 170a, respectively, leads to the familiar focus-directrix construction rules 169c and 170c. Remarkably, the construction rule for the hyperbola applies to both arms, even though the hybrid distance eq. does not.

derivation of focus-directrix construction rules
of central conics

hyperbolas

(a) $v = eu - ed + a(e^2-1)$ construction rule for the left arm of a hyperbola cast about the left focus and a normal line-pole to the right of the arm   (169)

(a') $v = eu + ed - a(e^2-1)$ eq. analogous to eq. 169a for the right arm and a line-pole to its left

(b) $d = a(e^2-1)/e$ condition for highest subfocal rank of a line-pole to the right of the left arm; identifies the traditional lt. directrix

(c) $v = eu$ most-symmetrical hyperbolas (both arms)

(d) $d = a(e^2+1)/e$ distance of rt. directrix from left focus

(e) $v = eu - 2a = eu - 2de/(e^2+1)$ construction rule of the lt. arm of a hyperbola about the lt. focus and rt. directrix

(e') $v = eu + 2a = eu + 2de/(e^2+1)$ construction rule of the rt. arm of a hyperbola about the lt. focus and rt. directrix

ellipses

(a) $v = eu - ed + a(1-e^2)$ construction rule of an ellipse cast about the left focus and a line-pole to the left of the curve   (170)

(a') $v = ed - eu + a(1-e^2)$ eq. analogous to eq. 170a for a line-pole to the right of the curve

(b) $d = a(e^2-1)$ condition for highest subfocal rank of a line-pole to the left of the curve; identifies the traditional lt. directrix

(c) $v = eu$ most-symmetrical ellipses

(d) $d = a(e^2+1)/e$ distance of rt. directrix from lt. focus

(e) $v = 2a - eu = 2de/(e^2+1) - eu$ construction rule of an ellipse cast about the left focus and right directrix

This unusual result ensues as follows. When the hybrid distance eq. for the right arm (eq. 143') cast about the left focus is combined with the appropriate relation between u, d, and x for this arm, $u = x - d$ (eq. 158b), the resulting construction rule for the right arm cast about the left focus and a line-pole to the left of the arm is 169a'. The only differences from the corresponding construction rule for the left arm are in the signs of both constant terms. But since the same condition applies for the vanishing of the constant terms from the eqs. of both arms, the same most-simple construction rule also applies for both arms. Thus, the left directrix is the unique line-pole in the plane of a rectangular hyperbola for which the polar-linear construction rule cast about the left focus is the same for both arms. Accordingly, this commonplace construction rule results from an unusual combination of circumstances.

To illustrate further the unusual nature of the above relation, the construction rules cast about the left focus and left directrix are contrasted with those cast about the left focus and right directrix. The same general construction rules that apply to the left and right arms of hyperbolas cast about the right directrix (eqs. 169a,a') also apply for the left directrix, since both directrices lie between the arms. For an ellipse, however, the right directrix is on the other side of the curve from the left directrix, and the relationship between u, d, and x becomes $u = d - x$ (eq. 158c), yielding the general construction rule 170a'. Substituting the distances between the left foci and the right directrices (eqs. 169d and 170d) in eqs. 169a,a' and 170a, a', yields the sought-after construction rules for the left foci and far directrices (eqs. 169e,e' and 170e); whereas a single construction rule applies for ellipses (see Problem 4), separate rules apply for each of the arms of hyperbolas.

Focus-Line-of-Symmetry Construction Rules

To derive construction rules about the left focus and the conjugate or minor axis, one needs to employ both eqs. 169a,a' for hyperbolas and eqs. 170a,a' for ellipses. Since portions of both loci lie to each side of the selected line-poles, two eqs. are needed to represent each locus. Letting $d = ae$ for both central conics (the focus-center distance) yields the degenerate construction rules 171a and 172a. These are expressed as products of factors in eqs. 171a' and 172a'. Note that the factors that apply to the segments in the half-planes opposite the foci (poles p_v) are identical for both curves, leading to a common construction rule, $v = eu + d/e$, for these segments.

focus-line-of-symmetry construction rules of central conics

(a) $(v-eu)^2 = d^2/e^2$ two-arm hyperbolas cast about a focus and the conjugate axis (an incongruent complementary locus exists)   (171)

(a') $(v-eu-d/e)(v-eu+d/e) = 0$ eq. 171a as the product of factors; each factor is identified relative to focus p_v (the point-pole)
    opposite   initial
      arm   arm

(b) $[x+2d/(e^2-1)]^2 - y^2/(e^2-1)$
    $= d^2(e^2+1)^2/e^2(e^2-1)^2$ incongruent hyperbolas whose arms complement the arms of the initial hyperbolas of eqs. 171a,a'

(a) $(v - d/e)^2 = e^2 u^2$ ellipses cast about a focus and the minor axis (an incongruent complementary locus exists)   (172)

(a') $(v-eu-d/e)(v+eu-d/e) = 0$ eq. 172a as the product of factors; each factor is identified relative to focus p_v (the point-pole)
    opposite   same
   half-plane half-plane

(b) $[x+2d/(1-e^2)]^2 + y^2/(1-e^2)$
    $= d^2(1+e^2)^2/e^2(1-e^2)^2$ incongruent ellipses complementing the initial ellipses of eqs. 172a,a'

In general, to investigate the existence of, and identify, a complementary polar-linear locus in the half-plane opposite to a given locus, one simply transforms the polar-linear construction rule derived for the given locus in the one half-plane to the corresponding rectangular eq. for the opposite half-plane. By this procedure, for example, one finds that the construction rules 171a,a' for an initial hyperbola also represent the incongruent hyperbola of of eq. 171b, one arm of which complements each arm of the initial curve (Fig. 31b_1,b_2; see Problem 12)

Vertex-Line-of-Symmetry Construction
   Rules

Some examples of *compound ellipses* are represented in Fig. 31a_1,a_2; the initial locus is the right half of an ellipse cast about its right vertex and minor axis (Fig. 31a_1). A complementary segment of a larger ellipse of the same eccentricity is represented in the left half-plane; this becomes continuous with the initial locus at the line-pole, forming a closed curve. The complementary situation is represented in Fig. 31a_2. The initial hemi-ellipse of the left half-plane is cast about the minor axis and the right vertex. This forms a closed curve with the portion of a complementary similar ellipse in the right

Quadratic Polar-Linear "Central Conics"

non-incident

compound ellipses

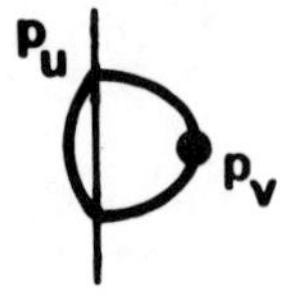

$$v^2 + 2du - e^2u^2 = d^2(2-e^2) \qquad a_1$$

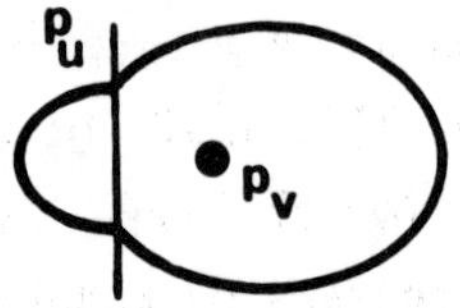

$$v^2 - 2du - e^2u^2 = d^2(2-e^2) \qquad a_2$$

disparate two-arm hyperbolas

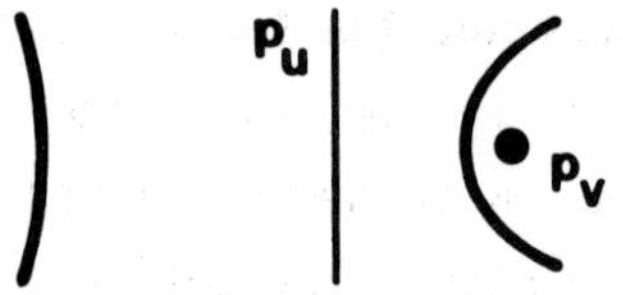

$$v + d/e = eu \qquad b_1$$

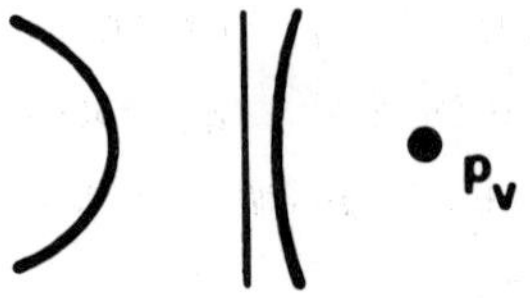

$$v - d/e = eu \qquad b_2$$

ellipses

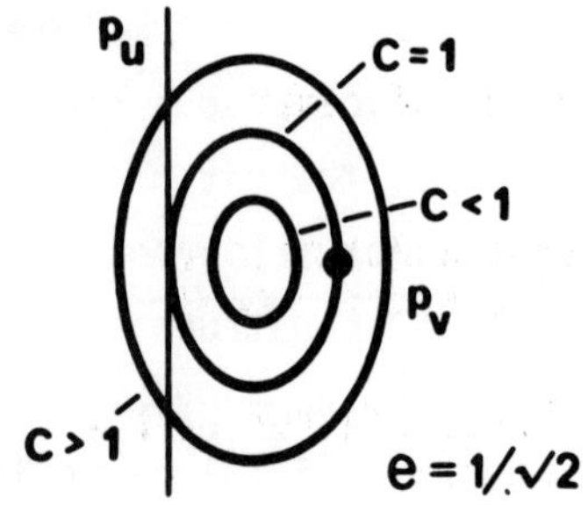

$$u^2 + v^2 = Cd^2 \qquad c$$

$$e = 1/\sqrt{2}$$

incident

reflective vertex-tangent two-arm hyperbola

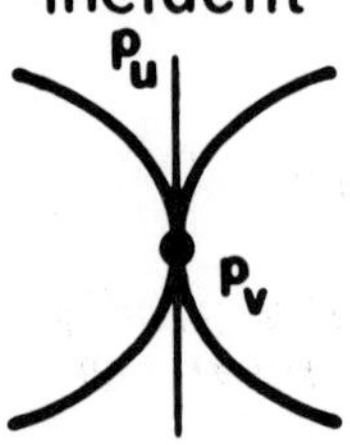

$$v^2 - e^2u^2 = 2au(e^2-1) \qquad d_1$$

ellipse

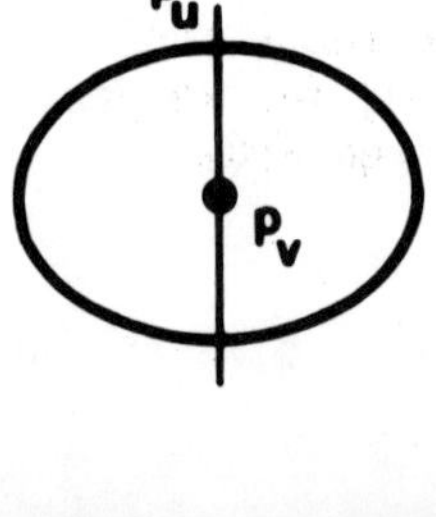

$$v^2 - e^2u^2 = a^2(1-e^2) \qquad d_2$$

Fig. 31

half-plane. Since these axial ellipses are non-point-focal (i.e., p_v is a vertex rather than one of the foci), it requires a 2nd-degree construction rule to represent them (and a 4th-degree eq. is needed to represent the complete loci).

Focus-Vertex-Tangent Construction Rules

Construction rules cast about a central conic focus and the near vertex-tangent now are considered (Fig. 30a$_1$,c). Eqs. 169a and 170a apply in this situation. Letting $a = d(e-1)$ for hyperbolas and $a = d(1-e)$ for ellipses, yields the common central conic construction rule 173a, which represents no complementary locus for ellipses (see Problem 5). When cast about a focus and the far vertex-tangent, however, the two construction rules are different; letting p_v be the left focus, the applicable initial eqs. are 169a and 170a'. With $d = a(e+1)$ for both curves, the construction rules become 173b,c. In both cases, the eqs. apply only to the initial arms of hyperbolas (whose foci are at p_v) but to entire ellipses. It readily is shown that there can be no complementary real locus in the half-plane beyond the vertex-tangent of an ellipse (see Problem 5) but a complementary locus does exist for the arm of a hyperbola.

focus-vertex-tangent construction rules of central conics

(a) $v - eu = d$ — ellipses and arms of hyperbolas cast about a focus and a near vertex-tangent (no complementary locus exists for ellipses) (173)

(b) $eu = v + d$ — arms of a hyperbola cast about a focus and far vertex-tangent (a complementary locus exists)

(c) $eu = d - v$ — ellipses cast about a focus and the far vertex-tangent (no complementary locus exists)

(a) $(x-ae)^2 - y^2/(e^2-1) = a^2$ — rectangular hyperbola of eq. 173a (174)

(b) $[x - ae(e-1)/(e+1)]^2 - y^2/(e^2-1) = a^2[(e-1)/(e+1)]^2$ — complementary rectangular hyperbola of eq. 173a cast about a focus and the far vertex-tangent

$(v-eu)^2 = d^2$ — dual two-arm hyperbolas cast about a given focus and a near vertex-tangent of one initial arm and far vertex-tangent of the other initial arm (derived from eqs. 173a,b) (175)

To identify the latter locus, eq. 173a is transformed to its rectangular equivalents, yielding two rectangular hyperbolas (see Problem 3). One of these (eq. 174a) represents both arms of the initial curve. The other (eq. 174b) is the complementary incongruent hyperbola, diminished in size in the ratio $(e-1)/(e+1)$. As represented by eq. 173a, the incongruent arm is cast about its vertex-tangent and the far focus.

In other words, for an initial right arm of a hyperbola cast about its vertex-tangent and focus, the complementary incongruent locus is a left arm cast about its vertex-tangent and the far focus (Fig. $30a_1$). Thus, p_v is at the focus of both right arms. Since the line-pole is the vertex tangent of both the initial right arm and the complementary left arm, these two arms are tangent to one another at their vertices.

On the other hand, the complementary locus, corresponding to construction rule 173b, $ev-u=d$, for an arm of a hyperbola cast about a focus and the far vertex-tangent, reveals a different situation. Whereas the complementary arm of eq. 173a was reduced in size and cast about its vertex-tangent and the far focus, the complementary arm for this case is magnified and cast about both the far focus and the far vertex-tangent (Fig. $30a_2$; see Problem 6). Consequently, the degenerate eq. 175, derived from eqs. 173a,b, is the construction rule for dual two-arm hyperbolas (composite of Fig. $30a_1,a_2$). Note that this construction rule differs from the corresponding eq. (171) cast about a focus and the conjugate axis only in that the right side of the latter eq. is divided by e^2.

The Shapes of Complementary Segments

Examination of eqs. 171 and 174 reveals that both the initial and the complementary hyperbolas have the same eccentricity, and this also is true of complementary ellipses (see eqs. 172). Similarly, the complementary segments of axial circles (see below) and parabolas (see above) are similar to one another, bringing to attention the following RULES. The basis for these RULES is understood readily in terms of the derivations of the rectangular eqs. of the complementary polar-linear segments of an initial curve in the opposite half-planes, as follows.

> RULE 43. *The complementary segments of axial polar-linear conic sections in the two half-planes possess the same eccentricity.*
>
> RULE 44. *The complementary segments of polar-linear curves (other than axial conics) in the two half-planes generally are dissimilar.*

RULE 45. *The complementary segments of curves in the two half-planes of two-element coordinate systems in which one element is a line (and the other element is wholly to one side of the line) generally are dissimilar.*

RULE 46. *In general, a "partial translation" of a rectangular cubic or curve of higher degree yields a dissimilar curve of the same degree.*

To obtain the rectangular eq. of the segment of a polar-linear locus in the p_v-containing half-plane, one substitutes $[(x-d)^2+y^2]$ for v, and x for u, whereas to obtain the eq. of the segment in the opposite half-plane, one substitutes $[(x+d)^2+y^2]$ for v, and x for u in the same polar-linear eq. If either of these segments is considered to be the initial rectangular one, the substitution employed to obtain the initial rectangular eq. from the polar-linear one is appropriate for that segment. However, the substitution employed to obtain the rectangular eq. of the segment in the opposite half-plane is the equivalent of performing a "partial translation" (see *Glossary of Terms*) of the initial rectangular curve. What influence does this have on the nature of the locus in the opposite half-plane?

[It is evident, for example, that if the substitutions for u and v given above yield the rectangular eq. of the segment of the polar-linear locus in the p_v-containing half-plane, then the substitutions for the opposite half-plane amount to an x + 2d translation of the locus of this rectangular eq. for the terms involved in the substitutions for v but no translation for the terms involved in the substitutions for u.]

Consider, first, the influences of such "partial translations" on the loci represented by linear and quadratic rectangular eqs. In the case of a linear eq., a "partial translation" cannot take place, since there is only one term in x. When the eq. is quadratic, two results are possible. The locus of the eq. may merely be translated (for example, for the eqs. $x^2 + y^2 = R^2$ and $y^2 = 4ax$) or it may be both translated and changed in size (for example, in replacing the x of the x^2-term of the eq. $x^2 + dx + y^2 = a^2$ by $x \pm h$). In both cases, the eccentricity is unaltered.

On the other hand, if the rectangular eq. of the locus is 3rd-degree or greater, three influences are possible--both of those mentioned above plus a change in shape. Thus, in general, a "partial translation" of a rectangular cubic or higher-degree curve gives a dissimilar curve of the same degree (RULE 46).

Returning to the polar-linear situation, whenever the rectangular eqs. of the segments of the initial locus are of greater than 2nd-degree, whether degenerate or not, the complementary segments in the opposite half-plane are dissimilar to the initial locus (since terms in both u and v occur in the polar-linear eq., a partial translation always occurs). For example, the reflective dual hemi-parabolas of Fig. 29d_1,d_1 have polar-linear construction rules of 4th-degree in each half-plane. The rectangular eqs. obtained for the initial curves in the p_v-containing half-plane are degenerate quartics that represent two parabolas, whereas those of the complementary loci in the opposite half-plane are non-degenerate quartics. The latter represent curves that possess two parabola-like segments. These relationships are dealt with in Problems 10 and 11.

The Incident System

Before proceeding to a consideration of circles, two additional construction rules for ellipses and hyperbolas in the incident polar-linear system are considered. The first, the non-degenerate quadratic eq., $v^2 - e^2u^2 = 2au(e^2-1)$, defines the *reflective vertex-tangent two-arm hyperbola* of Fig. 31d_1. Of particular interest is the fact that this is the first "hyperbolic" polar-linear construction rule we have encountered for (congruent) two-arm hyperbolas.
[Although two-arm hyperbolas with arms that are tangent to one another are a new concept, they are no less valid constructions than the familiar rectangular hyperbolas.]

The second example--an ellipse centered in the incident system--illustrates a polar-linear construction rule identical with the hybrid polar-rectangular distance eq. for the same curve, save for replacing x by u. Eq. 150 is the hybrid distance eq. of 0°-ellipses about their centers. If line-pole p_u of the polar-linear construction rule is taken to be the minor axis, then x of the distance eq. becomes ±u of the polar-linear construction rule. But since x occurs only to the 2nd-degree in the hybrid eq., no additional squaring is needed to eliminate the alternate signs. Thus, the resulting construction rule with ±u replacing x is not degenerate. The same considerations apply to all hybrid distance eqs. involving only even powers of x, of which those for ellipses and hyperbolas cast about their centers are the only examples in Table 8 (for another example, see Problem 25).

Circles

Most of the polar-linear construction rules for ellipses can be converted to those for the corresponding circles by letting e = 0. In this respect, the most interesting eqs. of those considered in the preceding section are those for the *compound ellipses* of Fig. 31a₁,a₂. Letting e = 0 yields eqs. 176a₁,a₂ for *compound circles* (Fig. 27s,t).

(a₁) $v^2 + 2Ru = 2R^2$ hemi-circle in the right half-plane cast about the diameter as line-pole and the right intersection with the polar-linear axis as the point-pole; the locus in the left half-plane is a segment of a circle of radius $\sqrt{5}R$ (176)

(a₂) $v^2 - 2Ru = 2R^2$ the complementary segment of the circle of eq. 176a in the opposite half-plane

Some additional examples of polar-linear circles and the degrees of their construction rules are given in Fig. 15, and further treatments are to be found in Kavanau (1980). Our additional considerations here are confined to some most interesting cases, namely *disparate* and *compound axial circles*. Although the following development could be carried out employing the hybrid distance eqs. for circles about an axial point (eq. 145), it is more easily comprehended in the following form.

The general eq. for axial rectangular circles is $(x-h)^2 + y^2 = R^2$. If one takes the standard polar-linear configuration of Fig. 27a, with the y-axis as line-pole and the point-pole at x = (d,0), the pertinent transformation eqs. are 155a,b,b'. Making the appropriate substitutions and combining terms leads to the general polar-linear construction rule 177 for axial circles. Note that this has "displaced parabolic" form (the form of the eq. of rectangular vertex parabolas translated along the x-axis), like eqs. 176a₁,a₂, but unlike any of the other construction rules for central conics. In fact, all polar-linear "vertex parabolic" and "displaced parabolic" construction rules represent axial circles (see Problem 8).

There are four obvious symmetry conditions on eq. 177. The condition d = h, for vanishing of the term in u, yields monopolar circles and is not enlightening. The condition for the vanishing of the constant term, however, defines the most-symmetrical polar-linear circles in the non-incident system, that is the polar-linear circles with the simplest construction rules (possessing only two terms--the minimum possible for an eq. in both variables).[These, in fact, are focus-directrix constructions; see RULE 47, below.] The most-symmetrical of these circles and, consequently, the most-symmetrical circle in the non-incident system, is the circle for which h = 3d/2 (Fig. 27i) and construction rule

179a takes its simplest form, $v^2 = du$ (eq. 179b). The curve that is polar-exchange symmetrical with this axial circle, $u^2 = dv$, is a *parabolic polar-linear conchoid* (see Kavanau, 1980, 1982).

construction rules for axial circles

	$v^2 \pm 2u(d-h) = R^2 - h^2 + d^2$	axial circles centered at $x = h$, i.e., a distance h to the right of pole p_u	(177)
	$h^2 = R^2 + d^2$	condition for vanishing of the constant term of eq. 177	(178)
(a)	$v^2 = 2u(h-d)$	most-symmetrical polar-linear circles	(179)
(b)	$v^2 = du$	most-symmetrical circle ($h = 3d/2$ in eq. 179a)	
(a)	$(x+h-2d)^2 + y^2 = 4d(d-h) + R^2$	rectangular eq. of the complementary circles in the left half-plane defined by eq. 177	(180)
(b)	$(x+h-2d)^2 + y^2 = (2d-h)^2 - d^2$	circles of eq. 180a for the most-symmetrical cases (condition of eq. 178)	
(a)	$h \geqslant 2d$, (a') $R^2 \geqslant 4d(h-d) \geqslant 4d^2$	conditions on eq. 180a for the existence of real complementary circles centered in the left half-plane	(181)
(b)	$(2d-h)^2 \geqslant d^2$, (b') $h \geqslant 3d$	conditions on eq. 180b for the existence of segments of real complementary circles	
	$v^2 = 2uR$	eq. 179 for the incident system	(182)
(a)	$v^2 + du = d^2$	the most-symmetrical circle with a 3-term eq.; includes point-pole p_v and is tangent to line-pole p_u	(183)
(b)	$v^2 - du = d^2$	the second-most-symmetrical circle with a 3-term eq.; is tangent to p_u	

Condition 178 requires that d be less than h but greater than h-R ($R \neq 0$). Since d is the distance between the point-pole and the line-pole, this means that the most-symmetrical polar-linear axial circles are those for which p_v is enclosed by the circle in the p_v-containing half-plane--those circles for which pole p_v lies to the left of the center of the circle at a distance from the line-pole greater than h-R (Fig. 27i-k). It is no coincidence that the same circumstances apply to the most-symmetrical bipolar circles--the circles of Apollonius. It is these most-symmetrical and generally disparate circles (eq. 179) which are considered next.

The upper alternate sign in eq. 177 allows for the portion of a locus to the right of the line-pole, and the lower sign for the portion to its left.

Since $h > d$, it is the upper alternate sign that applies to the circles in question, for which the polar-linear construction rule becomes eq. 179. However, eq. 179 does more than define the locus of the initial circle, $(x-h)^2 + y^2 = R^2$, for the stated conditions. Under certain conditions, it also defines complementary circles in the left half-plane.

The rectangular construction rule for these complementary circles is derived by transforming eq. 177 with the lower alternate sign, and with condition 178 satisfied, into rectangular coordinates, yielding eq. 180a. In order for these circles to be real, the right side of eq. 180a must be positive, leading to the left side of condition 181a' $[R^2 \geqslant 4d(h-d)]$.

The condition $d = h/3$ yields a complementary point-circle at the reflection of p_v in p_u (Fig. 27j), with the circle in the right half-plane being of radius $\sqrt{8}d$. As p_v approaches p_u, that is, as d diminishes from $h/3$, the complementary circle in the left half-plane enlarges. Concomitantly, the circle in the right half-plane also enlarges (Fig. 27k). In the limit, for $d = 0$, the incident system is attained, both circles attain the radius $h = R$ (Fig. 27h), and the construction rule achieves its simplest form, 182, which is identical with the hybrid distance eq. 144 for a circle about an incident point (with u replacing x).

The third symmetry condition on eq. 177 is that for which all three terms attain unit coefficients, namely $h = R = d/2$. This is the condition for a tangent circle that includes the pole (eq. 183a and Fig. 27m). This circle is less symmetrical than those of eq. 177 by virtue of the fact that its construction rule has three terms, as opposed to two (RULE 42). But it is the most-symmetrical circle with a three-term construction rule by virtue of the fact that it possesses only unit coefficients (see RULES 40 and 42). For $d = 0$, eq. 183a yields a point-circle, the most symmetrical monopolar circle of the incident system (Fig. 27f, $R = 0$).

[The reader is reminded that, in assessing the simplicity of construction rules, the characteristic distance is considered to have the status of a variable in the evaluation of the coefficients of terms, such as d, d^2, dv, du, dv^2, etc.]

The fourth symmetry condition on eq. 177 is that for which the coefficients of the v and d terms are unity and the coefficient of the term in du is -1 (eq. 183b). This is the construction rule for a tangent circle ($R = h = 3d/2$) that encloses the pole (Fig. 27n). Construction rule 183a is simpler than 183b, and its locus is more symmetrical, because both terms in the variables have positive coefficients (see also RULE 40). As in the case of eq. 183a, eq. 183b represents no real locus in the left half-plane.

Eq. 180a is the construction rule for complementary rectangular circles in the left half-plane of the general axial polar-linear circles of eq. 177. For the existence of real complementary circles that are centered in the left half-plane, conditions 181a,a' must be satisfied, where R is the radius of the circle in the right half-plane. For example, for the limiting case of $h = 2d$ and $R = 2d$, the circle in the right half-plane is tangent to p_u and the complementary locus is a point-circle at the point of tangency. If both inequalities 180a,a' are satisfied, with, say, $h = 3d$ and $R = 4d$, compound circles are obtained (see Fig. 27t), with the segment in the left half-plane centered at $x = -d$ and of radius $2\sqrt{2}d$.

On the other hand, the segments of real complementary circles in the left half-plane can have their centers in the right half-plane. For example, for $h = 0$ and $R = d$, the complementary circle with a segment in the left half-plane (radius $\sqrt{5}d$) is centered at $x = 2d$ in the right half-plane (Fig. 27s).

Lastly, we inquire as to whether it is possible for a circle in the right half-plane to possess pole p_u as a secant (i.e., for a circle in the right half-plane to overlap the line-pole) in the absence of a circle in the left half-plane that also possesses p_u as a secant. Otherwise stated, can we have a segment of a circle in the right half-plane (as in Fig. 27s,t) without also having a complementary segment of a circle in the left half-plane that joins it to form a compound circle? It is readily shown that this is not possible (see Problem 24).

Limacons

We consider only the salient features of the polar-linear eqs. of axial limacons (polar eq. $r = a - b\cos\theta = a - ae\cos\theta$; see Fig. 19). These are determined using the limacon hybrid distance eqs. about the two foci and the pole at $-b/2$ (eqs. 146, 148, and 154). Attention is confined primarily to the initial loci, that is, the identification of complementary loci in opposite half-planes is not pursued (but see Problems 15 and 19). Of particular interest is the matter of identifying the directrices of limacons and homologizing them with those of Cartesian ovals (see next section). From our treatment of polar-linear conic sections, it is evident that directrix lines are the line-poles of polar-linear construction rules for which the constant terms vanish (or for which no constant term exists).

RULE 47. *A directrix-line of an axial curve is a line-pole for which no constant term is present in the curve's polar-linear (or linear-circular) construction rule.*

Focal Axial Limacons

Consider, first, focal axial limacons cast about the DP as point-pole and having their lines of symmetry in coincidence with the polar-linear axis. It is evident from inspection of the hybrid distance eq. for the DP (eq. 146, repeated here as eq. 184) that, since no constant term is present, the normal line that includes the DP is the directrix for such limacons. Thus, when the line-pole is taken to be the limacon LR (the normal chord or secant through the DP), and one makes the substitution $x = \pm u$ (eq. 159a) in eq. 184, one obtains the degenerate 8th-degree construction rule 185a, which possesses no constant term (the eq. is 8th-degree because it must be squared to eliminate the alternate sign).

$v^2 + aex = \pm av$ hybrid distance eq. of limacons cast about the DP (upper sign for elliptical limacons and the large loop of hyperbolic limacons, lower sign for the small loop)   (184)

(a) $v^2(v \mp a)^2 = a^2 e^2 u^2$ construction rule for limacons in the incident system; p_v the DP and p_u the LR (lower alternate sign for the small loop)   (185)

(b) $(v^2+aeu-av)(v^2-aeu-av) \cdot (v^2-aeu+av) = 0$ eq. 185a as the product of factors; one factor has been discarded (since the small loop is entirely to the left of p_u)

Since eq. 185a is doubly degenerate (because it also is a perfect square), it can be expressed as the product of three factors (eq. 185b) that represent portions of various segments to the right and left of the directrix. The first (left-most factor) is for the portions of elliptical limacons and the large loop of hyperbolic limacons to the right of the directrix; the middle factor is for the portions of the same segments to the left of the directrix; and the last (lower) factor is for the entire small loop, which lies to the left of the directrix. A fourth factor, for which there is no real locus, has been discarded. Accordingly, for this selection of p_u and p_v (the incident polar-linear system), entire elliptical limacons can be represented by a degenerate 4th-degree construction rule, and entire hyperbolic limacons by a degenerate 6th-degree construction rule, both of which lack a constant term.

If line-poles other than the directrix are selected, say, normal lines entirely to the right of the curves, such as the tangents to the curves on the DP sides, the appropriate substitution eq. for x is $x = d-u$ (eq. 158c). For limacons of eccentricity equal to or greater than 1/2, the normal tangent on the DP side lies at a distance $a^2/4b = a/4e$ to the right of the DP, while for e less than or equal to 1/2, it lies at a distance $(a-b)$ to the right. Employing these values of d yields the two degenerate 4th-degree construction rules 187a,b.

(a) $x = a^2/4b - u = a/4e - u$ value of x in eq. 184a for a normal tangent line-pole on the DP side of limacons for $e \geq 1/2$ (186)

(b) $x = a - b - u = a(1-e) - u$ eq. 186a for $e \leq 1/2$

(a) $v(v \mp a) = aeu - a^2/4$ $e \geq 1/2$ limacons cast about the DP and the near normal tangent (lower sign for small loop)   (187)

(a') $v(v \mp 4ed) = 4e^2du - 4e^2d^2$ eq. 187a in terms of d and e

(b) $v(v \mp a) = aeu - a^2e(1-e)$ eq. 187a for $e \leq 1/2$

(b') $v[v \mp d/(1-e)] = deu/(1-e) - d^2e/(1-e)$ eq. 187b in terms of d and e

(c) $v(v \mp 2d) = d(u - d)$ eqs. 187a,b for $e = 1/2$

It is evident that for the combination of the DP and the normal tangent-line on the DP side of the curve as reference elements, entire rectangular elliptical limacons can be represented by a 2nd-degree polar-linear construction rule (upper alternate sign only, of eqs. 187), whereas it requires a 4th-degree construction rule to represent entire rectangular hyperbolic limacons (the lower alternate sign of eqs. 187a,a' also is needed to represent the small loop). The most symmetrical of these polar-linear limacons is the $e = 1/2$ representative (Fig. 19b), for which the construction rule simplifies to eq. 187c. Similar eqs. readily are derived employing the $-(a+b)$ vertex-tangent as line-pole (see Problem 13).

The hybrid distance eq. 148 for the focus of self-inversion, which is within the small loop of hyperbolic limacons and exterior to elliptical limacons (see Fig. 19), is repeated here as eq. 188a. Inspection of this eq. reveals that, for a line-pole appropriately positioned to the left of the point-pole (eqs. 157), the constant term of the resulting polar-linear construction rule vanishes, with simplification from four terms to three. From testing eqs. 157 and 188a for various values of e, it is found that this line-pole (directrix) sometimes lies to the left of the curve, sometimes intersects it, and sometimes lies to its right (but never lies to the right of the focus of self-inversion). Eq. 157a, rearranged as eq. 188b, gives the appropriate relation between u, x, and d (see Fig. 19a,b for the locations of these directrices in the e = 2 and e = 1/2 limacons).

derivation of polar-linear construction rules of
limacons cast about the focus of self-inversion

(a) $v^2 \mp aev + ax/e +$ hybrid distance eq. of limacons cast about the focus of self-inversion at (188)
$a^2(1-e^2)^2/4e^2 = 0$ $(a^2-b^2)/2b$ (plus sign for the DP of elliptical limacons)

(b) $u = \pm(x+d)$, $x = \pm u - d$ transformation eqs. 157 for x and u

(a) $v^2 - aev \pm au/e - ad/e +$ eliminant of eqs. 188a,b   (189)
$a^2(1-e^2)^2/4e^2 = 0$

(b) $v^2 - aev \pm au/e = 0$ limacons cast about the focus of self-inversion and the directrix of eq. 190a

(b') $(v^2-aev+au/e)(v^2-aev-au/e) = 0$ eq. 189b as the product of two factors
     portion to   portion to
       right   left

(c) $v^2 - 4de^2v/(1-e^2)^2 \pm$ eq. 189b in terms of d and e
$4du/(1-e^2)^2 = 0$

(a) $d = a(1-e^2)^2/4e$ axial position of the directrix to the left of the focus of s.-i.   (190)

(b) $x = a(1-e^2)/2e - a(1-e^2)^2/4e$ axial position of the directrix in relation to the DP

(a) $v^2 - aev + au/e = 0$ eq. 189b for $e \geq \sqrt{2}+1$ (191)
(a') $v^2 - dv + (3-2\sqrt{2})du = 0$ eq. 189c for $e = \sqrt{2}+1$
(b) $v^2 - aev - au/e = 0$ eq. 189b for $e \leq \sqrt{2}-1$
(b') $v^2 - dv - (3+2\sqrt{2})du = 0$ eq. 189c for $e = \sqrt{2}-1$
(c) $v^2 - av \pm au = 0$ eq. 189b for e = 1, the cardioid
(c') $v^2(v-a)^2 = a^2u^2$ eq. 191c squared to eliminate alt. signs

Eliminating x between eqs. 188a,b yields eq. 189a. The condition for vanishing of the constant term is eq. 190a. For this condition, eq. 189a becomes the degenerate 4th-degree focus-directrix construction rule 189b,b'. From the location of the directrix relative to the DP, it can be shown (see Problem 14) that the initial curve lies entirely to the right of the directrix for $e \geq \sqrt{2}+1$, and entirely to its left for $e \leq \sqrt{2}-1$, whence the quadratic eqs. 191 apply to these special cases (see Problem 15 for complementary loci).

The most symmetrical of the 4th-degree polar-linear limacons with directrices that intersect the curve is the cardioid, for which $e = 1$ and construction rule 189a simplifies to eqs. 191c,c'. For this case, $d = 0$ and p_v, the cusp-point, is incident upon the latus rectum as line-pole (the incident system). No eq. analogous to eqs. 191c,c' exists in the bipolar system, since, when both point-poles are at the cusp-point, the construction rule is trivial. Consonant with RULE 40, the most-symmetrical of the quadratic limacons with directrices that do not intersect the curve are the $e = \sqrt{2}+1$ and $e = \sqrt{2}-1$ members (eqs. 191 a',b'), with tangent directrices (see also Problem 16).

As examples of the directrices of non-focal axial limacons (following the definition of RULE 47), curves cast about the pole at $-b/2$ ($-ae/2$) are considered. The applicable hybrid distance eq. 154 is repeated here as eq. 192a and rearranged to 192a'. Examination of the latter eq. reveals that the constant term is positive for $e < 2$ but negative for $e > 2$. This change in sign at $e = 2$ means that the directrix is to the right of p_v for $e < 2$ but to its left for $e > 2$. Accordingly, since the directrix generally intersects the curve (the pole at $-b/2$ is interior to all limacons), both transformation eqs. 157a and 158a must be employed, yielding the degenerate 8th-degree construction rule 193a.

Expressed in terms of d and e, eq. 193a becomes 193a". It readily is shown (see Problem 17) that for $e = 4$, the directrix is tangent at the $-(a+b)$-vertex, while for $e > 4$ the curve lies wholly to the right of the directrix, whereupon the alternate sign can be eliminated, yielding the non-degenerate 4th-degree construction rule 193b. The most-symmetrical of these limacons (i.e., of those for which $e \geq 4$) is the $e = 4$ representative, for which eq. 193b greatly simplifies to eq. 193c (see RULE 40). Eq. 193a' is the incident polar-linear construction rule for the $e = 2$ limacon, for which p_v is the small-loop vertex and p_u is the vertex-tangent at this point. The simple form of the construction rule for these poles is consonant with RULE 40.

derivation of polar-linear construction rules of limacons cast about the pole at -b/2

(a) $(v^2-a^2e^2/4)^2 - a^2(v^2+a^2e^2/4) = -a^3ex$ hybrid distance eq. of limacons cast about the pole at -b/2, all cases   (192)

(a') $v^4 - a^2v^2(1+e^2/2) = -a^3ex + a^4e^2(1-e^2/4)/4$ eq. 192a rearranged

(b) $d_{F-D} = ae(1-e^2/4)/4$ directed distance between the -b/2 pole and the directrix; p_u lies to the right of p_v for positive values

(a) $v^4 - a^2v^2(1+e^2/2) = \mp a^3eu$ limacons cast about the pole at -b/2 and its directrix; the latter lies at a distance d (eq. 192b) to the right of p_v for $e \leq 2$ and to the left for $e \geq 2$ (the upper sign is for portions of the loci to the right of p_u), all cases   (193)

(a') $v^4 - 3a^2v^2 = \mp 2a^3u$ eq. 193a for e = 2, the incident system

(a") $v^4 - \dfrac{16d^2v^2(1+e^2/2)}{e^2(1-e^2/4)^2} = \mp eu|64d^3/e^3(1-e^2/4)^3|$ eq. 193a in terms of e and d (the upper external alternate sign is for portions of the locus to the right of p_u)

(b) $v^4 - \dfrac{16d^2v^2(1+e^2/2)}{e^2(1-e^2/4)^2} = 64ud^3/e^3(1-e^2/4)^3$ eq. 193a" for $e \geq 4$

(c) $v^4 - d^2v^2 = -4d^3u/27$ the most-symmetrical curve represented by eq. 193b, namely, the e = 4 representative

Parabolic Cartesians

Parabolic Cartesians are taken as examples of polar-linear constructions of Cartesian ovals, but the treatments easily are extended to all members of the group. Eq. 119a, cast about the monoval-bioval focus (repeated here as eq. 194), is taken as our initial hybrid distance eq. (with p_u relabelled as p_v, since the label "p_u" has been assigned to polar-linear line-poles).

It is evident from eq. 194 that the directrix must lie to the right of its focus, pole p_v, since the linear term in x on the right side of the eq. has the same sign as the constant term on the left. Making the substitution u = ±(x-d) of eq. 158a (assuming intersection with the curve), yields construction rule 195a, from which it can be seen that, for a line-pole lying midway ($d_{uv} = d_{vz}/2$) between the monoval-bioval focus and pole p_z (the -b/2 homologue), the constant terms cancel. The resulting simple construction rule 195b represents both monovular and biovular parabolic Cartesians (see Problem 19).

derivation of polar-linear construction rules of parabolic Cartesians cast about the monoval-bioval focus

$$v^2 - Cd_{vz}v + d_{vz}^2 = 2d_{vz}x \qquad \text{conventional (and hybrid) distance eq. for the monoval-bioval focus of parabolic Cartesians (d_{uz} of Chapter IV is relabeled d_{vz})} \qquad (194)$$

(a) $\quad v^2 - Cd_{vz}v + d_{vz}^2 = 2d_{vz}(\pm u + d_{uv}) \qquad$ eq. 194 with the substitution $x = \pm u + d_{uv}$ $\qquad (195)$

(b) $\quad v^2 - Cd_{vz}v = \pm 2d_{vz}u \qquad$ construction rule for parabolic Cartesians cast about p_v and its directrix, letting $d_{uv} = d_{vz}/2$ (plus sign for locus to right of p_u)

(c) $\quad v^2 - Cd_{vz}u = 2d_{vz}u \qquad$ construction rule for monovular parabolic Cartesians for $C \leqslant 1/2$

$$x = (d_{vz}/2)[2 \pm C \pm \sqrt{(C^2 \pm 4C)}] \qquad \text{locations of vertices of parabolic Cartesians} \qquad (196)$$

The vertices of parabolic Cartesians are defined by eq. 196, obtained by letting $y = 0$ in eq. 116 and solving for x. Testing this eq. for various values of C shows that the directrix is tangent to the $C = 1/2$ monoval but lies wholly to the left of the curve for all smaller values of C. Accordingly, the simple 2nd-degree construction rule 195c represents these monovals. For larger values of C, up to 4, this directrix intersects the monovals (Figs. 12f$_1$ and 21a). For $C = 4$, it intersects the large loop of the equilateral limacons (Fig. 12f$_1$), while for $C > 4$ it intersects the large oval (Figs. 12f$_3$ and 23a$_2$).

The locations of the directrices for the other two foci of self-inversion of biovular parabolic Cartesians are ascertained from the distance eqs. 119b,c (after conversion to the hybrid eqs.). The directrix for the other focus within the small oval (p_v of Chapter IV) lies to its left (fig. 23a$_2$), while the directrix for the exterior focus, p_w, lies to its right (Fig. 23a$_2$). Since these two foci approach one another with decreasing values of C--becoming coincident at the DP of the equilateral limacon ($C = 4$)--and since their directrices always lie between them, we see the basis for the LR of limacons being the directrix for the DP (and for the DP hybrid distance eq. lacking a constant term).

The same situation also occurs in other representatives of linear Cartesians. In every group, the directrices for the other two foci of self-inversion (i.e., other than the monoval-bioval focus) lie between them, both the foci and the directrices approach one another as the ovals approach one another, and, when the ovals come into contact and form a limacon (the curve of demarcation), the foci come into coincidence at the DP and the directrices come into coincidence at the LR (which includes the DP).

Since the small oval of parabolic Cartesians lies entirely to the right of both directrices (for foci p_w and p_v of Chapter IV), a 2nd-degree polar-linear construction rule suffices to represent it (cast about either of these directrices and the corresponding focus), and a 6th-degree polar-linear construction rule suffices to represent both ovals (see Problems 20 and 21). This is analogous to the situation for the limacon DP and its directrix (the LR; see eq. 185b and the accompanying text).

The fact that the directrices for the three foci are discrete lines is not unexpected, since no prior instance has been encountered in which two individual polar-linear point-foci of a curve have a common directrix. Accordingly, it is of great interest to find that the monoval-bioval focus, p_v (p_u of Chapter IV), and pole p_z possess a common directrix that lies midway between them in all parabolic Cartesians (i.e., in both monovals and biovals).

This, then, is another unusual feature of construction rules that involve these two particular poles. We already have noted that the simplest parabolic bipolar construction rule cast about these poles can represent both monovular and biovular parabolic Cartesians, and that the hybrid distance eq. takes its simplest form and represents both monovals and biovals when cast about the monoval-bioval focus employing the distance from pole p_z to this focus as its parameter.

To illustrate the coincidence of the directrices, the conventional distance eq. 120 (repeated here as eq. 197a) is converted to the hybrid eq. by letting $x_{M-B} = x_z + d_{vz}$, yielding eq. 197b (d_{vz} in this eq. and eqs. 197 and 198 is the distance between the monoval-bioloval focus and pole p_z). Inspection of this equation reveals that the directrix for pole p_z must lie to the left of p_z (since a negative constant term is needed on the right to cancel the negative constant term on the left). Substituting $x = \pm u - d$ from eq. 157a, yields eq. 198a.

254   THE POLAR-LINEAR SYSTEM

Letting d_{p_z-D} (the p_z-directrix distance) equal $d_{vz}/2$, which places the directrix for p_z in coincidence with the directrix for the monoval-bioval focus, yields the sought-after p_z-directrix construction rule 198b. Since the directrix for p_z always lies to the right of the small oval in biovular members, the 4th-degree construction rule 198c suffices to represent the small oval in its entirety (it also represents the left half-plane segment of the large oval).

derivation of polar-linear construction rules of
parabolic Cartesians cast about the pole at
$-b/2$

(a) $z^4/C^2 - d_{vz}^2 z^2 + d_{vz}^4 = 2d_{vz}^3 x_{M-B}$ conventional distance eq. for pole p_z of parabolic Cartesians (x_{M-B}, the abscissa of the monoval-bioval focus, p_v, of eqs. 194-196)   (197)

(b) $z^4/C^2 - d_{vz}^2 z^2 - d_{vz}^4 = 2d_{vz}^3 x_z$ hybrid eq. corresponding to eq. 197a ($x_{M-B} = x_z + d_{vz}$)

(a) $z^4/C^2 - d_{vz}^2 z^2 - d_{vz}^4 = 2d_{vz}^3(\pm u - d_{p_z-D})$ construction rule for parabolic Cartesians cast about point-pole p_z and a line-pole d_{p_z-D} units to its left (plus sign for p_z locus to right of point-pole)   (198)

(b) $z^4/C^2 - d_{vz}^2 z^2 = 2d_{vz}^3 u$ p_z-directrix construction rule of parabolic Cartesians cast about pole p_z and a directrix $d_{vz}/2$ units to its left (in coincidence with the directrix of the monoval-bioval focus)

(c) $z^4/C^2 + 2d_{vz}^3 u = d_{vz}^2 z^2$ eq. 198b for the small oval of biovals and the left half-plane portion of the lg. oval

Geometrical Correlates

With the knowledge that the directrices of pole p_z and the monoval-bioval focus are in coincidence in parabolic Cartesians, one realizes that this also must apply to the equilateral limacon (Fig. 12f$_2$), which is the curve of demarcation between monovular and biovular parabolic Cartesians. This is verified by substituting $e = \sqrt{2}$ in eqs. 190a and 192b (see Problem 22). This circumstance constitutes another example of a geometrical correlate in curves with the most simple construction rules. Thus, we already have noted that the $e = 2$ and $e = \sqrt{2}$ limacons are the most symmetrical limacons from several points of view (excluding circles; see page 163). Accordingly, another geometrical correlate of the
[The $e = \sqrt{2}$ limacon, for example, is the inverse of the equilateral hyperbola, which has the simplest rectangular construction rule for axial hyperbolas

($x^2-y^2 = a^2$), and its bipolar construction rule (eq. 97) about the DP and pole at $(a^2-b^2)/b$ has only three terms, as opposed to four terms for other limacons.] high symmetry of the equilateral limacon is the fact that the directrices of the pole at $-b/2$ and the focus of self-inversion are in coincidence.

Note, in this connection, that the most symmetrical polar-linear parabolic Cartesian cast about either pole p_z or the monoval-bioval focus as the point-pole is the $C = 1$ monovular representative (Fig.21-1a). Thus, for $C = 1$, two of the three coefficients of eq. 198b have unit value (see Problem 23). This curve also is the most-symmetrical bipolar parabolic Cartesian. A geometrical correlate of this high symmetry or great structural simplicity is that the curvature at the vertex nearest the monoval-bioval focus changes from convex ($C < 1$) to concave ($C > 1$) as C passes through unity (see Chapter IV, *Geometrical Correlates*).

Summary and Comparisons with the Bipolar and Other Coordinate Systems

Regarding the question of the degrees of eqs., the polar-linear system can hold few surprises, since it possesses the property of *partial degree-restriction*. This refers to the fact that polar-linear eqs. of axial curves cast about a given point and a normal line-pole are restricted to a degree not less than, and equal to or twice, that of the corresponding hybrid polar-rectangular eq. For example, since the hybrid focal eqs. of axial parabolic Cartesians (for monovals and both biovals) are 2nd-degree, no corresponding polar-linear eq. can be 1st-degree; they are either 2nd or 4th-degree. There is no possibility for reduction to a 1st-degree eq. for a selected line-pole, as there is for a second selected point-pole in coordinate systems that are not partially degree-restricted, such as the bipolar system and several other systems (for example, as in the polar-circular system of the next section).

These circumstances, of course, come about by virtue of the employment, and special properties, of a line-pole. When two line-poles are employed, complete degree-restriction results. [The highest polar-linear degree category of a rectangular curve, however, is the same as in the bipolar system--four times the corresponding rectangular degree.] Accordingly, as a consequence of partial degree-restriction, only conic sections can have 1st-degree polar-linear eqs. and, excluding incomplete-linears and the incident system, not even all conic sections. For example, there is no 1st-degree non-incident polar-linear eq. for a line, though there is for line-segments and rays.

The fact that reduction of construction rules from the degrees of the hybrid eqs. is impossible in the polar-linear system is accompanied by other consequences of the special properties of line-poles. Degree reduction in the bipolar system--attained through the use of particular combinations of poles--typically is accompanied by the loss of congruent multiple loci. In the polar-linear system, however, degree retention and doubling typically are accompanied by the presence of supernumerary similar incongruent loci or dissimilar loci (or segments thereof), with the formation of bizarre disparate and compound curves.

The origin of the latter curves depends on the fact that the line-pole of the non-incident system divides the plane into two non-equivalent half-planes. With few exceptions, a rectangular locus with segments in both half-planes is represented by a degenerate polar-linear construction rule that consists of the product of two factors equated to 0. Each factor, alone, equated to 0 represents the segment of the initial locus in one or the other half-plane.

But in addition to representing a segment of the initial locus in one of the half-planes, a given factor of the polar-linear eq. equated to 0 also represents a complementary locus in the other half-plane. The latter locus is obtained by a certain "partial translation" of the x-variable of the rectangular eq. of the segment of the initial locus. Only in the case of a focal axial hyperbola is the complementary locus in the other half-plane congruent with the initial locus, and, only exceptionally, is it similar but incongruent. In most cases it is dissimilar.

The occurrence of loci possessing similar but incongruent segments to an initial curve is not a monopoly of systems possessing a line-pole. As will be seen in Chapter VI, multiple but incongruent loci also occur in a system possessing a point and a circle as reference elements, and the possibilities are greatly increased when both a line-pole and a circle-pole are employed (see *Linear-Circular Coordinates*).

When attention is directed to the special line-poles that lead to the simplest construction rules within degree categories, namely, to eqs. in which all terms include one or both variables, unsuspected new relations emerge between related but different curves and between different point-poles. Thus, the special relation between the monoval-bioval focus and pole p_z of parabolic Cartesians (that leads to the simplest bipolar parabolic construction rule and the simplest hybrid polar-rectangular eq. for Cartesian ovals--both of which apply to both ovals of biovular members) also is found to extend into the polar-linear domain. Thus, these poles possess a common directrix.

Similarly, the convergence of point-foci into coincidence, as Cartesian biovals approach one another and make contact in curves of demarcation, is accompanied by a like convergence and coming into coincidence of the directrices for these foci. However, whereas the coincident foci of self-inversion become imaginary, the two coincident directrices become the directrix of the DP. As a consequence, the hybrid distance eq. for the DP possesses no constant term.

Lastly, two additional new properties were encountered in the polar-linear system. Polar-exchange symmetry is typical of almost all coordinate systems with disparate elements--whether dissimilar or merely incongruent--and more than two elements (it would be lacking, for example, in the equilateral tripolar system but not in the square tetrapolar system). The fundamental significance of this new method of relating curves to one another is seen from the fact that many curves are polar-exchange symmetrical with themselves and that 1st-degree polar-linear hyperbolas and ellipses are polar-exchange symmetrical with each other (through the relation of reciprocal eccentricities).

In addition, the property of one coordinate element including another coordinate element was encountered for the first time--the incident polar-linear system. In the case of the bipolar system, coincidence leads to triviality, whereas in the polar-linear system, incidence often leads to the most symmetrical locus for a given construction rule. For example, for the construction rule $u = v$, the line locus of the incident system is more symmetrical than the parabola of the non-incident system, inasmuch as it is symmetrically distributed in both half-planes, whereas the parabola is not. The same consequences apply to the axial circles $v^2 = ju$. This eq. represents the most symmetrical axial circle in the non-incident system (for $j = d$), which encloses the pole, whereas in the incident system it represents circles tangent to and reflected in the line-pole (directrix), and including the point-pole.

Problems

1. Find the locus in the right half-plane that complements the initial locus of eq. 161c-left in the left half-plane.

2. Derive the general construction rule of polar-linear parabolas for the line-pole in coincidence with the line of symmetry and an axial point-pole. Derive the special cases of the point-pole (1) at the traditional focus, (2) at the vertex, and (3) at the reflection of the vertex in the focus.

3. Assuming that construction rule 173a is cast about the left focus and left vertex-tangent of an initial hyperbola and represents the left arm, derive eqs. 174a,b for the corresponding two-arm rectangular hyperbola and the complementary rectangular two-arm hyperbola whose right arm is represented by the same construction rule. What is the corresponding eq. for an initial ellipse, assuming it to be cast about the right focus and right vertex?

4. Show that the polar-linear construction rule of an ellipse cast about the left focus and right directrix (eq. 170e) represents no locus in the right half-plane.

5. Show that the construction rules 173a,c represent no locus in the half-planes beyond the vertex-tangents of ellipses.

6. Derive the rectangular eq. of the hyperbola whose arm complements the arm of an initial hyperbola cast about its focus and the vertex of the other arm (eq. 173b).

7. Find the eq. of the complementary hyperbola of Fig. $31b_1,b_2$ from the eq., $v + d/e = eu$, for an arm of an initial hyperbola cast about its focus and conjugate axis. By what linear factor is the complementary hyperbola magnified in size? Is the line-pole located in a significant relation to the complementary arm (see Problem 12)?

8. Explain the absence of terms in u^2 in the polar-linear construction rules of axial circles (eq. 177)

9. Derive the polar-linear construction rule for the parabolas of Fig. $29d_1$. Then let $x_1 = 2a$ and let pole p_v be incident upon the curve. By comparing the resulting construction rule for the portions of the parabolas in the upper half-plane with that of the parabolas of Fig. $29d_2$, determine which parabolas are more symmetrical (about the reference elements).

10. Confirm the existence of a real complementary locus in the lower half-plane of the dual hemi-parabolas of Problem 9 by finding its point(s) of intersection with the 2a-chord.

11. The construction rule, $2d^2v^2 = u^4 - 4d^3u + 3d^4$, of Problem 10 represents reflective dual hemi-parabolas (Fig. $29d_1,d_2$) in the upper half-plane. What is the rectangular eq. of the complementary locus represented by the same construction rule in the lower half-plane? Does this eq. represent parabolas?

12. Derive the construction rule for the hyperbola (eq. 171b) that complements the hyperbola of eq. 171a'. Does pole p_u occupy an exceptional position in the plane of the complementary locus (i.e., is p_u either the conjugate axis, a vertex-tangent, an LR, or a directrix of the complementary locus)?

13. Derive the eq. of limacons (polar eq., $r = a - b\cos\theta = a - ae\cos\theta$) about the DP and the vertex-tangent at $h = -(a+b)$.

14. Find the eccentricity of the limacon for which the directrix of eqs. 190 is tangent to the -(a+b)-vertex.

15. What is the eq. of the complementary loci in the left half-plane for the initial limacons of construction rule 191a?

16. What is the construction rule and eccentricity of the limacon for which the coefficient of the du-term of eq. 189c is unity?

17. Find the value of e for which the directrix of the limacons (polar eq., $r = a - b\cos\theta = a - ae\cos\theta$) of eqs. 193a,a' is tangent to the curve at the -(a+b)-vertex.

18. Show that the locus complementary to the e = 4 limacon of construction rule 193c in the left half-plane is imaginary.

19. What is the rectangular construction rule for the complementary locus to the right of directrix p_u represented by eq. 195b? Find the intercept(s) of this locus with the polar-linear axis to the right of p_u for the C = 5 parabolic Cartesian biovals.

20. Derive the focus-directrix eq. for focus p_v of eq. 119b for (a) the small oval, (b) the large oval, and (c) both ovals of the C = 5 parabolic Cartesian.

21. Solve Problem 20 for focus p_w of eq. 119c.

22. Show that the directrices for the focus of self-inversion (eq. 190a) and the pole at $-b/2$ (eq. 192b) of the equilateral limacon are in coincidence.

23. Confirm eq. 198b for the vertices of the C = 1 parabolic Cartesian of Fig. 21-1a.

24. Show that if an initial circle in the p_v-containing half-plane possesses line-pole p_u as a secant, a complementary circle exists in the opposite half-plane that also possesses line-pole p_u as a secant. Show that these two circle-segments meet (come into contact with one another) at the line-pole.

25. What is the incident polar-linear construction rule of the curve of Fig. 8a (polar eq. $r = a + b\cos^2\theta$) cast about the true center as point-pole and the 90°-270°-axes as line-pole? Confirm your result by checking it at the vertices.

CHAPTER VI.   THE POLAR-CIRCULAR AND LINEAR-CIRCULAR
              COORDINATE SYSTEMS

THE POLAR-CIRCULAR COORDINATE SYSTEM

The most general two-element coordinate systems consisting of circle-poles are the *non-congruent bicircular systems*, with circles of arbitrary radii. Two special groups of these *bicircular systems* are the *congruent bicircular systems* and the *polar-circular systems* of the present Chapter, in which one of the circle-poles is a point-circle. The most specialized non-trivial bicircular system, of course, is the non-coincident bipolar system of Chapter IV. Similarly, the most general two-element systems consisting of a line and a circle-pole are the *linear-circular systems* of the present Chapter, of which the non-incident polar-linear system of Chapter V was merely a special case, and of which incident polar-linear systems are most specialized.

Inasmuch as all the bicircular systems consisting of two circle-poles, whether bipolar, polar-circular, congruent bicircular, etc., are related to one another through their common type of reference elements, and similarly for all systems consisting of a circle-pole and a line-pole, it can be anticipated that some of the properties of the systems in each group will be possessed in common, while others will be characteristic of or unique to specific systems. We already have considered the most-simple and most-general relationships that exist between two point reference elements   Exchanging one of these point-circles for a non-point-circle results in the loss of virtually all of these special relations, as follows.

1. The two elements no longer are the simplest possible.

2. The symmetry of position of the two elements with respect to one another no longer is the simplest possible. An infinite number of different distances exist between the point-pole and points of the circle-pole.

3. The relative location of the two elements no longer is completely general, inasmuch as the point-pole bears a different relation to the circle-pole in the four possible different locations--centered, interior, incident, and exterior. But the two elements do retain the most general orientations.

4. The amount of directional bias no longer is the least possible in a system possessing two poles, inasmuch as a moving point may be simultaneously receding from the point-pole and all points of the circle-pole, approaching all points of the circle pole but receding from the point-pole, approaching some points of the circle-pole while receding from others, while at the same time either approaching or receding from the point-pole, etc.

5. The system cannot be regarded as the only non-redundant multipolar system consisting of both point and circle-poles. For example, three-element systems consisting of two points and one circle, and one point and two circles, are neither equivalent to one another nor redundant to the polar-circular system.

Despite these differences in relations between the poles of the polar-circular system, as compared to those between the poles of the bipolar system, it is clear that the information (diagnostic features) about positional symmetry or structural simplicity that is explicit and obvious in bipolar construction rules also must be possessed, if only implicitly, by polar-circular construction rules. This is asserted because an appropriate transformation of variables converts polar-circular eqs. of given loci into the bipolar eqs. of the same loci (see Problem 25), of which more than one eq. generally exists (because different bipolar eqs. usually are required to represent greatest and least-distance constructions and/or portions of polar-circular loci that are exterior and interior to the circle-pole). The treatments of the present Chapter are concerned [Although letting the radius of the circle-pole, R, equal 0 in a polar-circular construction rule transforms it into a bipolar construction rule, the latter does not necessarily represent the same locus. Correspondingly, the locus represented by a polar-circular construction rule in which R does not appear (for example, u = v) usually is not the same locus represented by the identical bipolar eq.] primarily with illustrating the intriguing properties of polar-circular systems and their curves, rather than with transformations of polar-linear eqs. to bipolar eqs., and relations between diagnostic features in the two systems (but see Problem 25). Linear construction rules suffice for this purpose.

In polar-circular systems, one meets two new aspects of coordinate systems not encountered previously and one new type of locus. One new aspect is *extremal-distance symmetry* (defined below). This characterizes a relation between certain loci in coordinate systems that possess at least one closed reference element. The other, in essence, characterizes all other coordinate systems (i.e., other than those already treated). This is the fact that the identity of a locus defined by a given eq. generally is not unique, but depends upon the possession of specific information concerning relations between the reference elements.

In the non-coincident bipolar system and the non-incident polar-linear systems, any given construction rule uniquely defines a locus (or loci). In polar-circular systems that are neither central (with the point-pole at the center of the circle-pole) nor incident (with the point pole incident upon the circle-

262   THE POLAR-CIRCULAR SYSTEM

pole), the identity of a locus generally is not established until the ratio R/d is known.

The new type of locus consists of compound conics in which the complementary segments possess different eccentricities, a situation that is not possible for polar-linear loci (see Chapter V, *The Shapes of Complementary Segments*).

Transformation Equations

As the standard arrangement (Fig. 32a) for conversions between non-centered polar-circular coordinates and rectangular coordinates, the circle-pole, p_u, is taken with its center, x_c, at the origin, and the point-pole, p_v, is located on the x-axis at (d,0). With this arrangement, the distances u, from the circle-pole to a point of the locus, are given either by eq. 199a or 199a'. [If these two eqs. were to be combined into one, with two pairs of alternate signs, the disallowed directed-distance relation $u = -R - \sqrt{(x^2+y^2)}$ would be included.] Unless otherwise noted, u represents either the distance u_L (the least distance from the circle-pole) or u_G (the greatest distance). The distance v, from the point-pole, is given by eq. 199b.

transformation equations between polar-circular
and rectangular coordinates

(a) $u = \pm[\sqrt{(x^2+y^2)} - R]$ least distances from circle-pole p_u to point (x,y); upper sign for exterior pts.   (199)

(a') $u = \sqrt{(x^2+y^2)} + R$ greatest distances from circle-pole p_u for both interior and exterior points

(b) $v = \sqrt{[(x-d)^2 + y^2]}$ distance from point-pole p_v to point (x,y)

(a) $x = [(u \pm R)^2 + d^2 - v^2]/2d$ abscissa of point (x,y) in terms of u and v (upper sign for least distances and exterior points, lower sign for least distances and interior points or for greatest distances and all points)   (200)

(b) $y = \pm\sqrt{\{4d^2(u \pm R)^2 - [(u \pm R)^2 + d^2 - v^2]^2\}}/2d$ ordinate of point (x,y) in terms of u and v (alternate signs of radicand as in eq. 200a)

If eqs. 199a,a',b are squared and the difference, $u^2 - v^2$, is obtained and solved for x (see Problem 1), one obtains the expression 200a. If this is substituted in eq. 199b, and the resulting expression is solved for y, one obtains eq. 200b. Comparing eqs. 200a,b with the corresponding bipolar trans-

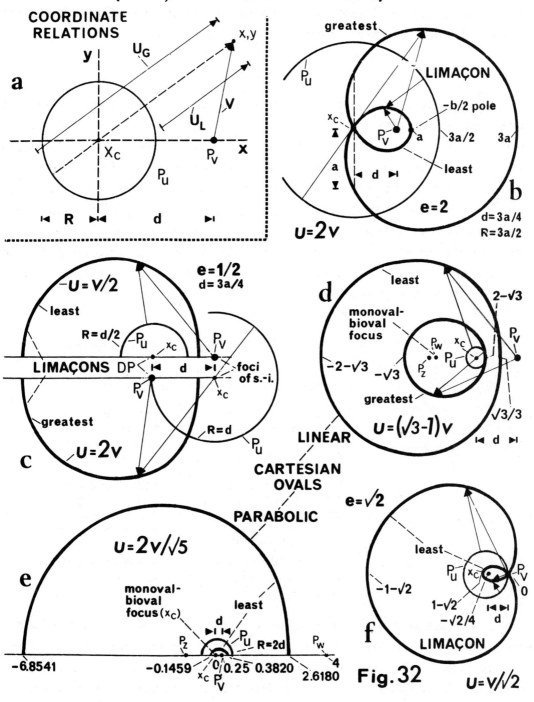

Fig. 32 Polar-Circular Constructions (limaçons & Cartesian ovals)

formation eqs. 22a,a',b, one finds that the former have twice the degree of the latter; 2nd and 4th-degree eqs. relate x and y to the bipolar u and v, whereas 4th and 8th-degree eqs. are required to express the corresponding polar-circular relations (since one additional squaring is needed to eliminate the alternate signs for greatest and least distances and interior and exterior points). This means that general polar-circular eqs. have eight times the corresponding rectangular degree, as opposed to only four times this degree for general bipolar eqs.

Of course, if one derives only eqs. for least distances and interior points, least distances and exterior points, or greatest distances, the general polar-circular and bipolar eqs. have the same degree. However, such a practice would be entirely arbitrary. First, it would exclude portions of loci exterior or interior to the circle-pole. Second, since both greatest and least distances generally are defined uniquely, there is no reason to favor one type of construction over the other (sometimes a greatest-distance construction does not exist; see Fig. 33a$_2$).

The Construction Rule u = v

Since our attention is to be concentrated on 1st-degree polar-circular construction rules and the loci that they represent (but see Problems 10 and 25), it behooves us to use the 1st-degree construction rules themselves as initial eqs., and transform them by means of eqs. 199a,a'b to ascertain their rectangular equivalents. But derivations employing initial hybrid eqs. also are given (see below and Problem 3).

Making the substitutions of eqs. 199a,a',b in $u = v$, squaring, cancelling the terms x^2 and y^2, regrouping with the radical on the left, squaring again, and putting in standard form for central conics, gives eq. 201. It is seen that the central conics in question have eccentricity, $e = d/R$, semi-major axis of length $R/2$, and semi-minor axis of length $\sqrt{[\pm(R^2-d^2)]}$ (or semi-transverse and semi-conjugate axes for hyperbolas). They are positioned with their left foci at x_c (the center of circle-pole p_u) and right foci at p_v (Figs. 33a$_1$,a$_2$; see Problem 2). Both arms of hyperbolas (Fig. 33a$_1$) are represented, with the left arm being a greatest-distance construction that intersects the circle-pole, and the right arm being a least-distance construction that is exterior to it. Ellipses (Fig. 33a$_2$) lie entirely within the circle-pole and are solely least-

Non-Incident Polar-Circular Conics

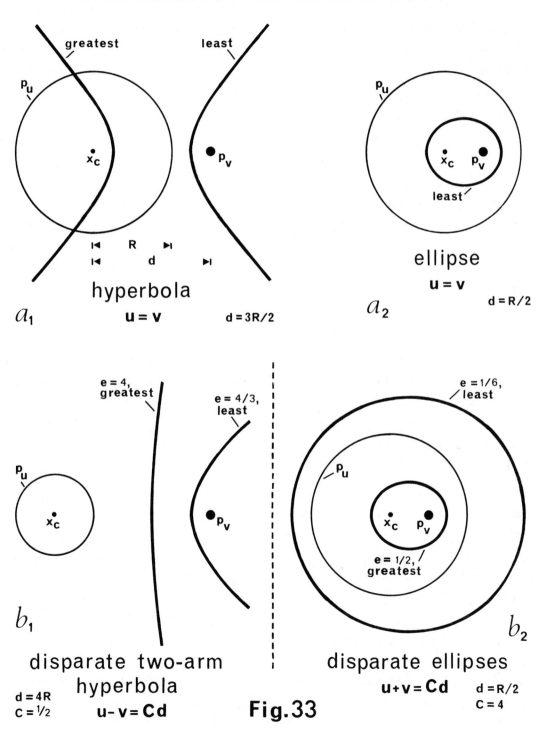

Fig. 33

distance constructions. A greatest-distance construction cannot exist, since with p_V located within the circle-pole, no point in the plane exists for which $u_G = v$.

[Note (Figs. 32b,c-lower, 33a₁, and 34a,d) that arms and segments of rectangular curves can intersect the circle-pole and continue as the same rectangular curve in the interior or exterior, as greatest-distance constructions. On the other hand, the rectangular locus of least-distance constructions changes in passing from one domain to the other (i.e., the first derivatives are continuous at the intersections for greatest-distance constructions but not for least-distance constructions).]

polar-circular constructions for $u = v$

	$(x-d/2)^2/(R^2/4) + y^2/[(R^2-d^2)/4] = 1$	rectangular equivalent of $u = v$ in standard form for central conics	(201)
(a)	$v = a(e^2-1) + e(x_c-d)$	conventional distance eq. of the right arm of a hyperbola cast about the polar-circular origin, x_c	(202)
(b)	$v = a(e^2-1) + e[(u\pm R)^2 - v^2 - d^2]/2d$	right arm of a hyperbola cast about p_V, its focus, and p_u, the circle-pole	
(c)	$(v + d/e)^2 = (u \pm R)^2$	eq. 202b with $d = 2ae$, rearranged	
(d)	$\pm(v + d/e) = u \pm R$	roots of eq. 202c	
(e)	$v = u$	eq. 202d with $e = d/R$ and correct signs	

The constructions for the hyperbolas are the first examples of *extremal-distance symmetry*, which characterizes all coordinate systems that employ one or more closed curves as reference elements. Any two curves or segments of a curve are said to be extremal-distance symmetrical if both are represented by the same construction rule (for the same R/d-ratio) but one is a greatest-distance construction and the other is a least-distance construction.

The three exceptional eccentricities, $e = 0$, 1, and ∞, for central conics are accounted for as follows. $e = 0$ corresponds to $d = 0$ (or R infinite), in which case p_V is at x_c (the center of p_u). This is the centered system; although it is not trivial like its bipolar equivalent, all the represented loci are circles. For this case, the locus of eq. 201 is a centered circle of radius R/2, for which $u_L = v$.

For $e = 1$, we have $d = R$; this yields the incident system, for which the equivalent of eq. 201 is $y^2 = 0$, representing a line through p_V and x_c. For the exterior ray on the side of p_V, and the near-radius segment, the construction rule is $u_L = v$, while for the far-radius segment, and the exterior ray on the far side, it is $u_G = v$. In other words, for the ray from the center including p_v, $u_L = v$, while for the opposite ray from the center, $u_G = v$.

For $e = \infty$, R is 0 (or d is infinite) and one passes over to the bipolar system. For $R = 0$, the equivalent of eq. 201 is $(2x-d)^2 = 0$, or two coincident normal lines lying midway between the poles--the most-symmetrical locus in the bipolar system. In this case, the locus represents the two coincident arms of a hyperbola of infinite eccentricity, with semi-transverse axis $a = 0$ and semi-conjugate axis $b = d/2$ (since the two poles can be regarded as the foci of the limiting hyperbola as $R = 2a$ approaches 0).

To illustrate a comparable derivation using hybrid distance eqs., let p_v be the focus of the right arm of a hyperbola (eq. 143 with the plus sign), in which case, $x_c = x_v + d$, and the conventional distance eq. for the right arm about the origin (x_c) is eq. 202a. Substituting for x_c from eq. 200a yields eq. 202b. If one lets $d = 2ae$, p_v becomes the focus of the right arm and, after cancelling and re-arranging, eq. 202c is obtained. For $e = d/R$ and the applicable combination of signs, this becomes the simplest linear construction rule, $u = v$, which is the least-distance construction for the right arm. A parallel derivation for the left arm (with x_c as its focus) gives the same eq. for the greatest-distance construction (Problem 3). [d must be less than R for ellipses, and only a least-distance construction exists.]

Polar-Circular and Bipolar Construction Rules

Although polar-circular construction rules lack the obvious explicit diagnostic features of bipolar eqs., it was pointed out above that all the information obtainable from a given bipolar eq. is implicit in the corresponding polar-circular eqs. (with the circle pole centered on either of the bipolar poles). Transformations to the bipolar eqs. of specific polar-circular loci require only that the ratio R/d be known. But this ratio always has to be known for polar-circular constructions; otherwise no unique locus is represented. This need to know the R/d-ratio for the non-centered and non-incident polar—circular systems lends emphasis to the more specific character of the bipolar system.

Having described the loci of the polar-circular eq. $u = v$, this most-simple case can be used for a correspondingly simple illustration of the polar-circular to bipolar transformation. Since the polar-circular variable u is simply a composite of another variable, say w--the distance from the center of the circle-pole--and a constant term R--the radius of this pole--one readily can transform the polar-circular construction rule to an ensemble of bipolar construction rules that represent the same loci.

For example, suppose one wishes to ascertain the bipolar eqs. of the polar-circular loci that lie within the circle-pole for the least-distance construction rule $u=v$. The substitutions for u then take the form given in the accompanying chart and lead to the bipolar eq., $w+v=R$. When $d=R$, this becomes $w+v=d$, for the bipolar connecting line-segment, corresponding to the near radius of the incident polar-circular system. When $d<R$, it becomes $w+v=d/e$ for an ellipse, whereas for $R<d$, there can be no bipolar locus, since $w+v$ cannot sum to less than d.

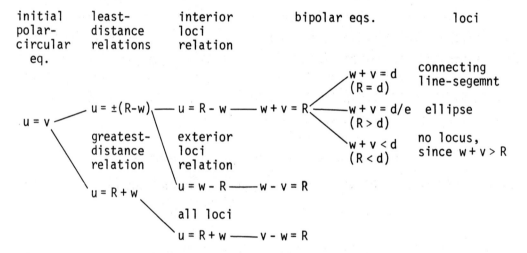

With the above relations in mind, one could have predicted that when the polar-circular construction rule $u=v$ represents ellipses and hyperbolas, R must equal d/e, and when it represents rays, R must equal d, etc. As will be evident from the treatments that follow, similar predictions can be made in the cases of the constant terms and parameters of other polar-circular construction rules, such as, $u=Cv$ and $u+v=Cd$, simply by considering the loci represented by the corresponding bipolar eqs. of the forms $\pm(w\pm R)=Cv$ and $\pm(w\pm R)+v=Cd$.

For example, it is shown below that for a certain polar-circular construction of limacons, $u=ev$. But since it already is known that the linear bipolar construction rules of limacons take the form $\pm ev \pm u = ed$ (and $\pm eu \pm v = ed$; see eqs. 94a'-c'), one can predict that R must be equal to ed for this polar-circular construction. Thus, letting $u = \pm(w\pm R)$ in the polar-circular eq. $u=ev$, yields $\pm(w\pm R) = ev$, from which, by comparison with the bipolar eq. $\pm ev \pm u = ed$, $R = ed$ [In the above eqs. with two sets of alternate signs, it is tacit that the combination of alternate signs that is valid only for directed distances is not allowed.]

The Construction Rule $u = Cv$

Limacons

Carrying out a derivation parallel to that for eq. 201 with the initial eq. $u = Cv$ ($C > 0$) yields eq. 203a. If one compares this construction rule with eq. 105a, repeated here as eq. 204, it is seen to have the same form as the eq. for linear Cartesians, which were found to be "master equations" for conic sections. Indeed, the polar-circular construction rule $u = Cv$, allowing for both greatest and least-distance constructions, encompasses the same loci encompassed by eq. 204 and its bipolar equivalent, $\pm Bu \pm Av = Cd$, but without the need for the alternate signs (emphasizing the fact that the variable u of the polar-circular eq. is composite, being the sum of a variable term and a constant term). This is shown with several representative derivations.

polar-circular constructions for $u = Cv$

(a) $[(x^2+y^2)(C^2-1)-2C^2dx+(C^2d^2-R^2)]^2 =$ rectangular equivalent of the   (203)
      $= 4R^2(x^2+y^2)$ polar-circular eq. $u = Cv$

(b) $[(x^2+y^2)-2C^2dx/(C^2-1)+(C^2d^2-R^2)/$ eq. 203a rearranged
      $(C^2-1)]^2 = 4R^2(x^2+y^2)/(C^2-1)^2$

(a) $[(x^2+y^2)(A^2-B^2)-2A^2dx+d^2(A^2-C^2)]^2$ linear Cartesians   (204)
      $= 4B^2C^2d^2(x^2+y^2)$

(b) $[(x^2+y^2)-2A^2dx/(A^2-B^2)+d^2(A^2-C^2)/$ eq. 204 rearranged
      $(A^2-B^2)]^2 = 4B^2C^2d^2(x^2+y^2)/(A^2-B^2)^2$

$(x^2+y^2+bx)^2 = a^2(x^2+y^2)$ rectangular limacons, DP at   (205)
                                                     origin

(a) $4C_{pc}^4 d^2/(C_{pc}^2-1)^2 = 4e^2R^2/(C_{pc}^2-1)^2 =$ equating the square of the coef-   (206)
      $4C_{pc}^2 d^2 e^2 (C_{pc}^2-1)^2$ ficient of the linear term in the
                                                                  brackets of eq. 203b to e^2 times
                                                                  the coefficient on the right, and
                                                                  letting $R^2 = C_{pc}^2 d^2$

(b)	$C_{pc} = e = R/d$	eq. 206a after factoring out common terms	(206)
(a)	$R = d$	conditions for limacons with the focus of self-inversion at the center of the circle-pole (x_c)	(207)
(b)	$C_{pc} = 1/e$		

Repeating eq. 81-upper for rectangular limacons as eq. 205 and comparing this with eq. 203b, it is seen that the condition $A^2 = C^2$ in eq. 204, which is one of the conditions that yields limacons, corresponds to the polar-circular condition $R = C_{pc}d$ (where C_{pc} is the C of the polar-circular eq. $u = Cv$). Furthermore, since $b = ae$ for the limacons of eq. 205, one can set up an eq. between the corresponding expressions for a and b of eq. 203b, yielding eq. 206a-upper. Making the substitution, $R = C_{pc}d$, yields eqs. 206a (upper left and lower) and 206b, $C_{pc} = e$, whence $e = R/d$. This is the reciprocal of the value of R/d for conics using the construction rule $u = v$.

For these limacons, the DP is at x_c (the center of circle-pole p_u) and the focus of self-inversion is at p_v. Although this leads to the standard orientation for elliptical limacons, hyperbolic limacons are reversed from the standard orientation (Problem 3). In the construction of these limacons, the elliptical members are represented by a least-distance eq. (which applies to the DP); the hyperbolic members are represented by both least-distance (for the small loop) and greatest-distance (for the large loop) constructions (see Fig. 32b,c). Hence, in this construction, the small and large loops are related by extremal-distance symmetry.

The second condition for which eqs. 204 represent limacons is $B^2 = C^2$. As might be guessed, since the condition $A^2 = C^2$ leads to a construction rule with the DP at x_c and the focus of self-inversion at p_v, the condition $B^2 = C^2$ leads to the reversed roles. For this circumstance, with the focus of self-inversion at x_c and the DP at p_v, one finds that $R = d$ and $C_{pc} = 1/e$ (Fig. 32f and Problem 4). Thus, for the two simplest linear polar-circular construction rules of limacons, C_{pc} has the value e (the eccentricity) in one case but 1/e in the other.

However, the most remarkable aspect of the second construction rule, with the focus of self-inversion at x_c, is that a least-distance construction applies to both loops of the hyperbolic members (Fig. 32f; no greatest-distance construction exists), whereas the construction rule for elliptical limacons now employs greatest distances. It also is found that, just as polar-linear central conics (u = Cv) of reciprocal eccentricities were polar-exchange-symmetrical with one another, so also are polar-circular limacons (u = Cv) of reciprocal eccentricities polar-exchange-symmetrical with one another.

Cartesian Ovals

The rectangular eqs. 203 for the polar-circular construction rule u = Cv are cast about the origin at x_c, and the rectangular linear Cartesians of eqs. 204 are cast about one of the three foci of self-inversion as origin (d, the distance to a second focus of self-inversion at x = d). Accordingly, if one solves for A_L, B_L, and C_L in terms of C_{pc} and R (for a common value of d), the relations obtained will be for the case of the circle-pole centered at one of the three foci, with a second focus located at x = d (the position of the bipolar and polar-circular poles p_v). Carrying through these solutions yields the parametric relations 208a,b.

polar-circular constructions for u = Cv

(a) $\quad C_{pc} = A_L/B_L$ $\qquad$ relations between polar-circular $\qquad$ (208)
parameters C, d, and R, and bi-
(b) $\quad R = C_L d/B_L$ $\qquad$ polar parameters A, B, and C of
linear Cartesian ovals

(c) $\quad u = C_{pc}v = A_L v/B_L$ $\qquad$ polar-circular Cartesian biovals in terms of A_L and B_L of the corresponding linear bipolar eqs.

(a) $\quad C_{pc} = A_L/B_L = (1+\sqrt{3})/(2+\sqrt{3}) =$ $\qquad$ conversions of the bipolar parame- $\qquad$ (209)
$\qquad\qquad (\sqrt{3}-1)$ $\qquad$ ters to polar-circular parameters
for the ovals of Figs.21-1b and 32d

(b) $\quad R = C_L d/B_L = d/(2+\sqrt{3}) = (2-\sqrt{3})d$

(c) $\quad u = (\sqrt{3}-1)v$ $\qquad$ polar-circular construction rule of ovals of Figs.21-1b and 32d

Thus, the Cartesian ovals represented by the linear bipolar eq. ±Bu ± Av = Cd are represented by the linear polar-circular eq. $u = A_L v/B_L$, where $R = C_L d/B_L$. This construction is illustrated in Fig. 32d (eq. 209c) for the Cartesian ovals of Fig.21-1b. Since the small oval is a greatest-distance construction and the large oval is a least-distance construction, the two ovals have extremal-distance symmetry. Any of the three foci of self-inversion may be at x_c and used in combination with either of the other two foci located at p_v.

Parabolic Cartesian Ovals

It was concluded in Chapter IV that members of the parabolic subgroup of linear Cartesians possess the greatest bipolar symmetry (among other things by virtue of possessing the simplest parabolic eq., $z^2 = Cdu$) as well as the greatest hybrid polar-rectangular symmetry (by virtue of possessing the simplest hybrid distance eq. for the group and the fact that this eq. applies to both ovals of the biovular members). Do these curves also possess the greatest polar-circular symmetry in the group?

To answer this question, the properties of the polar-circular equivalents of bipolar parabolic Cartesians are investigated. One property can be determined directly from eq. 208a, since the symmetry condition for the parameters of the subgroup in question is known, in the form of eq. 123d, $A^2C^2 + A^2B^2 = B^2C^2$. Making the substitutions of eqs. 208a,b leads to eq. 210a. Accordingly, whenever the constant C_{pc} of the polar-circular construction rule $u = Cv$ has the value of the ratio, $R^2/(R^2+d^2)$, the construction rule represents parabolic Cartesians.

polar-circular constructions for $u = Cv$

(a) $\quad C_{pc} = R^2/(R^2+d^2)$ $\qquad$ eliminant of eq. 123d and eqs. 207a,b for C_{pc} of eq. 209b $\qquad$ (210)

(b) $\quad u = C_{pc}v = R^2v/(R^2+d^2)$ $\qquad$ polar-circular construction rule for parabolic Cartesians

$\{(8-C^2+CE)(x^2+y^2)-16d_{uv}x + [8-16C/(C-E)]d_{uv}^2\}^2 = 64C^2d_{uv}^2(x^2+y^2)$ $\qquad$ rectangular eqs. of parabolic Cartesians in terms of d_{uv} and C (of the bipolar eq., $z^2 = Cd_{uz}u$) $\qquad$ (211)

(a) $C_{pc}^2 = 8/C(C-E)$, (b) $R = 4Cd/C(C-E)$ $\qquad$ conversion eqs. between parameters of eqs. 203 and 211 $\qquad$ (212)

In the general case of polar-circular construction rules for linear Cartesians (eq. 208c), the parameter C_{pc} cannot be expressed directly and simply in terms of R and d alone, but takes the form $C_{pc} = A_L/B_L = A_LR/C_Ld$. For limacons, however, a direct expression is possible, namely $C_{pc} = R/d = e$. But whereas C_{pc} may assume any value for this construction rule for limacons, it is always less than unity for the polar-circular construction rule of parabolic Cartesians.

We now find the relations between the polar-circular C (C_{pc} of eq. 203a,b) for parabolic Cartesians and the bipolar C (C of the eq. $z^2 = Cdu$) for parabolic

Cartesians. For this purpose eq. 125a is employed (repeated here as eq. 211). This recasts eqs. 204 for the parabolic subgroup. Using these results for the polar-circular construction of the C = 5 parabolic Cartesian of Fig. 32e, one finds a profound difference from the construction of the linear Cartesian ovals of Fig. 32d .

[For the $e = \sqrt{2}$ limacon curve of demarcation of Fig. 32f (C = 4), for which C_{pc} = $\sqrt{2}/2$ and R = d, one obtains the construction rule, $u = \sqrt{2}v/2$, with the focus of self-inversion at x_c and the DP at p_v.]

In the case of the linear Cartesian of Fig. 32d, the large oval is a least-distance construction and the small oval is a greatest-distance construction. In the present case, for the C = 5 parabolic Cartesian, both ovals are least-distance constructions and no greatest-distance construction exists (the applicable parameter values are R = 2d and C = $2\sqrt{5}/5$; note that d for this construction rule is the distance between x_c and p_v).

Accordingly, the parabolic subgroup of linear Cartesian ovals also has greater polar-circular symmetry than other linear Cartesians, since a single least distance construction rule suffices for both ovals, whereas both least and greatest-distance construction rules are required for the other members. That is, u_L = Cv produces both ovals of the parabolic subgroup (a 1st-degree eq.), whereas in general, it requires both of the construction rules, u_L = Cv and u_G = Cv, to produce the two ovals.

A noteworthy difference between the constructions of the Cartesian ovals of Fig. 32d and 32e is that the circle-pole is centered on the focus between p_v and p_w in Fig. 32d, but on the focus to the left of both p_v and p_w in the parabolic Cartesian of Fig. 32e. This latter focus is the monoval-bioval focus, whereas the former focus is not (p_w is the monoval-bioval focus in Fig. 32d). Accordingly, as the two sets of ovals approach one another (with changing parameter values) and come into contact to form limacons, the circle-pole of Fig. 32d comes to have the DP at its center, as in Fig. 32b (in which the orientation is reversed), while the circle-pole of the parabolic Cartesian of Fig. 32e comes to have the focus of self-inversion at its center, as in Fig. 32f--the limacon curve of demarcation.

One of the analytical bases for these differences is as follows. Of the three bipolar foci of the rectangular eqs. 204, one is at the origin, another is at x = d, and the third is at $x = (A^2 - C^2)d/(A^2 - B^2)$. For Cartesian ovals in general, this third focus may be either to the right or to the left of the focus at the origin, that is, it may be on the same side as the $x=d$ focus or on the opposite side. In the former case (same side), the focus at the origin is the monoval-bioval focus (i.e., the focus that remains real in the monovular members), but in the latter case (opposite side) it is not the monoval-bioval focus (the $x=d$ focus then is the monoval-bioval focus).

One of the characteristic features of the parabolic subgroup of linear Cartesians is that the focus at the origin is the monoval-bioval focus, and the third focus always is to the right of the origin (this is shown by means of the parameter symmetry condition, $A^2B^2 + A^2C^2 = B^2C^2$; Problem 6).

When the third focus of Cartesian ovals with the polar-circular construction rule $u=Cv$ is to the right, the focus at the origin is the monoval-bioval focus and the circle-pole is centered on it. This focus, then, becomes the pole about which the limacon curve of demarcation self-inverts. In this event, a least-distance polar-circular construction rule applies to both loops of hyperbolic limacons, regardless of whether the parent Cartesian biovals belong to the parabolic subgroup. But when the third focus of the Cartesian ovals is to the left, the circle-pole becomes centered at the DP of the limacon curve of demarcation (Fig. 32b). In that event, both least and greatest-distance constructions are required to represent both loops of hyperbolic members.

Accordingly, the property of parabolic Cartesian biovals that distinguishes them from other Cartesian ovals and confers greater polar-circular symmetry on them, is the fact that a single least-distance polar-circular construction rule applies to both ovals when x_c is at the monoval-bioval focus. The construction rule may be cast in conjunction with either of the other two foci as the point-pole (see Problems 7 and 8), but the monoval-bioval focus must be at x_c.

For example, the same $C=5$ parabolic Cartesian of Fig. 32d is represented by a least-distance construction rule for both ovals, if the exterior focus p_w is taken to be the point-pole (for $C_{pc} = \sqrt{5}/5$, rather than $2\sqrt{5}/5$, and $R=2j$, rather than $R=j/2$; $d=j/4$). But if the circle-pole is not centered on the monoval-bioval focus, the construction rules involve least distances for one oval and greatest distances for the other (see Problem 8).

This further illustrates the conclusion reached in Chapter IV, that the three bipolar foci of Cartesian biovals are equivalent only in the sense that they are all foci of self-inversion and that the use of any combination of two of them leads to a linear bipolar construction rule. In other respects, there

ELLIPSES AND HYPERBOLAS

are many differences between these three foci (although two of them are more closely related to each other than to the third). In the present coordinate system, any combination of two of them at x_c and p_v leads to a linear construction rule of the form $u = Cv$. But only when the monoval-bioval focus is at x_c does a single least-distance construction rule, $u_L = Cv$, apply to both ovals.

The Construction Rule $u \pm v = \pm Cd$

In the interests of simplicity, our knowledge of relations between bipolar and polar-circular construction rules is used to ascertain the nature of the loci represented by the polar-circular construction rule $u \pm v = \pm Cd$ (eq. 213a). It is known that $u = R + w$ represents all greatest-distance constructions (where w is the distance to a point of a locus from x_c, the center of p_u) and that $u = \pm(R - w)$ represents all least-distance constructions (eqs. 213b,b'). Accordingly, substituting the relations 213b,b' in eqs. 213a yields the corresponding bipolar construction rules 213c.

polar-circular constructions for $u \pm v = \pm Cd$

(a) $u \pm v = \pm Cd$, (a') $u + v = Cd$ initial eq. (a) as the composite (213)
of eqs. (a') and (a")

(a") $u - v = \pm Cd$

(b) $u = R + w$ relations between u and w (w, the distance to a point of a locus from x_c) for all loci of eqs. 213a; (b) represents greatest-distance constructions, (b') represents least-distance constructions

(b') $u = \pm(R - w)$

(c) $w \pm v = (Cd \pm R)$ bipolar eqs. corresponding to eqs. 213a, all cases

(d) $e = \pm d/(Cd \pm R)$ eccentricities of ellipses and hyperbolas of eqs. 213a,c

$$\frac{4(x-d/2)^2}{(Cd \pm R)^2} + \frac{4y^2}{[(Cd \pm R)^2 - d^2]} = 1 \quad \text{rectangular eqs. of the loci of eqs. 213a,c} \quad (214)$$

Eqs. 213c are recognized to be the construction rules for ellipses and one-arm hyperbolas whose foci are at point-poles p_w (x_c) and p_v, and for which the constant term is equal to d/e. This leads to the identification of the combination of parameters of the right side of eqs. 213c, as given in eqs. 213d. The corresponding rectangular eq. derived by making the substitutions of eqs. 199 a,a'b in eqs. 213a is eq. 214 (Problem 9). Since the center-focus distance of all the conics represented by eq. 214 is d/2, and all of them are displaced to the right a distance d/2, all of them have their left foci at the origin (x_c) and their right foci at p_v.

[The fact that the rectangular eqs. for both u and v (eqs. 199a,a'b) contain radicals usually makes the alternate signs of u and v in polar-circular eqs. superfluous, with the result that the corresponding rectangular eqs. often represent all possible combinations of signs of terms in the variables, as in the present case.]

Eqs. 214, of course, represent both ellipses and hyperbolas of the conventional type, whereas individual linear bipolar eqs. are known to represent only one-arm hyperbolas (in the polarized system). It will be found below that individual polar-circular eqs. can represent a great variety of curves possessing segments of hyperbolas (and ellipses), including the conventional hyperbolas of Fig. $33a_1$, with the construction rule u = v.

For a first examination of some specific loci represented by the polar-circular construction rule 213a, let Cd = 2R, leading to the eq. u ± v = ±2R. Taking first the case of an interior pole at x = d = R/2 yields two complete ellipses of eccentricities 1/6 and 1/2, represented by the construction rule u + v = 2R (Fig. $33b_2$). One ellipse (e = 1/2) is entirely interior to the circle-pole and is the product of a least-distance construction, while the other (e = 1/6) is entirely exterior and the product of a greatest-distance construction.

Next consider an exterior point-pole at x = d = 4R. This yields hyperbolas of eccentricities 4/3 and 4, represented by the construction rules u - v = 2R and v - u = 2R. The loci represented by the former eq. are illustrated in Fig. $33b_1$. While, indeed, only right arms of hyperbolas are represented by this construction rule, it represents two right arms; one of these (e = 4) is the product of a greatest-distance construction and the other (e = 4/3) the product of a least-distance construction. Neither arm intersects the circle-pole.

On the other hand, the eq. v - u = 2R represents the two left arms of hyperbolas of the same eccentricities of those of Fig. $33b_1$ (not illustrated). In this case, however, the e = 4/3 arm is the product of a greatest-distance construction (and intersects the circle-pole) and the e = 4 arm results from a least-distance construction. Note that the two segments, or curves, that are produced by each of these three construction rules possess extremal-distance symmetry; for identical construction rules, one ellipse or one arm of a hyperbola is represented by least distances and the other by greatest distances.

If the treatment of polar-circular curves were to be terminated at this juncture, one would be left with the impression that segments of curves that intersect the circle-pole are not disparate, since the examples of limacons and the arms of hyperbolas that have been encountered have been entirely conventional (which is possible only for greatest-distance constructions). A conventional arm of a hyperbola or a limacon simply intersects the circle-pole without having its size or shape altered in any way at the point of intersection, as usually occurs at the line-pole of polar-linear loci. The next examples reveal that the foregoing treatment has only scratched the surface of a rich variety of unconventional polar-circular curves, including types of loci not encountered in earlier chapters.

The loci of Fig. 34 are six examples of polar-circular constructions represented by the eq. $u \pm v = \pm Cd$. They were selected because each of two sets of three examples, taken as a composite, represent two complete central conics. On the left are segments of a hyperbola of eccentricity 2, and segments of an ellipse of eccentricity 2/3; on the right are segments of a hyperbola of infinite eccentricity (two coincident lines) and of an ellipse of eccentricity 1/2. All of these loci occur in the incident system ($R = d$) and all of them are displaced to the right of center an amount $d/2$, bringing their centers to the position $x = d/2$ and their foci into coincidence with x_c (on the left) and p_v (on the right).

[Note the interesting fact that the circle ($e = 0$), the $e = 2$ and $e = \sqrt{2}$ hyperbolas, and the parabola have emerged in several reference frames and in several connections as the most symmetrical conic sections (and so have the corresponding limacons). It already has been noted that the $e = \sqrt{2}/2$ polar-linear ellipse possesses unique symmetry (the loci of Fig. 31c, with construction rule $u^2 + v^2 = Cd^2$) and we note here that hyperbolas of infinite eccentricity and the $e = 1/2$ ellipse possess unique polar-circular symmetry (the loci of the eqs. $u \pm v = \pm d$). Accordingly, the most symmetrical curves from various points of view are found to fall into pairs with the reciprocal eccentricities, 0, 1/2, $\sqrt{2}/2$, 1, $\sqrt{2}$, 2, and ∞. Of course, the eccentricity of the parabola is its own reciprocal.]

The first of the examples on the left (Fig. 34a) is the left arm of a hyperbola with focus at x_c, with the construction rule $u - v = d/2$ ($C = 1/2$)--the product of a greatest-distance construction. There can be no corresponding least-distance construction, that is, $u_L - v = d/2$, because v always is greater than or equal to u_L.

The second example on the left (Fig. 34b) is a type of curve not heretofore encountered, and one whose construction rule has no bipolar analogue. Inasmuch

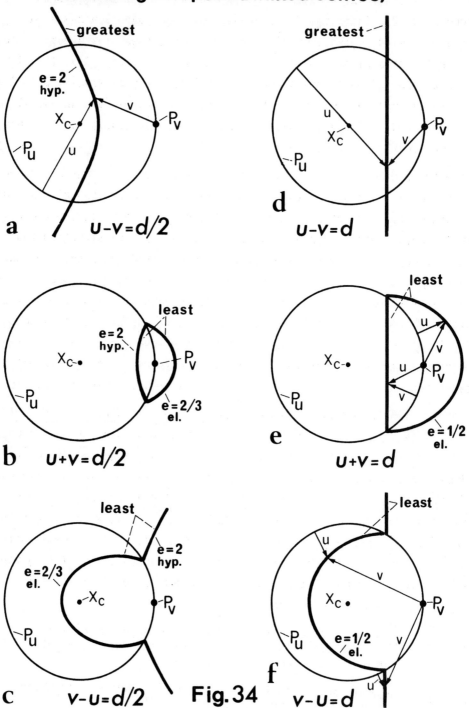

Fig. 34. Incident (R=d) Polar-Circular Conics (including compound mixed conics)

as it consists of a curve with joined incomplete segments, it meets the definition of a *compound conic* (see Chapter V). But since the exterior portion is a segment of an ellipse (e = 2/3) and the interior portion is a segment of a hyperbola (e = 2), it is the first example of a compound conic-- a *compound mixed conic*--with segments of different eccentricities. The hyperbola segment is a portion of a right arm congruent with the arm of Fig. 34a. This is confocal with and joined to a segment of an ellipse of eccentricity 2/3. A greatest-distance construction does not exist for $C = 1/2$, because u_G itself must equal or exceed d. There is no corresponding bipolar case of a construction $u + v$ = constant that yields a hyperbola.

The third example (Fig. 34c) is essentially the complement of the second one, in that it consists of the complementary portions of the ellipse and the right arm of the hyperbola of Fig. 34b. Again, this locus is the product of a least-distance construction and, again, it involves a construction that is non-existent among bipolar eqs., namely, a compound mixed conic and the representation of a portion of an ellipse by the construction rule $v - u = d/2$. This also is the first example of an open compound curve composed of portions of conics of different eccentricities. When the segments represented by this construction rule are combined with those of Fig. 34a,b, one obtains a complete rectangular ellipse and hyperbola. There can be no corresponding greatest-distance construction, $v - u_G = d/2$, since u_G always is greater than or equal to v.

Having illustrated three specific examples of incident polar-circular constructions that were not encountered previously, we now inquire as to which loci represented by the construction rule $u \pm v = \pm Cd$ are most symmetrical. Following RULES 1a,b, these are the loci represented by the eq. in its simplest form, namely $u \pm v = \pm d$. Letting $R = d$ and $C = 1$ in eq. 214 yields the ellipse of eq. 215, of eccentricity 1/2 (with $a = d$ and $b = \sqrt{3}d/2$, centered at $x = d/2$), and the hyperbola (two coincident lines) of eq. 216 of infinite eccentricity (with $a = 0$ and $b = d/2$, centered at $d/2$). Each corresponding locus for $C = 1$ is given in Fig. 34d-e to the right of the similar construction for $C = 1/2$. Again, the loci of Fig. 34e,f are compound mixed conics.

$$(x-d/2)^2 + 4y^2/3 = d^2 \qquad \begin{array}{l}\text{e = 2 rectangular ellipse, segments of which}\\ \text{make up the most-symmetrical loci with polar linear eqs. } u \pm v = \pm Cd\end{array} \qquad (215)$$

$$(x-d/2)^2 = 0 \qquad \begin{array}{l}\text{e = } \infty \text{ rectangular hyperbola, segments of which}\\ \text{complement those of eq. 215}\end{array} \qquad (216)$$

These latter three curves take on particular interest as examples of 1st-degree polar-circular construction rules for a line, a line-segment (plus a hemi-ellipse) and two rays (plus a hemi-ellipse). The conclusions reached above concerning the C = 1/2 constructions, as regards the non-existence of loci for the complementary types of constructions (least versus greatest distances) also hold for the loci of Fig. 34d-f.

Summary

The non-central, non-incident polar-circular system is the first example of an undirected-distance coordinate system in which the loci represented by construction rules are not uniquely defined until the relations between the reference elements are specified, in this case, the R/d-ratio. For the non-coincident bipolar system and the non-incident polar-linear system, such a specification is unnecessary. Whereas the coincident bipolar system is trivial, the centered polar-circular system is not, though all loci in it are circles.

Like the incident polar-linear system, the incident polar-circular system is non-trivial, but unlike the incident polar-linear system, no additional line of symmetry of the system is gained by virtue of the incidence. Like the situation in the polar-linear system, dissimilar curves can be represented by the same polar-circular construction rule, although only the simplest cases have been considered (where both dissimilar loci are conic sections).

At first sight, the relations between the bipolar and polar-circular systems appear to be very close. In the simplest cases one readily can interconvert between eqs. in the two systems, draw certain conclusions about the curves represented, and calculate the parameters in one system from those in the other. However, interconversions and analogizing must be carried out with care--even in the cases of linear three-term eqs. with unit coefficients for terms in the variables. Thus, the polar-circular construction rule $v - u$ = constant can represent a portion of an ellipse and the construction rule $u + v$ = constant can represent a segment of a hyperbola, whereas the corresponding bipolar eqs. cannot. The greatest difficulty attends predicting the precise makeup of polar-circular curves.

By having the variable u represent either greatest or least distances, one confers on polar-circular eqs. the ability to represent, with very few parameters, a great number of different loci whose eqs. require additional parameters in the bipolar system or that cannot be represented by a single eq. But this aspect of simplicity is largely offset by the need to specify the R/d-ratio. On the other hand, as is the case for the polar-linear system, very simple polar circular construction rules can represent numerout types of loci that are not representable by either degenerate or non-degenerate bipolar, polar, or rectangular construction rules.

By virtue of the fact that polar-circular construction rules can represent either least or greatest distances and interior and exterior loci, the transformation eqs. of x and y to polar-linear coordinates are twice the degrees of the corresponding bipolar transformation eqs. As a consequence, the degrees of general polar-circular eqs. are twice the degrees of general bipolar eqs. By limiting consideration to greatest distances only, or to least distances in a single domain, the degrees of the general construction rules in the two systems become equal, but the polar-circular eq. does not necessarily represent the entire locus represented by the bipolar eq. In the cases of curves in standard positions, the 2/1 degree ratio does not necessarily apply and polar-circular eqs. do not necessarily represent both greatest and least-distance constructions nor portions of loci in both domains.

THE LINEAR-CIRCULAR COORDINATE SYSTEM

Introduction

In the linear-circular coordinate system, another step is taken toward greater complexity of possible relations between curves and coordinate elements, beyond those that exist in the polar-linear and polar-circular systems. The possible relations now include both greatest and least distances, both interior and exterior segments, and portions of loci to both sides of the line-pole. A number of general properties of the system are immediately evident.

As in the polar-linear system, the presence of a linear reference element imposes partial degree-restriction. As in the polar-circular system, the existence of greatest and least-distance constructions leads to extremal-distance symmetry, while the possibility of different locational relations between the reference elements means that the loci represented by given construction rules are not uniquely defined until the R/d-ratio is specified (in this case, d is the distance between the center of the circle-pole and the line-pole).

As in both of the other systems, the fact that the reference elements are disparate leads to the existence of polar-exchange symmetry. One also expects to be able to interconvert between construction rules for given loci (or, at least, for segments of given loci) in the polar-linear and linear-circular systems, just as between the bipolar and polar-circular systems. Ready interconversion is, indeed, the case, as it always is for systems that differ only in the radius of a circular reference element.

Additionally, one expects linear-circular construction rules to be able to represent curves that are not representable by non-degenerate eqs. in previously treated systems, and this proves to be the case. Two new types of loci are encountered even for linear construction rules, and one can anticipate that additional ones would be encountered in an expanded treatment. Lastly, by reference to the hybrid distance eqs., one expects to be able to predict the types of loci represented by simple linear-circular eqs., and, in certain cases, the relations of the loci to the coordinate elements. In this connection, the linear-circular construction rule, $u^2 = Cdv$, is used to illustrate in some detail the diagnostic power conferred by knowledge of the hybrid eqs.

Transformations between Rectangular and Linear-Circular Coordinates

As our standard linear-circular system, x_c (the center of circle-pole p_u) is taken to be at the origin of the rectangular system, and line-pole p_v is taken to be normal to the x-axis at distance d to the right of x_c (Fig. 35b). But for the fact that p_v now is a line-pole normal to the x-axis, rather than a point-pole incident upon the axis, this arrangement duplicates the polar-circular arrangement of Fig. 32a.

With this arrangement, the distance u, from circle-pole p_u, is unchanged from the polar-circular values of eqs. 199a,a', repeated here as eqs. 217a,a'. The corresponding distances from line-pole p_v are given by eq. 217b. Solving these eqs. for the abscissa and the square of the ordinate of a point of a curve in terms of the distance u and v yields eqs. 217c,d,d'. Note that the expression y^2, for least-distance constructions (eq. 217d'), also applies to greatest-distance constructions. As in the case of polar-circular coordinates, the general linear-circular construction rule of a rectangular curve of degree n is 8n (see Problem 10).

transformation eqs. between rectangular and linear-circular coordinates

(a) $u = \pm[\sqrt{(x^2+y^2)} - R]$ least distance from p_u to point (x,y); upper sign for exterior points   (217)

(a') $u = \sqrt{(x^2+y^2)} + R$ greatest distance from p_u to point (x,y); applies for both interior and exterior points

(a") $(u \pm R)^2 = x^2 + y^2$ eq. encompassing both eqs. 217a,a'; minus sign for greatest distances and interior points of least distances, plus sign for exterior points of least distances.

(b) $v = \pm(x - d)$ distance from line-pole p_v, at x = d, to point (x,y); upper sign for points to right of p_v

(c) $x = (d \pm v)$ abscissa of point of curve; plus sign for points to right of p_v

(d) $y^2 = (u-R)^2 - (d \pm v)^2$ square of ordinate of point of curve for greatest distances (signs of v as in eq. 217c)

(d') $y^2 = (R \pm u)^2 - (d \pm v)^2$ square of ordinate of point of curve for least distances (signs of v as in eqs. 217c,d; plus sign of u for exterior points)

Since the treatment of linear-circular curves is confined to those represented by specific construction rules, the transformation eqs. 217a,a',b are used to convert these rules to their rectangular equivalents. For treatments employing hybrid distance eqs., the transformation eqs. are unchanged from those of eqs. 157-159 (except that v replaces u).

The Construction Rule $u = v$

The most-symmetrical linear-circular loci are represented by the construction rule $u = v$ (RULE 2a). It is clear that these loci must be composed of rectangular parabolas or segments of parabolas, because the corresponding polar-linear locus (Fig. 25a) is a parabola. Thus, since the polar-linear locus is derivable from the hybrid construction rule for a parabola cast about its focus, and because replacing w (used in place of the hybrid v) of the hybrid eq. by $u \pm R$ only alters the value of the constant term of the hybrid eq., the corresponding linear-circular loci also must be parabolas (or segments thereof).

Though one can anticipate that the foci of these linear-circular parabolas or partial parabolas will lie at x_c (just as the foci of the polar-linear parabolas lie at the point-pole), and that both greatest and least-distance constructions will be involved, the precise nature of the loci usually has to be ascertained by testing—through the medium of the construction rules of the corresponding rectangular loci.

Making the substitutions of eqs. 217a,a',b in the construction rule $u = v$ leads to the corresponding (degenerate) rectangular 4th-degree eq. 218 (Problem 11). For convenience, the locus for the positive alternate sign is labelled "arm 1" (eq. 218a') and that for the negative alternate sign is labelled "arm 2" (eq. 218a").

rectangular equations of the loci of $u = v$

(a) $\quad y^2 = -2(d \pm R)[x - (d \pm R)/2]$ $\quad\quad$ both arms $\quad\quad\quad\quad\quad$ (218)

(a') $\quad y^2 = -2(d+R)[x - (d+R)/2]$ $\quad\quad$ "arm 1" of eqs. 218a

(a") $\quad y^2 = -2(d-R)[x - (d-R)/2]$ $\quad\quad$ "arm 2" of eqs. 218a

It is evident from these eqs. that the vertex-focus distances for arms 1 and 2, respectively, are $(d+R)/2$ and $(d-R)/2$. Since arm 1 opens to the left and is translated $(d+R)/2$ units to the right, its focus is in coincidence with x_c.

Similarly, though arm 2 may open to the left (d > R) or right (R > d), it is translated a distance (d-R)/2 to the right in the former case, but (R-d)/2 to the left in the latter case, in both instances bringing the focus into coincidence with x_c. Accordingly, the two arms are confocal.

Examples for each of six illustrative cases are plotted in Fig. 35. Three are for the condition d > R, one for d = R, and two for d < R. In each case, two parabolic arms are represented; for d = R, however, arm 2 "collapses" into coincidence with the x-axis. Since the construction rule u = v possesses no contant term, the line-poles of all these loci are directrices. [But if a line-pole were to be shifted, it no longer would be a directrix for the identical parabolas, and the new eqs. would possess constant terms.]

The condition d > R (an exterior directrix) leads to both arms opening to the left. Since arm 2 is the product of a greatest-distance construction and arm 1 results from a least-distance construction, the two segments have extremal-distance symmetry. As d becomes very large relative to R, arm 2 approaches arm 1, whose vertex always lies midway between the near point of the circle-pole and the directrix (Problem 12). In the limiting case (R = 0), the greatest and least-distance constructions become one, and a single arm results. This is the familiar polar-linear construction, in which the vertex lies midway between the focus and the directrix.

As the directrix approaches the circle-pole (for a fixed value of R), arms 1 and 2 both decrease in absolute size, but the arm 1/arm 2 size-ratio increases without limit. For the directrix at d = 3R, arm 2 is tangent to the circle-pole (Fig. 35a) and arm 1 lies midway between arm 2 and the directrix. For progressively smaller values of d (approaching the value of R), the vertex of arm 2 approaches x_c, and the vertex of arm 1 approaches tangency to the circle-pole. Concomitantly, arm 2 rapidly diminishes in size (see Fig. 35c).

For d = R (Fig. 35d), the directrix and arm 1 are tangent to the circle-pole (a = R)--the *tangent linear-circular system*--but arm 2 collapses to the line (or two coincident lines) $y^2 = 0$ (a = 0), coincident with the x-axis. The ray extending from x_c to the left represents a greatest-distance construction and is the limiting greatest-distance construction of arm 2 (for this sequence of values of d relative to a fixed R). At this position, the arm 1/arm 2 size-ratio is infinite.

For d < R, the directrix penetrates the circle-pole as a secant. Concomi-

Linear-Circular Parabolas, u=v

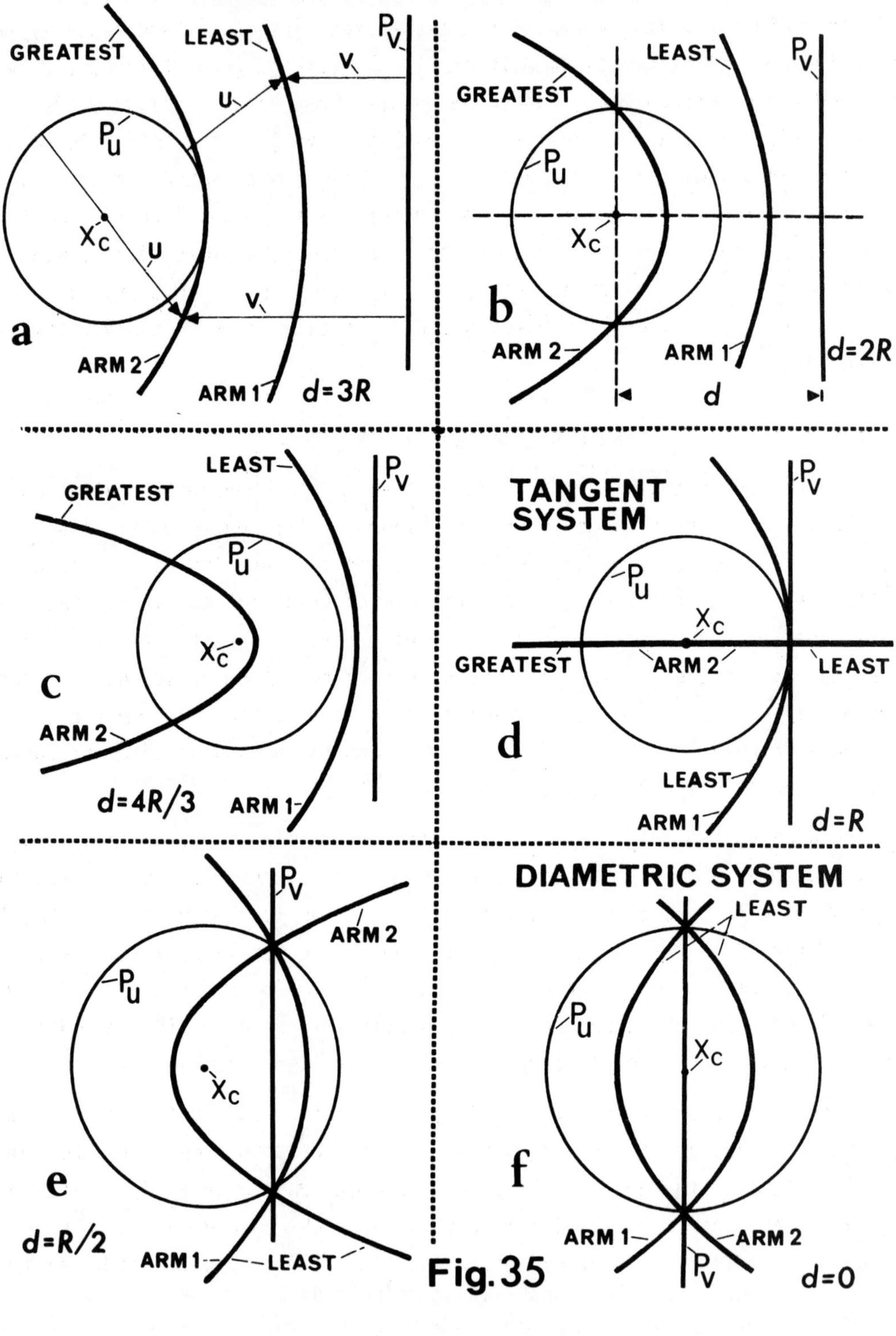

Fig.35

tantly, arm 1 intersects the circle-pole and continues to diminish in size (Fig. 35e). Arm 2--now a greatest-distance construction--enlarges absolutely (from a = 0) to a = (R-d)/2, its vertex progresses to the left of x_c, and its size relative to arm 1 also increases. Both arms now intersect each other, the directrix, and the circle-pole at two common points. When the directrix becomes a diameter (Fig. 35f), arm 1 attains its smallest size (a = R/2), which is equal to that of arm 2, with the arms being reflections of one another in the LR. For this position of the directrix, the system gains a second line of symmetry-- the *diametric linear-circular system*. These are the most-symmetrical linear-circular parabolas, inasmuch as they are reflections of one another, possess two lines of symmetry, and intersect p_u, p_v, and each other at diametrically-opposed points.

The Construction Rule u = Cv

One can predict that the construction rule u = Cv represents central conics cast about a focus at x_c, both from a knowledge of 1st-degree polar-linear construction rules for central conics (eqs. 169 and 170) and from a knowledge of the focal hybrid distance eqs. of central conics (eqs. 142 and 143). Thus, the polar-linear eqs. of central conics cast about a focus are of the form u = ev±Cd for a general orthogonal line-pole, with the constant term vanishing when the line-pole is the near directrix (the letters for the distances from the two poles have been exchanged in this eq. to conform to the linear-circular convention; p_u, the circle-pole, and p_v, the line-pole).

Accordingly, since the distance from the point-pole of polar-linear eqs. (the variable u of the above eq.) becomes u ± R for linear-circular eqs., the effect of using a circle-pole and extremal distances is simply to shift the location of the directrix. That is to say, when a circle-pole replaces the point-pole, the constant term of the polar-circular eq. vanishes for a different location of the line-pole than occurs in the polar-linear case. The reasoning based on the known hybrid distance eqs. is similar (see Problem 13).

Making the substitutions of eqs. 217a,a',b in the construction rule u = Cv leads to eq. 219a for the rectangular equivalents of this linear-circular construction. This is the eq. of central conics of eccentricity e = C, leading to eq. 219b. Since the linear-circular construction rules become u = ev, it follows that central conics of reciprocal eccentricities also have polar-exchange symmetry in the linear-circular system (recall that this was the case in the polar-linear system). Since the constant term of construction rules in the standard form of eq. 219b is equal to a^2 (the square of the semi-major axis), it follows that the displacement is by the amount ae from center. Consequently, all the represented central conics are focal (have a focus at x_c), and line-poles at x = d are their directrices.

[The fact that e, d, and R can be assigned values independently does not mean that an orthogonal line-pole at any location in the plane of a conic can be a directrix; once a value of d is chosen for given values of e and R, this fixes the location and size of the particular conic for which the selected line-pole is a directrix.]

rectangular equations of the loci of u = Cv

(a) $[x + C(Cd \pm R)/(1-C^2)]^2 + y^2/(1-C^2) = (Cd \pm R)^2/(1-C^2)^2$ central conics   (219)

(b) $[x + e(ed \pm R)/(1-C^2)]^2 + y^2/(1-e^2) = (ed \pm R)^2/(1-e^2)^2$ eq. 219a with C = e

Six cases of linear-circular ellipses have been selected for illustration. These are the generally disparate ellipses of Fig. 36a-f. Letting e = 1/2, six ratios of R/d that illustrate salient features are employed. It is evident from eqs. 219 that for large values of d (or small values of R), the ellipses approach congruence. The smaller ellipse is the product of a greatest-distance construction (the homologue of arm 1 of Fig. 35) and the larger one is the product of a least-distance construction. Accordingly, the two loci of each pair have extremal-distance symmetry. The smaller ellipse is not always interior to the circle-pole for large d. For example, it intersects the circle-pole for d = 4R (size-ratio 3/1) and becomes exterior and tangent to it for d = 5R (size-ratio of 7/3). For d = 5R/2 (Fig. 36a), the size-ratio is 9/1.

As d decreases from the value 5R/2, both ellipses decrease in size, but the value of the size-ratio increases. The smaller ellipse eventually becomes a point-locus at x_c for the value d = 2R (Fig. 36b). This point-locus is the homologue of the "linear parabolas" of Fig. 35d. Like them, it also has a greatest and least-distance construction, and like them, it demarcates between loci that have solely a greatest-distance construction (the smaller ellipses for d < 2R) and loci that have solely a least-distance construction (the smaller ellipses of Fig. 36c-e, for d > R). In general, this circumstance arises when the d/R-ratio equals 1/e. For example, for the e = 2 hyperbola, the *demarcating position* of the directrix is at d = R/2. In other words, the *demarcating locus*--the locus that has both a least and a greatest-distance construction--occurs for direc-

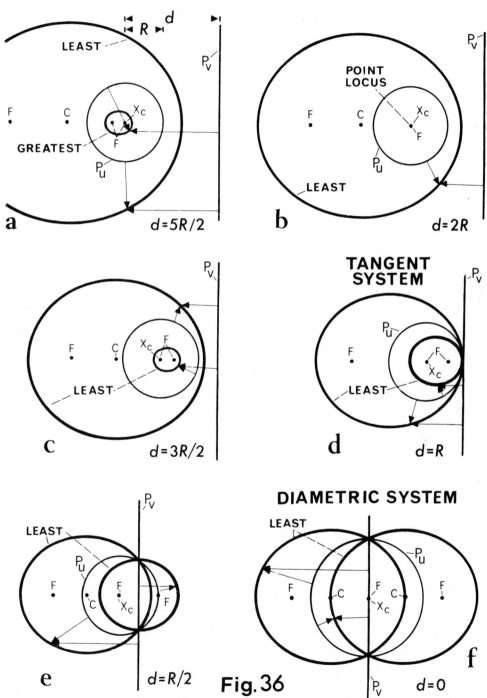

Fig. 36 Linear-Circular Ellipses, $u = ev = v/2$

trices that are secants of the circle-pole for "linear hyperbolas," exterior to the circle-pole for point-ellipses, and tangent to the circle-pole for "linear parabolas."

As the directrix approaches the circle-pole from its demarcating position, the point-ellipse enlarges; it reaches the same size at $d = 3R/2$ (Fig. 36c) that it had at $d = 5R/2$, but now it is the product of a least-distance construction (rather than a greatest-distance one). The two ellipses of the pair no longer have extremal-distance symmetry since both are least-distance constructions. Furthermore, x_c is at the left focus of the smaller ellipse rather than at its right focus, as it was for the smaller of the two ellipses of pairs that preceded the demarcation locus (Fig. 36a). Thus, the two ellipses (and hyperbolas) of a pair possess extremal-distance symmetry only when they are cast about similarly-located foci. The homologous circumstances for the parabola constructions of Fig. 35 were manifested by the curves opening in the same direction when they possessed extremal-distance symmetry but in opposite directions in the absence of such symmetry.

When the directrix is tangent to the circle-pole (the *tangent linear-circular system*), both ellipses also are tangent to it (Fig. 36d), and their size-ratio is 3/1. Once the directrix penetrates the circle-pole, the two ellipses intersect both it, each other, and the circle-pole at common points (Fig. 36e), just as in the case of the linear-circular parabolas of Figs. 35e,f. The directrix position $d = R/2$ finds the directrix coincident with the minor axis of the smaller ellipse (Fig. 36e). When the directrix is a diameter--the *diametric linear-circular system*--the system and all its loci gain a second line of symmetry (Fig. 36f). [The condition in which the center of one ellipse in this system is incident upon a vertex of the other is unique to the $e = 1/2$ ellipses.]

The Construction Rule $u \pm v = \pm Cd$

The treatment of 1st-degree linear-circular loci is concluded with examples drawn from the general construction rule $u \pm v = \pm Cd$. It is easy to predict that all the loci represented by this eq. will consist of rectangular parabolas or segments of such parabolas. Considering the identification of these loci only from the point of view of the derivation of the equivalent rectangular eqs., it is evident from a consideration of the transformation eqs. 217a,a',b that on substituting for u and v in the above eq., only one squaring is needed to eliminate the radical, yielding the construction rule 220a.

rectangular equations equivalent to $u \pm v = \pm Cd$

(a) $x^2+y^2 = x^2 \pm 2(Cd \pm d \pm R)x$ first step in forming the eqs.   (220)
        $+ (Cd \pm d \pm R)^2$

(b) $y^2 = \pm 2(Cd \pm d \pm R) \cdot$ eq. 220a in standard form (upper and lower alter-
              $[x \pm (Cd \pm d \pm R)/2]$ nate signs before the parenthetical expressions paired, and the same combinations of alternate signs for d and for R to be taken within parenthetical expressions)

(a) $y^2 = \pm 10R(x \pm 5R/2)$ two of the presumptive rectangular loci of $u \pm v = \pm d$ for $d = 2R$ (the locus for the plus signs does not participate)   (221)

(b) $y^2 = \pm 6R(x \pm 3R/2)$ two additional presumptive loci (the locus for the plus signs does not participate)

(c) $y^2 = \pm 2R(x \pm R/2)$ two additional presumptive loci (both loci participate)

Since the x^2-terms of eq. 220a cancel whenever the coefficients of the initial 1st-degree linear-circular eq. in u and v have the same absolute magnitude, there remains a degenerate three-term eq. in y^2, x, and a constant term. No matter how many squarings might be needed to eliminate alternate signs from eq. 220a, the resulting eq. merely will represent a multiple number of rectangular parabolas. Putting eq. 220a in standard form, yields eq. 220b. The most symmetrical linear-circular loci represented by these eqs. are those for which $C = 1$ (all coefficients of unit magnitude in the initial eq.).

The most-symmetrical linear-circular loci derived above, with the construction rule $u \pm v = \pm Cd$ are illustrated in Fig. 37 for the case $d = 2R$. Letting $C = 1$ and $d = 2R$ in eq. 220b yields the six rectangular parabolas of eqs. 221a-c (upper and lower alternate signs taken in combination with each other). Because of the profusion of alternate signs in the original rectangular eqs. and the number of possible constructions, these six parabolas must be tested against the initial eqs. to find out which are valid loci and what the precise constructions are.

When the six rectangular parabolas of construction rules 221a-c are plotted and tested, it is found that only four of them participate in the various constructions, that these constructions complement one another in including all parts of all four participating parabolas, and that only four of the six possible linear-circular construction rules apply (considering the eqs. involving

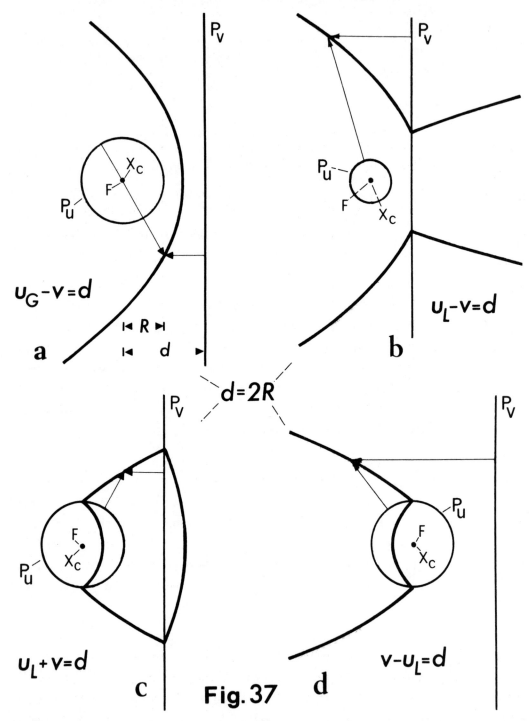

Fig. 37 Linear-Circular Parabolas, $u \pm v = \pm d$

u_L and u_G as different constructions). Thus, there is no construction $u_G + v = d$ nor $v - u_G = d$. Moreover, two of the four constructions in question are remarkably different from any previously encountered; one of them is a compound parabola made up of portions of three different parabolas, and the other is a compound parabola made up of both finite and infinite segments.

The construction rule $u_G - v = d = 2R$ yields the conventional parabola of Fig. 37a, which is of the same size relative to the circle-pole and occupies precisely the same location as arm 1 of Fig. 35b, with the construction rule $u_L = v$. If it were valid to compare two identical segments of two different loci (as opposed to the entire loci) in the same coordinate system, arm 1 of Fig. 35b would be regarded as being more symmetrical than the parabola of Fig. 37a, since it has a simpler construction rule. A more appropriate comparison would be between arm 1 of Fig. 35b and the compound parabola of Fig. 37b, since both are least-distance constructions. But the valid comparison is between the complete loci of Fig. 35b and the composite loci of 37a,b (since both loci being compared then are products of both greatest and least-distance constructions). The parabolas of Fig. 35b, of course, possess greater linear-circular symmetry than those of Fig. 37a,b, since they are represented by a simpler construction rule.

The compound parabola of Fig. 37b, produced by the construction rule $u_L - v = d$, is similar to the polar-linear parabola of Fig. 25d$_2$, produced by the construction rule $u - v = Cd$ (letting u be the distance from the point-pole for the sake of comparison). A difference between the two situations is that the absent segments of Fig. 25d$_2$ (for the construction rule $u + v = Cd$) make up the closed compound curve of Fig. 25d$_1$, whereas, in the present case, the segments produced by the construction rule $u_L + v = d$ do not participate in the same way. The absent segment of the larger locus in Fig. 37b participates in toto in Fig. 37c, but only portions of the absent segment of the smaller locus of Fig. 37b participate in Fig. 37c. The closing of the locus of Fig. 37c depends, instead, upon the participation of a segment of still a third parabola (eq. 221c with the minus signs). This is the first example that we have encountered in which segments of three similar rectangular curves participate in a compound curve (note that three of the four segments are from the two congruent parabolas of eqs. 221c).

The last illustration (Fig. 37d) is an open compound curve made up of segments of the two congruent parabolas that contribute to the locus of Fig. 37c. This also is a new type of locus, in that it is the only compound curve we have encountered that is made up of a segment of finite length of one locus and infinitely-long segments of another locus (all three segments are from the congruent rectangular parabolas of eqs. 221c).

The Construction Rule $u^2 = Cdv$

Our treatment of linear-circular construction rules is concluded with examples drawn from the general construction rule $u^2 = Cdv$, and this eq. is used to illustrate again, and in more detail, the diagnostic utility of hybrid distance eqs. It is apparent from perusing the hybrid eqs. 139-154 that the above construction rule must represent limacons (cast about either the DP or the focus of self-inversion) and Cartesian ovals (cast about a focus of self-inversion).

The basis for the above conclusion is the fact that the hybrid eqs. of the implicated curves (eqs. 146-149) are the only ones in which the variable for the distance from the hybrid pole (to a point of a locus) is present in both 1st and 2nd-degree terms. This is a prerequisite for the derivation of a linear-circular construction rule of the form $u^2 = Cdv$ (since otherwise there is no possibility for vanishing of the linear term in u, for the distance from circle-pole p_u). This is illustrated employing eq. 146 for limacons cast about the DP, repeated here as eq. 222a, where w (substituted for v of eq. 246) is the distance from the hybrid pole to a point of the curve.

deriving the equation $u^2 = Cdv$ for limacons from the hybrid DP-equation for limacons

(a) $w^2 + aex_w = \pm aw$ hybrid distance eq. for limacons cast about the DP (plus sign for large loop and elliptical ovals, minus sign for small loop)   (222)

(b) $(u\pm R)^2 + aex_c = \pm a(u\pm R)$ eq. 222a with $w = u\pm R$ (upper sign of R for least exterior distances)

(c) $u^2 + R^2 + aex_c = aR$ eq. 222b with linear terms in u cancelled by letting $a = 2R$

(c') $u^2 + R^2 + 2eRx_c = 2R^2$ eq. 222c with parameter a replaced by 2R

(d) $u^2 = -2eRx_c + R^2$ eq. 222c' rearranged (still a hybrid eq.)

(e) $u^2 = -2eR(d-v) + R^2$ eq. 222d for a line-pole to the right of x_c and the locus (a polar-linear eq.)

$\quad\quad = 2eRv + R(R-2ed)$

(f) $u^2 = 2eRv = 4e^2dv$ eq. 222e for the line-pole a directrix ($d = R/2e$)

Simplifying our treatment to consider only greatest distances and least distances to exterior points, the linear-circular variable u is equal to the hybrid variable $w \pm R$ (lower sign for least exterior distances), whence $w = (u \pm R)$. [For least distances to interior points one sets $w = R - u$ and arrives at the same final result.] If, after substituting for w in the hybrid eq. 222b, the resulting eq. is to possess only a term in u^2, there must originally be terms in both w^2 and w, to allow for cancellation of the linear terms. For example, these terms will cancel from eqs. 222b if $2uR = au$, leading to the first three conditions; $a = 2R$, least-distance constructions for elliptical ovals and large loops of hyperbolic limacons (upper signs), and greatest-distance constructions for small loops.

In other words, large loops and elliptical ovals of linear-circular limacons represented by the construction rule $u^2 = Cdv$, with their DP at x_c (Fig. 38a-left), will be the products of least-distance constructions, small loops will be the products of greatest-distance constructions, and the radius of circle-pole p_u must be half the value of the limacon parameter a (of the polar eq., $r = a - b\cos\theta$, and the hybrid eq. 222a). Fulfilling these requirements leads to eqs. 222c,c'.

Rearranging eqs. 222c,c' yields eq. 222d. In order for this to take the linear-circular form $u^2 = Cdv$, two additional conditions must be satisfied; the term in $-x_c$ must transform to a positive term in v--the distance from line-pole p_v--and the resulting constant term must vanish. Stated otherwise, the line-pole must be a directrix positioned to the right of x_c and the represented locus. [The line-pole cannot be positioned to the left of x_c (and be a directrix), with the curve to its left (eq. 157c), because in that event the constant term could not vanish.]

Fulfilling, first, the condition that the line-pole be positioned to the right of both x_c and the locus, one obtains $x_w = (d-v)$ (eq. 158c), leading to eq. 222e. Second, the condition that the line-pole be a directrix requires that $R = 2de$, leading to the final construction rule 222f. From this eq. it is seen that the eccentricity of the limacon with its DP at x_c is $e = \sqrt{C}/2$.

Accordingly, it has been established that the linear-circular construction rule $u^2 = Cdv$ represents certain limacons and Cartesian ovals cast about a directrix (by definition; RULE 45). By taking the hybrid distance eqs. of limacons cast about the DP as examples, and substituting $w = u \pm R$, the constant C

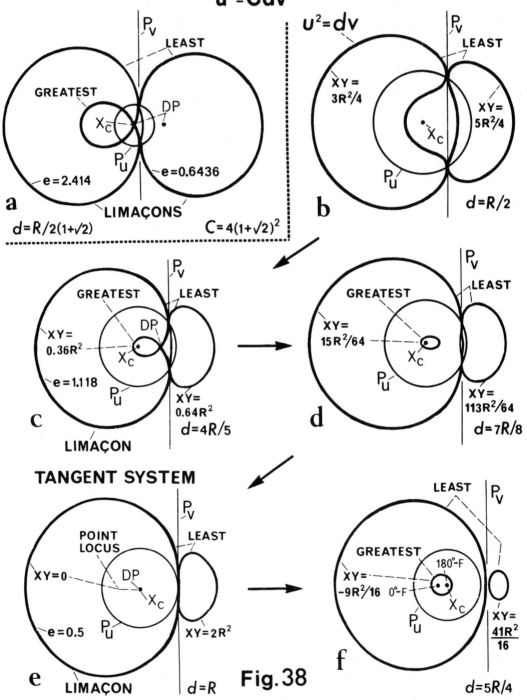

Fig. 38. Linear-Circular Limaçons & Cartesian Ovals, $u^2 = Cdv$

of the linear-circular eq. that represents these limacons (as well as certain Cartesian ovals; Fig. 38) was found to be equal to 2eR, with a = 2R. The resulting construction rules apply to least distances for elliptical limacons and large loops of hyperbolic limacons, and to greatest distances for small loops.

It was established further that the line-pole must lie to the right of both x_c and the locus, whereupon $x_c = d - v$. Finally, in order for p_v to be a directrix, we must have R = 2ed. The limacon constructions of Fig. 38a-left and 38e-left satisfy all these conditions. Similar derivations can be carried out for other limacons and Cartesian ovals (see Problem 13).

The rectangular eqs. of the loci represented by the construction rule $u^2 = Cdv$ are found to be eqs. 223 (see Problem 14). Inspection of these eqs. reveals that they represent Cartesian ovals (including parabolic Cartesians) and limacons (compare with eqs. 81, 105a, and 116). [The "d" of these eqs. is the distance between x_c and directrix p_v, rather than the distance between the bipolar foci p_u and p_v of eq. 105a (or focus p_u and pole p_z of eq. 116).]

One readily recognizes that when $R^2 = Cd^2$, eq. 223 represents a limacon curve of demarcation for the lower signs and Cartesian ovals for the upper signs. But the rectangular eqs. of limacons also take the form of eq. 223 when they are cast about the focus of self-inversion as origin, namely eq. 224 (see also eq. 17c). Comparing eq. 224 with eq. 223 reveals that the latter also represents limacon curves of demarcation for C = 4(R±d)/d (Problem 15).

rectangular equations equivalent to $u^2 = Cdv$

$[x^2+y^2 \mp Cdx+(R^2 \pm Cd^2)]^2 = 4R^2(x^2+y^2)$ upper and lower signs paired   (223)

$[(x^2+y^2)+a^2x/b+(a^2-b^2)^2/4b^2]^2 = b^2(x^2+y^2)$ rectangular eq. of limacons with the focus of self-inversion at the origin   (224)

Linear-circular Cartesian ovals represented by the upper signs of eq. 223 have curves of demarcation only for values of C > 4. For C ≤ 4, the curves are monovals for all R/d-ratios (Fig. 38b-f, ovals to right of p_v). The ovals represented by the upper signs, together with those represented by the lower signs, become curves of demarcation concomitantly for the unique value, C = $4(1+\sqrt{2})^2$ (Fig. 38a and Problem 17), whereupon e = $1 + \sqrt{2}$ for the limacon cast about its DP and e = $1/\sqrt{(1+\sqrt{2})}$ for the limacon cast about its focus of self-inversion.

Requiring that eq. 223 with the lower signs represent limacons cast about both the focus of self-inversion and the DP leads to the cardioid, the only limacon in which these foci are in coincidence and self-inversion is trivial (i.e., the product of two radii to the curve is identically 0 because one of the radii always has the value 0).

Accordingly, in general, eq. 223 represents two sets of Cartesian ovals. For some special cases (Fig. 38c,e) it represents only the first or second set and the limacon curve of demarcation of the second or first set, while in the unique case of Fig. 38a, it represents two limacon curves of demarcation. To illustrate these circumstances, consider the cases of the most-symmetrical 2nd-degree linear-circular loci, namely, the case for $C = 1$ ($u^2 = dv$, the simplest 2nd-degree construction rule), illustrated for $d = R, 4R/5, 7R/8, R$, and $5R/4$.

Since, from eq. 222f, $C = 4e^2$, the limacon cast about its DP and represented by the construction rule $u^2 = Cdv$, with $C = 1$, has an eccentricity of 1/2 (Fig. 38e; the limacon of Fig. 38c is cast about its focus of self-inversion). This is yet another example of the exceptional symmetry or structural simplicity of curves of eccentricity 1/2. Since one of the conditions leading to eq. 222f was $R = 2de$, it also is found that the most-symmetrical loci occur for the condition $d = R$ (Fig. 32e), which is the condition for the tangent system. Since both loci for this construction are tangent to both coordinate elements (as well as to each other), this gives further support to RULE 38.

Consider, first, the case $d = R/2$ (Fig. 38b). This leads to two opposed confocal Cartesian monovals bitangent to each other and the directrix at their points of intersection with circle-pole p_u. Both ovals are represented by a least-distance construction and both self-invert about the common focus, x_c. This is known as a *0°-intra-oval-mode focus of self-inversion* because each oval inverts to itself along coincident radii. Whereas the products of the paired radii for self-inversion of the two monovals about x_c (given by an xy-product for each curve) sum to $2R^2$ for all members of the ensembles, the individual products show progressive changes with the d/R-ratio (Problem 18).

Whenever the directrix intersects the circle-pole, the ovals (or loops) also make contact with it, and at the same points. This is required, because the construction rule possesses no constant term; when $u = 0$, v also must vanish. Furthermore, the points of contact must be tangent-points since, following the hybrid eq. analysis, the directrices do not intersect the loci. Accordingly, in these circumstances (Fig. 38b-e), real loci always exist and always are bitangent to one another.

For C = 1, the locus to the right of the directrix--represented by eq. 223 with the upper sign--always is a monoval, since the condition for a curve of demarcation to exist requires C > 4, and the latter curve defines the transition to biovals. For example, in Fig. 38a, for a case in which C exceeds 4, the limacon to the right of the directrix transforms to Cartesian biovals as the directrix recedes from x_c (moving to the right), and to a Cartesian monoval as it approaches x_c (moving to the left). On the other hand, as the directrix of Fig. 38b recedes from x_c (moving to the right), the monoval lying to the left of the directrix transforms first to a hyperbolic limacon curve of demarcation (Fig. 38c), then to biovals (Fig. 38d), then to an elliptical limacon curve of demarcation (Fig. 38e), and then again to biovals (Fig. 38f).

The directrix location d = 4R/5, of Fig. 38c, leads to the hyperbolic limacon curve of demarcation of the left ovals (e = √[5/2]) cast about its focus of self-inversion (corresponding to eq. 224), which also is the focus of self-inversion of the right monoval. The xy-products for self-inversion have increased in value for the oval on the right and decreased for the ovals on the left, but sum to $2R^2$. As shown above, the small loop of the limacon is a greatest-distance construction; the remainder of the loci are least-distance constructions. The ovals, of course, are bitangent to one another, and the concavities of both curves progressively flatten as the directrix recedes from x_c.

Beyond the demarcating position, d = 4R/5, the small loop of the hyperbolic limacon detaches at the DP, leading to the formation of the small interior oval and the large exterior oval of a biovular Cartesian. The situation is illustrated in Fig. 38d for the directrix located at d = 7R/8. Again, both curves (the left biovals and the right monoval) self-invert about the common focus at x_c, which now is a 0°-inter-oval mode focus for the biovals (i.e., one oval inverts to the other along coincident radii). The interior oval, derived from the small loop, is a greatest-distance construction, and the remaining segments are least-distance constructions.

As the directrix recedes farther from x_c toward the perimeter of p_u, the interior oval diminishes in size and the other oval flattens in the region near the directrix. Since the ovals remain bitangent to one another and to the directrix, the flattening leads to a progressive approximation of the points of tangency to one another.

When the directrix becomes tangent to the circle-pole, the interior oval "collapses" into coincidence with x_c to form the elliptical limacon curve of demarcation, whereupon x_c becomes the DP. On the other hand, x_c remains the focus of self-inversion of the monoval to the right of the directrix. The xy-products for the two ovals of Fig. 38e continue to sum to $2R^2$, but now one value is 0 (since the limacons self-invert only trivially about the DP) and the other is $2R^2$.

The former points of bitangency now have merged to a single point of tangency of the flattened ovals, which are tangent both to each other and to both coordinate elements. It already was noted that the $e = 1/2$ limacon defines the transition between elliptical limacons that are wholly convex and those that possess a concavity, and this property also characterizes the Cartesian monoval on the right of Fig. 38e (i.e., with shifts of position of the directrix). The ovals again are solely least-distance constructions.

As the directrix leaves its point of tangency with the circle-pole, the monovals also cease being tangent to one another and the coordinate elements. Concomitantly, an interior oval forms about x_c, again yielding biovular Cartesians (Fig. 38f). As in the case of the small oval of Fig. 38d, the interior oval is the product of a greatest-distance construction and the other two ovals arise from least-distance constructions.

A significant difference between the ovals of Fig. 38d and 38f, less conspicuous than the differences in form, is that x_c no longer is the 0°-interoval-mode focus of self-inversion of the ovals on the left. The latter focus lies to the left of x_c, as marked. Rather, the focus at x_c is the 180°-interoval-mode focus of self-inversion (i.e., the ovals on the left invert into each other about x_c along radii that are oppositely directed). In consequence, the algebraic xy-product for the biovals is negative. But since the xy-product for the monoval on the right now exceeds $2R^2$, the two products continue to sum to $2R^2$ (as they must for all loci represented by eqs. 223; see Problem 18).

One of the most noteworthy features of Fig. 38 is that it gives the first illustrated example of the manner in which elliptical limacon curves of demarcation form from, and give rise to, Cartesian ovals (with changing parameter values, or directrix locations). In both the hyperbolic and elliptical cases, the oval originates in the neighborhood of the DP (which is part of the locus of both groups of limacons).

In the transition from a hyperbolic limacon (passing from Fig. 38c to 38d), the small loop detaches to form the interior oval (or, in reverse, coalesces with the large oval to form two loops). Concomitantly, the 180°-inter-oval and 0°-intra-oval-mode foci of self-inversion come into coincidence to form the DP (or, in reverse, separate with the loss of the DP). In the transition from an elliptical limacon (passing from Fig. 38d to 38e), the 0° and 180°-inter-oval-mode foci of the biovals come into coincidence and the interior oval "collapses" onto them to form the DP (or the reverse occurs on passing from Fig. 38e to 38f, as the DP is lost and the two foci of self-inversion are formed).

Construction Rules as Relationships between Curves

Studies of curves in the linear-circular coordinate system, more than in any of the previously employed systems, emphasize the fact that the mere use of some coordinate systems to formulate even a single curve's construction rule also constitutes a study of relationships between curves. Having formulated the eq. of a given linear-circular locus, one could proceed to inquire as to the corresponding construction rules of each of the coordinate elements with respect to the locus in question paired with the other coordinate element. For example, what is the eq. of the circle-pole of Fig. 37a with respect to the parabola and the line-pole taken as reference elements, or of the line-pole with respect to the parabola and the circle-pole taken as reference elements? Explorations along these lines, though highly intriguing, would carry us too far afield from present goals.

Summary

Just as was the case with each new coordinate system studied previously, the investigation of linear-circular loci and their eqs. reveals fascinating new constructional relationships between curves and coordinate elements taken in different combinations and locations. Since the treatments have concentrated on conic sections and curves derivable from conic sections by the simple transformation of inversion, and since the coordinate elements employed also are conic sections, the simplicity of the construction rules is not entirely unexpected.

The very close relationship of Cartesian ovals to conics and their inverse loci (though Cartesian ovals are not related to conics by inversion) was evident already in the bipolar system, where linear construction rules represent conic sections, limacons, and Cartesian biovals, and from the fact that limacons are the curves of demarcation of the latter ovals [Of course, some of these relationships also are deducible from rectangular eqs., but generally pass unnoticed.]

The linear-circular system illustrates the close relationships between limacons and Cartesian ovals even more forcefully, since the simplest linear-circular eq. that represents these loci often represents them together in the same ensemble (with both curves represented by a least-distance construction in the tangent system). Ensembles of Cartesian ovals, alone, are most common (of course, strictly speaking, limacons are merely special cases of Cartesian ovals).

Again, one finds that the hybrid distance eqs. give immediate insight into the nature of loci in other coordinate systems that possess point or circle reference elements. Thus, it was simple to deduce that the eq. $u^2 = Cdv$ must represent Cartesian ovals and limacons, and that no 1st-degree linear-circular construction rule can represent either of these curves.

As anticipated in the *Introduction* to the linear-circular coordinate system, this system possesses the properties to be expected of a coordinate system in which a line and a circle participate as reference elements. Although the present treatment is restricted to very simple constructions, it is evident that this system possesses a rich variety of new curves, and that it illustrates new relations between curves, reference elements, and types of symmetry--most notably between lines and Cartesian ovals. The specific properties of individual loci is astonishing, and, though the rectangular curves from which some of these loci are derived are readily predicted, the origin of the properties only becomes evident upon close examination of the circumstances that give rise to each locus.

Other Coordinate Systems

Although a tremendous variety of additional types of curves and relations between curves exist in other coordinate systems consisting only of points, circles, and a line, these cannot be treated here (see Kavanau, 1980). One can predict that the congruent bicircular system will possess almost all the explicit diagnostic features of the bipolar system. Its linear eqs. not only will represent all of the loci representable by linear eqs. in the bipolar system, but many more. For example, because of the existence of both greatest and least-distance constructions, both arms of rectangular hyperbolas and both ovals of rectangular biovular Cartesians will be represented. Overlapping congruent bicircular systems lead to very complex new relationships.

When a third reference element is present, the situation becomes extremely complex, since now the relative locations of even point-poles influence the types of loci represented by given eqs. Of such systems, only those having solely point-poles have been considered (aside from the classical trilinear system). Use of a collinear tripolar system enables the identification of curves with three collinear foci, just as the bipolar system allows this identification for curves with at least two foci (since the construction rules of such curves must, by definition, be linear). In these systems one encounters the fascinating phenomenon of *harmonic parabolas*, wherein *non-degenerate* eqs. represent two-arm parabolas with arms of generally unequal size (these are comprised of the initial parabola and a *satellite* parabola).

Problems

1. Derive the transformation eq. 200a.

2. Show that the central conics of eq. 201b are positioned with their left foci at the center of the circle-pole and their right foci at p_v.

3. Letting circle-pole p_u be centered at x_c as the origin, and line-pole p_v be at $x = d$, use the hybrid distance eq. for the left arm of a hyperbola cast about its focus to find the greatest-distance construction for the left arm cast about poles p_u and p_v.

4. Show that each of the limacons of eq. 203b has its DP at x_c (the center of the circle-pole) and its focus of self-inversion at p_v.

5. Show that for $B^2 = C^2$ of eqs 204 for limacons, the common solution to eqs. 203b and 204b finds $R = d$ and $C_{pc} = A/B = 1/e$.

6. Show that the focus of Cartesian biovals (eqs. 204) at $x = (A^2-C^2)d/(A^2-B^2)$ always is to the right of the origin in the parabolic subgroup, for which the parameter symmetry condition is $A^2B^2 + A^2C^2 = B^2C^2$.

7. Using a vertex analysis (i.e., using only the vertices of the ovals as test points), show that a least-distance polar-circular construction rule of the form $u = Cv$ applies for both ovals when p_w, rather than p_v, of Fig. 32e ($C = 5$ parabolic Cartesians) is taken to be the point-pole.

8. Using a vertex analysis, show that both least and greatest-distance polar-circular construction rules of the form $u = Cv$ are required to represent the Cartesian biovals of Fig. 32e, if the circle-pole is not centered on the monoval-bioval focus.

9. Derive eq. 214 for the polar-circular loci $u \pm v = \pm Cd$.

10. Derive the rectangular eqs. of the polar-circular loci represented by the construction rule $u^2 = Cdv$ and of the polar-exchange-symmetrical curves. Relate these curves to parabolic Cartesians.

11. Derive eqs. 218, the rectangular eqs. of the linear-circular loci $u = v$.

12. Show that the vertex of arm 1 of the linear-circular parabolas $u = v$ (eq. 218a') lies midway between p_v and the nearest point of p_u.

13. Assume a linear-circular construction rule $u = ev = v/2$ for ellipses (p_u, the circle-pole, p_v, the line-pole) and a line-pole at $d = 5R/2$. What must the sizes of the ellipses be if this line-pole is to be the directrix? As the initial eq., use the hybrid distance eq. for an ellipse cast about its left focus (eq. 142a with the minus sign).

14. Find the condition for which the linear-circular eq. $u^2 = Cdv$ represents parabolic Cartesians cast about the monoval-bioval focus. Do least and/or greatest-distance construction rules apply? What is the value of C_{L-C} in terms of d_{L-C} and R? Show that your result applies to the monoval $L-C$ of Fig. 21-1a.

15. Derive eq. 223 for the linear-circular loci $u^2 = Cdv$.

16. Find the conditions for which eq. 223 with the upper signs represents limacons with their foci of self-inversion at x_c (the center of p_u).

17. Evaluate the parameters a and e of the two limacons of eq. 223 obtained when this eq. represents limacons concomitantly for both the upper and lower sets of signs.

18. Find the eccentricity of the limacon curve of demarcation cast about its focus of self-inversion for the lower signs of eq. 223.

19. Derive the expressions for the xy-products of the radii for self-inversion of the ovals of eq. 223 about the focus at x_c for the upper signs and for the lower signs, and show that they always sum to $2R^2$ (in other words, obtain the expressions for the products of the intercepts from the origin to the curves).

20. Derive the rectangular eq. of the linear-circular curves that are polar-exchange symmetrical with the curves represented by the construction rule $u^2 = Cdv$. Plot and describe the locus for $C = 1$ and $R = d$ that is polar-exchange symmetrical with the locus of Fig. 38e.

PART II.   STRUCTURE RULES AND STRUCTURAL SIMPLICITY

Introduction

Part I, *Construction Rules and Structural Simplicity*, introduces new concepts and approaches, new types of curves and coordinate systems, and new relationships between curves and coordinate systems. In the process, it provides extensive new raw materials and foundations for enhancing the beauty and fascination of Euclidean geometry for the student and scholar. However, the amplifications of analytic geometry of Part I lie entirely within classical boundaries; they deal primarily with visible aspects of structure. Thus, almost all the illustrations of Part I are those of the initial curves themselves.

Though all but the polar and rectangular reference frames employed in Part I either were little-used, little-known, or unknown, they are all coordinate systems, and all the new properties and relationships--no matter how novel-- are fully comprehensible to us from visual inspection of the curves and coordinate elements. Construction rules are merely explicit instructions for plotting curves within given reference frames, whose simplicities provide bases for assessing the corresponding structural simplicities of the encoded curves within these reference frames.

When one enters the domain of Part II, *Structure Rules and Structural Simplicity*, one transcends classical boundaries. First, coordinate systems and construction rules do not form the basis for assessing structural simplicity. Second, except in extreme cases, the structural features that are studied are not perceived on direct observation of the curves (compare the initial curves of Figs. 41 and 42b,c with their structural curves). Third, one comes to grips with fundamental properties of curves that enormously broaden our perspectives, yet, with very few exceptions, completely escaped previous notice.

The reference frames of this new domain possess only single fixed reference elements employed in conjunction with *probes*, characterized by fixed angles--the *probe-angles* (see Fig. 39 for probes with 90° and 180°-probe-angles). But, unlike coordinate reference angles, these probe-angles either are not referred to a fixed initial line (in the case that a point is the reference element) or the radii whose relative directions they specify do not emanate from a fixed point (in the case that a line or curve is the fixed reference element).

In consequence, the structural information that one obtains usually is not unique either to the initial curve or to the combination of the initial curve and the single reference element. This structural information is obtained in the form of eqs. termed *structure rules*. The corresponding features are named and characterized by interpreting these structure rules as the construction rules of rectangular curves (see Fig. 40).

In essence, structure rules describe relations that exist between pairs of points of curves and single reference elements, rather than relations that exist between each point of a curve and given coordinate elements (as with construction rules). The only familiar examples of the new relations are the classical symmetries, which deal with the simplest aspects of the structure of curves and figures as assessed from a point or a line.

The Classical Symmetries and Structure Rules

It is readily evident that the simplest structure rule, $X = Y$, describes these classical symmetries. Thus, stating the existence of a classical reflective symmetry in a line is the equivalent of stating that the intercept X, of a point of a curve to one side of the line, is equal to the intercept Y, of a point on the other side (Fig. 1a,c; repeated here with the addition of Fig. 1e).

Similarly, stating the existence of a classical rotational symmetry about a point at an angle α is equivalent to stating the following: when two radius vectors (hereafter referred to as *probe-lines*) that emanate from the point and are oriented at the fixed angle α to one another (the probe-angles of Figs. 2a and 39), are rotated about the point, the intercept, X, of one probe-line with the curve or figure, is equal to the intercept, Y, of the other probe-line, for all rotational positions (Figs. 1b,c,d).

Generalizing from the Classical Symmetries

It is only because the classical symmetries embody the simplest possible structural relation, $X = Y$, that they are detectable on visual inspection. Once the method of assessing structural simplicity with respect to single reference elements is generalized (generalized symmetry)--even when applied to a classically symmetrical curve or figure--one enters a domain in which structural relations very rarely are readily perceptible. In fact, the ability to perceive them occurs exclusively in trivial or near-trivial cases of linear structure rules with unit coefficients.

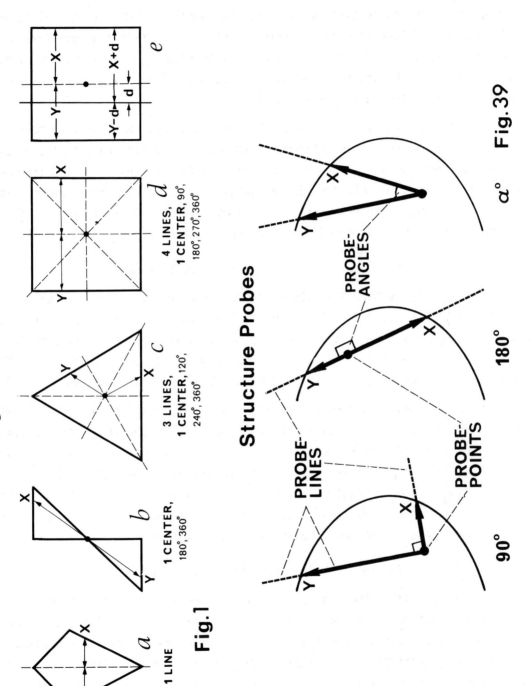

Fig. 39

One such trivial case serves as a convenient example. Thus, if one assesses the structural simplicity of a figure with reflective symmetry in a line, about a parallel line a distance d to its left (Fig. 1e), one is able to recognize the new structural relationship on sight. For example, if the displacement is in the direction of the Y-probe-lines, the structure rule describing the new relation is simply $X = Y + 2d$. This relation is clearly evident on inspection of Fig. 1e.

However, when non-trivial generalizations are considered, the relationships embodied in the new structure rules only very rarely can be perceived on inspection of the curve, which is the chief reason why these generalizations have remained for so long unknown. In assessing structural simplicity about a point by means of a pair of rotating probe-lines, one simply employs angles and/or points other than those of classical rotational symmetries--in fact, any angle and any point in the plane (Figs. 39, 40a, 41). [The points from which probe-lines emanate are referred to as *probe-points*, and a probe-point together with its probe-lines is referred to as a *probe*; see Figs. 2 and 39.]

When assessing structural simplicity about a line using probes that translate along the line, one simply orients the probe-lines at angles other than equal and opposite on each side of the line (Fig. 40b), and/or employs lines other than those of classical reflective symmetry (Fig. 42e)--in general, one employs any two angles and any line in the plane.

But having achieved these generalizations, there is no reason to limit reference elements for structural assessments to points and lines. Any curve in any position can be employed for this purpose (Figs. 40c and 42e), together with probe-lines set at any two fixed angles (fixed either with respect to a line of symmetry or to the instantaneous tangent to the curve, etc.).

The resulting three general methods of assessing structural simplicity with respect to single reference elements are illustrated in Fig. 40. When assessments are made with rotating probes (probe-lines emanating from and rotating about probe-points), the type of generalized symmetry embodied in the corresponding structure rules is referred to as *circumpolar symmetry* (literally, symmetry about a pole or point).

Fig. 40a shows the method of assessing the circumpolar symmetry of an ellipse about an unexceptional interior point, for a probe-angle of 90°. For each rotational position of the probe, a point is plotted along rectangular axes to

Symmetry Relative to Single Reference Elements

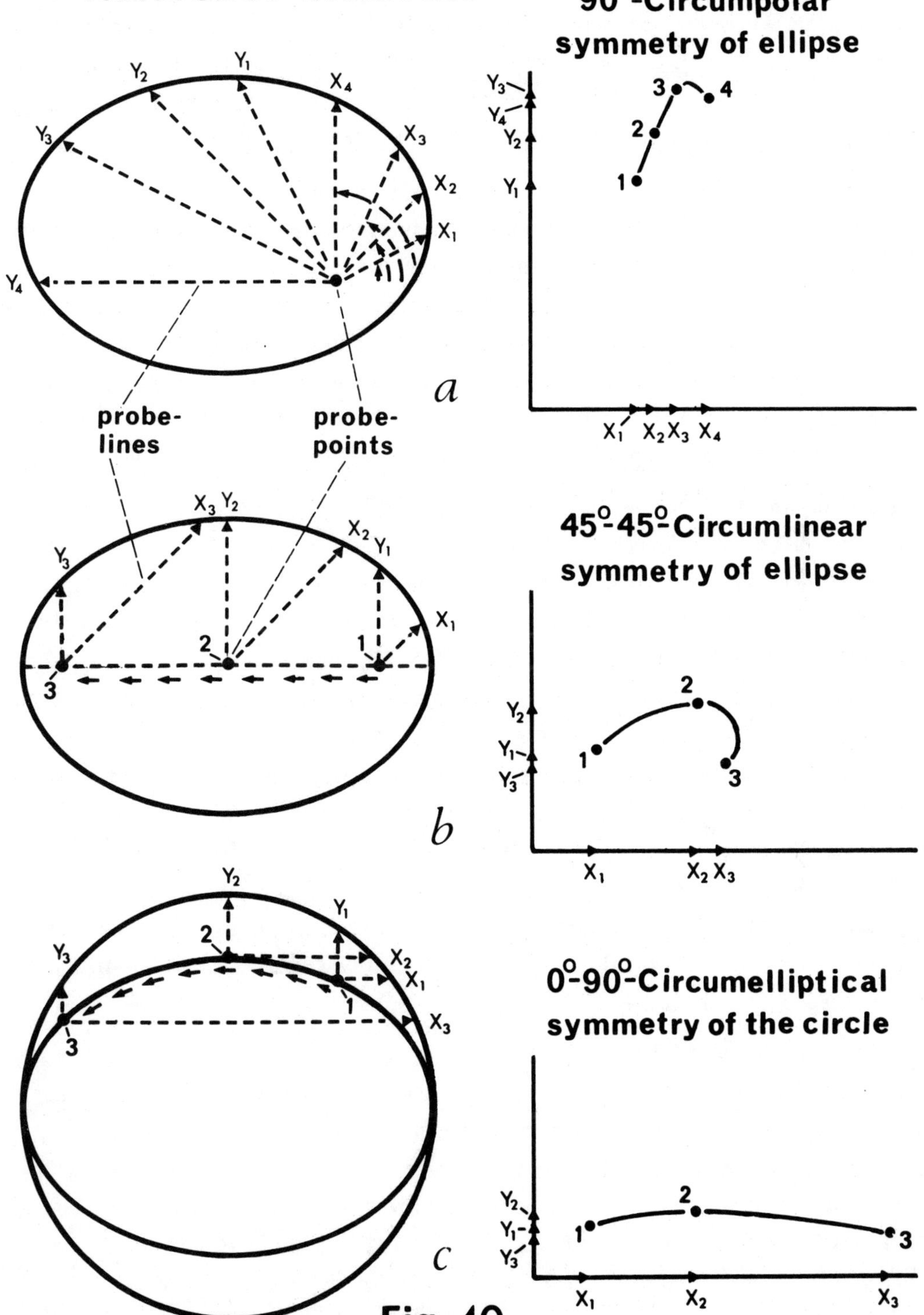
90°-Circumpolar
symmetry of ellipse
probe-lines
probe-points
a
45°-45°-Circumlinear
symmetry of ellipse
b
0°-90°-Circumelliptical
symmetry of the circle
c

Fig. 40

the right (see also Fig. 41), where the abscissa and ordinate for each plotted point are defined by the intercepts of the X and Y-probe lines (also referred to as the X and Y-intercepts).

Fig. 40b illustrates the method of assessing the structural simplicity of an ellipse about its major axis with 45°-probes translating along and oriented at 45° to the axis. Three probes in three positions are shown at the left, and the corresponding three points defined by the X and Y-intercepts of probe-lines are plotted at the right. Since one is employing 45°-probes, and these are being translated at a fixed angle of 45° to the axis, the type of generalized symmetry embodied in the corresponding structure rules is referred to as *45°-45°-circumlinear symmetry*, where the first figure gives the angle of the X-probe-line to the reference line, and the second one gives the angle between the probe-lines.

In Fig. 40c, the reference element is an ellipse, and the angles of reference are 0° and 90°. Hence, the type of symmetry embodied in the corresponding structure rules is referred to as *0°-90°-circumelliptical symmetry*. In this case, the initial curve is a circle, and the three points plotted to the right are defined by the intercepts of the X and Y-probe-lines shown at the left (see also Fig. 42e).

Structure Rules about Points and Inherent Structure

As mentioned above, structure rules express relations that exist between initial curves and single reference elements, that is, relations that are inherent to the combination of the initial curve and the reference element. Accordingly, if one wishes to obtain the structure rules that characterize the structure of initial curves alone--without being influenced by properties of a reference element--it is only necessary to employ a reference element that is itself devoid of structure.

Since the point is the only structureless element, in the ordinary sense, it follows that the only structure rules that describe properties inherent to an initial curve alone are its structure rules about points. Chapter VII and VIII are devoted almost exclusively to the analysis of the structural simplicity of curves and relationships between curves by means of structure rules about points (for treatments of structure rules about lines and other curves --*circumlinear* and *circumcurvilinear symmetry analyses*--see Kavanau, 1980).

Structure Rules about Points Versus Construction Rules about Points

The polar, hybrid polar-rectangular, and bipolar coordinate systems all have characteristic advantages for assessing structural simplicity of curves about points by means of *construction rules*. On the other hand, the study of *structure rules* about points--*circumpolar symmetry analysis*--stands alone as the method par excellence for assessing the structural simplicity of curves, and is by far the most powerful of the four methods.

The preeminent position of the circumpolar method hinges on the complete absence of bias or extraneous structure, such as that introduced by a second reference element or by a structured single reference element. Though one has as much freedom in positioning the polar and hybrid poles as in positioning the circumpolar probe-point, the resulting equations are biased by also being referred to a fixed half-line or line. [It is precisely the existence of this bias, of course, that produces a construction rule rather than a structure rule.]

Though one has complete freedom in positioning the bipolar poles, the resulting construction rule always represents the structural simplicity of the curve about the pole-pair, rather than about either pole alone. Even when the two poles are equivalent, the assessment still is based on the pole-pair.

There are only two circumstances in which the bipolar method begins to approach the diagnostic power of the circumpolar method for comparing structural simplicity about individual poles. These are when comparisons are made either between the poles of two bipolar eqs., both of which eqs. are cast about a pair of equivalent poles, or between two poles of a single bipolar eq., each of which is unique (in the sense that no equivalent pole exists for either of them).

However, there is no circumstance in which a bipolar analysis of an initial curve achieves the diagnostic power of the circumpolar approach. This is because circumpolar analyses possess the additional discrimination of an accounting for angular relations (i.e., inherent structure is analyzed by determining relations between ensembles of paired points of initial curves, where the points can be paired in an infinite number of ways).

CHAPTER VII.   STRUCTURE-RULE ANALYSIS, A SURVEY

Introduction

A polar construction rule defines one or more values of the distance r from the pole, for each value of the coordinate θ. Likewise, a specific circumpolar structure rule defines one or more values of the intercept Y from a given probe-point to the curve, for each value of the intercept X (and vice versa), for a given probe-angle (Figs. 2a, 39, 40a, 41). Generally speaking, the greater the simplicity of the standard polar construction rule of a curve about a pole (i.e., the polar eq. expressed as a function of the sine and/or cosine of the whole angle θ), tne greater the structural simplicity of the curve about the pole (RULE 8a).

[The ability to simplify polar eqs. by the use of multiple and fractional angles, though not without structural implications, is to be regarded primarily as a convenient shorthand method of writing eqs.]

A curve with a simple standard polar construction rule about a given pole also is expected to have relatively simple structure rules about the same pole (RULE 8a). Thus, the polar eqs. of curves provide the initial eqs. for the derivation of all structure rules--circumpolar, circumlinear, and circumcurvilinear (structure rules about points, lines, and curves, respectively). In general, this proves to be the case. Though structure rules usually are of much higher rectangular degree than construction rules, they are of lower degree when they represent cases of greatest structural simplicity (the classical symmetries).

The structure rules of the circle provide excellent illustrative examples, since the circle is the best-known curve and possesses uniquely high circumpolar symmetry. The circle is the only curve with the structure rules $X = Y$ and $XY = R^2$, about a true center, that apply for all probe-angles. For incident points, there are three trivial structure rules, $X = Y$ for 0°, and $X = -Y$ and $XY = 0$ for 180°. For 90°, the incident structure rule is "circular," $X^2 + Y^2 = 4R^2$ (Fig. 44c, outer curve), while for all other angles (except 0° and 180°) it is "elliptical" (Fig. 44a,b,d,e).

For unexceptional points in the plane, the lowest-degree structure rules of the circle are for probe-angles of 0° and 180° (Fig. 44f). These are the "equilateral hyperbolics," $XY = \pm(R^2 - h^2)$, where the applicable alternate sign depends upon whether the pole is interior or exterior (i.e., on whether h is less than or greater than R). These two structure rules, which include the special cases, $XY = 0$ and $XY = R^2$, for incident points and the center, respectively, are simply an expression of the fact that the circle self-inverts about any point in the plane. For all probe-angles other than 0° and 180°, structure rules about unexceptional points are of 6th-degree (see below and Figs. 41d and 44a-e). Only the line possesses structure rules of lower degree about unexceptional points in the plane; the non-incident structure rule of a line for an arbitrary angle α is of 4th-degree (see below and Fig. 41a). [No mention is to be found in the literature of either the 4th or 6th-degree structure rules of the line or circle, nor of any relation that can be regarded as the equivalent of these structure rules. On the other hand, the "circular" incident 90°-structure rule is implicit in the Pythagorean theorem.]

Structure Rules and Constructions

To give the reader a feeling for the significance of some simple *structure rules*, in relation to a known construction of a curve, the elliptical limacon of Fig. 8b is taken as an example. We have seen how this limacon can be constructed by the addition of two radius vectors of non-concentric circles. One radius vector is defined by the polar eq., $r = a$, of a centered circle of radius a, and the other by the polar eq., $r = b\cos\theta$, of an axial circle of diameter b that includes the pole. Following this construction of the curve, let us inquire as to its 90° and 180°-structure rules about the polar pole (the DP of the curve).

Considering, first, the 180° case, let the intercept of the X-probe-line be defined by the sum of the radius vectors for any angle θ, that is, $X = a + b\cos\theta$ (Fig. 8b). The intercept of the Y-probe-line then is the sum of the oppositely-directed radius vectors, $Y = a + b\cos(\theta + 180°) = a - b\cos\theta$. In other words, whereas the radius vector, $b\cos\theta$, of the second circle is added to the radius vector, a, of the first circle to obtain the intercept of the X-probe-line, it is subtracted to obtain the intercept of the Y-probe-line. This being the case, the sum of the intercepts of the two probe-lines is constant at $X + Y = a + b\cos\theta + a - b\cos\theta = 2a$, and the 180°-structure rule is $X + Y = 2a$.

For the 90°-structure rule about the DP, observe (Fig. 8b) that if the X-intercept is defined by the radius vector for any angle θ, that is $X = a + b\cos\theta$, the corresponding Y-intercept is defined by a radius vector 90° in advance of (or behind) the X-radius vector, namely, $Y = a + b\cos(\theta + 90°)$. In other words, it is in the nature of this construction of limacons from two circles, that if the parameter a is subtracted from both the X and Y-intercepts of radius vectors from the DP, the two quantities obtained, (X-a) and (Y-a), are the lengths of two radius vectors oriented at 90° to one another and extending from a point of the second circle to their points of intersection with this circle (see Fig. 8b). But it was just noted above that the squares of two such radius vectors of a circle sum to $4R^2 = b^2$ (the "circular" incident 90°-structure rule equivalent to the Pythagorean theorem). It follows that the sums of the squares, $(X-a)^2$ and $(Y-a)^2$, is b^2, and that the 90°-structure rule of limacons about the DP is the "circular" eq. $(X-a)^2 + (Y-a)^2 = b^2$.

Thus, it is seen how two particular structure rules of a limacon are determined both by the manner of construction of the curve from two circles, and by the properties of the circles (one property of which--the incident 90°-structure rule--is imposed on the limacon). These are among the simplest of many examples that could be given but even they alone suffice to emphasize the fundamental significance of structure rules in relation to the properties of curves.

Aside from extremely simple examples, such as the structure rule of the classical symmetries, $X = Y$, and the structure rule for self-inversion, $XY =$ constant, the 180°-structure rule, $X + Y = 2a$, of limacons about the DP describes one of only a very small number of such properties of curves that were known to classicists (an example for conic sections is given below). In this particular case, the customary classical description of the property is that "any chord or secant through the polar center of a limacon has a constant length." This also is one of the very few cases in which a relation described by a non-trivial structure rule is visible on inspection of the curve.

Comparing Structure Rules

Just as for construction rules, the exponential degrees of structure rules form the primary basis for comparisons between the underlying structural simplicities of curves; the lower the degree, the greater the simplicity. When one deals with structure rules, however, comparisons must include the specification of a probe-angle. Thus, one speaks of the 90°-structural simplicity (or 90°-circumpolar symmetry) of curves about given points, or the 0° or 180°-circumpolar symmetry, or even the α-circumpolar symmetry--that is, the structural simplicity for an arbitrary probe-angle α.

The most meaningful comparisons are made within common probe-angle categories, either for structure rules about homologous points of different curves, or about different points of the same curve. For example, the line has greater 90°-structural simplicity than the circle about an arbitrary point in the plane, since its structure rule is only of 4th-degree, compared to 6th-degree for the circle. A given curve that has greater structural simplicity than another curve about a homologous point for a given angle generally also has greater structural simplicity for other angles.

When making comparisons between structure rules on the basis of exponential degree, the primary index is the degree for positive intercepts. For compari-

[Generally, use of the term *positive intercepts* refers to the intercepts, X and Y, obtained from an intercept eq. for values θ and $\theta+\alpha$, respectively, without regard to sign; the corresponding X and Y *negative intercepts* are the oppositely-directed intercepts with the initial curve (at the angles $\theta + 180°$ and $\theta+\alpha+180°$, respectively. When used in conjunction with references to specific quadrants, however, the terms *positive* and *negative intercepts* are used in the conventional sense.]

sons within degree categories, one appeals to the comparative simplicities of the structure rules themselves, and to their versatility in representing other combinations of intercepts. For example, two 90°-structure rules of ellipses, one about their centers and the other about their traditional foci, are both of 4th-degree. The simplicity of the structure rule about the center, however, is greater than that about the foci.

[By way of comparison, this result is consonant with the corresponding assessment of comparative structural simplicities based on construction rules in the rectangular coordinate system but the opposite of the corresponding assessments in the polar, bipolar, and hybrid polar-rectangular systems.]

One indication of this greater structural simplicity is that the 90°-structure rule about the center is more symmetrical (in fact, symmetrical in an extraordinary way) than that about the foci (compare eqs. 231a and 237; see also Problem 6). Another is that the 90°-structure rule about the center applies for all combinations of intercepts--X and Y positive, X and Y negative, X positive and Y negative, and X negative and Y positive--whereas that for the foci applies only for X and Y positive.

The corresponding 90°-structure rule about the foci that applies for all combinations of intercepts is a degenerate eq. of 16th-degree. This consists of the product of four 4th-degree factors (each of which applies to one of the combinations of intercepts; see Problem 6).

Deriving Circumpolar Structure Rules

The 90°-Structure Rule of Lines

It is established in the foregoing (see also Chapter III) that the structure rules of a curve about a given point relate the intercept lengths, X and Y, of the probe-lines of a probe at that point, as these lines rotate about the probe-point and scan the curve. Accordingly, to obtain these structure rules, one need merely cast the polar eq. of the curve about the given point and find the relation between values of r that are out of phase with one another by the amount of the probe-angle.

In following this procedure, one lets X be the length of the intercept of the probe-line for the angle θ of the polar eq., and Y be the length for the angle $\theta + \alpha$, where α is the probe-angle. The probe-point always lies at the pole of the polar construction rule.

As a very simple example, consider the 90°-structure rule of a line about a non-incident point. Let the point be the polar pole and the line be normal to the 0°-axis at a distance $r = d$. The eq. of this line is $r\cos\theta = d$. Letting X be the value of r for the angle θ and Y be the value for the angle $\theta + 90°$, one obtains eqs. 225b,c. Equating the sum of the squares of $\sin\theta$ and $\cos\theta$ to unity yields the 90°-structure rule 226b (Fig. 41a). The curve represented by eq. 226b--the 90°-*structural curve*--is the most-symmetrical representative of the unicursal Cross Curves of ellipses, $b^2x^2 + a^2y^2 = x^2y^2$, namely the case for the circle, when $a = b = R = d$.

derivation of the 90°-structure rule of a line

(a) $r\cos\theta = d$ polar eq. of a line normal to the 0°- (225)
axis at a distance d from the pole.

(b) $X\cos\theta = d$,   (c) $-Y\sin\theta = d$ eqs. defining lengths of probe-lines from origin to line of eq. 225a for 90°-probe

(a) $\sin^2\theta + \cos^2\theta = d^2/X^2 + d^2/Y^2 = 1$ eliminant of eqs. 225b,c (simultaneous solution, eliminating θ) (226)

(b) $d^2(X^2 + Y^2) = X^2Y^2$ 90°-structure rule of line (unicursal Cross Curve of circle of radius d; applies for all combinations of intercepts)

Unicursal Cross Curves are constructed as follows, using an initial cen-

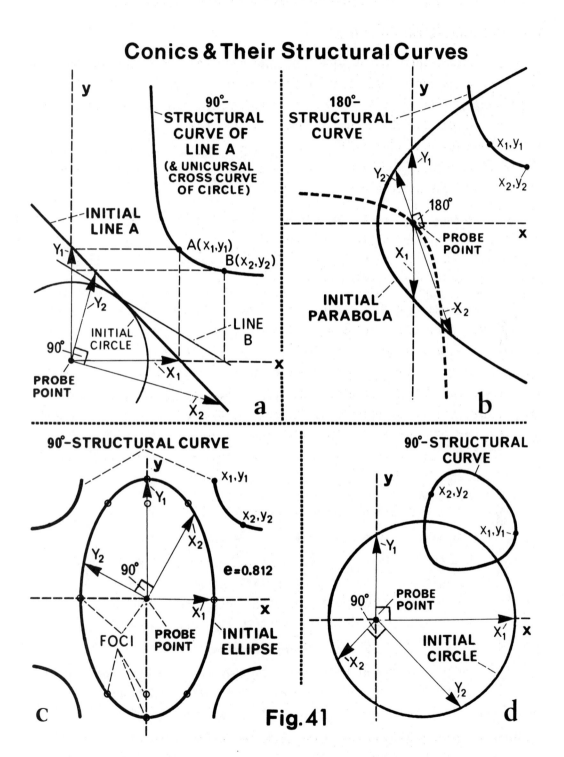

Fig. 41

tered rectangular 0°-ellipse. First, one draws tangents to the curve (for example, lines A and B of Fig. 41a, for a circle). At each point where a given tangent-line crosses one of the coordinate axes, a construction line is drawn parallel to the other axis (dashed lines of Fig. 41a). The points at which pairs of construction lines for a given tangent meet (points A and B of Fig. 41a) define the unicursal Cross Curve of the initial ellipse.

This manner of constructing the curve is shown together with the structure-probe derivation of the 90°-structure rule of an initial line (line A) in Fig. 41a, to emphasize the fact of the existence of two different transformations of a circle and line (the latter not previously known) that give the same 4th-degree curve (but the circle has no 4th-degree structure rule).

Letting the 1st-quadrant (-45°)-tangent-line of a centered circle be the initial line (initial line A of Fig. 41a) whose structure rule is being obtained, the X_1 and Y_1 probe-lines meet the tangent-line where it crosses the axes, in which case the construction lines for the unicursal Cross Curve are parallel to the probe-lines. In the case of tangent-line B, however, this is not the case. Point B of the curve is obtained from construction lines in the same way as for point A, but the probe-lines, X_2 and Y_2, that yield point B of the 90°-structure rule are not parallel to these construction lines.

[It should be noted that it is considerably more difficult to obtain the 90°-structure rule of a line from the initial line A of Fig. 41a (see Problem 1) than from a normal line at $X=d$.]

Non-Degenerate Structure Rules Can Represent All Combinations of X and Y-Intercepts-- Positive, Negative, Null, and Imaginary

It will be noted that the structural curve represented by eq. 226b is symmetrical about the x-axis (replacing y by -y does not alter the locus), the y-axis, the origin, and both coordinate bisectors. In other words, the curve has congruent segments in all eight octants (only the 1st-quadrant segments of which are shown in Fig. 41a) and its construction rule represents positive and negative intercepts in all possible combinations.

This result calls attention to the fact that the 90°-circumpolar structure rules derived in the manner described represent all combinations of positive and negative intercepts of the 90°-probe-lines with the initial curve (see Problems 5 and 6). Thus, for example, when both probe-lines are directed away from the initial line, both intercepts are negative; when only the X-probe-line is directed away, its intercept alone is negative, etc. Structure rules also represent probe-lines of length 0, which exist whenever a probe-point is incident upon a locus.

For purposes of simplicity and to facilitate comparisons, one either excludes negative intercepts in the derivation of structure rules or discards factors that represent them from degenerate construction rules; in many cases, such as the present one, the structure rule represents all combinations of intercepts for the given probe-angle. In some cases (see below) two positive and two negative intercepts exist at all rotational positions of a probe.

Structure rules also apply to imaginary intercepts (see Kavanau, 1980, pp. 12-57 to 60). In the present case, all intercepts are real (either finite or infinite), but in the cases of closed curves and exterior probe-points, imaginary intercepts exist for many probe orientations, and these also are represented by the structure rules (see Problem 7). Moreover, structure rules also can be obtained about imaginary probe-points, in which case they represent only relationships between imaginary intercepts. An intriguing aspect of the analysis of structural relationships by means of structure rules is that it applies with equal force--yielding entirely consistent results--in both the real and complex domains.

180°-Structure Rule of the Parabola About Its Traditional Focus

To derive the 180°-structure rule of the parabola about its traditional focus, the standard rectangular vertex parabola is translated the distance of parameter a to the left (x+a translation) to bring its focus to the origin (Fig. 41b and eq. 227a). In polar coordinates, the construction rule becomes 227b. Expressing eq. 227b in powers of $\cos\theta$ yields eq. 227c. The latter differs from the *simple intercept equations* 20-24 only in that the vertex is at a fixed axial point (-a) specified by a parameter of the curve, rather than at an arbitrary axial point (h or -h) specified by a purely positional parameter.

derivation of the 180°-structure rule of the parabola

(a) $y^2 = 4a(x+a)$ initial parabola (focus at origin)   (227)

(b) $r^2\sin^2\theta = 4a(r\cos\theta + a)$ eq. 227a in polar form

(c) $r^2\cos^2\theta + 4ar\cos\theta + (4a^2 - r^2) = 0$ eq. 227b in powers of $\cos\theta$

(d)	$\cos\theta = (-2a \pm r)/r$	simple intercept format of eqs. 227a-c	(227)
(e)	$\cos\theta = (-2a \pm X)/X$	eqs. defining the X and Y intercepts of the locus of eq. 227d for a 180°-probe	
(e')	$\cos(\theta+180°) = -\cos\theta = (-2a \pm Y)/Y$		
(a)	$a(X + Y) = XY$	180°-structure rule of the parabola about its focus for two positive intercepts	(228)
(a')	$X = -Y$	trivial 180°-structure rule of the parabola about its focus for one positive and one negative intercept	
(b)	$2XY/(X + Y) = 2a$	eq. 228a in the "harmonic mean" form	

Solving eq. 227c for $\cos\theta$ yields the $\cos\theta$-*simple intercept format* 227d. Like the simple intercept eq., this is one of the standard forms employed in structure-rule derivations and other analyses of Structural Equation Geometry. Letting $r = X$ for the intercept of the X-probe-line and $r = Y$ and $\theta = (\theta+180°)$ to obtain the intercept of the Y-probe-line, yields eqs. 227e,e'.

Eliminating $\cos\theta$ yields the sought-after 180°-structure rule 228a (Fig. 41b), which is seen to represent a 45°-equilateral hyperbola centered at (a,a). Only the arm in the 1st-quadrant (which is for pairs of positive intercepts) is the valid 180°-structural curve. Two orientations of the probe are shown, together with the location of the point contributed to the 180°-structural curve at each orientation.

Eq. 228a represents another of the very rare instances in which an invisible property of a curve embodied in a structure rule (other than self-inversion) was known previously. Structure rule 228a is an expression of the little-known fact that "the harmonic mean between the segment lengths (X and Y) of a focal chord of a conic is constant and equal to the hemi-latus rectum." For hyperbolas and ellipses the harmonic mean eqs. take the form 229a. On the other hand, although ellipses have only the trivial 0°-structure rule $X = Y$, hyperbolas also possess the previously unknown non-trivial 0°-structure rule 229b (Problems 3 and 4).

(a)	$2XY/(X+Y) = b^2/a$	180°-structure rule of central conics about a focus for two positive intercepts	(229)
(b)	$2XY/(X-Y) = b^2/a$	0°-structure rule of hyperbolas about a focus for two positive intercepts (X, the intercept with the far arm)	

90°-Structure Rules of Central Conics about Their Centers

An ellipse is used to illustrate the manner of derivation of the 90°-structure rule of a central conic about its center. The standard eq. (230a) of a centered ellipse is transformed to polar coordinates and expressed as the $cos^2\theta$-*simple intercept format* 230b (see also Problems 2-4). Letting $r = X$ for the intercept of the X-probe-line, and $r = Y$ and $\theta = \theta + 90°$ to obtain the intercept of the Y-probe-line, yields eqs. 230c,d. Eliminating θ by summing the two eqs. and equating to unity gives the beautifully symmetrical 90°-structure rule 231a. This applies for all combinations of positive and negative intercepts (see Problems 5 and 6). All four segments of the structural curve are plotted in Fig. 41c. The corresponding structure rule for hyperbolas (eq. 231b) is seen to have no real solution for $a > b$, that is, for $e^2 < 2$ (see also Problems 3 and 4).

derivation of a 90°-structure rule of ellipses

(a) $b^2x^2 + a^2y^2 = a^2b^2$ initial centered ellipse   (230)

(b) $cos^2\theta = a^2(r^2-b^2)/r^2(a^2-b^2)$ $cos^2\theta$-simple intercept format

(c) $cos^2\theta = a^2(X^2-b^2)/X^2(a^2-b^2)$ eqs. defining the X and Y intercepts of eq. 230b for a 90°-probe

(d) $sin^2\theta = a^2(Y^2-b^2)/Y^2(a^2-b^2)$

(a) $a^2b^2(X^2+Y^2) = X^2Y^2(a^2+b^2)$ 90°-structure rule of ellipses about their centers (applies for all combinations of intercepts)   (231)

(b) $a^2b^2(X^2+Y^2) = X^2Y^2(b^2-a^2)$ 90°-structure rule of hyperbolas about their centers (applies for all combinations of intercepts)

(c) $d^2(X^2+Y^2) = X^2Y^2$ 90°-structure rule of parallel line-pairs about points on their midlines (applies for all combinations of intercepts)

It will be noted that as parameter a tends to infinity, in the structure rule of the ellipse (eq. 231a), the eq. tends to the form $b^2(X^2+Y^2) = X^2Y^2$, the same as that derived above for a line. Actually, this same structure rule (repeated here as eq. 231c) also applies to a parallel line-pair for probe-points on the midline, in which case a pair of positive and a pair of negative intercepts exist for all orientations. These results call attention to the fact that the parallel line-pair can be regarded as the limiting form of an ellipse as parameter a tends to infinity in eq. 230a or b, for example, yielding the eq. of a parallel line-pair ($y^2 = b^2$ or $r^2\sin^2\theta = b^2$).

Although, as noted above, the 90°-structure rule of a line or parallel line-pair is the unicursal Cross Curve of a circle, the corresponding structure rules of other central conics are not unicursal Cross Curves or Bullet-Nose Curves, $x^2y^2 = a^2y^2 \pm b^2x^2$ (upper sign for unicursal Cross Curves), although they are closely related to them.

[It now is known (Kavanau, 1980) that, in addition to the classical constructions, the Bullet-Nose and unicursal Cross Curves also are the central inversions of the normal pedals of central conics about their centers.]

The General 90°-Structure Rule of
   the Circle

To derive the 90°-structure rule of the circle about a point in the plane, a distance h from center, the initial eq. is 232a and the polar eq. is 232b, leading to the $\cos\theta$-simple intercept format 232c. Substituting X for r to obtain the intercepts of the X-probe-line, and Y for r and $\theta + 90°$ for θ, to obtain the intercepts of the Y-probe-line, gives eqs. 232d,e (see Problem 7). Summing the squares and equating to unity, to eliminate θ, gives the non-degenerate, 6th-degree, 90°-structure rule 233a, which rearranges to 233a'.

            deriving a 90°-structure rule of the circle

(a) $(x-h)^2 + y^2 = R^2$ circle centered at $x = h$ (232)

(b) $r^2 - 2hr\cos\theta + h^2 = R^2$ polar eq. of eq. 232a

(c) $\cos\theta = (r^2+h^2-R^2)/2hR$ simple intercept format of eq. 232b

(d) $\cos\theta = (X^2+h^2-R^2)/2hX$ eqs. defining the X and Y intercepts of eq. 232c for a 90°-

(e) $\cos(\theta+90°) = -\sin\theta = (Y^2+h^2-R^2)/2hY$ probe

(a) $Y^2(X^2+h^2-R^2)^2 + X^2(Y^2+h^2-R^2)^2 = 4h^2X^2Y^2$ 90°-structure rule (eliminant of eqs. 232d,e; $h \neq 0$)   (233)

(a') $(X^2+Y^2)[X^2Y^2+(h^2-R^2)^2] = 4R^2X^2Y^2$ eq. 233a in another form

(a) $X^2Y^2(X^2+Y^2-4R^2) = 0$ eqs. 233a,b for $h = R$ (incident 90°-structure rule)   (234)

(b) $X^2 + Y^2 = 4R^2$ non-trivial 90°-structure rule

Structure rule 233a is seen to represent a locus symmetrical about the x-axis, the y-axis, the coordinate bisectors, and the origin. Only the segment of the structural curve in the 1st-quadrant is plotted in Fig. 41d (see also Fig.

44c). The structure rule applies for all combinations of intercepts--two positive, two negative, either one negative and the other positive (see Problems 5 and 6), two imaginary, and either one imaginary and the other real (see Problems 7 and 8).

For $h=R$, the general 90° structure rule 233a' simplifies to the degenerate 6th-degree eq. 234a, which is the incident 90°-structure rule. The two identical solutions $XY=0$ (i.e., $X^2Y^2=0$) are trivial ($XY=0$ is an incident structure rule of all curves) and are discarded, giving eqs. 234b. Since the derivation of eqs. 233a,a' involved multiplication and division by h, these eqs. are not valid for $h=0$, the case in which the probe-point is at the center.

Foci, Focal Loci, and Structural Simplicity at the Subfocal Level

Structural Equation Geometry provides incisive new tools for identifying and defining points (*point-foci*) and point-continua (*focal loci*) in the planes of curves, about which the curves possess exceptional structural simplicity (members in both categories are referred to as *foci*). Curves can be classified on the basis of their corresponding structural properties about these foci. Any point in the plane of a curve, for which the degree of a structure rule reduces from its value for all neighboring points, is a *circumpolar point-focus*.

Since classically-defined foci that lie on lines of symmetry are among the highest-ranking circumpolar point-foci, the new perspectives reveal each of them to be a point about which a curve possesses a very high degree of structural simplicity. Equivalently, each point can be said to have a very high degree of positional symmetry in the plane of the curve. In retrospect, the various classical definitions of foci, of which the best-known characterizes them as *"burning points" to which certain light rays striking the curve converge by reflection or refraction*, now are seen to have depended on various relations that merely were expressing different aspects of their high degrees of positional symmetry.

But, whereas only two real points in the finite plane of central conics were exceptional by classical criteria for identifying foci, analyses based on structure rules distinguish and define a hierarchy of 13 real point-foci for ellipses in the extended Euclidean plane (see Table 11 and the open circles and center in Fig. 41c). These include the center, the two traditional foci, the four conventional vertices, the four LR-vertices, and two point-foci of self-inversion on the ideal line at infinity (one each in the directions normal to the lines of symmetry). [There also are two imaginary point-foci on the minor axis.]

Other points in the planes of ellipses also are included in the hierarchy but do not possess point-focal rank. These are members of the point-continua that comprise *circumpolar focal loci*. [Paralleling the situation for bipolar foci, the *circumpolar focal rank* of a point-focus or focal locus is defined as the reciprocal of the degree of the corresponding structure rule.]

Circumpolar focal loci include the lines of symmetry, the initial curves themselves, certain asymptotes (for example, those of hyperbolas), and certain other exceptional loci referred to as *covert focal loci*. The latter bear no recognizable overt relationship to the initial curve; the most notable example is the line through the -b/2 focus of limacons normal to the line of symmetry (the 0° and 180°-structure rules for points on this line are of exponential degree 20, compared to 56 for neighboring points in the plane).

When one also takes these focal loci into account, ellipses possess a total of 16 real foci in the extended Euclidean plane; the two lines of symmetry and the curve itself supplement the 13 point-foci enumerated above. Hyperbolas possess the same total number of real foci, because the deficiency in two real point-foci--the b-vertex foci of ellipses--is compensated for by the presence of two additional focal loci, the asymptotes.

Classical approaches led to the recognition of the exceptional nature of most circumpolar foci on the basis of a number of different but more or less obvious and generally mutually exclusive criteria--lines and centers of symmetry, the traditional foci, poles or "centers" of self-inversion, the vertices (i.e., the points of intersection of curves with their lines of symmetry), points of infinite curvature, the asymptotes, double-points, etc.

Circumpolar symmetry analyses unify these previously diversely-categorized loci into a rigorously-defined, discrete, quantitative hierarchy on the basis of a single criterion--the degrees of their structure rules. Accordingly, a multiplicity of highly diverse classical criteria now can be seen merely to have expressed different aspects of an underlying theme of structural simplicity or positional symmetry. Additionally, the same new criterion identifies and quantitatively ranks other previously unrecognized exceptional points, lines, and loci, including the curves themselves.

It was pointed out in Chapter II that any otherwise unexceptionally-positioned rectangular curve that includes the origin is more-symmetrically positioned than a neighboring one that does not. This distinction is achieved solely by virtue of the curve including the unique symmetrically-positioned point in the coordinate system. The corresponding greater positional symmetry is manifested by the vanishing of a term from the construction rule. In the polar system, the corresponding curve includes the most-symmetrically-positioned point (the pole), and its construction rule not only simplifies by the loss of at least one term, its degree also reduces by at least one unit.

For circumpolar structure rules, the same type of relationship translates into a reduction of degree for a structure rule about *any* incident point of a curve. Moreover, this relation is angle-independent; reduction occurs at the level of the simple intercept eq. (see eqs. 20-24), which does not include a specification of angle. In other words, the basis for the fact of a curve itself being a circumpolar focal locus, that is, of points of the curve having greater positional symmetry than neighboring points off the locus, has its rectangular counterpart in the loss of a term by an eq. of a curve that includes the origin. It is because of this relationship that the latter circumstance is said to be "a crucial one from the point of view of generalized symmetry studies" in Chapter II.

But identifying and ranking foci on the basis of the degrees of the corresponding structure rules is only the most obvious aspect of a circumpolar symmetry analysis. Qualitative ordering of foci whose structure rules have the same degree is achieved on the basis of the relative simplicities of the corresponding structure rules, following the same criteria employed in Part I for construction rules that fell into common degree categories. Additionally, circumpolar symmetry analyses provide other analytical criteria for quantitatively ordering foci.

Additional quantitative analytical discriminations are achieved by relating $(ds/d\theta)^2$ [the square of the derivative of distance along an arc of the rectangular structural curve, with respect to the angle θ of the X-intercept of the initial curve] to a function of the X and/or Y-intercepts, that is, $(ds/d\theta)^2 = F(X,Y) = G(X)$. These eqs., known as the *XY* and *X-arc-increment equations*, respectively, enable the attainment of a unique characterization of the inherent structure of a curve for a given angle (see below, *Arc-Increment Equations and More Specific Characterizations of Inherent Structure*).

Only one curve, in combination with one probe-point in its plane (or an equivalent point) possesses any given *specific structure rule* (a structure rule for a specified probe-point and angle) in combination with the corresponding X-arc-increment-eq. In practice, sufficient specificity often is obtained by using only the exponential degrees of the XY and X-arc-increment-eqs. as criteria, in which cases the characterizations usually are not unique.

RULE 48. *A circumpolar point-focus of a curve is a point about which the curve has greater structural simplicity (a structure rule of lower exponential degree) than about neighboring points.*

RULE 49. *A circumpolar focal locus of a curve is a continuous point array, about every member point of which the curve has greater structural simplicity (a structure rule of lower exponential degree) than about neighboring points off the locus.*

The Classification of Curves

Aside from using similarities between construction rules, classical approaches to the classification of curves relied almost exclusively on aspects of overt structure, such as singular points (cusps, double-points, etc.), projective properties, and branches at infinity. This approach proved to be comparatively unproductive, however, and, to all intents and purposes, was abandoned in the 19th century. From the perspectives of Structural Equation Geometry, the basis for this lack of productivity is not far to seek. Since these overt criteria take into account only the most superficial aspects of structure, the resulting classifications are correspondingly superficial.

Applying these classical approaches to the classification of limacons and central conics would find the hyperbolic and elliptical representatives to have quite different properties, paralleling the differences in appearance. In contrast, a circumpolar symmetry analysis discloses very close parallels in almost all regards. When differences exist--as in many of the initial curves' construction rules--they usually take the form merely of differences in the signs of terms (see, for example, eqs. 250-256). In fact, as we saw above, the 180°-structure rules about the traditional foci of ellipses and hyperbolas are identical, despite the greatly different appearances of the curves.

Structural Equation Geometry introduces two fundamental, but largely independent, paradigms for characterization, classification, and comparative organization, namely, circumpolar symmetry analysis and *inversion analysis*. These yield much deeper insights into relationships between curves than the classical methods. These paradigms take quite different approaches to the problem of classification, and give results that not infrequently are unpredictable from a knowledge of the construction rules and appearances of curves.

For example, two looped cubics of very similar appearance are represented by the construction rules $y^2(a+x) = x^2(a-x)$ and $y^2(a+3x) = x^2(a-x)$. The former represents the equilateral strophoid (see Chapter IV and Figs. 24 and 56) and the latter the folium of Descartes. The construction rules of these curves are seen to differ in what seems to be only a trivial respect--the presence of the coefficient "3" in one but not the other. A difference of this nature has relatively little significance for the rectangular eqs. of conic sections. But for cubics, the difference is found to be crucial when the construction rules are subjected to either a circumpolar symmetry analysis or an inversion analysis.
[This change in the locus brought about by the change in the coefficient from 1 to 3 is slightly more complex than a "partial displacement." It already has been pointed out in Chapter V that "partial displacements" do not alter the shapes of rectangular curves with 1st or 2nd-degree construction rules but do alter loci with eqs. of higher degree.]

From the point of view of an inversion analysis, these two curves are unrelated. For example, the equilateral strophoid self-inverts through the loop vertex, and inverts through the DP to a curve with two lines of symmetry (the equilateral hyperbola), whereas the folium of Descartes does neither. On the other hand, a circumpolar symmetry analysis discloses both fundamental differences and similarities (see Kavanau, 1980, *Cubics*).

Inversion Analysis

The basis for the different results of the two approaches is as follows. An inversion analysis organizes curves according to their relatedness by the inversion transformation. This is achieved by obtaining the 1st-generation inverse loci of a given initial curve about *all points in the extended plane* (i.e., the Euclidean plane including the line at infinity) and ascertaining the relations between these loci.

The most remarkable of these relations is embodied in the recently-discovered fundamental property of *immediate closure* (see Kavanau, 1980, 1982). Immediate closure describes the following unique circumstance for a continuous non-trivial transformation: all the loci of a 2nd-generation inversion ensemble obtained by inverting any one of the loci of the 1st-generation ensemble about all points in the plane, already are represented in the 1st-generation ensemble. Thus, every member of the 1st-generation inversion ensemble inverts to every other member.

The study of relations between the inverse loci of curves inverted about all points in the plane is known as *Inversion Taxonomy*. The ensemble of inverse loci of all members of a group of curves, such as the conic sections or the Cartesian or Cassinian ovals, is known as an *inversion superfamily*; conic sections belong to the *quadratic-based inversion superfamily*. Of course, curves grouped according to this paradigm share all the properties that are conserved by inversion (for example, the magnitudes of angles of intersection or tangency are conserved).

By virtue of its property of immediate closure, the inversion transformation may be said to provide the first "natural" paradigm for classifying curves, since, by its very nature, it partitions curves into closed ensembles with unique memberships. This feature takes on all the more significance by virtue of the fact that many of the most important of the classically-known curves are inversions of one another.

[The inclusion of topics in Inversion Taxonomy would require a great expansion of length and carry us very far afield from the goals of this Chapter; the interested reader is referred to Appendix II and Kavanau (1980, 1982).]

Circumpolar Symmetry Analysis

But Inversion Taxonomy is of no use for the classification of curves that are not related by inversion. Accordingly, it is of no help for comparisons that have the objective of arriving at a more far-reaching classification of curves. To cross the boundaries of inversion superfamilies, other types of transformations--of which that for obtaining circumpolar structure rules about points is preeminent--must be employed.

Thus, the primary objective of a corresponding circumpolar symmetry analysis is not to probe the relatedness of curves from the point of view of whether they are derivable from one another by any transformation (although many such relations are disclosed). A circumpolar symmetry analysis asks questions about the types of inherent structure possessed by curves. Do they have similar, homologous, or identical circumpolar structure rules? A group of curves assembled by a circumpolar symmetry analysis (see Table 10) usually will include both members that are related to one another by inversion and members that are not so related. Specific examples of the relatedness of curves through their structure rules are given in Table 10. [It already has been noted that limacons and circles have "circular" symmetry in common.]

Design and Synthesis of Highly-Symmetrical Curves

Another topic that only can be mentioned here relates to the design and synthesis of curves that possess great structural simplicity. Once the factors that lead to a curve's possessing simple structure rules have been identified, new groups of curves with desired symmetry properties can be synthesized. The coordinate-system approach of Chapters IV-VI is primarily one of ascertaining the most-symmetrical curves with respect to specific coordinate elements. Approaches of a more synthetic nature are to be found in Kavanau (1980, Chapter XIV).

In one such approach, highly symmetrical curves are synthesized by employing individual terms in building-block style, such as $(x^2+y^2)^n$, which, taken in combination, do not violate design principles for high symmetry (i.e., high structural simplicity). A very fruitful approach stems from the finding that some very highly symmetrical curves belong to groups in which the focus of self-inversion is a component of a multiple-focus, that is, it always is in coincidence with one or more other foci (in contrast to other groups in which it is distinct from these other foci).

An example is the group of curves obtained by inverting central conics about their axial vertices, in which the focus of self-inversion comes to lie at the loop vertex (which is a circumpolar focus in its own right). In the example of Fig. 24a, the variable focus also comes to lie at the loop vertex, leading to a *triple-focus*. The precise paths that can be followed using this method are many and diverse.

Symmetry mimics provide another example of such syntheses. A symmetry mimic is a curve that has the identical 0° and 180°-structure rules as an initial curve about a homologous pole but has a rectangular construction rule of higher degree. The design capabilities for such curves even include making the 0° and 180°-structure rules identical with each other (i.e., having a single structure rule apply for both probe-angles). Some symmetry mimics have greater structural

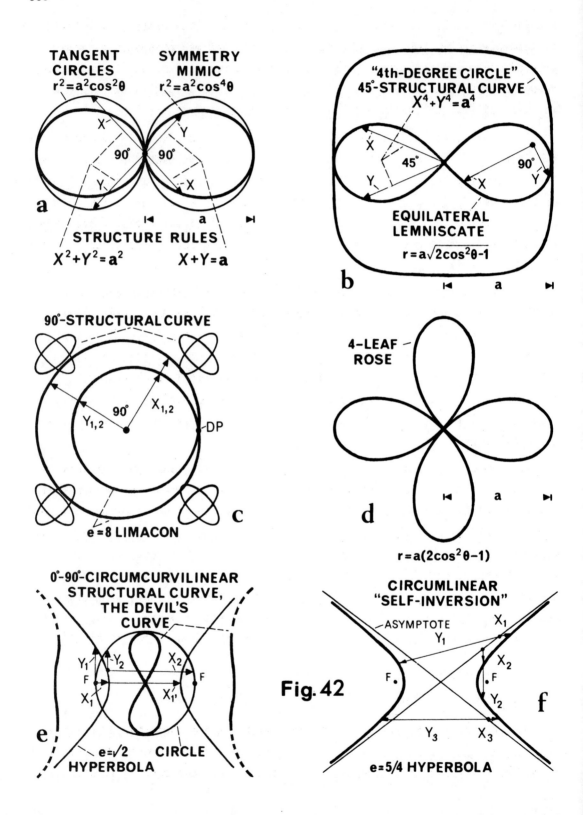

Fig. 42

simplicity than the initial curve. Thus, the sextic (6th-degree) symmetry mimic of tangent-circles (which are themselves of 4th-degree) possesses a higher degree of circumpolar symmetry than the tangent circles (Fig. 42a). The latter have linear 0° and 180°-structure rules and a "circular" 90°-structure rule, whereas all three structure rules of the symmetry mimic are linear.

The analytical basis for this circumstance is to be found in the comparative simplicities of the polar construction rules. Thus, although the rectangular eq. of the symmetry mimic, $(x^2+y^2)^3 = a^2x^4$, is of higher degree than that of the tangent-circles, $(x^2+y^2)^2 = a^2x^2$, its polar eq. is of lower degree; $r = a\cos^2\theta$ (see Fig. 8a) versus $r^2 = a^2\cos^2\theta$ (a root may not be discarded from the latter eq. without the loss of one of the circles). The above findings are consonate with RULE 8a.

In still another approach, curves are designed to self-invert at many angles about a double-focus (the example given above was for a single angle). Classical treatments of self-inversion recognize only two forms of this phenomenon--along probe-lines at 0° and 180°. However, the circumpolar structure rule for self-inversion, $XY = $ constant, is not limited to probe-angles of 0° and 180°. It can characterize the symmetry of a curve for any probe-angle. Consequently, Structural Equation Geometry also generalizes the classical concept of self-inversion.

Many curves are known to self-invert at 0° or 180°, and some are known to self-invert at both angles (for example, the biovular Cartesians and Cassinians of previous chapters). Only one classically-known group of curves has been found to self-invert at an angle other than 0° or 180°, namely, monovular Cassinians, which self-invert about their centers at 90° (see Kavanau, 1980, 1982). However, curves can be synthesized that self-invert about a single focus at all three angles and, in fact, one can synthesize highly symmetrical curves that self-invert at as many angles as desired (Kavanau, 1980). The circle, of course, is the only curve that self-inverts about a center at all angles.

Some Structure Rules of Unusual Interest

Most of the rectangular eqs. of conic sections are encountered relatively infrequently as structure rules. Even 3rd-degree structure rules are relatively rare. The most-frequently encountered exceptions are the eqs. $X = Y$, of classical symmetries, and XY = constant, of self-inversion. Aside from these, only the most highly-symmetrical curves, when characterized about their highest-ranking foci, have structural eqs. of 1st or 2nd-degree. 90°-structure rules typically are of much higher degree than those for 0° and 180°.

As we have seen, even for the line or parallel line-pair, the 90°-structure rule about an arbitrary point is 4th-degree, while for the circle it is 6th-degree. The corresponding structure rule of the equilateral hyperbola is of next-lowest degree at 16, while for other conic sections the degree is 20, and for limacons (and all other 4th-degree inverse loci of conic sections) it is 72. The structural curves defined by these structure rules almost always were previously unknown. In some cases, the new context of a structural curve constitutes the only known relation to another curve. Some structure rules of unusual interest and implications are considered in the following.

The '4th-Degree Circle"

An example of a case in which a structure rule of a given curve constitutes its only known relation to another curve is eq. 235a for the "4th-degree circle." Previously merely and interesting curiousity, this eq. now is known to produce the 45°-structural curve of the equilateral lemniscate about its cen-

(a) $\quad X^4 + Y^4 = a^4 \quad$ 45°-structure rule of the equilateral lemniscate about its center $\quad$ (235)

(b) $\quad (X^4+Y^4)[X^4Y^4 + d^8(C^2-1/16)^2]$ $\quad$ 45°-structure rules of Cassinian ovals about their centers (bipolar eq., $uv = Cd^2$)
$\quad\quad = 4C^2d^4X^4Y^4$

ter (Fig. 42b). The corresponding structure rule for Cassinian ovals (Chapter IV and Fig. 19), of which the equilateral lemniscate is the curve of demarcation, is 12th-degree (eq. 235b). Since eq. 235b reduces only for $C = 1/4$, the 3-fold difference in degree is a forceful illustration of the fact that curves of demarcation generally possess greater circumpolar symmetry than the curves that they demarcate.

RULE 50. *Curves of demarcation generally possess greater circumpolar symmetry than the curves they demarcate.*

Central and Focal 90°-Structure Rules
   of Conics

Two different 90°-structure rules for ellipses and hyperbolas were considered above (eqs. 231a,b). When expressed in terms of eccentricity and the parameter a, these may be combined into the single structure rule 236a. For $e^2 = 2$, for the equilateral hyperbola, this reduces to the eq. of a point-circle, $X^2 + Y^2 = 0$, which, as a structure rule, has only complex solutions (since no null intercept exists). For $1 < e^2 < 2$, the solutions also are complex but remain 4th-degree. For $e = 0$, the case for the circle, the structure rule simplifies to eq. 236b but, of course, this is trivial, since X and Y are identically equal to R. For e infinite but the parameter a finite, corresponding to a non-coincident parallel line-pair, structure rule 236a takes the form 236c, which already was encountered above for the parallel line-pair.

90°-structure rules of conics

(a) $(2-e^2)X^2Y^2 = a^2(1-e^2)(X^2+Y^2)$ central conics about centers ($e \neq 1$)   (236)

(b) $2X^2Y^2 = R^2(X^2+Y^2)$ eq. 236a for $e = 0$, $a = R$ (trivial)

(c) $X^2Y^2 = a^2(X^2+Y^2)$ eq. 236a for $e = \infty$, a, finite

(a) $(2-e^2)X^2Y^2 + e^2p^2(X^2+Y^2) = 2epXY(X+Y)$ conics about a focus (p, the focus-directrix distance)   (237)

(b) $2X^2Y^2 + R^2(X^2+Y^2) = 2RXY(X+Y)$ eq. 237a for $ep = a(1-e^2) = R$, $e = 0$ (trivial)

(c) $a(X^2+Y^2) = 2XY(X+Y)$ eq. 237a for $e^2 = 2$, $ep = a$, for the equilateral hyperbola

(d) $XY = 0$ eq. 237a for parameter a infinite, where $ep = a(e^2-1)$ (trivial)

It is most interesting that a single 90°-structure rule about the foci of conic sections (eq. 237a) applies for all values of eccentricity. In order to encompass the parabola using parameters possessed in common with ellipses and hyperbolas (by virtue of the polar-linear constructions of Chapter V), this must be formulated in terms of p, the focus-directrix distance. With $e = 0$ and $ep = R$, the values for the circle, structure rule 237a yields 237b (Problem 9). Again, this eq. is trivial, since X and Y are identically equal to R, but, like eq. 236b, it preserves the form possessed by the structure rules for other ellipses with very small values of e.

It will be noted that eq. 237a reduces to 3rd-degree for $e^2 = 2$, the value for the equilateral hyperbola, yielding eq. 237c. Accordingly, the equilateral hyperbola is the hyperbola with the greatest circumpolar symmetry about a traditional focus. For an infinite value of the parameter a, whereupon $ep = a(e^2-1) = b^2/a$, which represents the case for two coincident lines, the structure rule reduces to eq. 237d. Like the structure rule for the circle, this is trivial, since both X and Y are identically equal to 0 for all probe orientations for which both intercepts are defined (an intercept is not defined when a probe-line is in coincidence with the initial curve).

Unlike eqs. 235 and 236, which are valid for all combinations of positive and negative intercepts, real solutions of eqs. 237a-c exist only when both intercepts are positive. The structure rules (corresponding to eqs. 237a-c) that allow for all combinations of positive and negative intercepts are four times the degree (see Problem 6). The fact that the degree of the general 90°-structure rule 237a reduces only for the circle, the line, and the equilateral hyperbola, is the first example of a number of previously unknown homologies between these three curves revealed by Structural Equation Geometry.

[A most unusual second example occurs in the *equidistant collinear tetrapolar coordinate system*, in which the line, the circle, and the equilateral hyperbola are polar-exchange symmetrical (Kavanau, 1980). If the poles are aligned in the equidistant linear sequence p_u, p_s, p_t, p_v, then the construction rule uv = st represents an equilateral hyperbola, us = tv represents the midline (between poles p_s and p_t) and ut = sv represents both a circle and the midline.]

Eclipsing of Foci, and Inverse 4-Leaf Roses
    as the 90°-Incident Structural Curves
    of the Equilateral Hyperbola

The next example, which concerns the 90°-structure rule of the equilateral hyperbola about an a-vertex (eq. 238a) is most fundamental and has led to some extraordinary findings. Whereas the general 90°-structure rules of conics about the axial and LR-vertices are 8th-degree (eq. 239 and Fig. 43d,f), those for the equilateral hyperbola are only 4th-degree (eq. 238a and Fig. 43a,c). Both

incident 90°-structure rules of conics

(a) $(X^2+Y^2)^2 = 4a^2(X^2-Y^2)$ equilateral hyperbola about an a-vertex (the inverse 4-leaf rose)   (238)

(b) $[a^2(X^2-Y^2) + 4h\sqrt{(h^2-a^2)}XY]^2$
           $= 4(2h^2-a^2)^3(X^2+Y^2)$ equilateral hyperbola, general incident structure rule (h = abscissa of pole)

(c) $(X^2-Y^2+4\sqrt{2}XY)^2 = 108a^2(X^2+Y^2)$ equilateral hyperbola about an LR-vertex (the inverse 4-leaf rose)

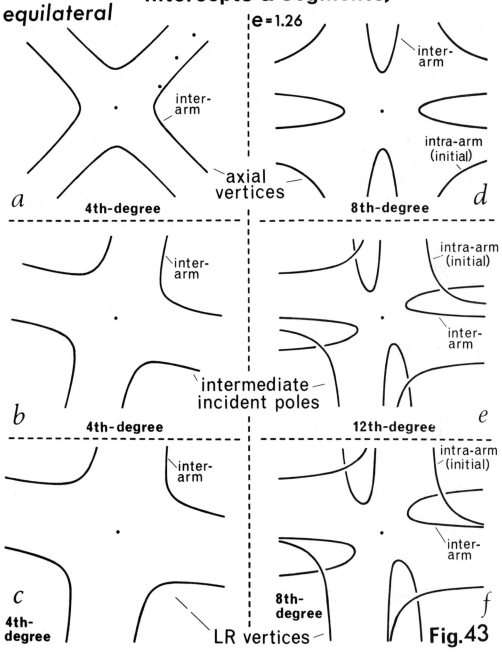

Fig. 43. Incident 90°-Circumpolar Intercept Transforms of Hyperbolas And Eclipsing of Vertex Foci (All Intercepts & Segments)

(d) $(X^2-Y^2)^2 = 4(2h^2-a^2)(X^2+Y^2)$ eq. 238b rotated CCW to bring the lines of symmetry into coincidence with the coordinate axes   (238)

$X^4Y^4 = (4a)^6(X^2+Y^2) + 3(4a)^4X^2Y^2$ 90°-structure rule of the parabola about its axial vertex   (239)

the axial and the LR-vertices of the parabola and central conics, in general, possess circumpolar focal rank, because their 90°-structure rules are four degree units lower than the 12th-degree structure rule for general incident points (Fig. 43e) (compared to 20th-degree for points in the plane; Fig. 52k).

The finding that the 90°-structure rules (eq. 238a) at the a-vertices and LR-vertices of the equilateral hyperbola only are 4th-degree raises the question of the degree of the equilateral hyperbola's general incident 90°-structure rule, since possession of focal rank by the vertices is dependent upon comparison with the degree of this general rule. Astonishingly, the general incident 90°-structure rule also is only 4th-degree (eq. 238b). This means that no incident point-focus of the equilateral hyperbola exists.

Among other known curves, this remarkable circumstance occurs only for the circle and, trivially, for the line (this constitutes a third homology between the three curves); the circle and the line are the only other known curves whose incident structure rules have the same degree for all points, that is, they are the only other known curves that possess no incident point-focus. Thus, just as the a, b, and LR-vertices are *eclipsable foci* of ellipses--because they lose their point-focal rank in the circle (when a=b)--so also are the a, and LR-vertices *eclipsable foci* of hyperbolas, because they lose their point-focal rank in the equilateral hyperbola (when a=b). The notable difference, of course, is that the a, b, and LR-vertices of ellipses also lose their identity (as vertices) in the circle, whereas they retain their identity in the equilateral hyperbola.

In the light of these remarkable circumstances, the question arises as to whether the a-vertices or LR-vertices of the equilateral hyperbola--which retain their non-point-focal rank as points of a focal locus--possess higher *subfocal rank* than other points of the curve. In other words, do they have simpler (as opposed to lower degree) structure rules? The answer to this question is "yes." By comparing eqs. 238a and c, for the 90°-structure rules of the a-vertices and LR-vertices, respectively, with eq. 238b for the general case, the structure rule of the a-vertices is found to be simplest and the structure rule of the LR-vertices the next-most simple.

[This conclusion is obvious for the a-vertices. For the LR-vertices, the value $h = \pm\sqrt{2}a$ in eq. 238b leads to the simplest expression for the coefficient of the XY-term, namely, $4\sqrt{2}$--which involves a radical with the smallest achievable integral radicand, together with the smallest achieveable integral value of the coefficient of the (X^2+Y^2)-term, namely 108.]

Identifying these incident 90°-structural curves of the equilateral hyperbola becomes of particular interest, because the homologous structural curves of the circle also are circles. Examination of these curves reveals that they all resemble four-arm equilateral hyperbolas (Fig. 43a-c), and that they progressively rotate and enlarge as the probe points move out along the arms from the a-vertices. Accordingly, to make meaningful comparisons between them, the axes of the curves must be rotated into coincidence with the x and y-axes.

Performing the required rotation yields eq. 238d, which embodies the following most remarkable finding: all the incident 90°-structural curves of the equilateral hyperbola are similar--they are all the central inversion (eq. 240b) of the well-known 4-leaf rose (eq. 240a and Fig. 42d). But the homology goes much further. Just as the incident 90°-structural curve of the circle has twice the radius of the initial curve, so also does the axis of the inverse 4-leaf rose for the a-vertex have twice the length of the transverse axis of the initial equilateral hyperbola.

(a) $r = a(\cos^2\theta - \sin^2\theta)/2$ 4-leaf rose (a, the parameter of eq. 238a)   (240)

(b) $r = 2a/(\cos^2\theta - \sin^2\theta)$ inverse 4-leaf rose (inverse of eq. 240a)

(c) $r = 2a/\sqrt{(\cos^2\theta - \sin^2\theta)}$ equilateral hyperbola (a, the parameter of eq. 238a)

In the light of these striking findings, it is natural to ask whether the inverse 4-leaf rose and its inversion about its center, the 4-leaf rose, also have remarkable circumpolar symmetry properties about their centers, like the circle and its central inversion (which also is a circle). The answers, again, are "yes." The 4-leaf rose has the 2nd-highest known linear circumpolar symmetry rank about a true center (for 90° and 180°-probes; Table 10)--second only to the circle (both arc-increment eqs. of the 4-leaf rose are of degree 2, compared to 0 for the circle; see Kavanau, 1980, 1982). Similarly, the inverse 4-leaf rose also has linear circumpolar symmetry about the center for

90° and 180°-probes, with arc-increment eqs. that are only 4th-degree (Table 9).

In view of these homologies, and because the incident 90°-structure rule of the circle also represents a circle, one expects some corresponding similarities between the eqs. of the equilateral hyperbola and the inverse 4-leaf rose. This also proves to be the case (compare eqs. 240b,c). Except for the square-root sign, the eqs. are identical. By taking the square-root of the function $(\cos^2\theta-\sin^2\theta)$, rather than taking the function itself, the two vertical arms of the inverse 4-leaf rose become imaginary and the horizontal arms are broadened slightly to become an equilateral hyperbola (compare the solid curve of Fig. 43a with the positions of the three solid-circle points).

These findings emphasize a third newly-discovered homology between the circle, the line, and the equilateral hyperbola; they are the only known curves that possess no incident point-focus (for a 90°-probe-angle), because all the incident 90°-structural curves are similar. In the case of the circle, the structural curves also are congruent, whereas in the case of the equilateral hyperbola, they enlarge progressively as the probe-point recedes from the a-vertex. For the line, the structural curve is a trivial point-locus, since all intercepts (when defined) are identically 0.

Focal loci of this nature, no point of which possesses point-focal rank, are known as *simple focal loci* (as opposed to *compound focal loci*). The line, circle, and equilateral hyperbola provide the only known examples of simple focal loci. The midline of the parallel line-pair, however, is a *simple subfocal locus*, as compared to the latera recta of conics, which are *compound subfocal loci*; thus, though the latera recta do not possess focal rank, the LR-vertices, which are incident upon them, are point-foci.

The 90°-Structure Rule of Limacons about the -b/2 Focus

One of the most remarkable eqs. known to me is the 90°-structure rule (eq. 241) of limacons about the -b/2 focus (though the pole at -b/2 does not possess focal rank in coordinate systems, it is a high-ranking circumpolar focus). Although this structure rule is not degenerate, when the eccentricity of the initial limacon is greater than 2 ($b > 2a$), it represents four sets of congruent loci. Each of these resembles a pair of concentric orthogonal ellipses--one set in each quadrant (Fig. 42c). This structure rule also represents all combinations of positive and negative intercepts.

90°-structure rules for the -b/2 focus of limacons

(a) $X^2Y^2[(X^6+Y^6)-2c^2(X^4+Y^4)+(c^4-2b^2d^2)(X^2+Y^2)$ 90°-structure rule (241)
of hyperbolic lima-
$+b^2(4c^2d^2-a^4)] + b^4d^4(X^2+Y^2) = 0$ cons about the -b/2 focus

$c^2 = (2a^2+b^2)$, $d^2 = (4a^2-b^2)/16$

(b) $(X^6+Y^6) - 6a^2(X^4+Y^4) + 9a^4(X^2+Y^2) = 4a^6$ eq. 241 for the e = 2 (b = 2a) limacon

(b') $(X^2+Y^2-4a^2)(X^2+Y^2+\sqrt{3}XY-a^2)(X^2+Y^2+\sqrt{3}XY-a^2) = 0$ eq. 241b factored

Perhaps the most remarkable property of the 90°-structure rule of hyperbolic limacons about the -b/2 focus is that the greater the parameter ratio b/a, that is, the more nearly the "small loop" of an initial hyperbolic limacon approaches the size of the large loop, the more nearly these orthogonal structural curves approximate true ellipses (for e = 8, in Fig. 42c, they are visually indistinguishable from true ellipses). For that reason, they are called *near-ellipses*. These eight near-ellipses are represented by an eq. of only 10th-degree, whereas a 16th-degree eq. is required to represent eight true ellipses.

Several cases were noted in Part I in which the e = 2 limacon has greater symmetry than other limacons, on the basis of simpler construction rules. Similar comparative structural simplicities are found in assessments based on structure rules. For e = 2 (b = 2a; $d^2 = 0$), eq. 241a reduces to the degenerate 6th-degree eq. 241b, which falls into three factors, each of 2nd-degree. These represent a 45°-ellipse, a concentric, congruent orthogonal ellipse, and a concentric circle, centered at the origin. The area of the circle ($4\pi a^2$) equals the sum of the areas of the ellipses.

These are not the only remarkable features of the circumpolar symmetry of limacons about the -b/2 focus. For example, the 0°-structure rule is a 4th-degree eq. that more nearly represents a line with increasing values of e (i.e., the segment of the curve produced by the structure rule, that is traced by the actual intercepts, becomes more nearly linear). For the e = 2 limacon, an XY-term factors out, leaving an "elliptical" structure rule.

The Devil's Curve--The 0°-90°-Structural Curve of the Circle about a Concentric Equilateral Hyperbola

Our last example of structure rules of unusual interest is drawn from the domain of circumcurvilinear symmetry. It illustrates: (a) a fourth homology between the circle and the equilateral hyperbola; (b) another case in which a structure rule provides the first known relation of a classical curve to other curves; and (c) *identical reciprocal circumcurvilinear symmetry*, that is, the circumstance in which the structure rule of one curve about another is identical with the structure rule of the other curve about the one (the 90°-probes employed for both types of structure-rule derivations are illustrated in Fig. 42e).

The corresponding structure rule--the 0°-90°-circumequilateral hyperbolic structure rule of the centered circle (see probe-lines X_2 and Y_2 of Fig. 42e), is the non-degenerate 4th-degree eq. 242a. This is the well-known Devil's Curve (Fig. 42e) first studied in the mid-18th century by G. Cramer and much employed as an example in analytic geometry and in presenting the theory of Riemann surfaces and Abelian integrals. However, the curve's relationship to the circle and equilateral hyperbola previously was unknown. A hint of a relationship to the equilateral hyperbola is obtained by putting eq. 242a in the form 242b.

0°-90°-circumequilateral hyperbolic structure rule
of the centered circle

(a) $(X^4+Y^4) - 2(R^2+a^2)X^2 + 2(R^2-a^2)Y^2 = 0$ the Devil's Curve   (242)

(b) $[X^2-(R^2+a^2)]^2 - [Y^2-(R^2-a^2)]^2 = 4a^2R^2$ eq. 242a in another form

It is most remarkable that the structure rule describing the 0°-90°-circumcircular symmetry of the centered equilateral hyperbola (see probe-lines X_1, X_1', and Y_1 of Fig. 42e) represents the identical Devil's Curve. This is the only known case of *identical reciprocal circumcurvilinear symmetry*, and constitutes a fourth example of a newly-discovered homology between the circle and the equilateral hyperbola.

Overt Symmetry of Structural Curves

Complete circumpolar structural curves and structure rules represent all combinations of positive and negative intercepts of the probe-lines of rotating probes with initial curves. The following descriptions apply primarily to such *complete* structural curves and structure rules, but in some cases they also apply in individual quadrants or in paired 1st and 3rd or 2nd and 4th-quadrants.

All structural curves possess an *ordinary* center, that is, a pole exists, on each side of which all lines that include it intersect the curve at equidistant points (see dashed curve of Fig. 13a). The rectangular eq. of such a curve, with the ordinary center at the origin, is unaltered if x and y, respectively, are replaced by -x and -y. Structural curves possess this property because the two negatively-directed probe-lines form the same probe-angle and sweep the curve in the identical manner as the two positively-directed probe-lines. In addition, any structural curve produced for a probe-angle, α, is congruent with the structural curve produced for a probe-angle ($360°-\alpha$). But, since the sense of the probe-lines is reversed, the two structural curves are reflections of one another in the coordinate bisector(s).

It will be noted that many of the structure rules encountered in the foregoing are "symmetric functions," that is, functions that are not altered by exchanging the variables X and Y. Geometrically, this means that the structural curves are symmetrical about the coordinate bisector, X=Y (and, similarly, for exchanging X and -Y with regard to symmetry about the coordinate bisector X=-Y). It is an inherent characteristic of structure rules of curves for probe-points on lines of symmetry that they are symmetric functions; thus, for any given pair of intercepts (X,Y), on one side of the line of symmetry, there exists an exchanged pair (Y,X) reflected in the line of symmetry and, consequently, reflected in the coordinate bisector.

On the contrary, structure rules about asymmetrically-located poles generally are not symmetric functions. For example, consider a given pair of positive intercepts (X,Y) of an asymmetrically-located 90°-probe (Fig. 42b, probe in right loop). No corresponding pair of positive intercepts necessarily exists in which the values of X and Y are interchanged, whereas the existence of all corresponding interchanged pairs is a prerequisite for symmetry about the X=Y coordinate bisector (see also Fig. 43b,c,e,f).

All 180°-structure rules are symmetric functions, because the positive X and Y-intercepts of a 180°-probe rotated about any point become exchanged each 180°. But what is true of the 180°-structure rule for two positive intercepts also must be true of the 0°-structure rule for one positive and one negative intercept (though not for 0°-structure rules restricted to two positive intercepts). In fact, complete 0° and 180°-structure rules are identical, save for the signs of terms; the corresponding structural curves simply are out of phase by 90°. Accordingly, 0°-structure rules also are symmetric functions.

All complete 90°-structure rules possess congruent segments in all four quadrants. This results from the fact that all combinations of positive and negative intercepts of 90°-probes form right angles. However, the congruent segments generally are not reflections of one another in the coordinate axes or coordinate bisectors, but only in the origin (the ordinary center; see Fig. 43b,c,e,f). If the probe-points of 90°-probes are incident upon a line of symmetry, the corresponding structural curve also possesses a *true center* (the point of intersection of two orthogonal lines of symmetry). This results because the combination of possessing an ordinary center (all structural curves), being a symmetric function (all structural curves about poles on lines of symmetry), and being a 90°-structural curve (congruent segments in all four quadrants) leads to the possession of a true center.

Table 9. Overt Symmetry Properties of Complete Structural Curves

Category of structure rules	Symmetry property of structural curves or structure rules
all	possess an *ordinary* center (replacing X by -X and Y by -Y leaves the structure rule unchanged)
for α and (360°-α)-probes	congruent, reflections of one another in the coordinate bisectors
for α and (180°-α)-probes	congruent and 90° out of phase
for probe-points on a line of symmetry	symmetric functions, symmetrical about coordinate bisector(s) (exchanging X and Y or X and -Y leaves the structure rule unchanged)
for 0° and 180°-probes	symmetric functions (see above)
for 90°-probes	congruent segments in all four quadrants
for 90°-probes with probe-points on a line of symmetry	possess a *true* center (the point of intersection of two orthogonal lines of symmetry)

Circumpolar Point-Foci and Probe-Angles

Generally speaking, a pole in the plane of a curve that possesses point-focal rank for a given probe-angle also is a point-focus for other probe-angles (but see below). Since 0° and 180°-structure rules are derived most readily, tests for foci preferably are carried out for these probe-angles. If a pole does not possess point-focal rank on the basis of a valid 0° or 180°-structure rule, then, with few exceptions, neither will it possess point-focal rank for other probe-angles. While tests with 0° and 180°-probes generally are feasible for points of curves of higher degree than 2, such tests are not valid when applied to the incident points of conic sections, since only trivial 0° and 180°-structure rules exist for these points.

A test for the presence of point-foci on a focal locus by means of 0°-structure rules is illustrated here for the *exterior* axis of the parabola, $y^2 = 4a(x-h)$, where h is the abscissa of the vertex ($h > 0$). Eq. 243 is the 0°-structure rule for a point h units to the left of the vertex (with the vertex translated to $x = h$). The fact that this eq. is only 4th-degree defines the exterior axis itself as a focal locus (allowing h to take on negative values also defines the interior axis as a focal locus), because the 0°-structure rules for neighboring points not lying on the axis are 8th-degree.

$$X^2Y^2 = h^2(X+Y)^2 + 4ahXY, \quad (h>0)$$
 0°-structure rule of the parabola for a point on the exterior axis h units from the axial vertex (243)

Examination of structure rule 243, however, reveals that further reduction of the eq. is not possible for any point of the exterior axis (i.e., for any positive value of h), since roots cannot be taken. Accordingly, since no point-focus exists on the exterior axis for a 0°-probe-angle, it follows that no point-focus exists for any probe-angle.

RULE 51. *In general, if a point in the plane of a curve does not possess point-focal rank on the basis of a non-trivial 0° or 180°-structure rule, it will not possess point-focal rank for any probe-angle.*

α-Structure Rules and Structure-Rule Formats

Although our examples employ 0°, 90°, and 180°-probe-angles almost exclusively, many structure rules readily are obtained for an arbitrary probe-angle α--the *α-structure rule*. The degrees of these structure rules often are twice as great as those for 90°-probe-angles and four or eight times as great as those for 0° and 180°, because of reductions of the latter and the need for additional squarings (to eliminate radicals) in the derivation of α-structure rules (Problem 10). Eq. 244 is the α-structure rule of central conics about their centers. Inspection of this eq. reveals why degree reductions are greatest for 0° and 180°-probe-angles.

α-structure rule of central conics about their centers for an arbitrary probe-angle α

$$X^4Y^4[(b^2\mp a^2)^2\sin^2\alpha - (b^2\pm a^2)^2\sin^2\alpha\cos^2\alpha] + a^4b^4(X^4+Y^4) = \tag{244}$$
$$2a^2b^2X^2Y^2(b^2\mp a^2)(X^2+Y^2)\sin^2\alpha + 2a^4b^4X^2Y^2(\cos^2\alpha-\sin^2\alpha)$$

The general possibilities for degree reduction of α-structure rules, however, are illustrated best by the *structure-rule formats*. These are the general forms for deriving structure rules obtained by eliminating θ from simple intercept formats. When a simple intercept format takes the form $\cos\theta$ or $\sin\theta = f(X)$, one simply substitutes $f(X)$ and $f(Y)$ in the $\cos\theta$-structure-rule format of eq. 245a [$f(Y)$ is obtained from $f(X)$ by replacing X by Y]. When it takes the form $\cos^2\theta$ or $\sin^2\theta = f(X)$, one uses the $\cos^2\theta$-structure rule format of eq. 245b.

(a) $f^2(X) + f^2(Y) - 2f(X)f(Y)\cos\alpha = \sin^2\alpha$ $\cos\theta$ and $\sin\theta$-structure-rule formats   (245)

(b) $[f(X) - \sin^2\alpha]^2 + [f(Y) - \sin^2\alpha]^2 -$ $\cos^2\theta$ and $\sin^2\theta$-structure-rule formats
     $2f(X)f(Y)(\cos^2\alpha-\sin^2\alpha) = \sin^4\alpha$

It is clear from inspecting the $\cos\theta$-structure rule format that for probe-angles of 0° and 180°, but not for 90°, there is immediate reduction (by root-taking) to half the degree of the α-structure rule. Thus, the eq. converts to a perfect square, yielding $f(X) = \mp f(Y)$. For the $\cos^2\theta$-structure rule format, a reduction to half the degree (by root taking) occurs for all three angles; for 0° and 180°, one obtains $f(X) = f(Y)$, while for 90°, one obtains $f(X) + f(Y) = 1$. The possibilities for further reductions, of course, are determined by the forms taken by $f(X)$ and $f(Y)$. But it is evident from these eqs. that fewer squarings generally will eliminate radicals for 0° and 180°-probe-angles than for 90°; one squaring often suffices for the former two angles but it never suffices for a probe-angle of 90°.

The Frequency of Occurrence of 1st and 2nd-Degree Structure Rules

Linear structure rules are extremely common (see Table 10), largely, but not exclusively, because many highly symmetrical surves have either two lines of symmetry or merely a center of symmetry. Two notable exceptions are the spiral of Archimedes and the logarithmic spiral, both of which have 1st-degree structure rules about the polar pole for all probe-angles ($Y = X + C$ and $Y = CX$, respectively; see Problem 11), but neither of which has either a line or center of classical symmetry.

A more common structure rule than that of the classical symmetries is the "equilateral hyperbolic" 0°-structure rule of self-inversion, $XY = $ constant. This follows because corresponding to every initial curve with one or more lines of symmetry there exist infinitely many inverse curves with at least one 0°-focus of self-inversion but no line of symmetry. In addition, for every line of symmetry of a curve, there exists a corresponding 0°-focus of self-inversion on the ideal line at infinity in the direction orthogonal to the line of symmetry. Loss of a line of symmetry and a corresponding 0°-focus of self-inversion on the line at infinity is accompanied by the gain of a 0°-focus of self-inversion in the finite plane.

[For several good and sufficient geometrical reasons, I employ the ideal line at infinity in the extended Euclidean plane, rather than the currently-favored expediency of a point at infinity in the inversive plane. Within the framework of a line at infinity, each 0°-focus of self-inversion of a curve with more than one (non-parallel) line of symmetry lies at a different position on the line at infinity (in the direction normal to the corresponding line of symmetry). This is an essential geometrical requirement. The currently-favored expediency places all such foci in coincidence, regardless of whether or not the lines of symmetry are in parallel (as they are, for example, for the curve $x = b\sin\{ay\}$).]

180°-foci of self-inversion (and corresponding structure rules) are known for very few curves, of which the circle is the best-known example. Unlike 0°-foci of self-inversion, 180°-foci of self-inversion exist only in the finite plane and are conserved in the finite plane of inverse curves. Thus, in the inversion of Cassinian biovals to parabolic Cartesian biovals, the 180°-foci of self-inversion at the centers of the former curves come to lie within the small ovals of the latter ones.

90°-foci of self-inversion also are very uncommon. Among classically-known curves, they occur only in Cassinian monovals (Fig. 19). They are not conserved by the inversion transformation; in their place, inverse loci possess *poles of congruent-inversion* (Kavanau, 1982), which do not possess focal rank (the latter are poles about which curves invert to a congruent but non-coincident locus).

In general, foci of self-inversion possess point-focal rank only by virtue of, and for the angle of, self-inversion. For other probe-angles, the positional symmetry of these foci is unexceptional. One of the reasons for formulating and qualifying RULE 51 for 0° and 180°-structure rules is that virtually all foci of self-inversion are for these angles. If 90°-structure rules were to be employed as the sole criteria for detecting point-foci, one would detect 0° and 180°-foci of self-inversion only when they were in coincidence with other foci.

Structure rules in other categories than "linear" (almost exclusively $X = Y$) and "equilateral hyperbolic" (almost exclusively $XY =$ constant) are comparatively uncommon. Only a few classically-known curves, most notably the circle and limacons, possess "circular" structure rules, but curves with circular structure rules are readily synthesized by the new approaches of Structural Equation Geometry (Kavanau, 1980).

Though "elliptical" structure rules also are relatively uncommon, curves that possess a circular structure rule for a given angle (0°, 45°, 90°, and 180° cases are known; see Table 10) possess elliptical structure rules for other angles. For example, limacons have linear structure rules about the DP for 0° and 180°-probes, circular ones for 90°-probes, and elliptical ones for all other probe-angles. The least common of the quadratic structure rules are "parabolic;" the only classically-known curves to possess them are the Norwich spiral (for all probe-angles but 0° for the polar pole) and the curves, $r = a + b\cos^4\theta$ (for 90°probes at the pole; Problem 12).

Arc-Increment Equations and More Specific Characterizations of Inherent Structure

Specific structure rules, unlike construction rules, do not define curves uniquely, but merely characterize particular types of inherent structural relations. In general, an infinite number of curves possess any given specific structure rule. Similarly, an infinite number of curves also possess given pairs of specific structure rules about the same or different points. Only when one reaches the more general level of point-specified (for arbitrary probe-angles) and angle-specified (for arbitrary probe-points) structure rules (see Table 2) does one attain one-to-one correspondences.

INVENTORY OF STRUCTURE RULES

Table 10. Inventory of Structure Rules of Low Degrees

Curve	probe-angle°	arc-increment eqs. XY-deg.	X-deg.	Pole
linear structure rules				
spiral of Archimedes	all	0	0	polar pole
logarithmic spiral	all	2	2	" "
limacons	0,180	2	2	DP
sextic symmetry mimic of tangent-circles	90,180	2	2	polar pole (true center)
4-leaf rose	90,180	2	2	" " " "
tangent-circles	180	2	2	" " " "
symmetry mimic of the parallel line-pair	180	3	3	" " " "
inverse 4-leaf rose	90,180	4	4	" " " "
parallel line-pair	180	4	4	" " " "
$r = \sqrt{(a^2 + b^2\cos^2\theta)}$	180	6	4	" " " "
central conics	0,180	6	6	" " " "
Cassinian biovals	0,180	18	10	" " " "
circular structure rules				
circle	90	2	0	incident point
limacons	90	2	0	DP
sextic symmetry mimic of tangent-circles	45	2	0	polar pole (true center)
4-leaf rose	45	2	0	" " " "
tangent-circles	90	2	0	" " " "
$r = \sqrt{(a^2 + b^2\cos^2\theta)}$	90	6	4	
hyperbolic structure rules				
parabola	180	3	6	traditional focus
symmetry mimic of the parallel line-pair	90	3	6	true center
hyperbolas	0,180	4	8	traditional focus
ellipses	180	4	8	" "
reciprocal spiral	all*	4	8	polar pole
equilateral strophoid	0	4	8	loop vertex (focus of self-inversion)
limacons	0	10	8	focus of self-inversion
Cartesian monovals and biovals	0	10	8	" " " "
Cartesian biovals	180	10	8	" " " "
circle	0,180	10	14	point in plane**
Cassinian biovals	0,180	18	8	true center
Cassinian monovals	90	18	8	" "

* only the trivial X = Y structure rule is obtained for a 0°-probe-angle
** non-central, non-incident

Thus, only central conics possess the point-specified α-structure rule of eq. 244; an α-structure rule about any other point in the plane of central conics also would be unique to central conics. Similarly, the 90°-structure rule 233a,a' of the circle about an arbitrary point in the plane is unique to the circle, as would be any other angle-specified structure rule, such as the 0° (and 180°) structure rules $XY = \pm(R^2 - h^2)$.

If attention is confined to an ensemble of known curves, one still does not necessarily attain the level of uniqueness by specifying more than one structure rule about a given point, more than one angle for which a given structure rule applies for a given point, different structure rules for specified or unspecified angles about different points, etc., But the field is narrowed. For example, only six of the curves in Table 10 with linear structure rules also have circular structure rules about the same point, and only two of them (Cassinian biovals and the *symmetry mimic of the parallel line-pair*; see Kavanau, 1980) also have hyperbolic structure rules about the same point.

Only seven of the twelve curves with linear structure rules listed in Table 10 possess the same linear structure rule for more than one probe-angle, and only five of the eleven curves with hyperbolic structure rules possess such rules at more than one angle. Likewise, only limacons, central conics, and Cassinian biovals possess a hyperbolic structure rule about one pole and a linear structure rule about another pole. In the same vein, only the circle and limacons possess a circular structure rule about one pole and a (non-trivial) hyperbolic structure rule about a different pole, and none with a circular structure rule also possesses a linear structure rule about a different pole.

On the other hand, in certain exceptional cases, very simple structure rules define curves uniquely. Only the circle possesses the circumpolar 0-degree structure rule $X = Y$ = constant (about its center) for all probe-angles and probe-orientations. Only the line possesses the 0-degree structure rule $X = Y = 0$ (about an incident point) for all probe-angles and all but a finite number of probe-orientations (the intercepts are not defined when a probe-line is coincident with the initial line itself). Among known curves, only the spiral of Archimedes possesses the 1st-degree structure rule $X = Y + C$ for all probe-angles and orientations, and only the logarithmic spiral possesses the 1st-degree structure rule $X = CY$ for all probe-angles and orientations.

But only in the case of a circle does a circumpolar structure rule contain explicit instructions or directions that enable one to draw the corresponding curve. It is only in the very simplest cases--through experience--that one can visualize the precise structure of a curve from knowledge of a structure rule. Information about the precise structure of a curve, even in the case where a single structure rule characterizes the curve uniquely, is implicit, not explicit. It is within this framework that we approach the next objective.

The question to be addressed here is not merely how to achieve a specification of inherent structural relations that applies uniquely, albeit implicitly, to a single curve. That already has been achieved in the form of the point-specified and angle-specified structure rules (Table 2). Except in a very few cases, however, these structure rules are very complicated and wholly impractical for characterization and classification. What is sought is an extension of the structure-rule approach that takes additional steps along the path toward unique specification, while remaining practical for characterizing and classifying curves.

The achievement of this goal followed the path of dealing with the most obvious generality of a structure rule, namely, the lack of specificity of the structure rule $X = Y$ for the classical symmetries. Assuming this to be a circumpolar 180°-structure rule, the knowledge that it applies to some specific curve yields absolutely no information about how X and Y change individually with changes in probe-orientation; the structure rule specifies only that the two intercepts always change in the same way.

Accordingly, it seemed desirable to take a further step in the direction of making the specification of the changes in X and Y less general (without leaving the domain of structure rules). To this end, instead of merely obtaining the intercepts X and Y directly, as a function of θ and the probe-angle α (two construction rules), and then eliminating θ (yielding a single structure rule), it appeared desirable to operate on X and Y with respect to θ in some manner, and then eliminate θ from the resulting eqs.

The Arc-Increment Equations

The obvious procedure was to obtain the derivatives $dX/d\theta$ and $dY/d\theta$ from the simple intercept format, eliminate θ from the resulting expressions, and then form the eq. for the changes in arc length with θ, that is, to obtain $ds/d\theta$ in terms of X and Y. For practical purposes, one employs $(ds/d\theta)^2$. This derivation (Problem 13) leads to an eq., $(ds/d\theta)^2 = F(X,Y)$, that describes how the square of the derivative of arc length along the structural curve (with respect to the angle of probe-orientation) is related to the lengths of the intercepts of the probe-lines with the initial curve. One also can phrase the description entirely in terms of $(ds/d\theta)^2$ and θ, that is, $(ds/d\theta)^2 = G(\theta)$.

For convenience, the eq. $(ds/d\theta)^2 = F(X,Y)$ is referred to as the *XY-arc-increment equation*. This merely is another type of structure rule of the initial curve that carries the specification of inherent structure a step further. As in the case of ordinary structure rules (those dealt with to this juncture), the degree of this eq. in X and Y supplies an index of structural simplicity, and is referred to as the *XY-degree*. If one refers to Table 10, it is seen that the nine curves with linear structure rules about a true center become divided into two groups when the XY-degrees are taken into account. Upon the same basis, the eleven curves with hyperbolic structure rules become divided into four groups.

It is in the nature of the trigonometric functions and their derivatives, however, that the XY-arc-increment eq. and, in consequence, the XY-degree are independent of the probe-angle (see Problem 14). But one can make the degree of [Since $(ds/d\theta)^2$ is the sum of $(dX/d\theta)^2$ and $(dY/d\theta)^2$, and since X and Y are simply out of phase with one another, their derivatives with respect to θ must be identical, whereupon their sum is independent of the probe-angle.] the resulting eq. specific for a given probe-angle simply by eliminating the Y-intercept, using the structure rule for that probe-angle (Problem 15). This gives the eliminant, $(ds/d\theta)^2 = H(X)$. The latter is referred to as the *X-arc-increment equation*; its degree in X is the *X-degree*.

In the cases of linear structure rules, the XY and X-degrees often are the same, as in eleven of the twelve cases in Table 10. But for higher-degree structure rules they generally differ. In the examples of Table 10, the use of the X-degrees results in two further subdivisions. However, one of these further subdivisions--that for the hyperbolic structure rules of circles and bi-ovular Cartesians--already was achieveable on the basis of the positions of the probe-points (at any exterior or interior point for the former, as opposed to being at a point-focus for the latter).

Taking all these criteria into account--the probe-angle, the type of pole at which the probe-point is located, the type of ordinary structure rule (linear, circular, etc.), the XY and X-degrees--only two curves share all categories in common. These are the 4-leaf rose (Fig. 42d) and the sextic symmetry mimic of tangent-circles (Fig. 42a), which possess all the criteria of Table 10 in common in both the linear and circular categories. These curves are distinguishable from one another, however, on the basis of their 45° and 90°-structure rules about their centers. Since this analysis leads to the conclusion that these curves are very closely related, it is instructive to pursue the characterizations to the levels of these structure rules (both 180°-structure rules are $X = Y$) and the arc-increments eqs., to illustrate the resolving power of this approach to inherent structure.

The initial polar eqs., structure rules, and arc-increments eqs. are eqs. 246 and 247. It is clear that one can distinguish between the inherent structure of these curves on the basis of structure rules in all the categories of eqs. 246 and 247. If one compares the symmetry of the curves about their centers on the basis of the simplicities of these eqs., the symmetry of the 4-leaf rose is greater based on all categories except the 45°-arc-increment eq., and this single discordant ranking is based on the category with the lowest priority for such rankings--the numerical coefficients.

polar eq. of curve	45° and 90° structure rules	XY-arc-increment eq. (all angles)	45°-X-arc-increment eq. (independent of X)	90°, 180°-X-arc-increment eq.
sextic symmetry mimic of tangent-circles				
$r = a\cos^2\theta$	45°: $(X-a/2)^2 + (Y-a/2)^2 = a^2/4$ 90°: $X + Y = a$	$\overset{\circ}{s}{}^2 = 4Y(a-Y) + 4X(a-X)$	$\overset{\circ}{s}{}^2 = a^2$	$\overset{\circ}{s}{}^2 = 8X(a-X)$ (246)
			$[\overset{\circ}{s}{}^2 = (dX/d\theta)^2 + (dY/d\theta)^2]$	
4-leaf rose				
$r = a(2\cos^2\theta - 1)$	45°: $X^2 + Y^2 = a^2$ 90°: $X = -Y$	$\overset{\circ}{s}{}^2 = 4(a^2-X^2) + 4(a^2-Y^2)$	$\overset{\circ}{s}{}^2 = (2a)^2$	$\overset{\circ}{s}{}^2 = 8(a^2-X^2)$ (247)

The Parabola Versus Central Conics

It will be recalled that the construction-rule approach to the symmetry of conic sections of Part I found the parabola to have greater structural simplicity than central conics in all coordinate systems with dissimilar reference elements, including the polar, hybrid polar-rectangular, polar-linear, and linear-circular systems. It is most notable that when comparisons of structural simplicity about a traditional focus are made in the circumpolar system, the parabola again is found to outrank central conics (with the exceptions of the circle and lines).

Not only are all the parabola's ordinary structure rules about the traditional focus simpler, its XY and X-arc-increment eqs. are both simpler (see also Problems 16 and 17) and of lower degree (see Table 10). The 180°-X-arc-increment eqs (248 and 249) for the traditional foci of the parabola and ellipses are given below for comparison, together with the 180°-structure rules.

structure rules and arc-increment eqs. of the
parabola and ellipses

(a) $XY = a(X+Y)$ focal 180°-structure rule, parabola   (248)

(a) $XY = (b^2/2a)(X+Y)$ focal 180°-structure rule, ellipses   (249)

(b) $a(X-a)^3 \overset{\circ}{s}{}^2 = X^2[(X-a)^4 + a^4]$ focal 180°-X-arc-increment eq., parabola   (248)

(b) $b^2(2aX-b^2)^4 \overset{\circ}{s}{}^2 =$ focal 180°-X-arc-increment eqs., ellipses   (249)
$X^2(2aX-b^2-X^2)[(2aX-b^2)^4 + b^8]$

It is clear from the above results that the structural simplicity of conic sections about their traditional foci in the circumpolar system does not achieve the exceptionally high status it attains in the hybrid polar-rectangular system. Nor does the 4-leaf rose achieve a high degree of structural simlicity about its center in the hybrid polar-rectangular system to compare with its exceptionally simple inherent structure in the circumpolar system. In fact, its hybrid distance eq. about its center is 3rd-degree ($v^3 + av^2 = 2ax^2$).

The circumpolar approach through structure rules clearly provides very extensive foundations for systematic characterizations, categorizations, and classifications of curves, very far removed, and fundamentally different, from the superficial classical approaches. The combination of the specific structure rule with the XY-arc-increment eq. to give the X-arc-increment eq. achieves complete specificity for the initial curve. In other words, only one curve exists that possesses the structure described by a given specific structure rule and its corresponding X-arc-increment eq.

The ability to achieve very simple descriptions of the inherent structure of the 4-leaf rose in terms of the 45°-structure rule and 45°-arc-increment eq. signifies that the curve possesses a very high degree of 45°-circumpolar symmetry (structural simplicity for a 45°-probe) about its center. It is clear from a comparison of eq. 247 (second entry from right) with eqs. 248b and 249b (see also Problem 15) that much more complex X-arc-increment eqs. are required to describe the inherent structure of curves (the parabola and ellipses) about poles of lesser positional symmetry, even for 180°-probes.

Simple Intercept Products--A Fundamental Property of Conic Sections

Simple Intercept Products

It is a fundamental but previously unrecognized or little-appreciated property of conic sections that the products of the X and Y-intercepts of 0° and 180°-probes positioned at a point in the plane (h,k) can be expressed very simply in the form $XY = f(\theta,h,k)$. A manifestation of this property already has been encountered in the case of self-inversion of the circle about a point in the plane, for which the 0° and 180°-structure rules are $XY = (h^2-R^2)$ and $XY = (R^2-h^2)$. The circle is the only conic section for which these products are independent of θ.

The derivation of these relationships is illustrated with central conics (eq. 250a). Here, and in Chapter VIII, the upper alternate signs of eqs. that represent both types of curves are for hyperbolas, and the lower signs are for ellipses. Eq. 250a represents x-k and y-k translations, which bring the origin to the point (-h,-k) in the planes of the initial rectangular curves. Transforming to polar coordinates gives the simple intercept eq. 250b. The two roots, r_1 and r_2, of this eq. represent the lengths of two probe-lines to the curve at the common angle θ. When both roots are positive (or negative), 0°-probe-lines are represented (coincident-radii), whereas when one root is positive and the other negative, 180°-probe-lines are represented (chord-segments).

Following the well-known rule, the product of the roots of eq. 250b is the ratio of the constant term to the coefficient of the r^2-term, given by eq. 250c. Accordingly, it is independent of the coefficient of the linear term in r. Reformulating the root-product eq. for the four different cases of intercept products of hyperbolas and ellipses in terms of X and Y, the lengths of the intercepts of the probe-lines (corresponding to r_1 and r_2), yields eqs. 251a,b and 252a,b.

derivation of the simple intercept products of central conics

(a) $\quad b^2(x-h)^2 \mp a^2(y-k)^2 = a^2b^2 \qquad$ initial central conics cast about $(-h,-k)$ $\qquad$ (250)

(b) $r^2[(b^2\pm a^2)\cos^2\theta \mp a^2] + 2(\pm a^2k\sin\theta - b^2h\cos\theta)r$ $\qquad$ polar form of eq. 250a
$\qquad + (b^2h^2 - a^2b^2 \mp a^2k^2) = 0$

(c) $r_1r_2 = [b^2h^2 - a^2b^2 \mp a^2k^2]/[(b^2\pm a^2)\cos^2\theta \mp a^2] \qquad$ product of roots of eq. 250b

(a) $XY = [b^2(h^2-a^2) - a^2k^2]/[(b^2+a^2)\cos^2\theta - a^2] \qquad$ coincident-radii intercept product of hyperbolas about point $(-h,-k)$ $\qquad$ (251)

(b) $XY = [b^2(a^2-h^2) + a^2k^2]/[(b^2+a^2)\cos^2\theta - a^2] \qquad$ same as eq. 251a but for chord-segments

(a) $XY = [b^2(h^2-a^2) + a^2k^2]/[(b^2-a^2)\cos^2\theta + a^2] \qquad$ coincident-radii intercept product of ellipses about point $(-h,-k)$ $\qquad$ (252)

(b) $XY = [b^2(a^2-h^2) - a^2k^2]/[(b^2-a^2)\cos^2\theta + a^2] \qquad$ same as eq. 252a but for chord-segments

(a) $\quad XY = (k^2 - 4ah)/\sin^2\theta \qquad$ coincident-radii intercept product of parabolas about point $(+h,-k)$ $\qquad$ (253)

(b) $\quad XY = (4ah - k^2)/\sin^2\theta \qquad$ same as eq. 253a but for chord-segments

(a) $XY = (k^2 - a^2)/\sin^2\theta, \quad k > a \qquad$ coincident-radii intercept product of the parallel line-pair about a point at distance k from the midline (line separation $= 2a$)

(b) $XY = (a^2 - k^2)/\sin^2\theta, \quad k < a \qquad$ same as eq. 254a for chord-segments

Inasmuch as central conics have two lines of symmetry, the positional parameters occur only as squares and represent all combinations of positive and negative displacements. The corresponding eqs. for the parabolas and parallel line-pair (eqs. 253 and 254, respectively) are given for comparison. [For the circle, one lets $a = b = R$ in eqs. 252, whereupon the $\cos^2\theta$-term vanishes.]

The simplicity of the above derivations is misleading, since it gives one the impression that the circumstances represented are of general occurrence. Thus, one could write scores of simple 2nd-degree polar eqs. that superficially resemble eq. 250b and that lead to simple products of roots involving $\sin\theta$ and/or cosθ. Nevertheless, the circumstances are not of general occurrence; they are unique to conic sections. The key factor is that eq. 250b is not a commonly-employed polar eq. (like the scores of others that could be written). It is a *simple intercept eq.* about an arbitrary point $(-h,-k)$; it is the *basic* polar construction rule arranged in the simple intercept form, whereas commonly-employed polar construction rules are cast about fixed points. The latter yield an intercept product in the simple form $XY = f(\theta)$ only for the fixed pole in question.

The corresponding simple intercept eq. for a rectangular curve of 3rd-degree or higher would not take the simple form of eq. 250b. Rather, it would be of 3rd-degree or higher (RULE 7a and Problem 18), in which case its general solutions would be extremely complex. Presentations of the XY-products of such solutions would require pages of space, rather than a mere fraction of a line.

Omnidirectional Circumlinear Self-Inversion of Hyperbolas

Having obtained the general expressions 251-254 for simple intercept products of conic sections about a point in the plane, it is of interest to examine the forms they take for certain exceptional points and lines. For this purpose, several "chord-segment" simple intercept products of hyperbolas and ellipses are listed below (expressions 255 and 256). As the corresponding denominators do not vary (see eqs. 251 and 252), they are omitted for convenience.

numerators of chord-segment intercept products of
central conics

transverse or major axis	conjugate or minor axis	center	traditional focus	real or imaginary asymptotes		
$b^2(a^2-h^2)$	$a^2(b^2+k^2)$	a^2b^2	$-b^4$	a^2b^2	hyperbolas	(255)
$b^2(a^2-h^2)$	$a^2(b^2-k^2)$	a^2b^2	b^4	a^2b^2	ellipses	(256)

A remarkable circumstance is that the intercept products for points on the asymptotes of hyperbolas are independent of h and k. Thus, hyperbolas have the extraordinary property that the product of one distance from a point on an asymptote to the curve along a line at an angle θ, and the second distance to the curve along the same line, is constant for all points of the asymptotes, along all lines at the same angle θ (Fig. 42f, examples for three points and three different angles).

The phenomenon described above is the *circumlinear* equivalent of circumpolar self-inversion, in which the product of two distances is constant for a fixed angle about all points (of a line), rather than for all angles about a fixed point. This aspect of circumlinear symmetry is referred to as *omnidirectional circumlinear self-inversion* of hyperbolas about their asymptotes (Fig. 42f; Kavanau, 1980). Although obscure, this is a classically-known property (see Salmon, 1879) and one of the rare examples of a known circumlinear structure rule (the structure rule X = Y of classical symmetries about lines being another). The same property applies to points on the imaginary asymptotes of ellipses, where these asymptotes are defined by the eq. $k^2 = -b^2h^2/a^2$.

Types of Circumpolar Point-Foci and Focal Loci

Detailed studies of the properties of circumpolar point-foci and focal loci in various groups of curves reveal many interesting divergent features, unsuspected from classical treatments. Some of these are reviewed briefly here. More detailed treatments are to be found in Kavanau (1980, 1982).

The fact that foci generally do not lie at fixed relative positions along axes, with respect to vertices, gives rise to many interesting variable relations between the locations of foci in the planes of curves. But the best-known foci--the traditional foci of conic sections--do not possess this variability. Thus, being located at $\pm\sqrt{(a^2-b^2)}$, with vertices at $\pm a$, the traditional foci of ellipses always are within the curve. Similarly, being located at $\pm\sqrt{(a^2+b^2)}$, with the vertices also at $\pm a$, the traditional foci of hyperbolas always lie beyond the vertices.

When one turns to the best-known inverse loci of conic sections--those about exceptional points on the lines of symmetry--quite different and much more varied circumstances prevail. The foci of limacons, the curves inverse to conics about the traditional foci, are taken as examples. The -b/2 focus, which is inverse to the reflection of the traditional focus of conics in the near directrix, always lies within the oval of elliptical limacons. But in the cases of hyperbolic limacons, it sometimes lies within the large loop, sometimes within the small loop (i.e., within both loops), and in one member is coincident with the small-loop vertex as a double-focus. Because of this variability of position, the -b/2 focus is designated as a *variable focus*.

Both of the other non-vertex axial foci of limacons also are variable; one of these, the focus of self-inversion--the point inverse to the opposite focus of the initial conic--is within the small loop of hyperbolic limacons, at the cusp-point in the cardioid (in which self-inversion is trivial), and outside the oval of elliptical limacons. Because the -b/2 focus and the focus of self-inversion can lie on either side of, as well as in coincidence with, a vertex, they are referred to as *penetrating variable foci*. The other non-vertex axial focus, the DP, while also variable, never penetrates a vertex; it is within the oval in elliptical members (but a DP of the locus), at the cusp in the cardioid, and at the point of intersection of the curve with itself in hyperbolic members. For this reason it is referred to as an *incident variable focus*.

[These relationships are readily understood in terms of the fact that limacons are focal inversions of conics, and their double-points always are the reciprocal inversion poles (the poles in the planes of the limacons about which inversion restores the initial conics). Since the focus of self-inversion of the limacon is the inverse of the opposite traditional focus of the conic (the reflection of the pole of inversion in the line of symmetry), it lies outside the oval of elliptical limacons (because the opposite traditional focus is inside the initial ellipse), within the small loop of hyperbolic limacons (because the opposite arm, within which the opposite focus lies, inverts to the small loop), etc.

The -b/2 focus, however, is the inverse of the reflection of the focus (the pole of inversion) in the near directrix. Accordingly, it is the inverse of a pole that may lie at any one of various positions in the planes of conics of different eccentricities. In ellipses, this reflected point always is outside the ellipse, since the directrix is outside; consequently its inverse comes to be situated within the elliptical oval of the inverse limacon. In hyperbolas, however, this reflected point may be beyond the vertex of the opposite arm (e > 2), at the vertex (e = 2), or between the two vertices (e < 2; it is at the center in the equilateral hyperbola). Accordingly, its inverse comes to be situated either within the small loop, at the small-loop vertex, or within the large loop.]

In some cases, penetrating variable foci never are incident upon the curve. For example, in Cartesian ovals, the monoval-bioval focus of self-inversion lies outside the monovals and within the small oval of the biovals. But it is not incident in the curve of demarcation; instead it is "captured" within the small loop as the indentations of the curve come into contact to form the two-looped limacon curve of demarcation (see Fig. 21-2). For this reason, it is known as a *conditionally-penetrating variable focus*.

Any curve that possesses variable foci also possesses *multiple-foci*. The simplest examples are those found when a variable focus is in coincidence with a vertex, such as the *double-focus* formed when the -b/2 focus of limacons is in coincidence with the small-loop vertex. In the cardioid, three otherwise discrete foci come into coincidence to form a *quadruple-focus*. These are the DP (itself a double-focus), the focus of self-inversion (though trivial), and the small-loop vertex (or one vertex of the elliptical oval). On the other hand, the -b/2 focus never comes into coincidence with either the DP, the focus of self-inversion, or the large-loop vertex.

Nor does the -b/2 homologue in Cartesian ovals ever come into coincidence with any of the three foci of self-inversion, though it can come into coincidence with a vertex of the small oval. The three foci of self-inversion, in turn, never come into coincidence with a vertex of an oval, though two of them come into coincidence with the DP of the limacon curve of demarcation (Fig. 21-2). From this point of view, the DP itself must be regarded as a quadruple-focus that includes two trivial foci of self-inversion (an analytically-justified point of view; see Chapter VIII and Kavanau, 1982).

We already have touched upon the distinction between *simple* and *compound curvilinear focal loci*, the latter being curvilinear focal loci (including lines) that include point-foci. In this connection, with the exception of some or all of the foci of self-inversion of curves that lack a line of symmetry, all point-foci lie on curvilinear focal loci.

We also have mentioned *covert focal loci*, which are focal loci that are neither lines of symmetry, asymptotes, nor the initial curve itself, such as the -b/2 axis of limacons (a line through the -b/2 focus normal to the line of symmetry). *Eclipsable foci* and *focal loci* have been encountered only in conic sections. In addition to the eclipsing of all the vertices (axial and LR) of ellipses and hyperbolas in the equilateral members (the circle and equilateral hyperbola), the major and minor axes of ellipses become eclipsed (i.e., diameters of the circle are not focal loci).

Summary

Although certain highly specific aspects of the inherent structure of curves were known classically, and the structure rule $X = Y$ is implicit in classical symmetries about points and lines, there lay unrecognized in associations between curves and points and lines of symmetry extreme examples of incisive and fundamentally different methods of analysis of curves. Whereas the principal classical approach has employed relations between distances from single points of curves to a pair of reference elements, the alternative method employs distances from paired points of curves to single reference elements. The classical approach yields construction rules, the alternative method yields structure rules, the topics of this introductory survey.

The classical symmetries of curves and figures are merely extreme cases of structural simplicity--involving inherent properties so simple that they can be detected by the most cursory visual inspection. By contrast, the majority of inherent structural properties of curves are not readily perceptible. The structure rules that embody them range from linear eqs., only slightly more complex than $X = Y$, to enormously more complicated eqs. that attain maximum exponential degrees in the range of 20 to 32 for conic sections and range very much higher for other curves.

Classical self-inversions at 0° and 180° (structure rule, XY = constant) also are found to be but special, albeit the most common, cases. Not only have curves been discovered that self-invert at other angles, self-inversion itself is merely a special case of a more general phenomenon--*congruent-inversion*. Self-inversion along coincident-radii (i.e., at an angle of 0°) retains a pre-eminent position, however, inasmuch as only 0°-foci of self-inversion exchange (on inversion) with lines of symmetry. Accordingly, from the point of view of inherent structure, the possession of a 0°-focus of self-inversion in the finite plane of a given curve is the counterpart of the possession of a line of symmetry in a certain curve inverse to the given curve.

But even a glimpse into the domain of structure rules and inherent structure exposes the fragmentary state of classical knowledge about inherent properties of curves. Even for the simplest of all curves, the line and the circle, the general 4th and 6th-degree structure rules, respectively, were unknown. Moreover, many fundamental properties of conic sections, in general, were recognized only recently--including the possession of *simple intercept products*; only two very simple instances of the latter were known to classical workers, one that characterizes the circle and another that characterizes hyperbolas (both of which cases involve types of self-inversion).

These limitations trace to the classical emphasis on analytical approaches that employ construction rules (i.e., coordinate eqs.) and that are targeted primarily at elucidating visible properties of curves (slopes, curvature, areas, intersections, etc.). With this emphasis, classical attempts to characterize, categorize, and classify curves systematically--which are essential stepping stones to a deeper understanding of structure—inevitably depended almost entirely on superficial visible properties. Accordingly, the resulting characterizations and schemes of classification gave correspondingly superficial insights and, being unproductive, were abandoned in the 19th century.

The same shortcoming is responsible for the failure to detect some of the most fundamental properties of the inversion transformation, probably the most important geometrical transformation of mathematics and physics. These include *immediate closure*, which makes possible the first "natural" paradigm for classification; it leads to a partitioning of curves into unique closed ensembles, every member of which inverts to every other member. The significance of this new tool is enhanced by the fact that a large fraction of all known curves fall into groups related in this manner.

Many of the structure rules are themselves construction rules for known curves; in some cases these structure rules provide the only known relation between two or more classical curves. The most intriguing example is a structure rule that describes relations between the circle and the equilateral hyperbola, which also is the construction rule for the Devil's Curve. Similarly, the eq. for the equilateral (a = b) unicursal Cross Curve—previously known only as a rectangular construction based on a circle—emerges as a structure rule of the line and parallel line-pair, while the eq. of the "4th-degree circle" is found to be a structure rule of the equilateral lemniscate. Additionally, a number of fascinating homologies between the line, the circle, and the equilateral hyperbola have been revealed.

Circumpolar structure rules provide the means both for identifying exceptional loci (points, lines, and curves) in the planes of initial curves, and ranking these loci in a quantitative hierarchy. This is achieved on the basis of the single criterion of the degrees of structure rules. The unifying concept underlying these rankings is that of the *structural simplicity* of the curves referred to these exceptional loci or, alternatively, the *positional symmetry* of these loci in the planes of the curves. On the other hand, classical approaches detected the exceptional nature of only certain of these loci, and required a number of diverse criteria to achieve even this incomplete recognition.

The essential distinction between the structure-rule and construction-rule approaches is that the latter approach merely enables one to construct the initial curves and carry out precise investigations of primarily overt features, whereas the former one, when employing a point reference element, penetrates beyond the level of superficial structure to yield characterizations of inherent properties. At the most general level, these characterizations provide the basis for broad schemes of classification based on commonly-held inherent properties, even for curves that differ greatly in their appearances. As the characterizations are pursued in ever greater depth, large groups sharing single properties become increasingly subdivided into smaller groups sharing greater numbers of properties, leading, ultimately, to unique characterizations in which the identities of the single members are implicit.

With recognition of these inherent properties of curves, and with knowledge of the structure rules that characterize specific inherent properties, it becomes possible to synthesize new curves that possess these properties in desired combinations, sometimes with surprising results. For example, whereas one cannot synthesize a curve with greater circumpolar symmetry (structural simplicity about a point) than a circle, one can synthesize a curve with greater circumpolar symmetry than two circles tangent to each other. The *sextic symmetry mimic of tangent-circles* fulfills this requirement; it possesses linear structure rules about its center for three different probe-angles, whereas only two of the corresponding structure rules of tangent-circles are linear (the third is circular).

The classical homology between the circle (the "equilateral ellipse") and the equilateral hyperbola is well known. Each curve is obtained from the corresponding general rectangular eq. by allowing the two parameters a and b, for the lengths of the orthogonal axes of the curves, to become equal. In the process, the major and minor axes of ellipses merge into an infinite number of diameters of the resulting circle, with the concomitant loss of identifiable axial vertices. Since the corresponding process in hyperbolas leaves the lines of symmetry identifiable, the essential classical geometrical basis for the homology between the two curves is equality of the axes. Circumpolar symmetry studies reveal a fascinating new homology.

Ellipses and hyperbolas, in general, possess incident circumpolar point-foci (the axial and LR-vertices). It is clear that these must lose their point-focal rank in the circle, where incident points are indistinguishable from one another. But a corresponding loss in the equilateral hyperbola was not anticipated, inasmuch as all exceptional incident points remain identifiable (with the axial vertices retaining their classical identity).

In a striking illustration of the great diagnostic resolving power of circumpolar symmetry analyses, it is found that all the incident point-foci of hyperbolas also are *eclipsed* in the equilateral member, that is, they lose their identity as circumpolar point-foci. While the axial and LR-vertices of the equilateral hyperbola retain an exceptional status (as compared to other incident points), this occurs at the *subfocal* level; their exceptional status is based on the comparative simplicities of corresponding structure rules, all of which fall in the same exponential degree category.

A notable correlation between analyses of structural simplicity based on construction rules and structure rules is found in comparisons between the parabola and central conics (ellipses and hyperbolas). On the basis of construction rules, the parabola has greater structural simplicity in coordinate systems based on a line and a point or a line and a circle (possessing one coordinate line of symmetry).

In corresponding analyses of the parabola and central conics based on structure rules, the parabola is found to have greater structural simplicity about its focus than do central conics about their foci (excluding the circle and lines). On the other hand, where the parabola has only one structure rule of classical symmetry (about its line of symmetry), central conics have three (one each about their two lines of symmetry and one about their true centers).

Problems

1. Obtain the 90°-structure rule (eq. 226b) of the rectangular line $x+y=\sqrt{2}d$ for the probe-point at the origin.

2. Derive the $\cos\theta$-simple intercept format of an ellipse about its left focus.

3. Obtain the simple intercept eq. and simple intercept format of hyperbolas for the probe-point on the x-axis at $x=-h$.

4. Obtain the simple intercept format of hyperbolas about their left foci from the result of Problem 3, and use it to derive the 0°-structure rules for all valid combinations of positive and negative intercepts.

5. Account for the fact that the non-degenerate eqs., 226b and 231a, for the 90°-structure rules of a line about a point in the plane and ellipses about their centers, are able to represent all combinations of positive and negative intercepts, as opposed to the frequent need for degenerate structure rules, as in Problems 4 and 6.

6. Derive the 90°-structure rule of an ellipse about the left focus for all combinations of positive and negative intercepts, and identify the factors that apply for each combination.

7. Letting $h = 3R/2$ and $\theta = 0°$, find the intercepts of the X and Y-probe-lines with the circles of eqs. 232a-c. Show that the intercepts, two of which are positive and two of which are imaginary, satisfy the 90°structure rule 233a'.

8. From an analysis of the 90°-structure rule 233a' for a circle about a point h units from center, determine the maximum value the two intercepts can take on simultaneously. What is the corresponding value of h? What is the maximum value an intercept can have for a real solution?

9. Derive eq. 237b, the trivial incident 90°-structure rule for the circle from the general eq. 237a.

10. Derive eq. 244 for the α-structure rule of ellipses about their centers (lower alternate signs).

11. Derive the α-structure rules of the spiral of Archimedes, $r = a\theta$, and the logarithmic spiral, $r = be^{\theta/a}$.

12. Derive the parabolic 90°-structure rule of the curve with the eq., $r = a + b\cos^4\theta$, about the polar pole.

13. Obtain the 90°-XY-arc-increment eq. for the curve with the eq. $r^2 = a^2 - b^2\cos^2\theta$ (the *root $cos^2\theta$-group*), about the polar pole. What is its degree in X and Y?

14. For the $\cos\theta$-simple intercept format, show that the XY-arc-increment eq. is independent of the probe-angle α.

15. Obtain the 90°-X -arc-increment eq. of the curve with the eq. $r^2 = a^2 - b^2\cos^2\theta$ from the 90°-XY-arc-increment eq. of Problem 13. What is its degree in X?

16. Using focal simple intercept formats for positive intercepts, derive and compare the XY-arc-increment eqs. of the parabola and ellipses. Evaluate $(dX/d\theta)^2$ at the axial vertices and the near LR-vertices.

17. Obtain and compare the 180°-X-arc-increment eqs. for the parabola and ellipses from the corresponding XY-arc-increment eqs. of Problem 16.

18. Explain why the simple intercept eq. for a 3rd-degree rectangular curve cast about a point in the plane cannot give rise to an intercept product about a point in the plane expressible in simple compact form, $XY = f(\theta,h,k)$, as in the cases of conic sections.

Chapter VIII.   Structure Rules of Curves About Points

                            Introduction

Having surveyed salient features of the circumpolar structure-rule approach in Chapter VII, this concluding Chapter addresses more technical aspects of deriving and employing structure rules of curves about points, and additional aspects of the inherent structures of curves uncovered by these analyses. The treatment begins with the relatively simple case of the circle (for additional treatments of the line and line-pairs, see Kavanau, 1980, 1982).

Only two curves have sufficiently great structural simplicity to permit explicit formulations of their structure rules about an arbitrary point in the plane for an arbitrary probe-angle α--the *general structure rule*. These are line-pairs (or a single line) and the circle. Because of their exceptional structural simplicity about single points, it is possible to express the general structure rules of the parallel line-pair (but not other line-pairs), the single line, and the circle in terms of a single positional parameter. For all other curves, two positional parameters are required and, except for line-pairs, the *general structure rules* of other curves have been expressed only in unsimplified or unreduced forms, or only in parametric form (the individual simple intercept eqs. or formats in X and Y).

                            The Circle

The general structure rule of the circle is derived from the initial eq. 257a. Converting to polar coordinates and solving for cosθ yields the cosθ-simple intercept format 257b. Letting X = r with θ unchanged, and Y = r with θ replaced by (θ+α), and eliminating θ (or substituting in the cosθ-structure rule format of eq. 245a), leads to the general structure rule 257c. As was pointed out in Chapter VII, this structure rule is unique to the circle. We already have considered the general eqs. for 0°, 90°, and 180°-structure rules of the circle (Chapter VII, *Introduction* and eq. 233a).

To obtain the general incident structure rule, one lets h = R in eq. 257c, giving the degenerate 6th-degree eq. 258a. The X^2Y^2-factor represents two identical trivial α-structure rules XY = 0, the rectangular eq. of the coordinate axes (an equilateral hyperbola with axes of null length), for two probe-lines at any angle α. For one of these structure rules, the X-intercept is identically

derivation of structure rules of the circle

(a) $(x-h)^2 + y^2 = R^2$ circle of radius R cast about a point in the plane at a distance h from its center   (257)

(b) $\cos\theta = (r^2+h^2-R^2)/2hr$ $\cos\theta$-simple intercept format of eq. 257a

(c) $X^2(Y^2+h^2-R^2)^2 + Y^2(X^2+h^2-R^2)^2$
$\quad -2XY(X^2+h^2-R^2)(Y^2+h^2-R^2)\cos\alpha$
$\quad = 4h^2X^2Y^2\sin^2\alpha$ general structure rule of the circle

(a) $X^2Y^2[X^2+Y^2-2XY\cos\alpha-4R^2\sin^2\alpha] = 0$ general incident structure rule of the circle   (258)

(b) $X^2(1-\cos\alpha)+Y^2(1+\cos\alpha) = 4R^2\sin^2\alpha$ ellipses of eq. 258a for the axes rotated into coincidence with the coordinate axes

0 (representing the intercept of the X-probe-line with the curve at the probe-point) and the Y-intercept scans the curve. For the other, the situation is reversed.

Since the incident structure rule $XY = 0$ exists for all curves, and gives no indication of inherent structure, it is trivial and is discarded. This leaves the bracketed 2nd-degree expression of eq. 258a equated to 0, as the valid general incident structure rule. Rotating the axes to obtain the eq. in standard form (Problem 1) yields eq. 258b, the standard eq. of an ellipse. For the two angles, $\alpha = 90°$ and $270°$, this becomes the eq. of a circle (Fig. 44c, 1/4-circle of radius 2R).

The general structure rule 257c is plotted in Fig. 44 for two positive intercepts, probe-angles of 30°, 60°, 90°, 120°, 150°, and 180°, and probe-points at distances from the center of 0, R/4, R/2, 3R/4, 7R/8, and R, as labelled in Fig. 44b,c (note that the technical term *intercept transform* or just *transform* is employed in all of the figures of this Chapter). We noted in Chapter VII that the 90°-structural curves for probe-points on the lines of symmetry of initial curves possess a true center, so the unplotted segments in the other three quadrants are congruent with those shown for the 1st-quadrant.

We also noted that the structural curves for an angle α are congruent with those for the angle $(360°-\alpha)$, so the plotted examples also represent angles of 210°, 240°, 270°, 300°, and 330°. Additionally, since complete structural

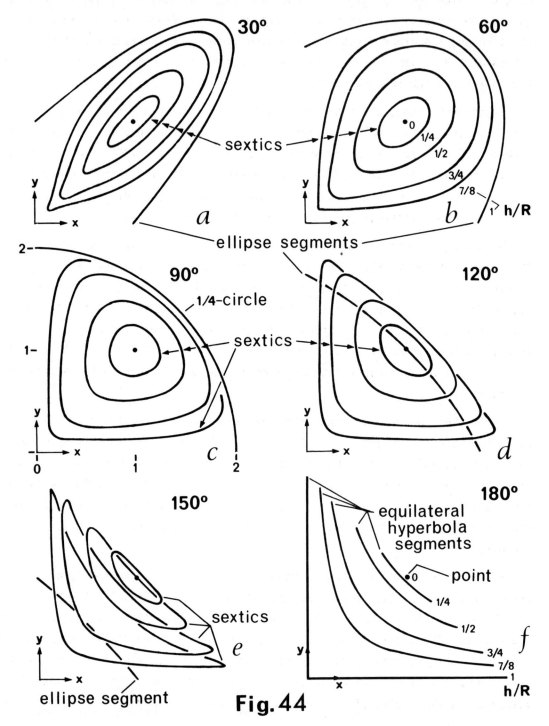

Fig. 44 Circumpolar Intercept Transforms of The Circle

curves for probe-angles that are 180° out of phase are themselves 90° out of phase (see Table 9), the 30° and 60°-segments for two positive intercepts are congruent with the 150° and 120°-segments, respectively, for mixed intercepts (i.e., the unplotted segments in the 2nd and 4th-quadrants for one intercept positive and the other negative), and vice versa. The incident structural curves of these supplementary plots are ellipses of a common eccentricity that are continuous between quadrants.

In essence, then, the plots of Fig. 44 are representative of all structural curves of the circle for interior and incident probe-points. Inasmuch as both the 0°-structure rules about exterior points and the 180°-structure rules about interior points are equilateral hyperbolas, Fig. 44f also is representative of the 0°-structure rules about exterior points (as well as 180°-structure rules about exterior points for mixed intercepts).

Structural curves of the circle about near exterior poles are very similar to those about interior poles equidistant from the perimeter. For example, the 90°-structural curve for $h = 0.875R$ and $h = 1.125R$ are almost in coincidence with each other. Although all structure rules about interior points have only real solutions, those about exterior points have both real and complex solutions; one real intercept exists for at least some orientations of all exterior probe-points, but the other intercepts for these particular orientations may all be imaginary.

Structural curves about the center are 0-degree for all probe-angles, and consist only of the point at (R,R), as labelled in Fig. 44f. The location of this point establishes the common scale, given in units of R along the abscissa and ordinate of Fig. 44c. As the probe-point moves out toward the perimeter from center, a series of gradually changing sextic structural curves are generated at all probe-angles except 0° and 180°, for which they are equilateral hyperbolas.

When the probe-point becomes incident upon the perimeter, the structural curves become quadratic. These are the ellipses of eq. 258a, for all probe-angles except 0°, 90°, 180°, and 270° (bracketed term equated to 0, and Fig. 44a, b,d,e). For 90° (and 270°)-probes, the curves are centered circles of radius 2R (Fig. 44c, outer curve), while for 180° (and 0°)-probes, they are 45°-equilateral hyperbolas of null axial length (Fig. 44f, limiting lines). The latter also are the asymptotes of the ensemble of equilateral hyperbolas.

The sextic structural curves obtained for probe-angles between 0° and 90° and between 270° and 360° (including 90° and 270°) are space-filling (i.e., the ensemble of these curves for all values of h, from 0 to R, fill the space within the elliptical or circular structural curves for incident points without intersection; Fig. 44a-c). On the other hand, the corresponding curves for probe-angles between 90° and 270°, with the exception of 180°, include intersecting members and possess segments exterior to the elliptical curves generated about incident points (Fig. 44d,e).

Similarly, the 180°-structural curves of Fig. 44f are space-filling, in the sense of nestling against one another, but these consist only of segments of equilateral hyperbolas. Although these equilateral hyperbolas gradually decrease in size as the probe-point moves out toward the perimeter (i.e., their axial lengths decrease), the lengths of the valid segments increase. This is evident in Fig. 44f, where only the valid segments are plotted.

Arc-Increment Equations

The *general XY-arc-increment eq.* (the XY-arc-increment eq. about a point in the plane) of the circle is obtained by finding $dr/d\theta$ from eq. 257b, eliminating θ by substituting from eq. 257b, and squaring, yielding eq. 259a (Problem 2). Since the X-intercept is the value of r for θ unchanged, $(dX/d\theta)^2$ is simply eq. 259a with X replacing r.

derivation of arc-increment eqs. of the circle

(a) $(dr/d\theta)^2 = r^2[4h^2r^2-(r^2+h^2-R^2)^2]/(h^2-R^2-r^2)^2$ $(dr/d\theta)^2$ for the polar eq. of a circle cast about a pole a distance h from center (259)

(b) $\overset{\circ}{s}{}^2 = X^2[4h^2X^2-(X^2+h^2-R^2)^2]/(h^2-R^2-X^2)^2 + Y^2[4h^2Y^2-(Y^2+h^2-R^2)^2]/(h^2-R^2-Y^2)^2$ 10th-degree XY-arc-increment eq. of the circle about a pole a distance h from center ($h \neq 0$)

(a) $\overset{\circ}{s}{}^2 = 8R^2 - (X^2+Y^2)$ XY-arc-increment eq. of the circle about incident points   (260)

(b) $\overset{\circ}{s}{}^2 = 4R^2$ X-arc-increment eq. for the incident 90°-structure rule of the circle (independent of X)

The finding of Chapter VII, that the XY-arc-increment eq. is independent of the probe-angle, also means that the value of $(dr/d\theta)^2$ is the same for any probe-angle. Accordingly, the expression for $(dY/d\theta)^2$ is identical to that for

$(dX/d\theta)^2$, but with Y replacing X, leading to eq. 259b. When cleared of fractions, this eq. is 10th degree in X and Y [The eq. is not valid for $h = 0$, however, since its derivation involved division by h.]

To obtain the incident XY-arc-increment eq., one lets $h = R$, whereupon there is great simplification and degree reduction, yielding the 2nd-degree eq. 260a. To derive the corresponding XY-arc-increment eq. for a given *specific structure rule*, that is, a structure rule for a specified point and angle (see Table 2), one simply eliminates Y from eq. 260a using the value of Y obtained by solving the specific structure rule in question.

Employing the solution for Y obtained from the "circular" incident 90°-structure rule (eq. 243b) yields the X-arc-increment eq. 260b which, being independent of X, is 0-degree (and provides the basis for the first circular structure rule entry in Table 10). The corresponding X-arc-increment eq. for an arbitrary angle and an elliptical structure rule is 4th-degree (Problem 3). The X-arc-increment eqs. for the trivial incident 0° and 180°-structure rules $X = Y$ and $X = -Y$ are 2nd-degree.

The Parabola

No circumpolar point-focus exists in the plane of the line or parallel line-pair, and only one exists in the plane of the circle. From another point of view one could say that these curves have such great circumpolar symmetry, compared to central conics in general, that all points in the plane acquire point-focal rank (either when $a = b$ or when a or b becomes infinite). Following the definition of RULE 46, however, only the center of the circle possesses point-focal rank.

At any rate, as soon as one leaves the domain of initial lines and an initial circle, the practicality of formulating general structure rules explicitly is lost. The closest we shall come to achieving such formulations involve either the corresponding parametric eqs. (in X, Y, α, and θ) or eqs. from which radicals have not been eliminated. It often is the case that only the maximum exponential degrees of these structure rules are known (see Table 11).

Intercept Equations and the General
Structure Rule

The standard parabola of eq. 261a is taken as the initial curve and translated to the left and upward (eq. 261b) to bring its vertex to the point (-h,k) [equivalent to translating the origin to the point (h,-k) in the plane of the initial curve]. Our object is to outline the procedures for obtaining the general structure rule about the point (h,-k) in the plane of the initial curve.

Transforming to polar coordinates and replacing r by X yields eq. 261c. Because translations have been employed along both axes, this eq. contains linear terms in both $\sin\theta$ and $\cos\theta$, whereupon it is designated as a *mixed-transcendental intercept equation*. As a consequence, it is impractical to obtain a simple intercept eq. in terms of $\sin\theta$ or $\cos\theta$ alone (nor $\sin^2\theta$ or $\cos^2\theta$). Accordingly, the structure-rule formats (eqs. 245) cannot be employed to obtain either the general structure rule or angle-specified structure rules.

Instead, eq. 261c is employed directly. The corresponding simple *mixed-transcendental* Y-intercept eq. for an α-probe is 261d. If, now, α were to be eliminated between eqs. 261c,d, a very complex eq. of 32nd-degree (maximum) in X and Y would be obtained, representing the general structure rule. Rather than pursue this extremely difficult goal, we work directly with the simple mixed-transcendental intercept eqs. to obtain *angle-specified structure rules*. These then are examined to ascertain focal conditions.

first steps in the derivation of the general structure rule of the parabola (261)

(a) $y^2 = 4ax$ initial parabola

(b) $(y-k)^2 = 4a(x+h)$ translated initial parabola with origin at (h,-k) and vertex at (-h,k)

(c) $X^2\sin^2\theta - 2kX\sin\theta + (k^2-4ah) = 4aX\cos\theta$ simple mixed-transcendental X-intercept eq.

(d) $Y^2\sin^2(\theta+\alpha) - 2Yk\sin(\theta+\alpha) + (k^2-4ah) = 4aY\cos(\theta+\alpha)$ simple mixed-transcendental Y-intercept eq. for $\alpha°$

(e) $Y^2\sin^2\theta \mp 2kY\sin\theta + (k^2-4ah) = \pm 4aY\cos\theta$ simple mixed-transcendental Y-intercept eq. for $\alpha = 0°$ and $180°$ (upper signs for $0°$)

(f) $Y^2\cos^2\theta - 2kY\cos\theta + (k^2-4ah) = -4aY\sin\theta$ simple mixed-transcendental Y-intercept eq. for $\alpha = 90°$

A focal condition, for example, is apparent immediately on examination of eq. 261c: if k^2 is allowed to take on the value $4ah$, the constant term vanishes and an X can be factored out, reducing the degree of the eq. in X from 2 to 1. Such a degree reduction at the level of an intercept eq. leads to great reductions in the degrees of the corresponding structure rules. Since the condition $k^2 = 4ah$ defines probe-points incident upon the curve, and the structure rules for these points reduce compared to those for neighboring points, the curve itself is a circumpolar focal locus. The incident 90°-structure rules, for example, are only 12th-degree, compared to 20th-degree for neighboring non-incident points.

0° and 180°-Structure Rules and Compound Intercept Formats

It was pointed out in the preceding section that it is impractical to obtain simple intercept formats in $\cos\theta$ or $\cos^2\theta$ from mixed-transcendental intercept eqs., and that different procedures of analysis are employed. The different procedures referred to involve *compound intercept formats*. These are solutions in functions of $\cos^n\theta$ or $\sin^n\theta$ for which the probe-angle is specified, and which, in consequence, are functions of both the X and Y-intercepts.

The procedure for obtaining the 0° and 180°-*compound intercept formats* for the parabola about a point in the plane is illustrated here. Both the 0° and 180°-formats are derived together because the procedures are identical for both and the formats differ only in sign.

[In fact, when all combinations of intercepts are taken into account in deriving structure rules, 0° and 180°-structure rules differ only in the signs of terms; see eqs. 326a,b and Problems 4, 7, and 8.]

Eq. 261c, the simple mixed-transcendental X-intercept eq., is divided by eq. 261e, the simple mixed-transcendental Y-intercept eq. for 0° and 180°, as below (or both eqs. are solved for $\cos\theta$ and the eliminant is obtained). By this simple procedure, the expressions $4a\cos\theta$ are eliminated in the quotient. On cross-multiplying the quotients, the terms in $\sin\theta$ also cancel, leaving the degenerate 3rd-degree eq. 262b, which rearranges to eq. 262b'. The factor $(Y \mp X)$ represents the trivial 0°-structure rule $X = Y$ (for two positive intercepts) and the trivial 180°-structure rule $X = -Y$ (for one positive and one negative intercept). Since these trivial factors are components of all complete 0° and 180°-structure rules, and since their existence is independent of the inherent structure of the initial curve, they are discarded, yielding the sought-after $\sin^2\theta$ compound 0° and 180°-intercept formats of eq. 262c.

These formats will be recognized as the simple intercept products treated

first steps in the derivation of general 0° and
180°-structure rules of the parabola

$$\frac{X^2\sin^2\theta - 2kX\sin\theta + (k^2-4ah) = 4aX\cos\theta}{Y^2\sin^2\theta \mp 2kY\sin\theta + (k^2-4ah) = \pm 4aY\cos\theta} \quad \text{quotient of eqs. 261c,e} \quad (261c,e)$$

(a) $\pm X^2Y\sin^2\theta \mp 2kXY\sin\theta \pm (k^2-4ah)Y =$ quotients 261c,e cross- (262)
multiplied
$XY^2\sin^2\theta \mp 2kXY\sin\theta + (k^2-4ah)X$

(b) $(Y\mp X)XY\sin^2\theta \mp (k^2-4ah)(Y\mp X) = 0$ eq. 262a simplified and rearranged

(b') $(Y\mp X)[XY\sin^2\theta \mp (k^2-4ah)] = 0$ eq. 262b rearranged

(c) $\sin^2\theta = \pm(k^2-4ah)/XY$ the $\sin^2\theta$-compound 0° and 180°-intercept format and simple intercept product (see eq. 253a)

in Chapter VII, whose existence in such simple forms is a fundamental property of conic sections. Note that the derivation in such simple forms--even for conics--hinges on the use of 0° and 180°-probe-angles. For other angles (see eq. 266), the linear terms in $\sin\theta$ (or $\cos\theta$) do not cancel. It is only because of the fact that X and Y are collinear that (regarding them as the two roots of a quadratic eq. in r, as in eq. 250b, in which intercept products were derived for central conics) their product becomes independent of the coefficient of the linear term in r.

Being in possession of the simple expression 262c, for $\sin^2\theta$ as a function of X and Y, this can be used to derive the 0° and 180°-structure rules of the parabola about a point in the plane. Thus, eq. 262c is used to eliminate θ from the mixed-transcendental X-intercept eq. 261c.
[Structure rules derived in this manner do not represent the trivial 0° and 180°-structure rules X = Y and X = -Y, since the factors representing them already were factored out and discarded in the derivation of eq. 262c.]

Eliminating functions of θ from eq. 261c by means of eq. 262c (Problem 4) leads to the 8th-degree 0° and 180°-structure rules 263 of the parabola about a point in the plane. This is arranged as shown to illustrate the derivation of focal conditions, that is, to illustrate the conditions for which degree reduction occurs. Reduction of this eq. clearly cannot be achieved by factoring an XY-term or an (X±Y)-term. Though the condition $A^2 = k^2 - 4ah = 0$ leads to reduction, this is unproductive, since this condition is for incident poles, for which no valid 0° or 180°-structure rule exists.

derivation of 0° and 180°-structure rules of the parabola

$16X^2Y^2[4a^2XY \mp A^2(4a^2+k^2)]^2 -$ angle-specified structure rule of   (263)
$8A^4XY(X\pm Y)^2[4a^2XY \mp A^2(4a^2-k^2)]$ the parabola about the point $(h,-k)$
$+ A^8(X\pm Y)^4 = 0$ [upper alternate sign for 0°]
   $[A^2 = (k^2-4ah)]$

(a) $X^2Y^2 = h^2(X+Y)^2 - 4ahXY$ 0°-structure rule for points on the line of symmetry   (264)

(b) $X^2Y^2 = h^2(X-Y)^2 + 4ahXY$ 180°-structure rule for points on the line of symmetry

(b') $XY = a(X+Y)$ 180°-structure rule about the traditional focus

    The only possibility for reduction of eq. 263 is to have the entire expression fall into a perfect square, which occurs solely for the condition $k = 0$. In other words, the line of symmetry of the parabola is the sole 0° and 180°-focal locus in the plane of the curve. Since the curve itself is not a valid locus for 0° and 180°-structure rules, but all other points in the plane are, it follows that any point-focus that exists in the plane of the parabola must be incident upon either the line of symmetry or the curve itself (RULE 51). By this simple means, all other possible candidates for focal loci, such as the directrix (for which $h = -a$), the LR (for which $h = a$), and the vertex-tangent (for which $h = 0$) are ruled out.

    Letting $k = 0$ in eq. 263 and taking roots yields the 0° and 180°-structure rules 264a,b for points on the line of symmetry. The conditions for the existence of a point-focus incident upon the line of symmetry (but excluding the vertex, since this also is incident upon the curve) are that eq. 264b reduce for interior axial points (h, positive) or that eq. 264a reduce for exterior axial points (h, negative). Both cases already were encountered in Chapter VII.

    Eq. 264b reduces for the probe-point at the traditional focus (for which $h = a$), yielding the 180°-structure rule 264b' (the "harmonic mean" eq.; see also eq. 228b). But there is no negative value of h for which eq. 264 reduces (see the treatment accompanying eq. 243; but since the latter eq. represents an x-h translation, the criterion for the existence of a focus is reduction of the degree of eq. 243 for a positive value of h).

    Plots of the 180°-structural curves for points incident upon the line of symmetry and the LR, and of 0°-structural curves for six exterior points, in-

0° & 180°-Circumpolar Intercept Transforms of The Parabola

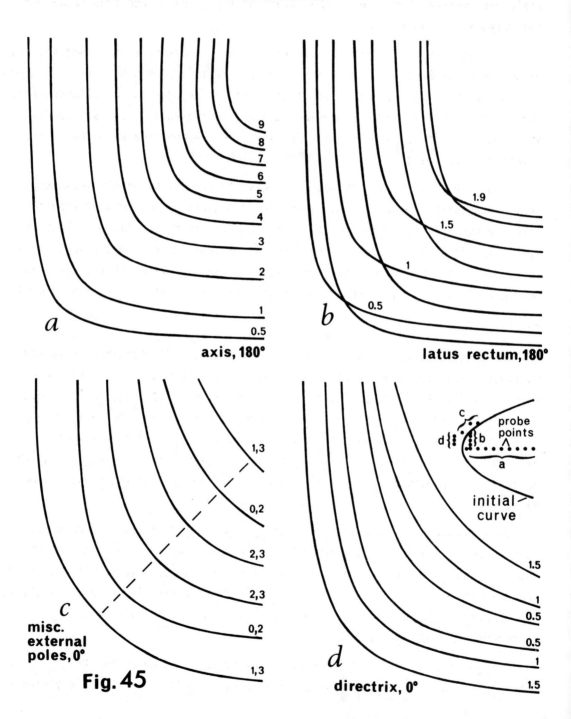

Fig. 45

cluding three points incident upon the directrix, are given in Fig. 45a-d, respectively. Note the symmetry of the structural curves about the coordinate bisector (see Table 9). The plots about axial points are in their proper relative positions, whereas those for probe-points on the LR have been translated for the purpose of illustration.

The 0°-structural curves about exterior poles possess paired inner and outer segments. These are plotted in the correct relative positions for each pair, but the pairs themselves have been translated for the purpose of illustration. The locations of the poles (probe-points) are indicated in the inset of Fig. 45d, and each structural curve is identified by the coordinates of the corresponding probe-point.

Though the 180°-structural curves appear to consist of single segments, they are recurrent each 180°, so that each represents two congruent superimposed segments. In all other cases, the probe-points do not lie on a line of symmetry, with the consequence that the paired segments are incongruent. Structural curves for non-axial interior poles typically possess intersecting segments, as in Fig. 45b, whereas those about exterior poles do not (Fig. 45c,d).

90°-Structure Rules

The procedures for obtaining the general incident 90°-structure rule are outlined first, and then specific structure rules are obtained for the axial vertices (Fig. 46, segments labelled "0") and LR-vertices (Fig. 47c). To obtain the polar eq. of the parabola cast about an incident point, one lets $k^2 = 4ah$ in eq. 261c, yielding the X-intercept eq. 265a (X replacing r). Replacing θ by $\theta+90°$ and X by Y in eq. 265a yields the corresponding Y-intercept eq. The 90°-XY-intercept product (eq. 266) is obtained readily from these expressions by solving for X and Y and taking their product.

Even though limited to incident points, the 90°-XY-intercept product is seen to be much more complicated than the general intercept products for 0° and 180° (eq. 262c) and is of no utility for deriving structure rules. Instead, the 90°-compound intercept eq. 267a is obtained by eliminating $\cos\theta$ (or $\sin\theta$) in the manner described above (see eq. 261). Solving for $\sin\theta$ yields the 90°-compound intercept format 267b.

The general incident 90°-structure rule is obtained by eliminating θ from the X-intercept eq. 265 by means of this solution 267b for $\sin\theta$. This yields a

90°-Circumpolar Axial Intercept Transforms of The Parabola

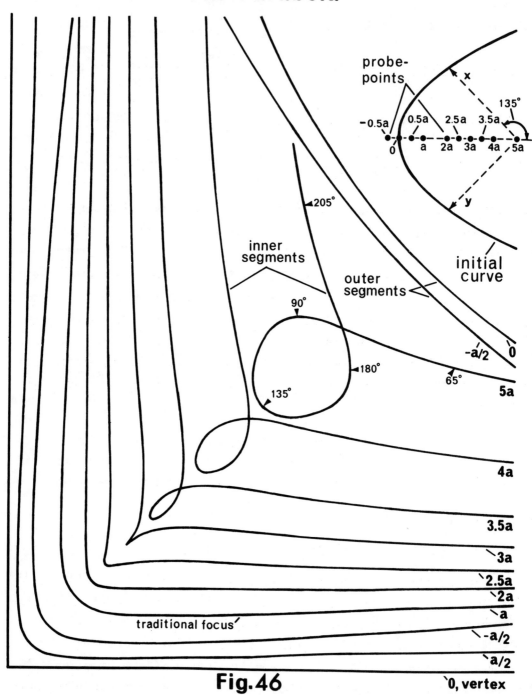

Fig. 46

derivation of incident 90°-structure rules of the parabola

(a) $X\sin^2\theta - 2(k\sin\theta + 2a\cos\theta) = 0$ X-intercept eq. for incident points   (265)

(b) $Y\cos^2\theta - 2(k\cos\theta - 2a\sin\theta) = 0$ 90°-Y-intercept eq. for incident points

$$XY = \frac{4(k\sin\theta + 2a\cos\theta)(k\cos\theta - 2a\sin\theta)}{\sin^2\theta\cos^2\theta}$$ 90°-XY-intercept product for incident points   (266)

(a) $(kX+2aY)\sin^2\theta - 2(k^2+4a^2)\sin\theta - 2aY = 0$ 90°-compound intercept eq. for incident points   (267)

(b) $\sin\theta = \dfrac{(k^2+4a^2) \pm \sqrt{[(k^2+4a^2)^2 + 2aY(kY+2aY)]}}{(kX+2aY)}$ 90°-compound intercept format for incident points

12th-degree eq. that is much too complex to reproduce here (but see Kavanau, 1980). Accordingly, the focal condition for incident points is that the 90°-structure rule reduce from 12th-degree.

For the axial vertex, one lets $k = 0$ in eq. 265a, whereupon the linear term in $\sin\theta$ vanishes. The resulting eq. then can be solved readily for the $\cos\theta$-simple intercept format. Substitution of this solution in the $\cos\theta$-structure-rule format 245a yields the 8th-degree structure rule 268 for the axial vertex (Fig. 46, segments labelled 0; Problem 5). Since this represents a reduction from the 12th-degree general incident 90°-structure rule, the axial vertex is a point-focus.

90°-structure rules of the parabola about its axial and LR-vertices

$X^4Y^4 = (4a)^6(X^2+Y^2) + 3(4a)^4X^2Y^2$ axial vertex   (268)

$X^4Y^4 = 8(4a)^6(X^2+Y^2) + 32a^2X^2Y^2(X^2+Y^2)$
$\quad - (4a)^4(Y^4+4Y^3X+6Y^2X^2-4YX^3+X^4)$ LR-vertices   (269)

In the cases of the LR-vertices, one lets $k = 2a$ (for the lower LR-vertex) in eq. 265. Since the linear term in $\sin\theta$ does not vanish, one eliminates the linear terms in $\cos\theta$ (or $\sin\theta$) as described above (see eq. 261c,e) and obtains the 90°-$\sin\theta$-compound intercept format (eq. 267b with $k = 2a$). Employing this to eliminate $\sin\theta$ from eq. 265 (with $k = 2a$) leads to the 90°-structure rule 269 for the LR-vertices (Fig. 47c; Problem 6). Since this eq. also is only 8th-degree, the LR-vertices also are point-foci.

Non-Axial Incident And Non-Incident 90°-Circumpolar Intercept Transforms of The Parabola (All Segments, All Intercepts)

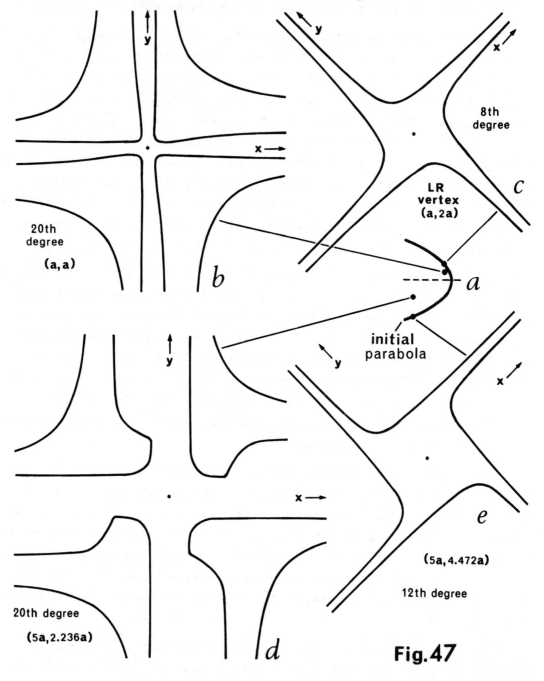

Fig. 47

A comparison of eqs. 268 and 269 reveals that the parabola possesses greater structural simplicity about the axial vertex for a 90°-probe than about the LR-vertices, since the 90°-structure rule about the axial vertex is much the simpler of the two eqs. However, this greater structural simplicity is at the subfocal level, since both structure rules are 8th-degree in the variables.

It is noteworthy that one term of eq. 269 differs from the polynomial expansion of $(Y+X)^4$ only in the sign of the $4YX^3$-term. This is a fascinating aspect of circumpolar structure rules that is of relatively common occurrence, namely, the possession of polynomial expressions that differ from perfect squares, cubes, etc., only because of the dissonant sign or value of a single coefficient. It is by virtue of this single difference in the sign that the 90°-structural curve for an LR-vertex lacks the symmetry about the coordinate bisectors (Fig. 47c) that is possessed by the 90°-structural curves about points on lines of symmetry (Fig. 46 and Table 9). Like all complete structural curves (Table 9), the full four-quadrant plot (Fig. 47c) for all combinations of intercepts possesses an ordinary center of symmetry.

To obtain the general 90°-structure rule for points on the line of symmetry (Figs. 46 and 48), one lets $k=0$ in the simple mixed-transcendental X-intercept eq. 261c. The linear term in $\sin\theta$ then vanishes, yielding the simple intercept eq. 270a, from which one can obtain the $\cos\theta$-simple intercept format 270b. Substituting in the $\cos\theta$-structure-rule format 245a yields a complex α-structure rule of 16th-degree about points on the line of symmetry. For a 90°-probe, this simplifies to eq. 271 but remains 16th-degree (Fig. 46). In fact, it is clear from inspecting the α-structure rule that it will reduce, in general, only for probe-angles of 0° and 180°.

In order for a point-focus to exist on the line of symmetry, the 90°-structure rule 271 must reduce. The left member of this eq. already is a perfect square, and it is evident that for $h=a$ the right member also becomes a perfect square, with reduction to the 8th-degree structure rule 272a. On regrouping terms (transferring the right term on the left side to the right side), eq. 272a also becomes a perfect square, reducing to the 4th-degree 90°-structure rule 272b for the traditional focus (Fig. 46, curve marked "a;" Fig. 48, heavy segments).

It will be noted from eq. 272b' that there is no real solution (nor is there for eq. 272b) if either X or Y alone is negative, while from eq. 272b

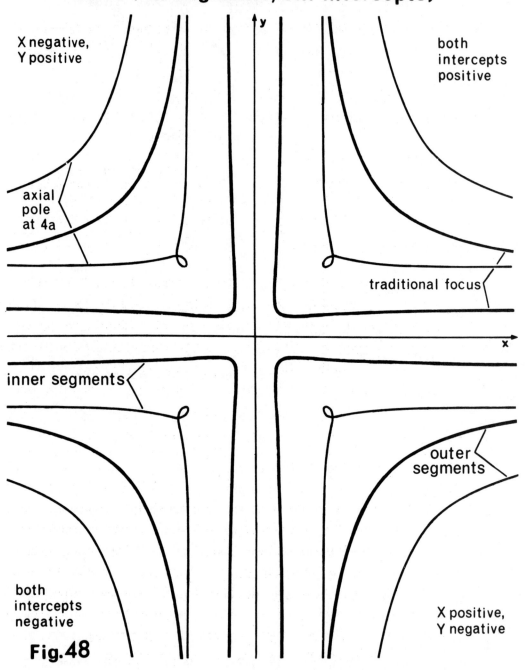

Fig. 48 90°-Circumpolar Intercept Transforms of The Parabola About The Traditional Focus And The Axial Pole At 4a (All Segments, All Intercepts)

derivation of 90°-structure rules of the parabola about points on the line of symmetry

(a) $X^2\sin^2\theta - 4aX\cos\theta - 4ah = 0$ simple intercept eq. about axial points (270)

(b) $\cos\theta = [-2a+\sqrt{(4a^2-4ah+X^2)}]/X$ $\cos\theta$-simple intercept format of eq. 270a

$\{[(8a^2-4ah)(X^2+Y^2)+X^2Y^2]^2 - 16a^2(A^2Y^4+B^2X^4)\}^2$
$= 32^2 a^4 A^2 B^2 X^4 Y^4$

90°-structure rule for points on the line of symmetry (271)

$[A^2 = (4a^2-4ah+X^2)]$, $[B^2 = (4a^2-4ah+Y^2)]$

(a) $[4a^2(X^2+Y^2)+X^2Y^2]^2 - 16a^2X^2Y^2(X^2+Y^2)$
$= 32a^2X^3Y^3$

positive root of eq. 271 for $h=a$ (272)

(b) $4a^2(X^2+Y^2) + X^2Y^2 = 4aXY(X+Y)$ 90°-structure rule about the traditional focus (positive root of eq. 272a)

(b') $[2a(X+Y)-XY]^2 = 8a^2XY$ eq. 272b in another form

(c) $X = [2aY^2 \pm 4aY\sqrt{(aY-a^2)}]/(Y-2a)^2$ eqs. 272b,b' solved for X

it is evident that there can be no real solution for both X and Y negative. This also is evident from eq. 272c, the solution of eqs. 272b,b' for X. On the other hand, the 90°-structure rule 271 for points on the line of symmetry is valid for all combinations of positive and negative intercepts. The explanation for the absence of the structure-rule components for other combinations of intercepts in eqs. 272 is that, in taking only positive roots in obtaining eqs. 272a,b, the other factors of the complete 90°-structure rule about the traditional focus were discarded.

Degenerate 90°-Structure Rules and Modular Segments

For the substitution $h=a$ in the 16th-degree general 90°-structure rule 271 for points on the line of symmetry, it becomes the degenerate product of four 4th-degree factors equated to 0 (see Problems 4-6 in Chapter VII). Each factor represents a quadrant of the complete structural curve of Fig. 48 (heavy segments). When the first positive root is taken (eq. 272a), the resulting 8th-degree structure rule represents the segments of the complete structural curve in the 1st and 3rd-quadrants; the discarded negative root represents the segments in the 2nd and 4th-quadrants. When the second positive root is taken, the resulting 4th-degree structure rule represents the segment in the 1st-quadrant only; the discarded negative root represents the segment in the 3rd-quadrant.

Accordingly, the complete 90°-structure rule for the traditional focus, representing all combinations of intercepts, is of degree 16, the same as about unexceptional axial points (for example, about the axial pole at $h = 4a$, for which the complete structural curve is represented by the light segments in Fig. 48). The essential difference is that the eq. becomes degenerate for points of exceptional positional symmetry, and the form of the complete structural curve can be deduced merely from the 4th-degree eq. of any one of its *modular* segments in one of the four quadrants (the segments in the other three quadrants being congruent with the "modular" segment).

This representation of *modular* segments by degenerate structure rules does not occur for unexceptional points; a non-degenerate 16th-degree structure rule is required to represent the forms of structural curves for such points. Nor does it occur for all foci. Thus, the 8th-degree 90°-structure rule of the parabola about the axial vertex represents all four quadrants of congruent segments of the complete structural curve, but no eq. of lesser degree can represent the modular form of a single quadrant (see Fig. 46).

These circumstances illustrate another aspect of the circumpolar structure-rule approach to analysis of the structural simplicity of curves about points and point-continua in their planes. Complete structural rules about exceptional points do not always reduce. But, in lieu of reducing, they become degenerate, with the result that the modular form of the complete structural curve can be represented by a structure rule of lesser degree. It is because of these circumstances that circumpolar symmetry analyses at the level of the degrees of structure rules employ the lowest-degree structure rules that can represent modular segments (of 90°-structural curves).

The Structural Curves

All of the 90°-structural curves of Fig. 46 for interior points are for positive intercepts and possess both inner and outer asymptotic segments, but only two of the outer segments (the segments farthest from the origin of the plot) are shown. The segments are for progressively displaced probe-points in the sequence shown, but the precise locations of the segments have been shifted for the purpose of illustration. The initial curve and probe-point locations are shown at the upper right, together with a representative probe at $h = 5a$, with its X-probe-line at $\theta = 135°$.

The 90°-structural curves of Fig. 47 are complete. The probe-point locations are shown in the plane of the initial curve (Fig. 47a). Since no probe-point is incident upon a line of symmetry, no structural curve possesses a true center (see Table 9). The exponential degrees of the structure rules and the probe-point coordinates accompany each structural curve. The two 90°-structural curves of Fig. 48 also are complete; being for axial poles, both possess a true center.

Central Conics

From the point of view of structural simplicity about points in the plane, ellipses and hyperbolas bear such close resemblances to one another that it is preferable to treat both together, drawing parallels and pointing out differences as they arise. Virtually every structure rule and construction rule of one of these conic sections is identical to a corresponding rule for the other, excepting only the signs of terms (differences in sign, of course, can have marked structural consequences).

In the following, the steps of structure-rule derivations generally are carried out either for ellipses alone or hyperbolas alone, but the final eqs. generally are given for both. The latter generally are in the form of a single expression for both structure rules, with the upper of a set of alternate signs applying for hyperbolas and the lower for ellipses. In cases where eqs. with alternate signs represent only hyperbolas or ellipses, the upper of a set of alternate signs is for 0° and the lower for 180°-probe-angles.

0° and 180°-Structure Rules

Eq. 273 is the general rectangular construction rule of hyperbolas cast about a point in the plane. The employment of x-h and y-k translations displaces the centers upward and to the right (assuming h and k to be positive) to the point (h,k). This is equivalent to placing the origin and probe-point at (-h,-k) in the plane of the initial curve. Accordingly, the structure rules that are obtained apply for the point in the plane of the initial curve at (-h,-k), in which case the intercept eqs. and formats generally are cast about the left focus, left vertex, points on the left LR, lower left LR-vertex, etc.

Transforming to polar coordinates and replacing r with X yields the simple intercept eq. 273b. As expected, the vanishing of the constant term, which occurs for any incident point of the initial curve, allows an X to be factored out of the eq. and identifies the curve itself as a focal locus; this factoring out of an X always leads to reductions of the degrees of the structure rules about incident points, as compared to those about neighboring points.

first steps in the derivation of 0° and 180°-
structure rules of central conics

(a) $\quad b^2(x-h)^2 - a^2(y-k)^2 = a^2b^2 \quad$ rectangular eq. of hyperbolas with centers at (h,k) $\quad$ (273)

(b) $\quad X^2(a^2+b^2)\cos^2\theta - 2b^2hX\cos\theta - a^2X^2 + (b^2h^2-a^2k^2-a^2b^2) = -2a^2kX\sin\theta \quad$ simple mixed-transcendental intercept eq. for hyperbolas

(c) $\quad XY = \dfrac{\pm(b^2h^2-a^2k^2-a^2b^2)}{[(a^2+b^2)\cos^2\theta - a^2]} \quad$ simple 0° and 180°-intercept products for hyperbolas (upper sign for 0°)

(c') $\quad \cos^2\theta = \dfrac{[\pm(b^2h^2-a^2k^2-a^2b^2)+a^2XY]}{(a^2+b^2)XY} \quad$ eq. 273c solved for $\cos^2\theta$

RULE 52. *In a simple (mixed transcendental) intercept equation cast about a point in the plane, the vanishing of the constant term is the focal condition for points incident upon the curve.*

RULE 53. *In a simple intercept equation cast about points on a line of symmetry, the vanishing of the constant term is the focal condition for the vertices.*

One obtains the simple mixed-transcendental Y-intercept eq. for 0°-probes simply by replacing X by Y in eq. 273b; for 180°-probes, one also replaces θ by (θ+180°). The linear terms in sinθ then are eliminated between the X and Y-intercept eqs., yielding the familiar XY-intercept products of eq. 273c (see also eqs. 251a,b). Alternatively, the product-of-the-roots method of Chapter VII can be used, but, as discussed earlier, this is useful for point-in-the-plane derivations only in the cases of conic sections. Solving eq. 273c for $\cos^2\theta$ (eq. 273c') and using the 180°-solution (lower alternate sign) to eliminate θ from eq. 273b, yields the 8th-degree 180°-structure rule of hyperbolas about a point in the plane (eq. 274).

Corresponding 180°-structural curves of the $e = 0.745$ ellipse for interior probe-points and two positive intercepts are given in Fig. 49. These are plotted to the same scale, but their positions have been shifted arbitrarily for

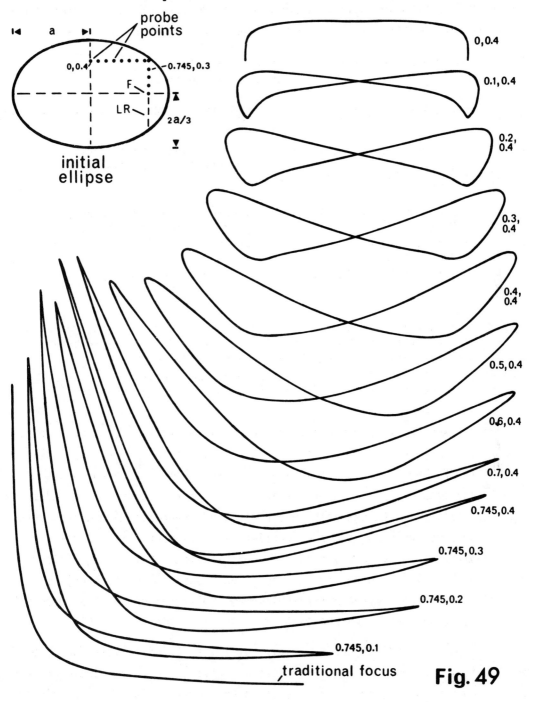

Fig. 49

the purpose of illustration. The probe-point positions in the plane of the initial curve are identified at the upper left.

For a focus to exist at a specific probe-point $(-h,-k)$ in the plane of the initial curve, eq. 274 must reduce in degree for displacements of the curve that bring the origin to this point. For example, $(-h,-k) = (0,0)$, correspond- to a null translation, would be for the center; $(-h,-k) = (-a,0)$ would be for the left vertex (x-a translation); $(-h,-k) = (-h,0)$ would be for a pole on the transverse axis at $x = -h$ (x-h translation); $(-h,-k) = (-2a,-2a)$ would be for a pole on the coordinate bisector in the 3rd-quadrant (x-2a and y-2a translations); etc.

Eq. 274 for the general 180°-structure rule of hyperbolas is arranged as shown to facilitate an investigation of the possibility that the entire polynomial is a perfect square, analogous to eq. 263 for the parabola. The first two terms already are perfect squares, while the third term is the potential cross-product term.

180°-structure rule of hyperbolas about a point
in the plane

$$16X^2Y^2[a^2b^2(b^2h^2-a^2k^2)XY-A^4(b^4h^2+a^4k^2)]^2 + A^{16}(b^2+a^2)^2(X-Y)^4 \qquad (274)$$

$$- 8A^8(b^2+a^2)(X-Y)^2XY[a^2b^2(b^2h^2+a^2k^2)XY-A^4(b^4h^2-a^4k^2)] = 0$$

$$A^4 = (b^2h^2-a^2k^2-a^2b^2)$$

general 180°-structure rules for central conics
about points on the transverse and major axes

$$4a^2h^2X^2Y^2 = (a^2 \pm b^2)(h^2-a^2)^2(X-Y)^2 \pm 4b^2h^2(h^2-a^2)XY \qquad \text{(upper signs for hyperbolas)} \qquad (275)$$

general 180°-structure rules for central conics
about points on the conjugate and minor axes

$$4b^2k^2X^2Y^2 = (b^2 \pm a^2)(k^2 \pm b^2)^2(X-Y)^2 \pm 4a^2k^2(k^2 \pm b^2)XY \qquad \text{(upper signs for hyperbolas)} \qquad (276)$$

Just as for the parabola, the entire eq. falls into a perfect square for $k = 0$. In the present cases for central conics, however, a perfect square also is formed for $h = 0$, yielding the 4th-degree 180°-structure rules 275 and 276 about points on the lines of symmetry (Fig. 50). The same eqs., with the alternate signs of the last terms reversed and terms in $(X-Y)^2$ replaced by $(X+Y)^2$,

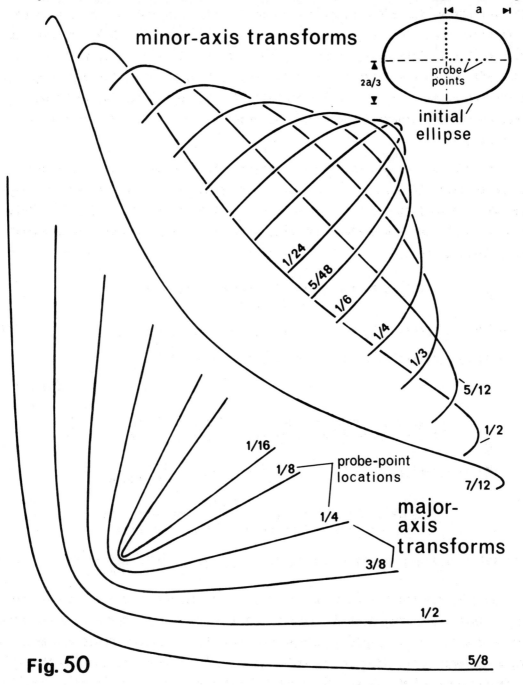

Fig. 50 180°-Circumpolar Intercept Transforms of The e=0.745 Ellipse About Interior Poles On The Lines of Symmetry

apply for 0°-probes. In order for an axial point-focus to exist, the corresponding eq. for points on the line of symmetry must reduce. Inspection of eqs. 275 and 276 reveals that structure rules for points on the transverse and major axes reduce for the conditions $h^2 = a^2 \pm b^2$ (upper sign for hyperbolas; Problem 7), yielding the 180°-"harmonic mean" eq. 229a for both hyperbolas and ellipses (see Fig. 49, traditional focus), and the analogous 0°-structure rule for hyperbolas.

Eqs. 276 also reduce for imaginary poles (the imaginary foci) at $k^2 = b^2 - a^2$ on the minor axis of ellipses and $k^2 = -(a^2+b^2)$ on the conjugate axis of hyperbolas (Problem 8). For the center, at $(h,k) = (0,0)$, both structure rules reduce to $X = Y$, the 180°-structure rule for classical rotational symmetries. Excluding the 0-degree structure rule ($X = R$) of the circle about its center, these are the lowest-degree structure rules of central conics (see Table 11).

These examples do not exhaust the possibilities for focal loci of central conics for 0° and 180°-probes. There is another focal condition for reduction of eq. 274 (and for the homologous eq. of ellipses). For $b^2 h^2 = a^2 k^2$, that is, for a probe-point incident upon an asymptote, the term $a^2 b^2 (b^2 h^2 - a^2 k^2) XY$ vanishes from eq. 274 (from the left-most squared term) leaving the 6th-degree cross-product term as the term of highest degree.

Accordingly, the asymptotes are circumpolar focal loci, with the 6th-degree 0° and 180°-structure rules 277. [We noted in Chapter VII that the asymptotes also are circumlinear focal loci.] These circumstances also apply for $b^2 h^2 = -a^2 k^2$, representing probe-points on the "imaginary asymptotes" of ellipses.

0° and 180°-structure rules of hyperbolas about
points on the asymptotes

$$16h^4(a^2+b^2)X^2Y^2 + a^4(a^2+b^2)(X \pm Y)^4 = 8a^2h^2XY(X \pm Y)^2[2XY \pm (a^2-b^2)] \qquad (277)$$

In summary, structure-rule analyses of central conics to this juncture have revealed the curves themselves to be focal loci--from mere inspection of the simple mixed-transcendental intercept eqs.--before the specification of a probe-angle. Subsequent 0° and 180°-analyses revealed that the lines of symmetry, the traditional foci, the centers, and the asymptotes of hyperbolas, also have focal rank in the real plane. Thus, in all, 6 of the 16 real foci of ellipses (see Fig. 41c) and 8 of the 16 real foci of hyperbolas have been identified.

Intercept Equations and Formats

For both the parabola and central conics, we have noted that it is possible to detect the existence of focal loci (the curve itself, the line of symmetry, and the asymptotes) through an examination of the simple (mixed-transcendental) intercept eqs. about a point in the plane, before the specification of a probe-angle (these are known as *angle-independent focal conditions*). Similar deductions sometimes can be made from the simple intercept eqs. and formats for points on other linear and curvilinear focal loci, before the specification of a probe-angle. This matter was touched upon briefly in Chapter III (*Simple Intercept Equations*).

Eqs. 278 and 279 are the simple intercept formats for points on the major and minor axes of ellipses. [Note that they have exchange symmetry in both a and b and h and k (x-h and y-k translations).] A few simple manipulations of the radicand of eq. 278 reveal that it becomes a perfect square (an indubitable focal condition) for $h^2 = a^2-b^2$ and simplifies greatly for $h = 0$ and $h = \pm a$ (presumptive focal conditions; see Problem 9), representing the abscissae of the traditional foci, center, and axial vertices, respectively. All of these conditions lead to reduction (from 16th-degree) of the 90°-structure rule for arbitrary points on the major axis (see Table 11). The corresponding conditions

simple intercept formats for ellipses about points
on the lines of symmetry

$$\cos\theta = \frac{b^2h \pm \sqrt{[b^4h^2-(b^2-a^2)(b^2h^2+a^2X^2-a^2b^2)]}}{X(b^2-a^2)} \quad \text{major axes} \quad (278)$$

$$\sin\theta = \frac{a^2k \pm \sqrt{[a^4k^2-(a^2-b^2)(a^2k^2+b^2X^2-a^2b^2)]}}{X(a^2-b^2)} \quad \text{minor axes} \quad (279)$$

on the radicand of the simple intercept format 279 are $k^2 = b^2-a^2$, $k = 0$, and $k = \pm b$, representing the imaginary foci on the minor axis, the center, and the b-vertices, all of which also lead to reduction of the 90°-structure rule of ellipses about points on the minor axes (see Table 11).

Some representative 180°-structural curves for interior probe-points on the major and minor axes (two positive intercepts) are illustrated in Fig. 50 for the $e = 0.745$ ellipse. The probe-points are identified in the plane of the initial curve at the upper right. All plots are to the same scale; they illustrate the full extent of the structural curves (as traced out by the inter-

Table 11. Degrees of Structure Rules of Conic Sections about Points[1]

Curve	Parabola				Ellipses				Circle			
Probe-angle	$\alpha°$	90°	180°	0°	$\alpha°$	90°	180°	0°	$\alpha°$	90°	180°	0°
Locus												
point in plane	32*	20	8	8	32*	20	8	8	6	6	2	2
x-axis	16	16	4	4	16	16	4	4	-	-	-	-
y-axis	16	16	4	4	16	16	4	4	-	-	-	-
incident point	32*	12	-	-	32*	12	-	-	2	2	-	-
a-vertex	16	8	-	-	16	8	-	-	-	-	-	-
b-vertex	-	-	-	-	16	8	-	-	-	-	-	-
LR-vertex	32*	8	-	-	32*	8	-	-	-	-	-	-
focus	4	4	2	2	4	4	2	2	-	-	-	-
center	-	-	-	-	8	4	1	1	0	0	0	0

Curve	Hyperbolas					Equilateral Hyperbola				Parallel line-pair			
Probe-angle	$\alpha°$	variable	90°	180°	0°	$\alpha°$	90°	180°	0°	$\alpha°$	90°	180°	0°
Locus													
point in plane	32*	-	20	8	8	32*	16	8	8	4	4	1	1
x-axis	16	14	16	4	4	16	12	4	4	-	-	-	-
y-axis	16	14	16	4	4	16	12	4	4	-	-	-	-
asymptotes	32*	-	-	6	6	32*	16*	6	6	-	-	-	-
incident point	32*	-	12	-	-	32*	4	-	-	4	4	-	-
a-vertex	16	14*	8	-	-	16	4	-	-	-	-	-	-
LR-vertex	32*	-	8	-	-	32*	4	-	-	-	-	-	-
focus	4	3	4	2	2	4	3	2	2	-	-	-	-
center	8	6	4	1	1	8	2^2	1	1	-	-	-	-

[1] dash indicates degree unknown, structure rule trivial or non-existent, or focus and structure rule non-existent
[2] solution is imaginary
*maximum

cepts) and are in proper relative positions in each set, but the sets have been displaced relative to one another for the purpose of illustration. Whereas the major-axis curves nestle, the minor-axis curves generally intersect.

Eqs. 280 and 281 are the simple intercept eqs. for points on the left LR and for general incident points of ellipses. Eq. 280 simplifies for two conditions: (a) $k=0$, which places the probe-point at the left traditional focus, and for which the $\sin\theta$-term vanishes; and (b) $k=b^2/a$, which places the probe-point at the lower-left LR-vertex ($y-b^2/a$ translation), for which the constant term vanishes and an X factors out. Both conditions are focal, but this cannot be established without further analysis (the factoring out of an X for the left-lower LR-vertex is only a presumptive focal condition, since an X would factor out for any incident point).

simple intercept eq. for points on the left LR of ellipses

$$(b^2-a^2)X^2\cos^2\theta - 2b^2\sqrt{(a^2-b^2)}X\cos\theta + (a^2k^2-b^4+a^2X^2) = 2a^2kX\sin\theta \qquad (280)$$

simple intercept eq. for incident points of ellipses

$$(b^2-a^2)X\cos^2\theta - 2hb^2\cos\theta + a^2X = 2ab\sqrt{(a^2-h^2)}\sin\theta \qquad (281)$$

simple intercept eq. for points incident upon the asymptotes of hyperbolas

$$(b^2+a^2)X^2\cos^2\theta - 2b^2hX\cos\theta - a^2(b^2+X^2) = -2abhX\sin\theta \qquad (282)$$

There are only three conditions for which the simple intercept eq. 281 for incident points simplifies (the designation *mixed-transcendental* is omitted hereafter for simplicity). All three, $h=0$, $h^2=a^2$, and $h^2=a^2-b^2$, are found to be focal conditions through analyses with 90°-probes (see below), identifying as point-foci the two b-vertices, the two a-vertices, and the four LR-vertices. The same circumstances apply for hyperbolas, but the "b-vertices" for them are imaginary.

Accordingly, with the addition of eight incident foci for ellipses and six incident foci for hyperbolas, the total number of real foci in the finite plane is brought to 14 for both central conic types. Two additional point-foci on the line at infinity bring the total number of real foci to 16. Lastly, the simple intercept eq. for points on the asymptotes of hyperbolas (and the homologous eq. for points on the imaginary asymptotes of ellipses) adds no new presumptive point-focus; the sole condition for simplification, h = 0, is the condition for the center.

90°-Structure Rules

To obtain the 90°-structure rule of ellipses about points in the plane, the initial simple intercept eq. 283a (see eq. 273b for the hyperbola) is complemented by the corresponding 90°-Y-intercept eq., formed by replacing X by Y and θ by (θ+90°). After converting the latter to functions of θ alone [by replacing cos(θ+90°) by -cosθ, and sin(θ+90°) by cosθ], and rearranging, one obtains eq. 283b.

simple X and 90°-Y-intercept eqs. of ellipses about
a point in the plane

(a) $\quad X^2(b^2-a^2)\cos^2\theta - 2b^2hX\cos\theta + a^2X^2 + (b^2h^2+a^2k^2-a^2b^2) = 2a^2kX\sin\theta \quad$ (283)

(b) $\quad Y^2(b^2-a^2)\sin^2\theta - 2a^2kY\cos\theta + a^2Y^2 + (b^2h^2+a^2k^2-a^2b^2) = -2b^2hY\sin\theta$

90°-compound intercept format of ellipses about a
point in the plane

$$\cos\theta = [-XY(b^4h^2+a^4k^2) \pm \sqrt{G^{16}}]/XY(b^2-a^2)(a^2kY-b^2hX), \quad h \ \& \ k \neq 0 \quad (284)$$

$$G^{16} = X^2Y^2(b^4h^2+a^4k^2)^2 - XY(b^2-a^2)(a^2kY-b^2hX) \cdot$$

$$[(a^2kX+b^2hY)(a^2b^2-b^2h^2-a^2k^2)-a^2b^2XY(hX+kY)]$$

Eliminating the linear terms in sinθ between eqs. 283a,b, replacing $\sin^2\theta$ by $(1-\cos^2\theta)$, and solving for cosθ using the quadratic-root formula, yields the 90°-compound intercept format 284 for ellipses about points in the plane. Though this intercept format is complicated, it gives only a hint of the complexity of the corresponding 20th-degree 90°-structure rule.

A number of 90°-structural curves for points in the plane of the e = 0.745 ellipse are plotted in Figs. 51 and 52. Whereas those of Fig. 51 are for non-axial and non-incident probe-points, all but one of those of Fig. 52 are for incident and axial probe-points. All of the structural curves are arbitrarily located for purposes of illustration. The three representatives of Fig. 52 for incident probe-points (Fig. 52i,g,j) are four-quadrant plots (all combinations of intercepts), whereas the others are for the 1st-quadrant only (two positive intercepts).

The only conditions for which the radicand G^{16}, of eq. 284 becomes a per-

Non-Axial Non-Incident 90°-Circumpolar Intercept Transforms of The e=0.745 Ellipse (Positive Intercepts)

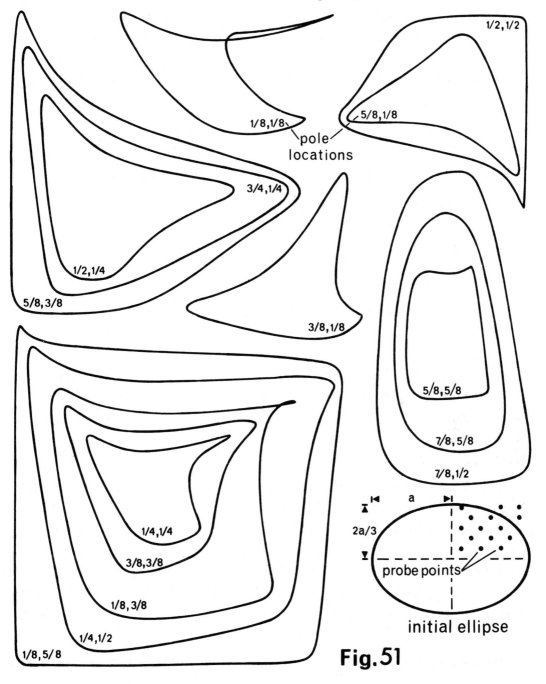

Fig. 51

fect square are $(h,k) = [\pm\sqrt{(a^2-b^2)},0]$, for the traditional foci. For the left traditional foci (eq. 284 does not apply for centers or for the special case of the circle) one obtains eq. 285. Note that, whereas eq 284 is a 90°-compound intercept format in both X and Y, eq. 285 is a simple intercept format in X alone. This calls attention to the fact that conversion of compound intercept formats in X and Y to simple intercept formats in X alone is accompanied by the loss of a specification of angle. [The angle specification is retained, however, if the compound intercept format is converted to a simple intercept format in Y alone (see eq. 287b and Problem 14).]

$$\cos\theta = (\pm aX+b^2)/X\sqrt{(a^2-b^2)} \qquad \text{simple intercept format of ellipses for the left traditional foci} \qquad (285)$$

RULE 54. *A condition for which the square-root radicand of a circumpolar intercept format is converted to a perfect square in the variables reveals the location of a high-ranking circumpolar focus.*

Conditions on the Radicand of the 90°-Compound Intercept Format

An examination of the conditions for h and k that lead to a simplification of the radicand G^{16} of eq. 284 reveals all the focal conditions for central conics, as follows. [Although the condition for the center leads to vanishing of the radicand, the format becomes indeterminate and the corresponding simple intercept format must be derived from eq. 283a.] For $b^2h^2 + a^2k^2 = a^2b^2$, or $k = \pm(b/a)\sqrt{(a^2-h^2)}$, the condition for points incident upon the initial ellipses, eq. 284 simplifies markedly to eq. 286 (from which k has not been eliminated completely). Elimination of θ between eqs. 283a and 286 leads to a complicated 90°-structure rule of 12th-degree for ellipses about unexceptional incident points. Further reductions occur for exceptional incident points (see below).

$$\cos\theta = \frac{(b^4h^2+a^4k^2) \pm \sqrt{[(b^4h^2+a^4k^2)^2 - a^2b^2(a^2-b^2)(a^2kY-b^2hX)(hX+kY)]}}{(a^2-b^2)(a^2kY-b^2hX)} \qquad (286)$$

90°-compound intercept format of ellipses about incident points

Substitution of $k = 0$ or $h = 0$ in eq. 284 simplifies the compound intercept formats to the simple X and 90°-Y-intercept formats 287a,b, for probe-points on the major and minor axis, respectively (including the centers). The

Ellipse 90°-Circumpolar Intercept Transforms

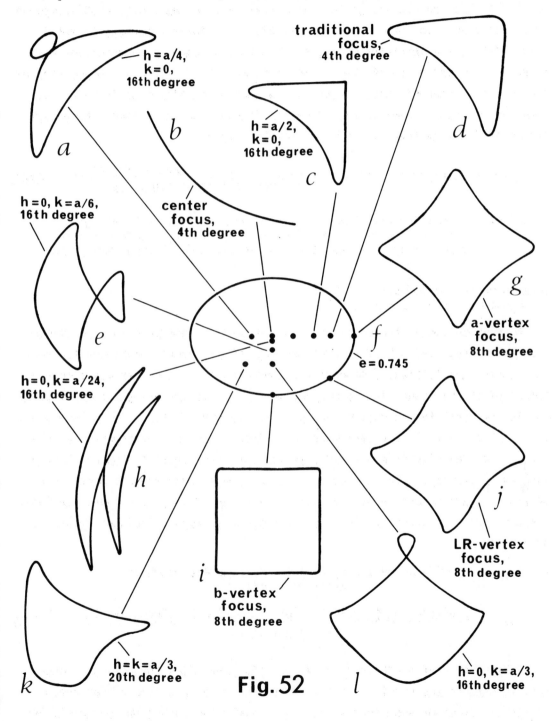

Fig. 52

corresponding 90°-structure rules are complicated and 16th-degree, compared to 12th-degree for incident points.

[To obtain the 90°-structure rule for ellipses about points on the major axis, one eliminates $\cos\theta$ between eq. 287a and the structure-rule format 245a (for $\alpha = 90°$). In the case of the corresponding structure rule for points on the minor axis, one lets $h = 0$ in eq. 283 and eliminates θ using eq. 287b, or converts eq. 287b to an X-intercept format and employs eq. 245a.]

simple intercept formats of ellipses about points
on lines of symmetry

(a) $\quad \cos\theta = \dfrac{b^2 h \pm \sqrt{[b^4 h^2 + (b^2 - a^2)(a^2 b^2 - b^2 h^2 - a^2 X^2)]}}{X(b^2 - a^2)} \quad$ X-intercept format, major axis $\qquad$ (287)

(b) $\quad \cos\theta = \dfrac{-a^2 k \pm \sqrt{[a^4 k^2 - (b^2 - a^2)(a^2 k^2 - a^2 b^2 + b^2 Y^2)]}}{Y(b^2 - a^2)} \quad$ 90°-Y-intercept format, minor axis

Some representative 90°-structural curves of the $e = 0.745$ ellipse and the $e = 1.118$ and $e = \sqrt{2}$ hyperbolas about points on lines of symmetry are illustrated in Figs. 52-55. In Fig. 52, the exponential degrees of the corresponding structure rules and the coordinates of the probe-points accompany each curve. The entire locus for two positive intercepts is plotted. In Fig. 53, for the $e = 1.118$ hyperbola, the plots also are for two positive intercepts; only the *inner segments* are shown (the segments nearest the origin). Both intercepts of these plots are with the same arm. Again, the locations have been shifted for the purpose of illustration.

Fig. 54 depicts the 90°-structural curve of the $e = \sqrt{2}$ hyperbola for a probe-point at the traditional focus, including all segments and all combinations of intercepts (as labelled). The plots of Fig. 55 are solely for a probe-point at the traditional focus of the $e = 0.745$ ellipse, but are for a series of probe-angles from 0° to 180°. The curves are to scale but have been shifted in positions.

Conics are the only known curves for which the general incident 90°-structure rules (12th-degree) are of lower degree than the general structural rules about points on the lines of symmetry (16th-degree). For comparison, the general incident 90°-structure rules of limacons are 60th-degree (Table 13), compared to only 24th-degree for points on the line of symmetry. 90°-structure rules of curves about axial vertices always are of lower degree than their structure rules about points on the two intersecting focal loci (the curve and

90°-Circumpolar Axial Intercept Transforms of The e=1.118 Hyperbola (Intra-Arm, Inner Segments)

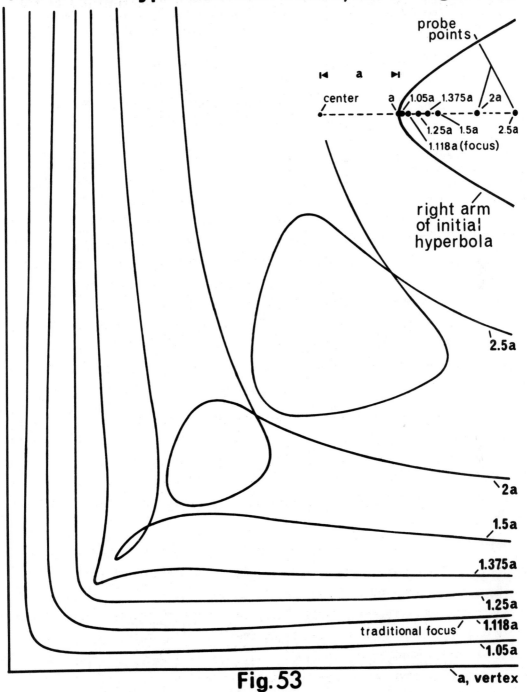

Fig. 53

a line of symmetry). Thus, for central conics, the 90°-structure rules about the axial vertices are only 8th-degree (Figs. 52 and 53; Table 11), while for limacons they are only 16th-degree (and 2nd-degree for the DP; Table 13).

RULE 55. *The exponential degrees of structure rules of curves about axial vertices reduce from the values for points incident upon the intersecting focal loci--a line of symmetry and the curve itself.*

The Latus-Rectum Vertices

For the lower-left LR-vertices, one lets $h = \sqrt{(a^2-b^2)}$ and $k = b^2/a$ in eq. 284 or 286, leading to the 90°-compound intercept format 288. The homologous format for the equilateral hyperbola greatly simplifies to eq. 289.

[The 90°-structure rules are obtained by eliminating θ between eqs. 288 and 283 a or b, leading to complicated 8th-degree structure rules (see Figs. 43f and 52j). These reduce to 4th-degree for the $e = \sqrt{2}$ hyperbola (eq. 283c and Fig. 43c), in which all of the incident 90°-point-foci are eclipsed (see Chapter VII, Fig. 43, and Table 11).]

90°-compound intercept format, left-lower LR-vertices of ellipses

$$\cos\theta = \{b^2(2a^2-b^2) \pm \sqrt{[b^4(2a^2-b^2)^2 + a(a^2-b^2)(ahX+b^2Y)(hX+aY)]}\}/(b^2-a^2)(hX+aY) \quad (288)$$

$$\cos\theta = \pm\sqrt{[9a^2+2(\sqrt{2}X+Y)^2]}/2(\sqrt{2}X+Y) \quad \text{90°-compound intercept format, left-lower LR-vertex of the } e = \sqrt{2} \text{ hyperbola} \quad (289)$$

The a and b-Vertices

Letting $h = a$ in eq. 287a, and $k = b$ in eq. 287b leads to the simple X and Y-intercept formats 290a,b for ellipses about their vertices (Problem 14). Eqs. 291 (upper alternate signs for hyperbolas) are the 90°-structure rules for the a-vertices of central conics (see Figs. 43a and 52g). These eqs. apply for both a-vertices, since the specificity of intercept eqs. and formats for a left or right vertex is lost when θ is eliminated.

simple intercept formats of ellipses

(a) $\cos\theta = \{ab^2 \pm a\sqrt{[b^4-X^2(b^2-a^2)]}\}/X(b^2-a^2)$ left a-vertices (for any probe-angle) (290)

(b) $\cos\theta = \{-a^2b \pm b\sqrt{[a^4+Y^2(b^2-a^2)]}\}/Y(b^2-a^2)$ lower b-vertices (90°-probe-angle)

90°-structure rules of central conics about their a-vertices

$$X^4Y^4(a^4-b^4)^2(a^2\mp b^2)^2 \mp 8a^2b^6(a^4-b^4)(a^2\mp b^2)(X^2+Y^2)X^2Y^2 + \quad (291)$$

$$16a^4b^8[(a^2\mp b^2)^2-4a^4]X^2Y^2 + 16a^4b^{12}(X^2+Y^2)^2 = 64a^6b^{12}(X^2+Y^2)$$

90°-Circumpolar Intercept Transform of The Equilateral Hyperbola About A Traditional Focus (All Segments, All Intercepts)

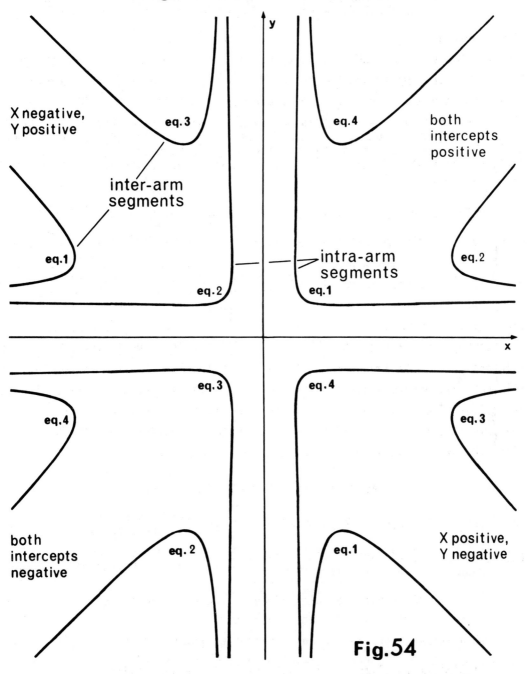

Fig. 54

The Traditional Foci and Centers

The only remaining foci to be considered are the traditional foci and centers, for which some intercept formats and 90°-structure rules already have been given. These are repeated here for convenience. The simple intercept formats of ellipses and hyperbolas for probe-points at the left traditional foci are eqs. 292a,b, respectively. Those for the centers are eqs. 294a,b.

simple intercept formats of ellipses (a) and hyperbolas (b) about the left traditional foci

(a) $\cos\theta = (aX-b^2)/X\sqrt{(a^2-b^2)}$ (b) $\cos\theta = (b^2 \pm aX)/X\sqrt{(a^2+b^2)}$ (292)

90°-structure rules of central conics about the traditional foci

(a) $b^4(X^2+Y^2) + (a^2 \mp b^2)X^2Y^2 = 2ab^2XY(X+Y)$ hyperbolas (upper sign) and ellipses   (293)

(b) $a(X^2+Y^2) = 2XY(X+Y)$ the equilateral hyperbola

simple intercept formats of ellipses (a) and hyperbolas (b) about their centers

(a) $\cos^2\theta = a^2(X^2-b^2)/X^2(a^2-b^2)$ (b) $\cos^2\theta = a^2(X^2+b^2)/X^2(a^2+b^2)$ (294)

90°-structure rules of central conics about the centers

(a) $X^2Y^2(b^2 \mp a^2) = a^2b^2(X^2+Y^2)$ hyperbolas (upper sign) and ellipses   (295)

(b) $X^2 + Y^2 = 0$ the equilateral hyperbola

Substitutions from eqs. 292 and 294 in the $\cos\theta$ and $\cos^2\theta$-structure-rule formats 245a,b, respectively, lead directly to the 90°-structure rules 293 and 295 (Problem 10), all of which are 4th-degree. The former is for positive intercepts only, whereas the latter is a *complete* structure rule (which applies for all combinations of intercepts; see Problem 5, Chapter VII). The corresponding complete 90°-structure rules about the traditional foci are degenerate 16th-degree eqs. (see Problem 6, Chapter VII).

The 90°-structure rule for the equilateral hyperbola about the traditional foci reduces to 3rd-degree, because of the vanishing of the 4th-degree term (as shown below, this also occurs for other hyperbolas at other probe-angles). At

Circumpolar Symmetry of The e-0.745 Ellipse About A Traditional Focus (Positive Intercepts)

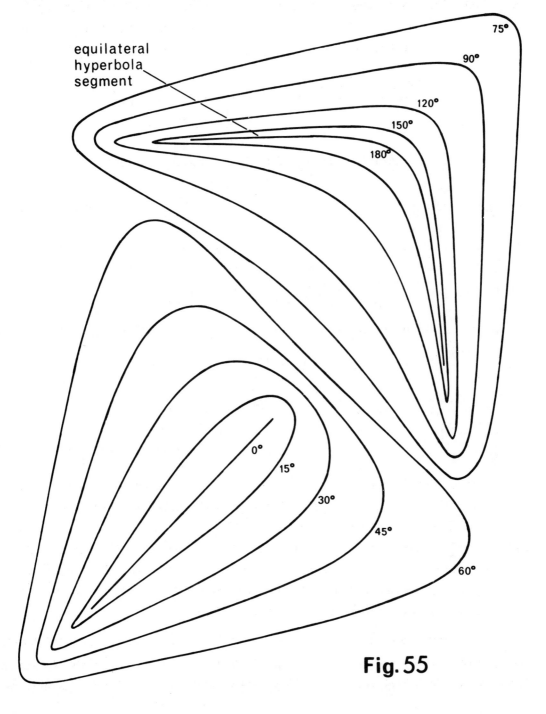

Fig. 55

the same time, the overall degree (including parameters) reduces from 6 to 3. In the cases of the centers, the degree of the 90°-structure rule reduces from 4 to 2 for the equilateral hyperbola--again through the vanishing of the 4th-degree term--while the overall degree reduces from 6 to 2. In this case, the resulting "point-circular" structure rule has only imaginary solutions, inasmuch as there is no real null solution. For every real intercept of value X, the corresponding Y-intercept is imaginary and has the value $Y = iX$ ($i^2 = -1$).

α-Structure Rules and Focal Conditions for Variable Probe-Angles

If one examines the 8th-degree α-structure rules 296 for central conics about their centers (Problem 11), one finds the several focal conditions (conditions for reduction) already considered, plus an additional condition of exceptional interest. When $\alpha = 180°$ or $0°$, one obtains the structure rules of classical symmetry, $X = Y$ and $X = -Y$, respectively (the former for two positive or two negative intercepts, the latter for one positive and one negative intercept). For $\alpha = 90°$, one obtains eqs. 295, which were considered above. But in addition to these focal conditions for specific angles, focal conditions also exist for the X^4Y^4-term alone (eqs. 297a,b) that are dependent upon the value of the probe-angle α. Thus, when $\cos\alpha = \pm(2-e^2)/e^2$, the 8th-degree term vanishes from structure rule 296, with consequent reduction of degree to a maximum of 6.

α-structure rules of central conics about their centers

$[(b^2 \mp a^2)^2 - (b^2 \pm a^2)^2 \cos^2\alpha] X^4Y^4 \sin^2\alpha + a^4b^4(X^4+Y^4) =$ (upper alternate signs for (296)
$2a^2b^2(b^2 \mp a^2)(X^2+Y^2)X^2Y^2 \sin^2\alpha + 2a^4b^4X^2Y^2(\cos^2\alpha - \sin^2\alpha)$ hyperbolas)

(a) $\cos\alpha = \pm(b^2+a^2)/(b^2-a^2)$ reduction condition on eq. 296 for ellipses (297)

(b) $\cos\alpha = \pm(b^2-a^2)/(b^2+a^2) = \pm(2-e^2)/e^2$ reduction condition on eq. 296 for hyperbolas

structure rule of hyperbolas about their centers for a probe-angle equal to the asymptote-angle

$$8(b^2-a^2)(X^2+Y^2)X^2Y^2 = 16a^2b^2X^2Y^2 + (a^2+b^2)^2(X^2-Y^2)^2 \qquad (298)$$

The probe-angle condition 297a cannot be fulfilled for ellipses because the magnitude of the cosine function cannot exceed unity. For hyperbolas, however, condition 297b defines the angle between the asymptotes (Problem 12). In other words, the structure rule of hyperbolas about their centers reduces to a maximum of 6th-degree whenever the probe-angle equals the asymptote-angle. Accordingly, focal condition 297b is referred to as a *variable probe-angle focal condition*. The corresponding structure rule, obtained by substituting condition 297b in eq. 296, is eq. 298. This is 6th-degree for hyperbolas of every eccentricity except $\sqrt{2}$ (for a 90°-probe-angle), for which it reduces to 4th-degree.

But the center is not the only circumpolar focus of a hyperbola for which a variable probe-angle focal condition exists. Consider, for example, the α-structure rule 299 for hyperbolas about their traditional foci (in which all the alternate signs are true alternates). In addition to the conventional reductions for probe-angles of 0°, 90°, 180°, and 270°, this eq. reduces from 4th-degree to 3rd-degree when the probe-angle is equal to the asymptote-angle --one of the relatively infrequent occurrences of a 3rd-degree structure rule (Problem 13).

In fact, the α-structure rule of hyperbolas about an arbitrary point on the transverse axis is found to reduce from 16th-degree to a maximum of 14th-degree for a probe-angle equal to the asymptote-angle, and it seems likely that the α-structure rules about all foci and focal loci of hyperbolas reduce for this same condition.

α-structure rule of hyperbolas about the traditional foci

$$b^4(X^2-2XY\cos\alpha+Y^2) + [2a^2(1\pm\cos\alpha)-(a^2+b^2)\sin^2\alpha]X^2Y^2 = \pm 2ab^2XY(X+Y)(1\pm\cos\alpha) \quad (299)$$

This phenomenon illustrates one of the very few known cases where differences in signs of terms of eqs. representing properties of hyperbolas and ellipses lead to conditions for the one type of curve that have no real or imaginary homologue for the other type (because the cosine function cannot exceed unity).

Hyperbola Axial-Vertex Inversion Cubics

Introduction

As the name implies, the *hyperbola axial-vertex inversion cubics* are the cubic curves obtained by inverting hyperbolas about their axial vertices. A fascinating recent finding is that all the vertex inversions of any curve obtained by inverting a hyperbola about a point on its transverse axis yield similar curves. In other words, similar curves are obtained by inverting, for example, the equilateral limacon about its axial vertices, the equilateral lemniscate about its axial vertices, the equilateral strophoid about its axial vertex, and the equilateral hyperbola about its axial vertices, namely, equilateral strophoids. It follows that the equilateral strophoid self-inverts about its axial vertex. This phenomenon is simply an expression of the recent finding embodied in RULE 56.

RULE 56. *Similar curves are obtained by inverting an initial curve A about a given pole B and by inverting any inverse curve A' about the pole B', inverse to B (or by inverting A' about a pole inverse to any pole equivalent to B).*

Consider the initial equilateral limacon of Fig. 56 ($r = a - \sqrt{2}a\cos\theta$), of which sixteen incident inversions are shown. When the limacon is inverted about its DP (the pole for $\theta = 45°$), the equilateral hyperbola is produced (the curve labelled "45°"). The curve inverse to the equilateral hyperbola about its axial vertices is known to be the equilateral strophoid. But the axial vertices of the equilateral hyperbola are inverse to the axial vertices of the initial equilateral limacon. It follows from RULE 56 that the curves inverse to the initial equilateral limacon about its axial vertices--curves labelled "0°" and "180°" (for $\theta = 0°$ and $180°$)--also are equilateral strophoids.

If one inverts a hyperbola about its left axial vertex (see Problem 40 of Chapter IV), letting the parameter a be the unit of linear dimension, one obtains the equilateral strophoid of eq. 300a (in the orientation depicted in Figs. 24a and 56-180°). For convenience, this is expressed in the standard form 300b, in which the ratios of the parameters are related as in eq. 300c. The vertex cubic parameter b is the diameter of the loop, that is, the distance from the DP to the loop vertex, while the parameter a is the distance from the DP to the *asymptote-point* (the foot of the asymptote on the line of symmetry; see Fig. 24a).

(a) $y^2(x+a^3/2b^2) = x^2(a/2-x)$ equilateral strophoid (conic parameters)   (300)

(b) $y^2(x+a) = x^2(b-x)$ equilateral strophoid (hyperbola axial-vertex inversion cubic parameters)

(c) $a/b = (a^3/2b^2)/(a/2) = a^2/b^2$ relations between the ratios of the parameters
    eq. 300b   eq. 300a   eq. 300a

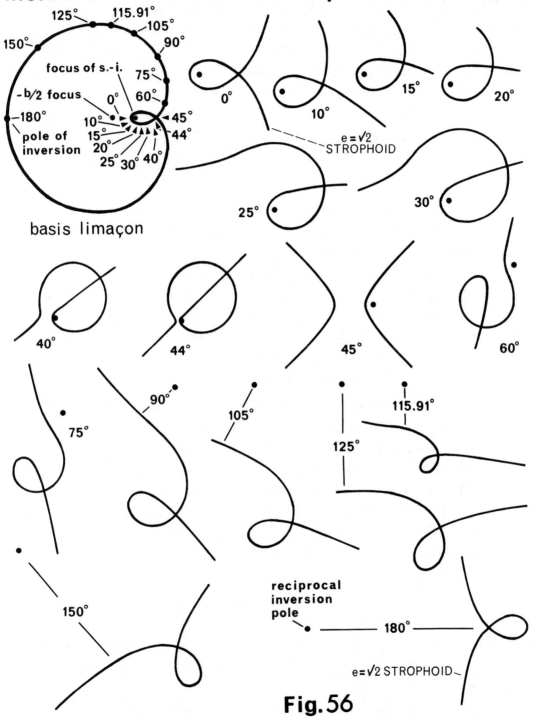

Fig. 56

The Simple Intercept Equation and Format

Following the usual procedures (employing an x+h translation), the simple intercept eq. and format of hyperbola axial-vertex inversion cubics about an arbitrary point on the line of symmetry are found to be eqs. 301 and 302. These appear to be formidable on first sight. However, one of the most intriguing aspects of circumpolar symmetry analyses is that the most formidable-appearing eqs. undergo remarkable reductions and simplifications for the points about which curves possess great structural simplicity. Since these are the points of greatest interest, extensive reductions and simplifications in structure-rule analyses are the rule, rather than the exception.

simple intercept eq. of hyperbola axial-vertex inversion cubics about an arbitrary point on the line of symmetry

$$(2h-a-b)X^2\cos^2\theta + [X^2+h(3h-2b)]X\cos\theta + (a+h)X^2 - h^2(b-h) = 0 \qquad (301)$$

simple intercept format derived from equation 301

$$\cos\theta = \frac{-(3h^2-2bh+X^2) \pm \sqrt{(3h^2-2bh+X^2)^2 - 4(2h-a-b)[(a+h)X^2-h^2(b-h)]}}{2(2h-a-b)X} \qquad (302)$$

Inspection of the simple intercept eq. 301 immediately reveals three focal conditions. For $h = b$, which translates the curve of eq. 300b to the left and places the origin at the loop vertex (which also is a focus of self-inversion), the constant term vanishes, an X can be factored from the eq., and the corresponding structure rules reduce. Similarly, for $h^2 = 0$, which leaves the origin at the DP, not only does the constant term vanish, but all the coefficients simplify and an X^2-term can be factored from the eq., with correspondingly greater simplifications and reductions of the DP-structure rules. Since h occurs as the square in the reduction condition, the DP is a *double-focus*.

The third evident focal condition is specified by the vanishing of the coefficient of the $\cos^2\theta$-term, which occurs for $h = (a+b)/2$. This is the *$\cos^2\theta$-condition*, the same condition that specifies the -b/2 focus of limacons and the p_z pole of Cartesian ovals (see Chapters IV-VI). With the vanishing of the $\cos^2\theta$-term, eq. 301 can be solved directly for the $\cos\theta$-simple intercept format without the need for a quadratic-root solution. The absence of a radical in this format leads to structure rules of greatly reduced degree. The focus at $h = (a+b)/2$ is termed the *variable focus*. Since, like the limacon -b/2 focus, it is within the loop in some members (b > a, e > √2), incident upon the loop vertex in the equilateral strophoid (b = a, e = √2), and exterior to the loop in the other members (b < a, e < √2), it is a *penetrating variable focus*.

Another focal condition that is evident from inspecting the simple intercept eq. is of a type not previously encountered. This is the condition for the vanishing of the coefficient, (a+h), of the X^2-term. This condition, h = -a, translates the curve of eq. 300b to the right and places the origin at the *asymptote-point focus* (see Fig. 24a). Accordingly, conditions revealed by inspecting the simple intercept eq. 301 account for four discrete axial foci--the DP, the loop vertex and focus of self-inversion (which are in coincidence), the variable focus, and the asymptote-point focus (see Table 12).

A fifth discrete pole, with only subfocal rank (the structure rules simplify but do not reduce) is the loop-pole at h = 2b/3. This is defined by the vanishing of the constant portion of the coefficient (3h-2b) of the cosθ-term of eq. 301. This pole achieves focal rank in the trisectrix of Maclaurin (e = 2), in which the variable focus comes into coincidence with it. Thus, we find again, as for hyperbolas and limacons, that the member of the group whose eccentricity is 2 has exceptional structural simplicity (see also below) in some respect.

Structure Rules

A cogent illustration of the implications of the above-described focal and subfocal conditions for the structural simplicity of hyperbola axial-vertex inversion cubics about various points in their planes is readily available. This is provided by an examination of the consequences of these conditions for the forms of the 8th-degree 0° and 180°-structure rules (303) of these curves for an arbitrary point on the line of symmetry.

0° and 180°-structure rules of hyperbola axial-vertex inversion cubics about an arbitrary point on the line of symmetry (upper alternate sign, 0°)   (303)

$(a+h)X^4Y^4 \mp h[h(b-h)+2(a+h)(3h-2b)]X^3Y^3 - h^2(b-h)(X^2+Y^2)X^2Y^2 + h^2(3h-2b)^2(a+h)X^2Y^2 \pm$
$h^3(3h-2b)(b-h)(X^2+Y^2)XY \pm h^4(3h-2b)^2(b-h)XY + h^4(b-h)^2(2h-a-b)(X\pm Y)^2 = 0$

$$X^2Y^2(XY \mp b^2)^2 = 0 \qquad \text{eq. 303 for h = b} \qquad (304)$$

For $h = -a$, the position of the asymptote-point, the 8th-degree X^4Y^4-term vanishes--with immediate reduction of eqs. 303 to 6th-degree--the X^2Y^2-term vanishes, and a portion of the coefficient of the X^3Y^3-term vanishes. For $h = 0$, the position of the DP, all terms of eqs. 303 but the first (left-most) vanish, leading to the trivial incident 0° and 180°-structure rule $XY = 0$. For $h = b$, the position of the loop-vertex and focus of self-inversion, four of the seven terms of eqs. 303 vanish and the coefficient of the X^3Y^3-term simplifies, leading directly to the degenerate 8th-degree structure rules 304. This represents both the trivial structure rule $XY = 0$, for the loop-vertex considered merely as an incident point, and the 0° and 180°-structure rules of self-inversion, $XY = \pm b^2$, for the loop-vertex considered as a focus of self-inversion (the 180°-structure rule is for one positive and one negative intercept).

For the $\cos^2\theta$-condition $h = (a+b)/2$, the condition that defines the variable focus, the $(X \pm Y)^2$-term vanishes. This leads to immediate reduction of structure rules 303 to 6th-degree by the factoring out of an XY-term. Lastly, for $h = 2b/3$, the position of the loop-pole, the X^2Y^2-term, the (X^2+Y^2)XY-term, and the XY-term vanish, and the coefficient of the X^3Y^3-term simplifies, but the structure rules (303) do not reduce. It is clear from these latter influences that the loop-pole at $h = 2b/3$ possesses exceptional positional symmetry, albeit at the subfocal level. In fact, the homologue of this pole possesses focal rank in certain other cubics (see Table 12 and Kavanau, 1980).

The Double-Point Focus

The DP, about which hyperbola axial-vertex inversion cubics possess the simple intercept format 305, is the highest-ranking circumpolar focus. The 0° and 180°-structure rule $XY = 0$, yielded by this format, is not trivial in the same sense as for the incident poles of quadratic curves. On the contrary, since the DP is the only incident point that possesses solely a trivial 0° and 180°-structure rule, this feature reveals a very special symmetry property of the curves about this point.

It is the 90°-structure rule 306a that identifies the DP as the highest-ranking focus. This structure rule is only 6th-degree, compared to a minimum of 8th-degree for the other foci, and 22nd-degree for a point in the plane. The DP attains its highest subfocal rank in the equilateral strophoid ($e = \sqrt{2}$), for which the 90°-structure rule attains the simple form 306b; it attains its second-highest subfocal rank in the trisectrix of Maclaurin (eq. 306c; $e = 2$).

$$\cos\theta = [X \pm \sqrt{(X^2+4a^2+4ab)}]/2(a+b) \quad \text{simple intercept format for the DP} \quad (305)$$

90°-structure rules of hyperbola axial-vertex inversion cubics about the DP

(a) $(X^2+Y^2)X^2Y^2 - [(a-b)^2-4a^2]X^2Y^2 - b^2(X^2+Y^2)^2$ general 90°-structure rule about the DP   (306)
$\quad + 2b(a^2-b^2)(a-b)(X^2+Y^2) - (a^2-b^2)^3 = 0$

(b) $\quad (X^2+Y^2)X^2Y^2 = a^2(X^2-Y^2)^2$ eq. 306a for the equilateral strophoid

(c) $(X^2+Y^2)X^2Y^2 + 96a^4(X^2+Y^2) = 9a^2(X^2+Y^2)^2 + 256a^6$ eq. 306a for the trisectrix of Maclaurin

The Loop-Vertex Focus

Next-highest ranking is the focus at the loop-vertex, for which the simple intercept format is eq. 307. The 0° and 180°-structure rules of self-inversion for this focus already have been noted (eqs. 304). The 90°-structure rule, which generally is 14th-degree, greatly simplifies and reduces to 8th-degree for the equilateral strophoid (eq. 308). This is accompanied by a great simplification of the simple intercept format to eq. 307b (the latter must be derived from eq. 301, since the denominators of eqs. 302 and 307a vanish). [Recall that as regards structural simplicity about more than one reference element, that is, as assessed by construction rules, the loop-vertex of the equilateral strophoid outranks the DP, though at the subfocal level (compare the hybrid distance eqs. 151 and 152).]

(a) $\cos\theta = \dfrac{\{-(b^2+X^2) \pm \sqrt{[(b^2-X^2)^2+4a^2X^2]}\}}{2(b-a)X}$ simple intercept format about the loop-vertex   (307)

(b) $\cos\theta = -2aX/(a^2+X^2)$ eq. 307a for the equilateral strophoid

90°-structure rule of the equilateral strophoid about the loop-vertex

$$X^4Y^4 + a^4(X^4+Y^4) + a^8 = 2a^6(X^2+Y^2) + 12a^4X^2Y^2 + 2a^2(X^2+Y^2)X^2Y^2 \quad (308)$$

The Variable Focus

The variable focus (the $\cos^2\theta$-condition focus) at $h = (a+b)/2$, is next-highest ranking on the basis of 0° and 180°-structure rules but generally is higher ranking than the loop-vertex for 90°-structure rules (Table 12). As noted above, its simple intercept format 309 involves no radical and leads to

Table 12. Degrees of Structure Rules of Cubics about Points

Focal locus	DP		Loop-vertex focus[5]		Variable focus		Asymptote-point		Loop-pole at 2b/3		Line of symmetry	
Probe-angle	0° & 180°	90°	0° & 180°	90°	0° & 180°	90°	0° & 180°	90°	0° & 180°	90°	0° & 180°	90°
Initial curve												
hyperbola axial-vertex cubics	1[1]	6	2	14	4	12	6	22	8	22	8	22
cissoid of Diocles	1[1]	6	*	*	**	12	**	22	*	22	8	22
equilateral strophoid	1[1]	6	2	8[3]	2[2]	8[2]	6	22	8	22	8	22
trisectrix of Maclaurin	1[1]	6	2	14	4	12	6	22	4[3]	12[3]	8	22
folium of Descartes	1[1]	-	6	-	none		10	-	8	-	12	-
Tschirnhausen's cubic	1[1]	-	12	-	-	-	12[4]	none	-	-	-	-

* same as cusp (DP) by virtue of coincidence, **no real solution except trivial
[1] trivial, $X = Y$, [2] double-focus by virtue of coincidence with loop-vertex
[3] reduces by virtue of coincidence with the variable focus
[4] structure rule for polar center; there is no asymptote-point
[5] a double-focus of self-inversion in the conic axial-vertex inversion cubics

---

Table 13. Degrees of Structure Rules of Limacons about Points

	probe-angle	0°	180°	90°
a point in the plane		56	56	72
a point incident upon the curve		44	44	60
a point on the $h = -b/2$ linear focal locus		20	20	72
the $-b/2$ peripheral vertices		12	12	< 60
a point on the line of symmetry		10	10	24
the $-(a+b)$ and $(a-b)$ axial vertices		6	6	16
the $-b/2$ focus		4	4	10
the $-b/2$ focus in the $e = 2$ member		2	2	2
$(a^2-b^2)/2b$ focus of self-inversion (two positive intercepts)		2	3	6
the DP		1	1	2

the relatively simple 0° and 180°-structure rules 310.

$\cos\theta = [h^2(b-h)-(a+h)X^2]/X[X^2+h(3h-2b)]$ simple intercept format about the variable focus   (309)

$4(a+b)(b^2-a^2)(X^2 \pm XY+Y^2) \mp 4(a+b)(9a^2-b^2)XY$
$+ (a+b)^2(b^2-a^2)(3a-b) - 16(3a+b)X^2Y^2 = 0$ 0° and 180°-structure rules about the variable focus (upper alternate sign, 0°)   (310)

$X^6Y^6 + 24a^4X^2Y^2(X^4+Y^4) =$
$\qquad 9a^2(X^2+Y^2)X^4Y^4 + 16a^6(X^6+Y^6)$ 90°-structure rule for the trisectrix of Maclaurin   (311)

The 90°-structure rule is only 12th-degree, compared to 14th-degree for the loop-vertex. It greatly simplifies for the trisectrix of Maclaurin, in which the variable focus is in coincidence with the loop-pole at 2b/3. This difference in the relative focal ranks of the loop-vertex and the variable focus, depending on the probe-angle, emphasizes the fact that circumpolar structure rules have the power to differentiate structural simplicity on the basis of a criterion of angular relations, whereas construction rules lack the ability to make such discriminations.

On the other hand, the distinction in this instance is not a good example of discriminations on the basis of structure rules for specific angles. As noted in Chapter VII, foci of self-inversion generally possess point-focal rank only for the angle of self-inversion; for other probe angles they usually are unexceptional.

[In fact, classicists included poles of self-inversion in the category of foci only when they were in coincidence with classically-defined foci; otherwise they were referred to merely as "centers."]

The Focus at the Asymptote-Point

Next-highest ranking is the focus at the asymptote-point, $h = -a$. The 0°-structure rule is only 6th-degree (eq. 312), compared to 8th-degree about arbitrary axial points (eq. 303). On the other hand, reduction of the 90°-structure

$X^3Y^3 + (X^2+Y^2)X^2Y^2 = a(3a+2b)(X^2+Y^2)XY$
$+ a^2(3a+2b)^2XY - a^2(a+b)(3a+2b)(X+Y)^2$ 0°-structure rule about the asymptote-point   (312)

rule from the value for arbitrary points on the line of symmetry (22nd-degree) has not been demonstrated (Table 12). Accordingly, it is possible that the asymptote-point does not possess point-focal rank for 90°-probes. Were this to be

412 STRUCTURE RULES ABOUT POINTS

confirmed, it would provide a much better example than the one mentioned above of the ability of circumpolar structure rules to differentiate between structural simplicity on the basis of a criterion of angle.

It is evident from eq. 312 that there can be no member of the hyperbola axial-vertex inversion cubics whose 0°-structure rule about the asymptote-point reduces from 6th-degree. This is because all the coefficients involve only parameter *sums* (in which case no term can vanish for any special value of eccentricity). Furthermore, only minor increases in subfocal rank are possible.

The Loop-Pole at $h = 2b/3$

The loop-pole at $h = 2b/3$ has focal rank in some other looped cubics (see Table 12). Among the hyperbola axial-vertex inversion cubics it is a focus only in the trisectrix of Macluarin, in which it is in coincidence with the variable focus. The 8th-degree 0° and 180°-structure rules 313a are obtained by letting $h = 2b/3$ in eq. 303. Only for the trisectrix of Maclaurin ($b = 3a$, $e = 2$) does it reduce to 4th-degree (eq. 313b). The corresponding structure rule for the equilateral strophoid (313c) is unexceptional. Note that the terms on the right of all three structure rules lack being a perfect square because the coefficient of the XY-term is unity, instead of 2. The general 90°-structure rule is complicated, and 22nd-degree. These results are summarized in Table 12 together with those for several other cubics of interest.

0° and 180°-structure rules of hyperbola axial-vertex inversion cubics about the loop-pole at $h = 2b/3$

(a) $16(b/3-a)b^6(X \pm Y)^2/27^2 + (2b/3+a)X^4Y^4 = 4b^3X^2Y^2(X^2 \pm XY + Y^2)/27$ general   (313)

(b) $\qquad 3X^2Y^2 = 4a^2(X^2 \pm XY + Y^2)$ eq. 313a for the trisectrix of Maclaurin

(c) $5 \cdot 3^6 X^4Y^4 = 32a^6(X \pm Y)^2 + 4 \cdot 3^4 a^2 X^2 Y^2 (X^2 \pm XY + Y^2)$ eq. 313a for the equilateral strophoid

Limacons

Introduction

Limacons are a remarkable and fascinating group of curves, closely related to conics by inversion, and having many homologies with conics, but possessing, in addition, extraordinary circumpolar symmetry properties that cannot exist in curves of 2nd-degree. Unfortunately, limacons have been a neglected group of curves from the standpoint of traditional analytical geometry; until recently (see Kavanau, 1980, 1982), there existed in the literature no treatment of them that could be characterized as anything but cursory. The most extensive classical treatment (Zwikker, 1950) devoted 9 of 11 pages to the cardioid. The present treatment outlines the most salient circumpolar symmetry properties of limacons.

Simple Intercept Equations and Formats

Analysis of the circumpolar symmetry of limacons is begun with the standard rectangular eq. 314a of the curve about the DP. This is subjected to x+h and y+k translations. Assuming h and k to be positive, these place the origin at the point (h,k) and displace the DP downward and to the left, to the position (-h,-k). After converting to polar coordinates, one obtains eq. 314b. To obtain [In employing x+h and y+k translations, emphasis is shifted to the location of the probe point in the plane of the initial curve, namely (h,k), rather than the the new location of the DP. The favored classroom practice is to employ x-h and y-k translations, which move the curves into convenient locations in the 1st-quadrant, with the DP at (h,k).]
the simple (mixed-transcendental) intercept eq., r is replaced by X, the left member is squared, and terms are regrouped, yielding eq. 315.

(a) $(x^2+y^2-bx)^2 = a^2(x^2+y^2)$ initial limacon with the DP at the origin   (314)

(b) $[r(b+2h)\cos\theta+2kr\sin\theta+r^2+(h^2+bh+k^2)]^2$
$= a^2[r^2+2r(h\cos\theta+k\sin\theta)+(h^2+k^2)]$ eq. 314 after x+h and y+k translations, conversion to polar coordinates and grouping of terms

simple intercept eq. of limacons about a point in the plane

$$X^2[(b+2h)^2-4k^2]\cos^2\theta + 2X[(b+2h)(X^2+h^2+bh+k^2)-a^2h]\cos\theta +$$
$$4kX(X^2+h^2+bh+k^2-a^2/2)\sin\theta + 4kX^2(b+2h)\sin\theta\cos\theta +$$
$$X^4 + X^2(2h^2+2bh+6k^2-a^2) + (h^2+bh+k^2)^2 - a^2(h^2+k^2) = 0$$ (315)

simple intercept eq. of limacons about a point on the line of symmetry (k = 0)

$$X^2(b+2h)^2\cos^2\theta + 2X[X^2(b+2h)+h(2h^2+3bh+b^2-a^2)]\cos\theta +$$
$$X^4 + [2h(b+h)-a^2]X^2 + h^2[(b+h)^2-a^2] = 0$$ (316)

The constant terms of eq. 315 define the first (and weakest) focal condition, for they make up the eq. of the initial curve in h and k, that is, with h and k replacing x and y; these terms vanish when the point (h,k) is incident upon the curve, whereupon an X factors out of the eq. and all structure rules reduce. For the condition h = -b/2 (for a point on a line normal to the line of symmetry at the level of the -b/2 focus), the cross-product term in $\sin\theta\cos\theta$ vanishes, greatly simplifying subsequent analytical operations and leading to reductions in the degrees of structure rules of the curve about probe-points on the line in question. Accordingly, this line is a focal locus. [Because the line bears no apparent exceptional relationship to the curve (i.e., it is neither a line of symmetry, a tangent, nor an asymptote, etc.), it is referred to as a *covert linear focal locus*.]

For k = 0, for points on the line of symmetry (the x-axis or 0°-axis), eq. 315 simplifies and both the $\sin\theta$ and $\sin\theta\cos\theta$-terms vanish, yielding eq. 316 and still further reductions in the degrees of structure rules. For this condition, a simple intercept format (the solution of eq. 316 for $\cos\theta$) can be obtained in the explicit form of eq. 317.

simple intercept format of limacons about a point
on the line of symmetry

(a) $\quad \cos\theta = \dfrac{a^2h - (b+2h)(X^2+h^2+bh) \pm a\sqrt{[b(b+2h)X^2+h^2(a^2-2bh-b^2)]}}{X(b+2h)^2}$ (317)

(b) $\quad \cos\theta = (a-X)/b \quad\quad$ simple intercept format about the DP

(c) $\cos\theta = [a^2(X^2+b^2/4)-(X^2-b^2/4)^2]/a^2bX \quad$ simple intercept format about the -b/2 focus

(d) $\cos\theta = [4b^3X-4b^2X^2-(a^2-b^2)^2]/4a^2bX \quad$ simple intercept format about the focus of self-inversion

As examples of some of the structure-rule degree reductions, the general 0°-structure rule about incident points is 44th-degree, compared to 56th-degree for the general structure rule about a point in the plane. About a point on the covert linear focal locus, the structure rule reduces to 20th-degree, while the reduction is to 10th-degree about a point on the line of symmetry (see Table 13). Since the covert linear focal locus intersects the curve, and the curve itself is a focal locus, one expects these points of intersection to be point-foci, in which case the structure rule would have to reduce to below 20th-degree. This proves to be the case, since the 0°-structure rule in question is 12th-degree.

Attention now is directed to the simple intercept eq. 316 and format 317a about points on the line of symmetry. For h = 0, the constant term of eq. 316 vanishes and the coefficients of all other terms simplify. This is the location of the DP, the highest-ranking circumpolar focus. It will be noted that the parameter h occurs as the square in the constant term, which means that the DP ranks at least as a double-focus (counting as a single-focus for each intersection with the line of symmetry). The corresponding simple intercept format is eq. 317b, which will be recognized as the common form of the polar construction rule for limacons.

Another high-ranking focus is defined for h = -b/2. This is the $\cos^2\theta$-condition, which defines the -b/2 focus, which has been referred to frequently in previous analyses. The corresponding simple intercept format is eq. 317c, which also involves no radical. Note that the (b+2h) portion of the coefficient of the $\cos^2\theta$-term of eq. 316 is squared, which defines the -b/2 focus as a double-focus. Whereas one readily perceives the geometrical basis for the DP having at least the status of a double-focus (actually it is a quadruple focus, since two trivial foci of self-inversion are coincident with it), the basis for the -b/2 focus also being a double-focus is obscure.

To understand the multiple focal status of the -b/2 double-focus, one must consider the ensemble of curves obtained by inverting central conics about all non-incident axial poles, the *conic axial-inversion quartics*. For example, the curves obtained by inverting hyperbolas about all poles lying between the vertex and the traditional focus of the same arm possess two $\cos^2\theta$-condition foci (recall also that there are two such foci--the loop-foci--in the central inversions, for example, the equilateral lemniscate). As the poles of inversion approach the traditional focus, these foci also approach one another in the inverse curves. They come into coincidence in hyperbolic limacons, which are the curves obtained in inversions about the traditional foci. For poles of inversion beyond a focus, two $\cos^2\theta$-condition foci also exist in the inverse curves, but they are imaginary.

Similar circumstances apply to the inverse loci of ellipses, in which the two foci are imaginary for poles of inversion of the initial ellipses between the center and a focus, real and in coincidence for poles of inversion at the focus, and real and diverging from one another for increasingly distant post-focal poles (at the vertices, one focus goes off to infinity).

A third point-focus is defined by the radicand constant-condition, which is the condition for the vanishing of the constant term, $h^2(a^2-2bh+b^2)$, of the radicand of eq. 317a. Equating the parenthetical expression to 0 defines the focus of self-inversion at $h = (a^2-b^2)/2b$. The corresponding simple intercept format is eq. 317d, which, like the formats 317b,c, involves no radical.

[Angle-independent foci of self-inversion of other quartics that lack a true center also typically are defined by radicand constant-conditions (recall that in the corresponding simple intercept format of central conics, this type of condition defines the traditional foci). On the other hand, foci of self-inversion that are not angle independent—those that possess focal rank solely by virtue of the property of self-inversion of the curve about them—typically are defined by structure-rule conditions, i.e., conditions for which the structure rule reduces to $XY = $ constant.]

When the constant term of the radicand of eq. 317a is considered in the context of being the term that defines the foci of self-inversion of limacons, its h^2-component is seen to define the two foci of self-inversion (of the corresponding biovular Cartesians) that come into coincidence (see Fig. 22e') and become trivial at the DP of the limacon curve of demarcation (see Fig. 21-2). Since the DP already is a double-focus by virtue of the two intersections with the line of symmetry, it is most properly regarded as a *quadruple-focus*, as mentioned above and in Chapter VII.

[Similar circumstances apply to the DP of the equilateral lemniscate, considered as a curve of demarcation of Cassinian ovals. This DP forms by the coming into coincidence of two vertices with the center. But since the center is already a triple-focus (it is a focus by virtue of each mode of self-inversion of the biovals plus being the intersection of two lines of symmetry), a *quintuple-focus* is formed. The two foci of intersection of the curve with the axes are defined by the constant-condition of the simple intercept eq., the center by the $\cos\theta$-condition of the same eq. (h is a multiplicative component of the coefficient of the $\cos\theta$-term), and the trivial foci of self-inversion by the h^2-component of the radicand constant-condition (see Problem 15).]

Two additional conditions that define foci are obtained by equating the bracketed portion of the constant term of eq. 316 to 0. This condition defines the locations of the vertices, $h = (a-b)$ and $-(a+b)$. Eq. 318 is the corresponding simple intercept format (upper sign for the vertex at $h = -(a+b)$). These complete the identification of the foci on the line of symmetry of a limacon. All eqs. apply to both elliptical and hyperbolic members, since the initial eq. 314a applies to all representatives (without the need for alternate signs).

simple intercept format about the axial vertices

$$\cos\theta = \frac{[(b\pm 2a)X^2 \pm a(a\pm b)^2] \pm a\sqrt{[(a\pm b)^4 - b(b\pm 2a)X^2]}}{(b\pm 2a)^2 X} \qquad (318)$$

Structure Rules

Points on Focal Loci

Eliminating θ from eqs. 316 and 317a leads to complicated 12th-degree degenerate eqs. for the 0° and 180°-degree structure rules about points on the line of symmetry. A term that represents the trivial 0° and 180°-structure rules, $(X \mp Y)^2$, can be factored out of these eqs., leaving a structure rule of 10th-degree. An additional XY-term will factor out, with reduction to a maximum of 8th-degree, if a certain 6th-degree eq. for the vanishing of a coefficient is satisfied. Since this eq. represents a focal condition at the level of a structure rule (as opposed to the level of a simple intercept eq. or format), it is referred to as a *structure-rule condition*.

When the coefficient referred to above is equated to 0 and factored, one obtains the degenerate eq. 319. On equating the various factors to 0, this is seen to represent four of the five axial foci, each with its proper weight as a single or double-focus. The fact that the focus of self-inversion is not defined by this eq. is yet another indication that foci of self-inversion possess a somewhat different status than other circumpolar point-foci.

$$h^2[h-(a+b)][h+(a+b)](h+b/2)^2 = 0 \quad \text{0°-structure-rule condition for axial foci of limacons} \quad (319)$$

$$36b^8 - 76b^6a^2 + 53b^4a^4 - 14b^2a^6 + a^8 = 0 \quad \text{structure-rule condition for the non-focal pole at } (b^2-a^2)2b \quad (320)$$

A highly interesting and illustrative condition on structure rules for a pole that is not a focus can be derived from the 0° and 180°-structure rules for the pole at $(b^2-a^2)/2b$. This is the point conjugate to the focus of self-inversion (the reflection of the focus of self-inversion in the DP). With changes in the eccentricity of elliptical limacons, this pole ranges all along the line of symmetry from the (a-b)-vertex to the -(a+b)-vertex, and exterior to it to the axial point at infinity. Concomitantly, the pole of self-inversion ranges from the (a-b)-vertex and exterior to it in the other direction to the axial point at infinity.

Since the structure rule for the conjugate pole must reduce at each position at which it comes into coincidence with one of the axial foci, the roots of eq. 320 must define the limacons of different eccentricities in which these coincidences take place. The eight roots of eq. 320 are found to be $b = \pm a$,

±√2a/2, ±√2a/2, and ±a/3 (Problem 16). Only the positive roots are valid, since parameters a and b are positive by definition. The first of these roots, b = a, defines the cardioid, in which the conjugate pole is in coincidence with three of the five axial foci--the DP, the (a-b)-vertex, and the otherwise non-trivial focus of self-inversion, all of which are in coincidence at the cusp-point.

The double-root b = √2a/2 defines the coincidence with the -b/2 double-focus in the e = √2/2 limacon, and the root b = a/3 defines the coincidence with the -(a+b)-vertex focus in the e = 1/3 limacon. These same structure-rule conditions for reduction (eqs. 319 and 320) can be extracted from the 90°-structure rule about points on the line of symmetry, for which the minimum reduction is from 20th to 16th-degree.

In the case of the -b/2 covert linear focal locus, the 0° and 180°-structure rules are derived by eliminating $\sin\theta$ from the simple intercept eq. (eq. 315 with h = -b/2) by means of the 0° and 180°-compound intercept formats (Problem 17). These are complicated eqs. of 20th-degree, compared to the 56th-degree structure rules about a point in the plane, making this a high-ranking focal locus. For the -b/2 peripheral vertices (the points of intersections of this focal locus with the curve), of which there are four real representatives for e > 2, and two real and two imaginary representatives for e < 2, the structure rules reduce to 12th-degree. The corresponding 90°-structure rule for this focal locus is very complex, and a reduction from 72nd-degree (the same value as for a point in the plane) has not been established (see Table 13).

The Double-Point Focus

Employing the simple intercept format 317b for the DP yields the α-structure rule 321a, which is one of the simplest in existence for curves of finite degree. The corresponding 0°, 180°, and 90°-structure rules for two positive intercepts are eqs. 321b,c,d, respectively, of which that for 0° is trivial. (see note on page 315 regarding positive and negative intercepts). As pointed

structure rules of limacons about the DP

(a) $(X-a)^2 + (Y-a)^2 - 2(X-a)(Y-a)\cos\alpha = b^2\sin^2\alpha$ α-structure rule   (321)

(b) $X = Y$ (c) $X + Y = 2a$ 0° and 180°-structure rules

(d) $(X-a)^2 + (Y-a)^2 = b^2$ 90°-structure rule

out earlier, eq. 321c represents one of the rare cases in which an inherent property of a curve embodied in a structure rule (aside from the classical symmetries) can be detected on visual inspection of the curve. It also is one of the examples of curves of finite degree that have the greatest linear structural simplicity about a point (see Table 10 and Problem 19). For a 90°-probe, the structure rule is circular (eq. 321d), while for all other probe-angles the structure rules are elliptical (Problem 20).

The Focus of Self-Inversion

Eqs. 322a,b,c, respectively, are the 0°, 180°, and 90°-structure rules of limacons about the focus of self-inversion for two positive intercepts (obtained by substituting from the simple intercept format 317d into the structure-rule format 245a). The 0°-structure rule represents the relation of self-inversion. The right term of eq. 322a is the square of the distance of the focus of self-inversion from the DP, an obligatory value for self-inverting curves with double-points.

Aside from the structural simplicity of the circle about its center--which ranks highest--the degree of structural simplicity of the limacon about this focus in the equilateral-hyperbolic category for foci of self-inversion (see Table 10) ranks second only to that of conic axial-vertex inversion cubics about their foci of self-inversion. It even exceeds the degree of structural simplicity of the circle about a point in the plane and that of Cassinian ovals about their centers (of course, the latter ovals are more symmetrical about their centers in an absolute sense, since they possess the linear 180°-structure rule of classical rotational symmetries).

$$\text{structure rules of limacons about the focus of self-inversion}$$

(a) $XY = (a^2-b^2)^2/4b^2$ 0°-structure rule   (322)

(b) $XY(2b-X-Y) = (a^2-b^2)^2(X+Y)/4b^2$ 180°-structure rule

(c) $16b^4(X^2+Y^2)X^2Y^2 - 8b^3[(a^2-b^2)^2+4b^2XY](X+Y)XY$ 90°-structure rule
$+ 16b^4(3b^2-2a^2)X^2Y^2 + (a^2+b^2)^4(X^2+Y^2) = 0$

The 180°-structure rule is unusual in being of 3rd-degree. The 6th-degree 90°-structure rule is a symmetric function. The corresponding complete structure rule represents either an oval in each quadrant (initial elliptical limacons) or two non-congruent intersecting ovals (initial hyperbolic limacons). In the elliptical members, the oval becomes progressively more circular as the focus approaches the (a-b)-vertex and e approaches 1. At the vertex, it becomes a circle (eq. 321d), since the initial curve becomes the cardioid, in which the focus of self-inversion (now trivial) is in coincidence with the cusp-point.

The $Cos^2\theta$-Condition Double-Focus

From the point of view of the inherent structure of limacons about this focus, it is the most extraordinary focus known to me, with structure rules and structural curves (see Fig. 42c) of compelling interest. Since limacons are the curves of demarcation of Cartesian ovals, which possess a homologous focus, the latter curves possess similar inherent structural properties (see Kavanau, 1980 and below). The structure rules 241 and 323a-c are obtained in the usual manner from the simple intercept format 317 and structure-rule format 245a.

$0°$ and $180°$-structure rules of limacons about the
$-b/2$ double-focus

(a) $(a^2+b^2/2)XY + (b^2/4 - a^2)b^2/4 = (X^2+XY+Y^2)XY$ $0°$-structure rule   (323)

(b) $(a^2+b^2/2)XY - (b^2/4 - a^2)b^2/4 = (X^2-XY+Y^2)XY$ $180°$-structure rule

(c) $0°$, $X^2+XY+Y^2 = 3a^2$, $180°$, $X^2-XY+Y^2 = 3a^2$ eqs. 323a,b for $e = 2$

The $-b/2$ double-focus is within the large loop for $e < 2$ (Fig. 12f$_2$, p$_z$), at the small-loop vertex for $e = 2$ (Fig. 12d$_2$), and within the "small" loop for $e > 2$ (Fig. 42c). The structural curves take their simplest forms for $e = 2$. The 4th-degree $0°$-structure rule 323a represents recurrent-arc segments of curves that resemble ellipses (the *near-lines*). For $e = 2$ (eq. 323c) the curve is an ellipse of eccentricity $e = \sqrt{(2/3)}$.

The remarkable property of the $0°$-structural curves is their very close approach to linearity for values of e greater than 2 (a location of the double-focus within the small loop). This phenomenon is dependent upon the facts that the midpoint of the segment traced by the structural curve is an extremal point, at which the slope changes from negative to 0 to positive (a minimum), and that the changes become more and more gradual with increasing e. For large e, the entire segment traced by the structural curve comes to occupy the very much flattened region of the extremal point.

Concomitantly with these changes in the 0°-structural curves, the corresponding segments traced by the 180°-structural curves increasingly resemble circles (the *near-circles*). For e = 20, the curve appears circular to the naked eye, with an axial ratio of 1.0006. However, unlike the near-lines and near-ellipses, which continue to approximate the forms of lines and ellipses with increasing e, the near-circles do not become circular in the limit.

Quite remarkable and much more complex changes occur in the 90°-structural curves (eqs. 241) with changes in eccentricity. As e increases to 1, three-sided ovals evolve toward a boomerang shape with two cusps. For e > 1, the two cusps become double-points with satellite loops, that progressively enlarge as the focus approaches the vertex of the small loop. For e = 2, the structural curves are a circle and two concentric, orthogonal ellipses (eq. 241b'), already discussed in Chapter VII. When the focus enters the small loop, these curves break up into a pair of intersecting ovals in each quadrant, and these increasingly take on the shapes of ellipses with increasing e. At e = 8, for example, the naked eye cannot distinguish these curves from true ellipses (Fig. 42c).

The Axial-Vertex Foci

Use of the simple intercept format 318 with the structure-rule format 245a yields the structure rule for the -(a+b) and (a-b)-vertices. Eqs. 324 are the 0°-structure rules (upper signs for the -[a+b]-vertices), which are seen to be 6th-degree. The 180°-structure rules differ only in the signs of terms. The 90°-structure rules are 16th-degree and are very complicated. Table 13 summarizes the degrees of limacon structure rules.

$$0°\text{-structure rules of limacons about the} \\ -(a+b) \text{ and } (a-b)\text{-vertices}$$

$$(b\pm2a)^3(X-Y)^2X^2Y^2 + 4a(b\pm2a)^2[abXY\mp(a\pm b)^2(X-Y)^2]XY + \quad (324)$$

$$4a^2(b\pm2a)(a\pm b)^2[(a\pm b)^2(X^2-3XY+Y^2)\mp 2abXY] + 4a^3(a\pm b)^4[ab\pm 2(a\pm b)^2] = 0$$

Parabolic Cartesians

The initial rectangular construction rule of bipolar parabolic Cartesians, $z^2 = Cdu$, with the monoval-bioval focus of self-inversion p_u at the origin, is eq. 325. Converting to polar coordinates, replacing r with X, and solving for $\cos\theta$, yields the simple intercept format 325b about focus p_u. The corre-

deriving structure rules of parabolic Cartesians

(a) $(x^2+y^2-2dx+d^2)^2 = C^2d^2(x^2+y^2)$ rectangular eq. of parabolic Cartesians, $d = d_{uz}$ (325)

(b) $\cos\theta = (X^2 \pm CdX+d^2)/2dX$ simple intercept format about focus p_u

(c) $\cos\theta = [X^4-C^2d^2(X^2+d^2)]/2C^2d^3X$ simple intercept format about focus p_z

complete 0° and 180°-structure rules of parabolic Cartesians about the monoval-bioval focus of self-inversion, p_u

(a) $(X-Y)^2(d^2-XY)^2[(X-Y)(d^2-XY)-2CdXY][(X-Y)(d^2-XY)+2CdXY] = 0$ for 0°   (326)

(b) $(X+Y)^2(d^2+XY)^2[(X+Y)(d^2+XY)-2CdXY][(X+Y)(d^2+XY)+2CdXY] = 0$ for 180°

(c) $XY = d^2$ 0°-structure rule of self-inversion for two positive intercepts

(d) $(X+Y)(d^2+XY) = 2CdXY$ 180°-structure rule for two positive intercepts

sponding simple intercept format about focus p_z is eq. 325c. Format 325b includes both roots of the polar eq. obtained from eq. 325a (representing two plots of the ovals 180° out of phase). Since this format allows for both positive and negative intercepts, employing it in conjunction with the structure-rule format 245a leads to the complete 0° and 180°-structure rules 326a,b, which differ only in the signs of terms.

In the case of eq. 326a, the first squared term represents the trivial 0°-structure rule, the second squared term represents the 0°-structure rule of self-inversion (eq. 326c) for two positive or two negative intercepts, and the other two terms represent the 0°-structure rules for one positive and one negative intercept (the first for X negative and Y positive, and the second for X positive and Y negative). If the sign of either X or Y is changed, these become the corresponding 180°-structure rules (326b), since complete 0°-structure rules with the sign of one variable changed are identical with complete 180°-structure rules, and vice versa.

The first squared term of eq. 326b represents the trivial 180°-structure rule, the second squared term represents the 180°-structure rule of self-inversion for either intercept positive and the other negative (in which case they are directed at the same angle), the third represents the structure rule for two positive intercepts, and the fourth term represents that for two negative intercepts. Conventionally, the structural simplicity of a curve about a point is characterized by the structure rule for two positive intercepts, whereupon eq. 326c is the 0°-structure rule of self-inversion and eq. 326d is the 3rd-degree 180°-structure rule.

For the $\cos^2\theta$-condition-focus p_z (the homologue of the limacon $-b/2$ double-focus), the structure rules 327a-c are the homologues of the limacon near-conic structure rules treated in the preceding section and Chapter VII. The corresponding structural curves have unusual properties that parallel those of the limacon near-conic structural curves (see Kavanau, 1980, and compare with eqs. 323a,b and 241, respectively).

structure rules of parabolic Cartesians about focus p_z

(a) $\qquad XY(X^2+XY+Y^2) = C^2d^2XY - C^2d^4 \qquad$ for 0° (327)

(b) $\qquad XY(X^2-XY+Y^2) = C^2d^2XY + C^2d^4 \qquad$ for 180°

(c) $X^2Y^2[(X^6+Y^6)-2C^2d^2(X^4+Y^4)+C^2d^4(C^2-2)(X^2+Y^2)] + C^4d^8(X^2+Y^2) = 0$, for 90°

Of course, biovular parabolic Cartesians are simply special cases of the linear Cartesians treated below, for which limacons are the curves of demarcation (recall, though, that linear bipolar eqs. cannot represent monovular Cartesians, since the latter possess only one real bipolar point-focus, whereas two such foci are required for linearity). Accordingly, the corresponding structure rules and structural curves of linear Cartesians about the $\cos^2\theta$-condition focus (it is shown below that this also is a double-focus) also are homologues of those of limacons. Though more complex, these structure rules of linear Cartesians include the parabolic Cartesian representatives as special cases.

Linear Cartesians

Taking the bipolar construction rule of linear Cartesians, $\pm Bu \pm Av = Cd$, as the initial eq., converting to rectangular coordinates (eq. 105a), performing an x+h translation, converting to polar coordinates, and replacing r by X yields the simple intercept eq. 328 about points on the line of symmetry. When this is expanded and arranged in powers of $\cos\theta$, the condition for vanishing of the $\cos^2\theta$-term is found to be $h = A^2d/(A^2-B^2)$, which gives the location of the $\cos^2\theta$-condition focus, p_z. Substituting this value in eq. 328 and solving for $\cos\theta$ yields the corresponding simple intercept format 329d. This is the format that generates the linear Cartesian near-conic homologues.

simple intercept eq. of linear Cartesians about
points on the line of symmetry

$$[(X^2+2hX\cos\theta+h^2)(A^2-B^2) - 2A^2d(X\cos\theta+h) + d^2(A^2-C^2)]^2 = \qquad (328)$$
$$4B^2C^2d^2(X^2+2hX\cos\theta+h^2)$$

simple intercept format of linear Cartesians about
the foci of self-inversion

(a) $\quad \cos\theta = [(A^2-B^2)X^2+(A^2-C^2)d^2 \pm 2BCdX]/2A^2dX \qquad$ focus at $x = 0 \qquad (329)$

(b) $\quad \cos\theta = [(A^2-B^2)X^2+(C^2-B^2)d^2 \pm 2ACdX]/2B^2dX \qquad$ focus at $x = d$

(c) $\quad \cos\theta = \dfrac{[(A^2-B^2)^2X^2-(C^2-A^2)(B^2-C^2)d^2 \pm 2AB(A^2-B^2)dX]}{2(A^2-B^2)C^2dX} \qquad$ focus at $x = (A^2-C^2)d/(A^2-B^2)$

simple intercept format of linear Cartesians about
the $\cos^2\theta$-condition focus

(d) $\quad \cos\theta = \dfrac{[d^2(A^2B^2+A^2C^2-B^2C^2) - X^2(A^2-B^2)^2]^2 - 4B^2C^2d^2[(A^2-B^2)^2X^2 + A^4d^2]}{8A^2B^2C^2d^3X(A^2-B^2)}$

The simple intercept formats for the foci of self-inversion are obtained in a similar manner by letting $h = 0$, d, and $(A^2-C^2)d/(A^2-B^2)$, as per the derivation leading to eqs. 111 of Chapter IV. The 0°-structure rules obtained from these formats are eqs. 330a-c, respectively. The alternate signs depend both upon the specific values of the parameters, A, B, and C, and on whether the ovals self-invert about the focus in question in the 0° or 180°-mode (for two positive intercepts).

structure rules of self-inversion for linear
Cartesians

(a) $\quad XY = \pm(A^2-C^2)d^2/(A^2-B^2) \qquad$ about focus at $x = 0 \qquad (330)$

(b) $\quad XY = \pm(B^2-C^2)d^2/(B^2-A^2) \qquad$ about focus at $x = d$

(c) $\quad XY = \pm(B^2-C^2)(A^2-C^2)d^2/(A^2-B^2)^2 \qquad$ about focus at $x = (A^2-C^2)d/(A^2-B^2)$

(d) $\quad (B^2-A^2)^4(X^2\pm XY+Y^2)XY = 2d^2(A^2B^2+A^2C^2+B^2C^2)\cdot \qquad$ 0° and 180°-structure rules about the $\cos^2\theta$-condition focus
$\qquad (B^2-A^2)^2XY + d^4[(A^2B^2+A^2C^2-B^2C^2)^2 - 4A^4B^2C^2]$

Eq. 330d is the 0°-structure rule for the $\cos^2\theta$-condition focus. The fact that it is more complex than eqs. 330a-c and of 4th-degree shows that the structural simplicity of linear Cartesians (see also eqs. 326c,d versus 327a,b) is much greater about the foci of self-inversion than about the $\cos^2\theta$-condition focus. This also was found to be true of the corresponding structural simplicities treated in Chapters IV-VI.

The finding that the three foci of self-inversion of Cartesian ovals are circumpolar foci in their own right, rather than merely for the angles of self-inversion, is implicit in the fact that their simple intercept formats 329a-c involve no radical, and in RULE 22. The latter RULE states that "all bipolar foci also are circumpolar foci." These foci, like the foci of self-inversion of limacons, are specified by radicand constant-conditions. Since no probe-angle is specified at this level of analysis (as opposed to the circumstances for foci of self-inversion that are defined by *structure-rule conditions*), the status of foci so-defined is not restricted to the angle of self-inversion.

The derivation of these poles as circumpolar foci is of interest as another example of the power of a circumpolar symmetry analysis to define foci whose definition (and detection) presented difficulties to classicists. This was the case, for example, for a third focus of self-inversion of Cartesian ovals. Its discovery by Chasles (without an existence proof) in the 19th century surprised classicists, who had been unaware of its existence from the time of discovery of the ovals by Descartes in the 17th century. In fact, as illustrated in Chapter III for the traditional foci of conics, these foci can be defined at the level of polar eqs., expressed in the form of $\cos\theta$-simple intercept formats. Thus, if eq. 328 is expanded and solved for $\cos\theta$ (see Kavanau, 1980, and below), the term involving the radical in the quadratic-root solution is expression 331.

radical-term of solution of eq. 328 for $\cos\theta$

(a) $\pm 2ABCd\sqrt{\{[A^2d-(A^2-B^2)h]dX^2+(A^2-B^2)dh^3+(B^2+C^2-2A^2)d^2h^2+(A^2-C^2)d^3h\}}$ (331)

(b) $(A^2-B^2)dh^3+(B^2+C^2-2A^2)d^2h^2+(A^2-C^2)d^3h = 0$ radicand constant-condition of expression 331a

$h = 0$, $h = d$ the three roots of eq. 331b defining the   (332)
                      three circumpolar foci of self-inversion
$h = (A^2-C^2)d/(A^2-B^2)$

When the constant term of the radicand vanishes, a root can be taken, with the vanishing of the radical and reduction of the corresponding structure rules.

Hence the three roots of the radicand constant-condition 331b define the three circumpolar foci of eqs. 332. As with all known circumpolar foci of 4th-degree curves defined in this manner, these are foci of self-inversion (either trivial or non-trivial).

The three foci of self-inversion, together with the $\cos^2\theta$-condition focus, and the four axial vertices (defined by the vanishing of the constant term of eq. 328) are the eight discrete circumpolar foci defined by conditions on the simple intercept eqs. and formats of Cartesian ovals (the radicand constant-condition of Cartesian monovals also define three foci of self-inversion but two of them are imaginary; Problem 23 and below).

RULE 57. *Radicand constant-conditions on simple intercept formats, for which the radical vanishes, define foci of self-inversion of 4th-degree rectangular curves.*

It will be noted that a common reduction condition exists for both the simple intercept format and the 0°-structure rule of linear Cartesians about the $\cos^2\theta$-condition focus. Thus, when the constant term of the numerator of eq. 329d vanishes, an X can be factored out of the numerator and denominator--a certain reduction condition. Similarly, when the constant term of the 0°-structure rule 330d vanishes, an XY can be factored out (representing the trivial incident structure rule XY = 0), with reduction of the degree from 4 to 2. This common condition (eq. 333) is the *vertex-coincidence symmetry condition*, that is, the condition that defines a coincidence between the $\cos^2\theta$-condition focus and an axial vertex of the small oval. When this condition is satisfied, the 0° and 180°-structure rules become eqs. 334a. A 45°-rotation of axes (eqs. 334 b,c) reveals the structural curves to be ellipses.

$(A^2B^2+A^2C^2-B^2C^2)^2 = 4A^4B^2C^2$	the vertex-coincidence symmetry condition	(333)
(a) $(X^2 \pm XY+Y^2) = \dfrac{2d^2(A^2B^2+A^2C^2+B^2C^2)}{(B^2-A^2)^2} = G^2$	0° and 180°-structure rules for condition 333	(334)
(b) 0°, $3X^2 + Y^2 = 2G^2$, 180°, $X^2 + 3Y^2 = 2G^2$	eq. 334a after 45° CCW rotation of axes	

Accordingly, striking homologies exist between biovular Cartesians and hyperbolic limacons as concerns their structure rules for the $\cos^2\theta$-condition foci. Since the biovals in which the focus becomes coincident with a vertex of the small oval are homologues of the e = 2 limacon, the corresponding structural curves also are homologues. In evidence of this, it is found that the complete

0° and 180°-structure rules of these linear Cartesians represent congruent, orthogonal, centered ellipses, just as do those of the $e = 2$ limacon (eq. 323c), even to the point of the ellipses having the same eccentricity [$e = \sqrt{(2/3)}$] and also being represented only by recurrent-arc segments.

The 0° and 180°-structural curves of eqs. 330d are completely homologous with those of limacons. For example, the former also are the eqs. of near-lines that become more nearly linear as the two ovals become more nearly equal in size and the $\cos^2\theta$-condition focus becomes more nearly centered in them. Similarities also are found for the homologous 90°-structural curves. But, since Cartesian biovals do not come into contact, the structural curves cannot become unicursal (continuous) and only "partial ellipses" form.

Central Cartesians

Introduction

As mentioned earlier, aside from *central quartics* (the curves obtained by inverting central conics about their centers), central Cartesians (Figs. 20-1, 2; 22a,c,f) are the quartic curves most closely related to central conics. Until recently, these central curves were known only through their most-symmetrical representatives (members with the greatest bipolar and circumpolar symmetry), the Cassinian ovals.

Central Cartesians are the curves that invert to Cartesian ovals about certain axial poles (see Problems 24-26), and Cassinian ovals are the subgroup whose members invert to parabolic Cartesians (see Fig. 22). The latter, in turn, are the most-symmetrical Cartesian ovals, whose curve of demarcation, in turn, is the most-symmetrical limacon (the equilateral limacon).

Central Cartesians, central conics, and central quartics are grouped together into a broad new category of curves that share a common property, itself indicative of a very high degree of circumpolar symmetry. These are the *central curves with 4th-degree-maximum inverse loci* (see Kavanau, 1982). In other words, no inverse locus of these curves possesses a rectangular eq. of higher degree than 4. By virtue of these homologies with central conics, the circumpolar symmetry properties of central Cartesians are of compelling interest.

Central Cartesians not only self-invert about their centers in both 0° and 180°-modes (the biovular members), they fall into three groups whose members invert to one another about the four axial poles that are at *modular* distance from center (i.e., at the distance of the unit of linear dimension for the self-inversions about the center). This heretofore totally unknown phenomenon is illustrated in Fig. 22, in which curves a, c, and f derive from one another by inversions about their four modular poles on the lines of symmetry.

Axial Vertices, Products of Roots, and
　　Self-Inversion

The initial eq. for central Cartesians centered at the origin, with their lines of symmetry in coincidence with the rectangular axes (eq. 126a, repeated here for convenience as eq. 335a), takes the polar form 335b. Solving for r^2 yields eq. 335c. The axial vertices now are located directly by letting $\theta = 0°$ for the x-axis and $\theta = 90°$ for the y-axis, yielding eqs. 336a,b.

It is readily evident from eqs. 336 that, when D^4 is negative, there are two real and two imaginary vertices on both axes, in which case the eq. represents one real oval and one imaginary oval. In order that four real vertices exist on both axes (annular ovals--one oval inside the other; Fig. 22a), B^2 and C^2 must be negative and D^4 must be positive but less than both $B^4/4$ and $C^4/4$. If the other conditions apply but only B^2 is negative, the two ovals are opposed and intersect the x-axis (Fig. 22c), and similarly for the y-axis if C^2 is negative (Fig. 22f). [If any of the prohibited conditions of eq. 335a applies, the curves become two real or imaginary circles (see Problem 31).]

[Recall that the exponents of capital-letter symbols denote dimensionality; for example, if the coefficient is B^2, B has the dimension of length squared; B^4 then represents the square of B^2, with the dimension of length to the 4th-power.]

(a) $(x^2+y^2)^2 + B^2x^2 + C^2y^2 + D^4 = 0$　　central Cartesians ($B^4 \neq 4D^4$, $C^4 \neq 4D^4$, $B^2 \neq C^2$, B^2, C^2, $D^4 \neq 0$)　(335)

(b) $r^4 + r^2(B^2\cos^2\theta + C^2\sin^2\theta) + D^4 = 0$　　polar form of eq. 335a

(c) $r^2 = \{-(B^2\cos^2\theta + C^2\sin^2\theta) \pm \sqrt{[(B^2\cos^2\theta + C^2\sin^2\theta)^2 - 4D^4]}\}/2$　　eq. 335b solved explicitly for r^2

(a) $r^2 = [-B^2 \pm \sqrt{(B^4-4D^4)}]/2$　　x-axis roots (vertices) of eqs. 335　(336)

(b) $r^2 = [-C^2 \pm \sqrt{(C^4-4D^4)}]/2$　　y-axis roots (vertices) of eqs. 335

$r_1r_2 = \sqrt{D^4} = \pm D^2$, for $D^4 > 0$　　product of the roots of eq. 335c of different absolute values　(337)

If one considers the products of the two roots of eq. 335c, they are found to be constant at $\sqrt{D^4}$ (eq. 337). Accordingly, if D^4 is positive, the curves self-invert in the 0°-mode (along coincident radii) about their centers. Since they then consist of either opposed or annular ovals, a further consequence is that they also self-invert in the 180°-mode (along oppositely-directed chord-segments). Furthermore, since 180°-foci of self-inversion are conserved in all inversions, it also follows that every curve inverse to a biovular central Cartesian also self-inverts in the 180°-mode (any inverse curve also must possess at least one, but not more than three, 0°-foci of self-inversion).

[Any modification of eq. 335a, aside from those producing translations and/or rotations, destroys the property of self-inversion about the center, even though the new curves also possess a true center. Thus, addition of a term in x^2y^2 does not alter the locus from being a central curve, but the self-inversion property is lost; the product of the roots no longer is constant, and the curve inverts to an 8th-degree locus. The addition of a term in Axy^2, $Ax(x^2+y^2)$, etc., destroys the central self-inversion property because a line of symmetry is lost.]

The Simple Intercept Equation and
Format, and the Cartesian Term

The simple intercept eq. 338a and format 338b of central Cartesians about points on the x-axis are derived from the initial eq. 335a in the usual manner, using x+h and y+k translations. Examination of eq. 338a reveals the existence of three focal conditions that define seven discrete foci as follows. The coefficient of the $\cos^2\theta$-term vanishes for $h^2 = (C^2-B^2)/4$, defining two $\cos^2\theta$-condition foci at $h = \pm\sqrt{(C^2-B^2)}/2$. The $\cos\theta$-condition, $h = 0$, for which the $\cos\theta$-term vanishes, defines the center as a focus. The vanishing of the constant term, which allows an X to be factored out, defines the four x-axis vertices (not all of which necessarily are real) as foci.

simple intercept eq. and format of central Cartesians
about points on the x-axis

(a) $(4h^2+B^2-C^2)X^2\cos^2\theta + 2hX[2(X^2+h^2)+B^2]\cos\theta +$ (338)

$$X^4 + (C^2+2h^2)X^2 + (h^4+h^2B^2+D^4) = 0$$

(b) $\cos\theta = \dfrac{-h[2(X^2+h^2)+B^2] \pm \sqrt{[(C^2-B^2)X^4 + (C^2-B^2)(C^2-2h^2)X^2 + (C.T.)]}}{(4h^2+B^2-C^2)X}$

(c) $(C.T.) = h^4(C^2-B^2) + h^2(B^2C^2-4D^4) + D^4(C^2-B^2)$ the *Cartesian term*

(d) $h^2 = \dfrac{(4D^4-B^2C^2) \pm \sqrt{[4D^4-B^2C^2]^2 - 4D^4(B^2-C^2)^2}}{2(C^2-B^2)}$ poles of the *Cartesian condition*

When the simple intercept format is examined, a radicand constant-condition is found, but it does not satisfy RULE 57 or define circumpolar foci. This is because even though the constant term vanishes, terms in both X^4 and X^2 remain. Accordingly, a root cannot be taken, and the radical does not vanish (simultaneous vanishing of both the constant term and either the X^4 or X^2-term leads to the initial curve being composed of circles). Even though an X^2 can be factored from the radicand when the constant term vanishes, an X cannot be factored from the numerator and denominator of the format. Accordingly, the vanishing of the constant term is not a focal condition on either account.

The constant term of the radicand, expression 338c, has been designated as the *Cartesian term*, because, when equated to 0 (the *Cartesian condition*), it defines the four axial poles (see Figs. 22a,c; 23a_1,a_3) about which central Cartesians invert to Cartesian ovals. For the reasons just given, these generally are only subfocal, that is, the poles defined by the Cartesian condition generally are not circumpolar foci.

[The corresponding radicand constant-term for central conics, the *limacon term*, when equated to 0, defines the traditional foci, about which conics invert to limacons.]

The Cartesian condition also can be derived by inverting central Cartesians about an axial pole h and translating the origin to any one of the foci of self-inversion of the inverse locus. When the corresponding simple intercept format of this inverse curve is examined, the radicand is found to contain terms in X^4, X^2, and a constant. In order for this intercept format to represent Cartesian ovals, it must take the form of one of the eqs. 329a-c (in which no radical is present). But this cannot occur unless both the X^4 and constant terms of the radicand vanish. The condition for this to occur is found to be the same for both terms, namely, the Cartesian condition (see Problems 25 and 26). Through an analysis of eqs. 335 and 338d, it can be shown that every central Cartesian, whether monovular or biovular, inverts to a Cartesian oval.

The Cartesian Condition, Circumpolar Symmetry Analyses, and the Inversion Transformation

The existence of the Cartesian condition reveals an implicit relation between the manner of employment of the polar coordinate system in circumpolar symmetry analyses, and the transformation of inversion. Thus, poles defined by the vanishing of the Cartesian term are exceptional in only two respects; (a) central Cartesians have greater structural simplicity about them at the subfocal level (the structure rules about them are simpler but not of lower degree than those about neighboring axial points), and (b) the curves obtained by inverting central Cartesians about them are Cartesian ovals, the most symmetrical of all the inverse loci about non-focal poles.

In other words, the most notable respect in which the poles identified through a circumpolar symmetry analysis are exceptional is that the inverse curves about them are Cartesian ovals. This is the implicit relation referred to above. Viewed from another perspective, circumstance (b) emphasizes the discriminative power of circumpolar symmetry analyses at the subfocal level.

Coincidences of Cartesian-Condition
    Poles with Foci

Inspection of the Cartesian term 338c reveals that the pairs of poles it defines are modular to one another, that is, the product of their two distances from center is the square of the unit of linear dimension for self-inversion about the center. Thus, from eq. 337, the product of the roots is $\sqrt{D^4}$, while from the expression 338c equated to 0, $h_1^2 h_2^2 = D^4$ (derived by taking the ratio of the constant term to the coefficient of the h^4-term), whence $h_1 h_2 = \pm\sqrt{D^4} = \pm D^2$, provided that $D^4 > 0$ (which is the condition for the existence of two ovals). Since the poles are modular to one another and cannot be coincident with the vertices (if they were, inversions about them would produce cubics; Problem 21), the only foci with which they can be coincident are those of the $\cos^2\theta$-condition.

We now consider the identity of the central Cartesians in which these coincidences occur. The curves in question are found to be none other than Cassinian ovals (Problem 22), the most-symmetrical members of the group. This finding calls attention to RULE 58, according to which, in groups of curves that possess variable foci, members in which multiple-foci are formed or the variable foci are in coincidence with poles of high subfocal rank typically are highly symmetrical. Numerous examples of this phenomenon already have been encountered (it occurs, for example, in the $e = 2$ limacon; in all limacons, considered as curves of demarcation of Cartesian ovals; in the equilateral strophoid; in the trisectrix of Maclaurin; in Cartesian ovals characterized by the vertex-coincidence symmetry condition; etc.).

[We already found in Chapter IV that Cassinian ovals are the only central Cartesians in which the $\cos^2\theta$-condition foci are in coincidence with the poles at $x_1^2 = C^2/2$, which are poles of high bipolar (and hybrid polar-rectangular) subfocal rank (see eqs. 129b,c and 131). The latter poles generally are not Cartesian-condition poles (Problem 31), so each $\cos^2\theta$-condition focus of Cassinian ovals now is seen to be in coincidence with two poles of high subfocal rank. Neither coincidence occurs in any other central Cartesian.]

RULE 58. *In groups of curves with variable foci, members in which multiple-foci occur and/or the variable foci are in coincidence with poles of high subfocal rank typically are highly symmetrical.*

Cartesian Ovals and $\cos^2\theta$-Condition
  Double-Foci

It was noted that central Cartesians possess two discrete $\cos^2\theta$-condition foci (eq. 338), whereas Cartesian ovals possess only one (as found by expanding eq. 328). To elucidate the basis for this difference, central Cartesians are inverted about a pole at $x = h$ (the origin of eq. 338) and the simple intercept eq. of the inverse curve is obtained about an axial pole at $x = H$ (Problem 24). Expression 339a is the $\cos^2\theta$-term of this simple intercept eq. It follows from equating expression 339a to 0 that the $\cos^2\theta$-condition foci of the curves obtained by inverting central Cartesians about an axial pole at $x = h$ lie at the two axial points H, so defined (i.e., the two solutions for H obtained in this way).

coefficient of the $\cos^2\theta$-term of curves inverse to
central Cartesians (about a pole at $x = h$) cast
about a pole at $x = H$

(a) $\qquad [4(h^4+h^2B^2+D^4)H^2 + 4hHj^2(2h^2+B^2) + j^4(4h^2+B^2) - C^2j^4]X^2\cos^2\theta \qquad$ (339)

locations of the $\cos^2\theta$-condition foci defined by
eq. 339a

(b) $\qquad H = [-hj^2(2h^2+B^2) \pm j^2\sqrt{(C.T.)}]/2(h^4+h^2B^2+D^4)$

When eq. 339a is solved for H, the radicand is found to consist solely of the Cartesian term (expression 338c). In consequence, both the radical and the radicand vanish whenever the pole of inversion of a central Cartesian is a Cartesian-condition pole. That is, for any x-axis pole of inversion of a central Cartesian other than a pole of the Cartesian condition, the inverse curve possesses two $\cos^2\theta$-condition foci (although one of these goes off to infinity in the vertex cubics). But when a central Cartesian is inverted about a Cartesian-condition pole, these two foci become coincident in the inverse curve, forming a double-focus. Inasmuch as the inverse curves in question are Cartesian ovals, the $\cos^2\theta$-condition foci of Cartesian ovals are double-foci, just as in limacons.

Accordingly, the homologies between central conics and central Cartesians, in this respect, are complete. Of all the inverse loci of central conics about points on the x-axis, only those about the traditional foci--limacons--have their $\cos^2\theta$-condition foci in coincidence as double-foci. These are defined by the radicand constant-condition. Similarly, of all the inverse loci of central Cartesians, only those about the Cartesian-condition poles--Cartesian ovals--have their $\cos^2\theta$-condition foci in coincidence as double-foci. These also are defined by the radicand constant-condition.

Angle-Independent Versus Angle-Dependent
   Foci of Self-Inversion

We consider now the simple intercept format of all curves inverse to central Cartesians (eq. 335a) about poles on the x-axis. To this end, the curves of eq.338a (the simple intercept eq. of central Cartesians about a pole at $x = h$) are inverted and an $x+H$ translation is performed on the inverse loci (bringing the origin to the pole at $x = H$ relative to the pole of inversion). The simple intercept eq. then is derived (Problem 24) and the simple intercept format is formed by solving for $\cos\theta$ (Problem 27). Examination of the radicand of this format reveals that only when the pole of inversion is a Cartesian-condition pole can the radicand define a focus of self-inversion (by the vanishing of the radical through the conversion of the radicand to a perfect square).

In other words, even though all the curves obtained by inverting central Cartesians about non-central points on the x-axis possess three foci of self-inversion (two in the 0°-mode and one in the 180°-mode), only those of Cartesian ovals are defined by a condition on the radicand of the simple intercept format. The foci of self-inversion of the other inverse curves (other than Cartesian ovals) can be divided into two groups; (a) those that are incident, and (b) those that are not incident. A discussion of the curves of group (a) and their foci of self-inversion calls attention to RULE 59 (see Kavanau, 1982). Inasmuch as the axial vertices of central Cartesians occur in modularly-inverse pairs (since biovular central Cartesians self-invert about their centers), it is a consequence of RULE 59 that all the vertex cubics obtained by inverting a central Cartesian about its x-axis vertices are similar.

    RULE 59. *The curves obtained by inverting an initial curve about a pair of modularly-inverse poles (two poles that invert, one to the other, about a focus of self-inversion) are similar.*

Vertex Cubics

In the inversions of central Cartesians about non-central points on the x-axis, the axial vertices of the inverse curves are inverse to the axial vertices of the initial curve (inversions about axial poles carry axial vertices into axial vertices). But according to RULES 56 and 59, similar curves are obtained by inverting an initial curve about a given pole and by inverting any inverse curve about the pole inverse to the given pole (or inverse to a modularly-inverse pole). It follows that curves obtained by inverting the inverse loci of central Cartesians about their axial vertices are similar to the curves obtained by inverting the central Cartesians themselves about their axial vertices.

Thus, all the axial-vertex inversion cubics of all curves inverse to central Cartesians about points on the x-axis (whether incident or not) are similar. This is equivalent to saying that only one x-axis vertex-inversion cubic exists (i.e., the various axial vertex-inversion cubics differ only in their sizes, positions, and orientations) for any given central Cartesian and all the curves obtained by inverting it about points on the x-axis (in fact, similar arguments apply generally to central curves; see Kavanau, 1982).
[Although space does not permit treating the many intriguing properties of the inversion transformation that have been elucidated recently, some of these are included in Appendix II.]

It follows from these considerations that if a central Cartesian is inverted about an x-axis vertex to yield a vertex-inversion cubic, inversions of the vertex-inversion cubic about its own axial vertices must yield the identical vertex-inversion cubic. That is, all three of the axial vertices of vertex-inversion cubics of central Cartesians are foci of self-inversion. Returning to the main theme of our discussion, since the curves in question belong to group (a), it is a consequence of RULE 53 that their vertex foci are defined by the vanishing of the constant term of the simple intercept eq., just as were those of hyperbola axial-vertex inversion cubics. [In fact, no known vertex-focus of self-inversion is defined by a radicand condition on a simple intercept format.]

> RULE 60. *Foci of self-inversion that are in coincidence with vertices are defined by the constant-condition on simple intercept equations, not by conditions on the radicands of simple intercept formats.*

Accordingly, the foci of self-inversion of these vertex cubics are angle-independent circumpolar foci in their own right, even though they are not defined by conditions on the radicands of their simple intercept formats. The implication of these findings is that the foci of self-inversion of the curves of group (b), that is, quartic curves inverse to central Cartesians about x-axis poles other than the centers, the vertices, or those of the Cartesian condition, are not angle-independent (with the exception of the *modular* poles, as discussed below). They appear to be circumpolar foci solely by virtue of the property of, and for the angle of, self-inversion.

Again, there is a close parallel with central conics and their inverse loci. The only angle-independent foci of self-inversion of quartic curves inverse to conics are those of limacons, the homologues and curves of demarcation of Cartesian ovals. These are defined by a radicand constant-condition. The foci of self-inversion of all other self-inverting quartic inversions of conics are angle-dependent, and are defined solely by structure-rule conditions, that is, conditions for which the structure rules themselves reduce. The foci of self-inversion of conic axial-vertex inversion cubics lie at the vertices. Accordingly, they are angle-independent and defined by constant-conditions on their simple intercept eqs.

Locating the Foci of Self-Inversion
    of Cartesian Ovals

Returning to the simple intercept format of the above derivation, and expressing it for the Cartesian condition, that is, for the case in which the initial central Cartesian is inverted about a Cartesian-condition pole, the radicand constant-condition is found to be eq. 340a. Although this eq. is a cubic,

radicand constant-condition of Cartesian ovals
obtained by inverting central Cartesians
about x-axis Cartesian-condition poles

(a) $2hH^3[2h^2(C^2-B^2) + (B^2C^2-4D^4)] + H^2j^2[6h^2(C^2-B^2) + (B^2C^2-4D^4)]$ (340)

  $+ 4hHj^4(C^2-B^2) + j^6(C^2-B^2) = 0$ h, any solution of (C.T.) = 0

locations of the foci of self-inversion of the
Cartesian ovals of eq. 340a

(b) $H = -j^2/2h$, $\quad H = \dfrac{hj^2(B^2-C^2) \pm j^2\sqrt{[(B^2-C^2)(B^2C^2-4D^4) - h^2(B^2-C^2)^2]}}{[(B^2C^2-4D^4) - 2h^2(B^2-C^2)]}$

eq. 340b for parabolic Cartesians (C, of the bipolar eq. of Cassinian ovals uv = Cd²)

(c) $H = -j^2/2h$, $H = [-hj^2 \pm j^2\sqrt{(d^2/2 - h^2 - 4C^2d^2)}]/(2h^2 - d^2/2 + 4C^2d^2)$ (340)

it can be solved readily by virtue of the fact that the location of one of the three foci of self-inversion already is known (first solution in eqs. 340b,c). This follows from RULE 61 (which is a consequence of RULE 56).

RULE 61. *The focus of self-inversion of an inverse locus that replaces a line of symmetry of the initial curve is the pole inverse to the reflection of the pole of inversion of the initial curve in its line of symmetry.*

To derive the known location of this focus of self-inversion, consider that the reflection of the pole of inversion in the line of symmetry (for h positive) is at the distance 2h to the left of the pole of inversion. Accordingly, if j is the unit of linear dimension, this reflected pole inverts to a pole at a distance $j^2/2h$ to the left of the pole of inversion. Accordingly, to bring this pole to the origin for an x+H translation, we need $H = -j^2/2h$. Dividing eq. 340a by the factor $(H+j^2/2h)$ and solving the resulting quadratic eq. in H yields eq. 340b (Problem 28).

For parabolic Cartesians derived from initial Cassinian ovals [h = d/2, B^2 = $-d^2/2$, $C^2 = d^2/2$, $D^4 = d^4(1/16-C^2)$], eq. 340b becomes eq. 340c. It is consistent with our earlier interpretation that, for the equilateral limacon curve of demarcation of parabolic Cartesians (C = 1/4), eq. 340c yields the focus of self-inversion at $(a^2-b^2)/2b$ plus two coincident foci of self-inversion at the DP (Problem 29). These, of course, are trivial.

Ensembles of Inverse Curves of Central Cartesians
   that Possess at Least One Line of Symmetry

The most fascinating properties of central Cartesians are those by which they differ from central conics. Perhaps the most remarkable of these concern the mappings of poles that yield prescribed ensembles of inverse curves (as explained below). In the interests of simplicity, attention is confined to ensembles of inverse curves that possess at least one line of symmetry. In the case of a centered central conic, the inverse curves that possess at least one line of symmetry can be divided into two ensembles. One ensemble is obtained by inverting about all points on the non-negative x-axis, and the other by inverting about all points on the non-negative y-axis.

These two ensembles possess two members in common. One is the central quartic resulting from inversions about the center. The other involves the two loci obtained by inverting about the points where the x and y-axes meet the line at infinity. Though both of these inverse loci are identical with the initial central conic, they are the products of different mappings. These duplications are not to be regarded as trivial, since the two points where the axes intersect the line at infinity are the only points in the plane about which the initial central conic self-inverts. For all other points on the line at infinity, the inverse curves are *congruent-inversions*, that is, inverse curves that are congruent but not coincident with the initial curve.

[The significant distinctions referred to above are lost in systems such as the inversive plane, that replace the line at infinity with a point at infinity.]

As a consequence of the principles embodied in RULE 59, namely, the fact that the curves obtained by inversions about a pair of modularly-inverse poles are similar, axial inversions of biovular central Cartesians give radically different mapping results from those of central conics. One difference is that a complete ensemble of the curves obtained by inverting about all points on the x-axis of central Cartesians can be obtained by inverting only about points on a segment of the x-axis (of finite length). Either segment, from the center to the pole at modular distance to either side of the center along the x-axis, fulfills this requirement (each of these poles at modular distance is modularly inverse to itself in the 0°-mode and to the other in the 180°-mode). The same circumstances apply for the y-axis (Fig. 22a,c,f; modular poles marked on each axis).

Another departure is that only one curve is shared by the x and y-axis ensembles, rather than two, namely, the curve obtained in inversions about the center, which is a self-inversion. The same circumstances apply to the x and y-axis ensembles obtained by inverting about points from the modular poles to the line at infinity, which also comprise a complete set of axial inversions. In this situation, however, the points at which the lines of symmetry meet the line at infinity are modularly-inverse to the center, since the center is a focus of self-inversion. Accordingly, the curves obtained by inverting about the center and about these points are similar (and, in fact, identical).

The two loci obtained in inversions about the modular poles--the one for the x-axis and the other for the y-axis--are neither congruent nor similar. The extraordinary feature in which they resemble one another is that both of them also possess a true center (considering curve a of Fig. 22 as the initial curve, these would be curves c and f). This finding was unprecedented, in being the first known instance in which a curve with a true center did not lose at least one line of symmetry in an inversion about a non-central point. [Equally, or more, startling was the finding that a curve with a true center can self-invert about a point other than its center; see Kavanau, 1982.]

A third departure is a consequence of the principles embodied in RULE 56. This previously unknown inversion-mapping property concerns the locus about which a non-axial curve that is inverse to an axial curve inverts to axial curves. In other words, when a curve with, say, a single line of symmetry loses that line of symmetry in an inversion, where do the points lie, in the plane of the inverse curve, about which the inverse curve inverts to curves with a single line of symmetry?

RULE 56 tells us that the points in question are inverse to the points in the plane of the initial curve about which the initial curve itself inverts to curves with a single line of symmetry. But these are known to be the points incident upon the single line of symmetry of the initial curve. Since these points are known to lie on a circle in the plane of the inverse curve, it follows that the sought-after points lie on the circle inverse to the line of symmetry of the initial curve.

Returning to central Cartesians, it also is a consequence of RULE 56 that a circle must exist in the plane of every biovular central Cartesian (but not of a central conic) about which it inverts to curves with a single line of symmetry. Since one type of central Cartesian (say, Fig. 22a) is obtained from the other two types (say, Fig. 22c,f) by inversion about one of the modular poles, the circle in question is the inverse of the other line of symmetry (the line of symmetry connecting the other pair of modular poles). In fact, this circle is the *modular circle* of the inverse central Cartesian, that is, it is a circle at modular distance from center (shown in Fig. 22f).

Accordingly, the inverse loci of central Cartesians that possess only a single line of symmetry fall into three groups. One group is obtained by inverting, say, Cassinian ovals (Fig. 22c) about non-central, non-modular poles on the x-axis (Fig. 22e,e'), a second group is obtained by inverting them about non-central, non-modular poles on the y-axis, and a third group is obtained by inverting them about poles on the modular circle (excluding the axial modular poles). Cartesian ovals (in this case biovals) belong only to the first group. An ensemble that includes the members of all three groups is obtained without duplication by inverting the given Cassinian bioval about a continuous figure that includes the quarter of the modular circle in the 1st-quadrant plus two radii consisting of segments of the x and y-axes from the origin to the corresponding modular poles.

The Symmetry Condition of Central Cartesians
   Inverse to Cassinian Ovals

It was noted above that Cassinian ovals are the most-symmetrical subgroup of central Cartesians by virtue of the fact that one pair of poles of the Cartesian condition is in coincidence with the $\cos^2\theta$-condition foci. Accordingly, one also expects the two corresponding subgroups of central Cartesians (obtained by inverting Cassinian ovals about their axial modular poles; all three of which groups are encompassed by eq. 335) to be the most-symmetrical members of their corresponding groups.

For these other two subgoups of central Cartesians, however, the poles of the Cartesian condition do not lie on the line of symmetry that includes the $\cos^2\theta$-condition foci. In one case--in which the ovals are annular (Fig. 22a)--they lie on the orthogonal axis (the y-axis), while in the other case--in which the ovals are opposed, like the Cassinians themselves (Fig. 22f)--they lie off the axis on the modular circle. Accordingly, it is not possible for the Cartesian-condition poles of the two subgroups of curves in question to be in coincidence with the $\cos^2\theta$-condition foci.

What, then, are the symmetry conditions in these other subgroups of central Cartesians that are the homologues of the coincidence of the Cartesian-condition poles and $\cos^2\theta$-condition foci of Cassinian ovals? Or, stated otherwise, by virtue of what properties, following RULE 58, are the subgroups of central Cartesians inverse to Cassinian ovals more symmetrical than the members of the groups inverse to other central Cartesians?

The answer is the same for both subgroups. The $\cos^2\theta$-condition foci of the members of the two subgroups of central curves inverse to Cassinian ovals are at modular distance from center, that is, they are in coincidence with the modular poles. For this reason, the curves are called *central modular-focal quartics*. Moreover, the members of these two subgroups invert to one another about these modular-focal poles, whereas they invert to Cassinians about their unexceptional modular poles on the other axis (see Fig. 22a,c,f).

An equally fascinating but remarkably different situation characterizes the inversion properties of monovular Cassinians. Since these self-invert about their centers at 90° (and 270°), a given pair of modularly-inverse poles lie on different lines of symmetry, that is, the modular poles are not in coincidence with themselves (for a treatment of these groups, see Kavanau, 1982). At any rate, the analyses of the circumpolar symmetry properties of central Cartesians give us a glimpse into a vastly different and far richer domain of geometrical relations and properties than those that characterize central conics.

But even these new vistas are only a prelude to the extraordinary features that emerge in studies of central and non-central curves with more than two lines of symmetry, including those with infinitely many lines of symmetry (Kavanau, 1980, 1982).

Problems

1. Identify the bracketed portion of structure rule 258 by rotating the axes to eliminate the XY-term. Determine the eccentricities of the represented structural curves as a function of the probe-angle α.

2. Derive the XY-arc-increment eq. of the circle (eq. 259b) about a point in the plane.

3. Derive the X-arc-increment eq. of the circle about an incident point for an arbitrary probe-angle α (i.e., for the incident structure rule that represents ellipses; eq. 258a, bracketed expression).

4. Derive eq. 263 for the 0° and 180°-structure rules of the parabola about the point (h,-k).

5. Derive the 90°-structure rule (eqs. 239 and 268) for the axial vertex of the parabola.

6. Derive the 90°-structure rule (eq. 269) for the LR-vertices of the parabola.

7. Derive the harmonic-mean eqs. 229a and the analogous 0°-structure rule (229b) for hyperbolas about the traditional foci from eqs. 275 and 276.

8. Derive the 0° and 180°-structure rules of ellipses about the minor-axis imaginary foci at $k^2 = b^2-a^2$, given that the corresponding simple intercept eq. is $X^2(b^2-a^2)\sin^2\theta + 2a^2\sqrt{(b^2-a^2)}X\sin\theta - b^2X^2 + a^4 = 0$. Show that the two imaginary intercepts for $\theta = 90°$ satisfy the 0°-structure rule.

9. Derive the $\cos\theta$-simple intercept formats of ellipses for the conditions, $h^2 = a^2-b^2$, $h = 0$, and $h = a$, from the general major-axis $\cos\theta$-simple intercept format of eq. 278.

10. Derive the 90°-structure rule of ellipses (eq. 293a) about the traditional foci employing the $\cos\theta$-simple intercept format 292a.

11. Derive the α-structure rule 296 for central conics about their centers.

12. If α is the angle between the asymptotes of a hyperbola in the sectors containing the arms, show that $\cos\alpha = \pm(2-e^2)/e^2$.

13. Show that the α-structure rule of a hyperbola about a traditional focus (eq. 299) reduces from 4th to 3rd-degree for a probe-angle equal to the asymptote-angle.

14. Show that the simple Y-intercept format 290b is for a 90°-probe-angle.

15. Show that the constant term (the term not involving X) of the radicand of the simple intercept format of the equilateral lemniscate (eq. 6) about points on the x-axis contains a multiplicative factor h^2.

16. Outline a method of obtaining the roots of eq. 320.

17. Obtain the compound 0°-intercept format for points on the -b/2 covert linear focal locus of limacons from eq. 315.

18. Account for the fact that the 0°-structure rule of limacons about the DP, as obtained from eq. 321, is trivial, whereas a radius vector from the DP to the curve at any angle generally has two different intercepts, one on each loop.

19. Derive the XY-arc-increment eq. of limacons about the DP. What are the X-arc-increment eqs. for 180° and 90°?

20. Show that, in general, eq. 321a for the α-structure rule of limacons about the DP represents ellipses. What is the expression for their eccentricities?

21. Show that the poles of the Cartesian condition of central Cartesians cannot be in coincidence with a vertex.

22. Show that the central Cartesians in which a pair of poles of the Cartesian condition come into coincidence with the $\cos^2\theta$-condition foci are Cassinian ovals. What is the location of the other pair of Cartesian-condition poles?

23. Find the location of the three foci of self-inversion of parabolic Cartesians and show that two of these foci are imaginary in the monovular members. Confirm your results for the C = 5 parabolic Cartesian of Fig. 23.

24. Letting the unit of linear dimension be j, invert the curves of eq. 338a about pole h (which is at the origin) and perform an x+H translation of the resulting loci. What is the corresponding simple intercept eq. about the pole H (which is at the origin)?

25. In the simple intercept eq. of Problem 24, let $H = -j^2/2h$, which places the origin of the inverse curve at the location of the focus of self-inversion that replaces the line of symmetry. Show that the radicand of the resulting simple intercept format becomes independent of X for the Cartesian condition. Derive the simple intercept format that applies when the initial curve is inverted about a Cartesian-condition pole. [h need not be expressed explicitly in the fianl solution. Simply let h define the location of a Cartesian-condition pole.]

26. Evaluate the $\cos\theta$-simple intercept format of Problem 25 on the assumption that the initial central Cartesians are Cassinian ovals (eq. 6h). Compare the simple intercept formats obtained with that of eq. 325b. Assuming that the initial Cassinian oval was the curve of demarcation ($C_{bipolar} = C_{bp} = 1/4$, the equilateral lemniscate), show that the resulting parabolic Cartesian simple intercept format becomes that of the equilateral limacon curve of demarcation cast about its focus of self-inversion.

27. Derive the $\cos\theta$-simple intercept format from the simple intercept eq. of Problem 24.

28. Obtain the quadratic eq. for the other two roots of the radicand constant-condition of eq. 340a by dividing by $H+j^2/2$.

29. Evaluate eq. 340c for an initial equilateral lemniscate curve of demarcation of Cassinian ovals ($C_{bp}=1/4$), that leads to an equilateral limacon curve of demarcation of the resulting parabolic Cartesians, and locate the positions of the foci of self-inversion in the plane of the equilateral limacon.

30. Solve Problem 29 for the $C_{bp}=1/5$ Cassinian ovals, which invert to the $C_{bp}=5$ parabolic Cartesians. Confirm your results from Fig. 23.

31. Show that Cassinian ovals are the only central Cartesians (with non-degenerate eqs.) in which the poles at $x_1=\pm\sqrt{2C}/2$ (see eqs. 129b,c and 130a) are in coincidence with the Cartesian-condition poles. Allowing the eq. (126a or 335a) of central Cartesians to be degenerate, in what other group of curves does this coincidence also occur?

APPENDIX I.   LISTING OF RULES

1a. The greater the generalized symmetry (structural simplicity) of a curve or figure with respect to one or more reference elements, the simpler its structural equations (either *structure rules* or *construction rules*) and the lower their degrees.

1b. The greater the symmetry of a curve or figure with respect to given reference elements, the simpler its structural equations and/or the lower their degrees.

1c. For measurements or instructions of comparable simplicity, the fewer the number needed to determine or specify all distances to points of a curve or figure from one or more reference elements, the simpler and/or lower the degrees of the structural equation that relates these distances.

1d. For measurements or instructions of comparable simplicity, a 50% decrease in the number needed to determine or specify all distances to points of a curve or figure from one or more reference elements, frequently results in a 50% or 75% decrease in the degree of the structural equation that relates these distances. A decrease in number by less than 50% is insufficient to lead to degree reduction.

1e. For curves specified by measurements or instructions comparable or equal in numbers, the greater the simplicity of the measurements or instructions, the simpler and lower the degree of the equation needed to specify them.

2a. The curve with the greatest symmetry of form and position with respect to two reference elements is the locus of the construction rule $X = Y$.

2b. A curve is most-symmetrically positioned with respect to one or more reference elements when its structural equation with respect to the same element(s) assumes its simplest and lowest-degree form.

2c. Of a given group of curves with structural equations with the same form, for example, $Ax^2 + By^2 = Cd^2$, the one with the simplest equation, for example, $x^2 + y^2 = d^2$, has the greatest symmetry of form and position (excluding loci consisting of a finite number of points).

3a. The greater the positional symmetry of a curve, the lower the exponential degree of, and the simpler, its structural equation.

3b. The greater the positional symmetry of a curve, the simpler its construction rule.

4. A curve that includes, or is tangent to, a reference element generally is more-symmetrically positioned in the coordinate system (has a simpler construction rule) than a congruent neighboring curve that does not include, or is not tangent to, the reference element.

5. The magnitude of the slope of a tangent to a conic (in standard position) at its latus rectum vertices (the intersections of the latera recta with the curve) is equal to the conic's eccentricity.

6. The degree in r of the polar construction rule of a curve of finite rectangular degree is defined by the lowest-degree, non-degenerate equation that gives a single complete trace of the curve for a single sweep of values of θ ranging between the limits set by two limiting tangent lines from the polar pole to the curve.

7a. A polar construction rule of a curve derived directly from the curve's rectangular equation about an arbitrary point in the plane (the *basic* polar construction rule) has the same exponential degree in r as the degree of the rectangular equation in x and y.

7b. The maximum exponential degree in r of the polar construction rule of a curve is the exponential degree of the curve's rectangular construction rule.

8a. Simpler polar construction rules tend to yield simpler structure rules.

8b. In general, the simpler the 90°-structure rule of a curve about a point (the structure rule for a 90°-angle between the probe-lines), the simpler the polar construction rule about the same point.

9. Among loci with comparably simple construction rules in coordinate systems with two congruent elements, the most-symmetrical locus is:

   (1) the one for which the coefficients of terms in u and v are identical; and/or
   (2) the one that is symmetrical about the most symmetrical point in the system; and/or
   (3) the one that includes one or more points of the reference elements; and/or
   (4) the one that retains its symmetry in the identical but polarized system.

10. Bipolar point-foci of curves fall on lines of symmetry (as defined in the rectangular and polar systems).

11. Bipolar construction rules typically reduce in degree in steps of 50% or 75%, for example, from 8 to 4 to 2 to 1 and from 16 to 4 to 2 to 1.

12. Bipolar construction rules reduce in degree for each of the following conditions.

   (1) One pole an independent point-focus, the other an arbitrary point in the plane.
   (2) Both poles otherwise unexceptional, but incident upon a line of symmetry of the curve.
   (3) One pole an independent point-focus, the second incident upon the same line of symmetry as the first but otherwise unexceptional.
   (4) Both poles independent point-foci incident upon the same line of symmetry.

13. An independent bipolar point-focus is an individually identifiable point on a line of symmetry which (a) when paired with an otherwise unexceptional pole leads to a construction rule of lower degree than the general bipolar equation, or (b) when paired with an otherwise unexceptional pole on the same line of symmetry leads to a construction rule of lower degree than the general 0°-axial bipolar equation.

14. A line connecting the poles of a bipolar construction rule of exponential degree 1 or 2 is a line of symmetry of the corresponding rectangular curve and a focal locus, that is, a locus upon which any pole-pair has bipolar focal rank.

15. For an array of similarly-positioned congruent bipolar loci, those members that lie between a member of highest symmetry (eq. $u = v$) and a member of 2nd-highest symmetry of the array, possess greater symmetry than the members that do not.

16. Bipolar construction rules of circles are weighted heaviest in favor of the pole nearest the center, which is the most-symmetrically positioned of the two poles.

17. First-degree bipolar construction rules of curves other than lines and circles identify two independent bipolar point-foci and one individually identifiable rectangular line of symmetry. Second-degree construction rules define at least one independent point-focus, or two dependent foci, and one individually-identifiable rectangular line of symmetry.

18. General bipolar construction rules of curves with one line of symmetry undergo major simplifications--without reduction--for both poles located on a line either normal or parallel to the line of symmetry, and for one pole on the line of symmetry.

19. The coefficients of the variables of bipolar construction rules cast about poles on a line normal to a line of symmetry are weighted heaviest in favor of the pole nearest the line of symmetry, that is, the pole nearest the line of symmetry has the greater subfocal symmetry rank and the curve is more-symmetrically positioned about it.

20. In a bipolar construction rule, the focal rank of a pole represented by the square of the resultant of a linear term augmented or diminished by a constant--for example, $(u+d)^2$--generally is greater than the focal rank of a pole for which u or v occurs only as the pure square--for example, u^2.

21. The pole with the greater *weight* in a bipolar construction rule, in which the highest degree of both variables is the same, generally is the pole of higher focal rank.

22. All bipolar foci are circumpolar foci, but not vice versa.

23. The degree of a *general distance equation*--the equation for the distance from a point in the plane to an arbitrary point of an initial curve (with the abscissa of the arbitrary point as the independent variable)--is twice the rectangular degree of the curve.

24. A general distance equation reduces to the corresponding rectangular degree of the initial curve for an arbitrary point on a line of symmetry, giving the *general axial distance equation*.

25. General axial distance equations reduce in degree only through root-taking.

26. Independent bipolar point-foci are defined for any condition for which the degree of a distance equation reduces from the rectangular degree of the initial curve.

27. Independent point-foci are defined by any condition for which the degree of a distance equation reduces from the rectangular degree of the initial curve in coordinate systems employing two points, a point and a line, a point and a circle, a circle and a line, or two circles as reference elements.

28. Vertices of bipolar curves (single intersections of a curve with a line of symmetry) have only subfocal symmetry rank.

29. General bipolar construction rules of curves with a true center undergo major simplifications--without reduction--for both poles located on a line parallel to either line of symmetry & for one pole incident upon either line.

30. The coefficients of the variables of bipolar construction rules cast about non-focal poles on lines parallel to a line of symmetry of a curve with a true center are weighted heaviest in favor of the pole nearest the center, that is, the pole nearest the center has the greater subfocal symmetry rank, and the curve is more-symmetrically positioned about it.

31. Bipolar construction rules that are symmetrical in terms involving the variables u and v represent curves that are either symmetrical with respect to the bipolar midline or for which a center of symmetry lies midway between poles p_u and p_v.

32. If the bipolar construction rule of a curve cast about equivalent poles (poles reflected in a line of symmetry) is of 1st or 2nd-degree, a true center of the curve lies midway between poles p_u and p_v (also frequently true of equations of 4th and 8th-degree).

33. Comparative subfocal ranking of the poles of a bipolar construction rule is not necessarily valid unless the initial curve possesses an equivalent pole for either both or neither of the poles being compared. When the latter conditions are not fulfilled, the ranking is biased against the singular pole.

34. If a bipolar construction rule is linear in v but not in u, pole p_v is a point-focus and outranks pole p_u.

35. The presence of a linear variable in a bipolar construction rule indicates point-focal rank of the corresponding pole.

36. Bipolar point-foci do not exist in cubics (neither dependent nor independent).

37. Dependent bipolar point-foci exist only as equivalent pairs.

38. The degree of the general polar-linear construction rule of a rectangular curve is four times the rectangular degree.

39. The degree of the general axial polar-linear construction rule of a rectangular curve is twice the rectangular degree.

40. Within given degree categories of given curves in given orientations and general locations in two-element coordinate systems with undirected distances, the most-symmetrical representatives (i.e., the representatives with the simplest construction rules) frequently are those that include and/or are tangent to the coordinate elements. In the cases of two loci with indistinguishably simple construction rules, the locus that includes and/or is tangent to both coordinate elements is the more symmetrical.

41. The second-most symmetrical loci in two-element coordinate systems with dissimilar elements are those represented by the construction rule $u = Cv$.

42. In discriminating between simplicities of construction rules (in given degree categories) in both variables of two-element coordinate systems with dissimilar elements, priority goes to the equations with fewest terms.

43. The complementary segments of axial polar-linear conic sections in the two half-planes possess the same eccentricity.

44. The complementary segments of polar-linear curves (other than axial conics) in the two half-planes generally are dissimilar.

45. The complementary segments of curves in the two half-planes of two-element coordinate systems in which one element is a line (and the other element is wholly to one side of the line) generally are dissimilar.

LISTING OF RULES

46. In general, a *partial translation* of a rectangular cubic or curve of higher degree yields a dissimilar curve of the same degree.

47. A directrix-line of an axial curve is a line-pole for which no constant term is present in the curve's polar-linear (or linear-circular) construction rule.

48. A circumpolar point-focus of a curve is a point about which the curve has greater structural simplicity (a structure rule of lower exponential degree) than about neighboring points.

49. A circumpolar focal locus of a curve is a continuous point array, about every member point of which the curve has greater structural simplicity (a structure rule of lower exponential degree) than about neighboring points off the locus.

50. Curves of demarcation generally possess greater circumpolar symmetry than the curves they demarcate.

51. In general, if a point in the plane of a curve does not possess point-focal rank on the basis of a non-trivial 0° or 180°-structure rule, it will not possess point-focal rank for any probe-angle.

52. In a simple (mixed-transcendental) intercept equation cast about a point in the plane, the vanishing of the constant term is the focal condition for points incident upon the curve.

53. In a simple intercept equation cast about points on a line of symmetry, the vanishing of the constant term is the focal condition for the vertices.

54. A condition for which the square-root radicand of a circumpolar intercept format is converted to a perfect square in the variables reveals the location of a high-ranking circumpolar focus.

55. The exponential degrees of structure rules of curves about axial vertices reduce from the values for points incident upon the intersecting focal loci-- a line of symmetry and the curve itself.

56. Similar curves are obtained by inverting an initial curve A about a given pole B and by inverting any inverse curve A' about the pole B', inverse to B (or by inverting A' about a pole inverse to any pole equivalent to B).

57. Radicand constant-conditions on simple intercept formats, for which the radical vanishes, define foci of self-inversion of 4th-degree rectangular curves.

58. In groups of curves with variable foci, members in which multiple foci occur and/or the variable foci are in coincidence with poles of high subfocal rank typically are highly symmetrical.

59. The curves obtained by inverting an initial curve about a pair of modularly-inverse poles (two poles that invert, one to the other, about a focus of self-inversion) are similar.

60. Foci of self-inversion that are in coincidence with vertices are defined by the constant-condition on simple intercept equations, not by conditions on the radicands of simple intercept formats.

61. The focus of self-inversion of an inverse locus that replaces a line of symmetry of the initial curve is the pole inverse to the reflection of the pole of inversion of the initial curve in its line of symmetry.

APPENDIX II.   INVERSION-MAPPING MAXIMS*

6-4-4. One obtains similar curves by inverting an initial curve (or any figure) about a given pole and by inverting any inverse curve or figure about the pole inverse to the given pole.

6-4-6. The equivalent inversion poles for an ensemble of inverse curves obtained by inverting an initial curve B about all points on a given locus L comprise the locus L' inverse to L in the plane of any curve B' inverse to B.

6-4-7. Inversion of a given initial curve B about all points in the plane generates an ensemble of inverse curves B', each member of which inverts to every other member but to no non-member. This is the property of *immediate closure* or *one-step closure*.

6-4-8. Inverse loci of 4th-degree inversions of conics about unexceptional incident poles are of 3rd-degree (exceptional incident poles are double-points, about which all such curves invert to quadratics).

6-4-9. Inverse loci of 3rd-degree inversions of conics about unexceptional incident poles also are 3rd-degree (inverse loci about exceptional incident poles are quadratics).

6-4-10. Inverse loci of cubics about non-incident poles in the finite plane are of higher degree than 3.

6-4-11. A non-axial inversion of a conic of non-unit eccentricity inverts to axial inversions of conics about points incident upon its two (orthogonal) circles of inversion (circles of modular radius centered upon its foci of self-inversion), excepting the two points of intersection of these circles. In these transformations, the circle upon which the pole of inversion is incident inverts to the line of symmetry of the inverse curve.

6-4-12. A non-axial inversion of a conic of non-unit eccentricity inverts to a central conic and a central quartic about the two points of intersection of its two circles of inversion. In these transformations, the two circles of inversion invert to the two lines of symmetry of the central curves.

6-4-14. The pole of inversion to a similar curve (but not a focus of self-inversion) of a non-axial inversion of a conic of non-unit eccentricity lies midway between the two poles for inversions to central curves.

6-4-15. An axial inversion of a conic of non-unit eccentricity inverts to a central conic and a central quartic about the two points of intersection of its circle of inversion with its line of symmetry. The circle of inversion inverts to the orthogonal line of symmetry of the inverse central curve.

6-4-17. A non-axial inversion of a parabola inverts to the entire ensemble of axial inversions of parabolas about points on its circle of inversion.

6-4-19. Equivalent inversion poles on the line of symmetry of axial inversions of central conics lie in modularly-inverse pairs, both members of a given pair lying either to the right or the left of the focus of self-inversion.

6-5-1. The two inverse poles at which a given radius through a 0°-focus of self-inversion (or given chord-segments through a 180°-focus of self-inversion) intersect a curve are equivalent inversion poles.

*Reworded and otherwise simplified from the correspondingly-numbered MAXIMS of Kavanau (1982) to conform to the usage and restricted contents of this work.

6-6-3. For the appropriate value of the unit of linear dimension, the inverse of an initial asymptotic arm of an axial curve, obtained by inverting about a pole incident upon the arm at points increasingly distant from a center, DP, vertex, etc., approaches coincidence with the arm. In the limit, coincidence is achieved between the inverse arm and the initial arm.

6-6-5. A non-axial inverse curve of a central conic or central quartic possesses three finite collinear poles of congruent-inversion. Two of these poles are foci of self-inversion. They lie on lines parallel to the lines of symmetry of the initial curve that pass through the pole of inversion. The third is a pole of inversion to a similar but non-coincident curve**. It lies at the point of intersection of the line-segment from the pole of inversion to the center of the initial curve, and the line segment connecting the foci of self-inversion.

6-7-1. Inversions of a closed initial curve or finite segments of an open initial curve about distant poles yield inverse curves that increasingly approach similarity to the initial curve, the more distant the pole.

6-7-5. Inversion of an initial curve or figure about an infinitely distant pole yields a similar curve or figure.

6-8-3. For the appropriate value of the unit of linear dimension, the inverse locus B' of an axial closed initial curve, or of finite segments of an axial open initial curve B about increasingly distant poles on a line orthogonal to the line of symmetry, increasingly resemble a reflection of B, or a reflection of finite segments of B, in the line of symmetry. In the limit, B' is a coincident reflection of B, whereupon B self-inverts.

6-8-4. For the appropriate value of the unit of linear dimension, the inverse locus B' of an axial closed initial curve, or of finite segments of an axial open initial curve B about increasingly distant poles on a line L not orthogonal to the line of symmetry, increasingly resemble a reflection of B, or a reflection of finite segments of B, in a line L* orthogonal to L. In the limit, B' is a congruent reflection of B in L*.

6-8-8. The possession of a line of symmetry by a curve is equivalent to the possession of a 0°-focus of self-inversion on the ideal line at infinity in the direction orthogonal to the line of symmetry.

6-8-10. The gain (or loss) of a line of symmetry in an inversion is accompanied by the transfer of a 0°-focus of self-inversion to (or from) the ideal line at infinity from (or to) the finite plane (see also MAXIM 8-15-4a).

6-8-11. In the finite plane, curves of odd degree self-invert only about incident foci; contrariwise, curves of even degree self-invert only about non-incident foci.

8-10-1. There are two axial poles about which a self-inverting monovular initial curve with a line of symmetry inverts to curves with true centers. These poles are reflections of one another in the focus of self-inversion at the modular distance.

8-10-2. The two curves with true centers, obtained by inverting monovular self-inverting curves with a line of symmetry, are central inversions of one another. These curves may be similar or dissimilar.

8-10-3. If an axial self-inverting locus inverse to a conic of non-unit eccentricity is inverted about the reflection of the DP in the focus of self-inversion, a central quartic is obtained.

**In references to inversion to a similar but non-coincident locus, it is tacit that the locus is congruent but not coincident with the initial curve for the appropriate value of the unit of linear dimension.

8-10-4. There are two pairs of axial poles about which Cartesian ovals invert to central Cartesians. The members of one pair of poles are modular reflections of one another in one 0°-focus of self-inversion, and the members of the other pair are modular reflections of one another in the other 0°-focus of self-inversion.

8-15-1. Upon inversion of points in the plane about a pole P, all point-pairs inverse at 0° in a circle of inversion C_1 become inverse at 0° in a circle of inversion C_2 centered on pole P', the inverse at 0° of pole P in circle C_1, provided that poles P and P' are not coincident (i.e., provided that pole P is not incident upon circle C_1).

8-15-2. Upon inversion of points in the plane about a pole P incident upon a circle of inversion C, all points inverse at 0° in C become reflected in a common line.

8-15-3. Upon inversion of points in the plane about a pole P, all point-pairs inverse at 180° in a circle of inversion C_1 become inverse at 180° in a circle of inversion C_2 centered on pole P', the inverse at 180° of pole P in circle C_1.

8-15-4. In the inversion of curves to dissimilar non-central inverse loci:

(a) the gain (or loss) of a 0°-focus of self-inversion in the finite plane is accompanied by the loss (or gain) of a line of symmetry and a focus of self-inversion on the ideal line at infinity in the orthogonal direction (see also MAXIM 6-8-10);

(b) 180°-foci of self-inversion are conserved, i.e., they are neither gained nor lost;

(c) 0° and 180°-foci of self-inversion that are coincident are conserved but are non-coincident in the inverse locus.

(d) foci of self-inversion at 90° and 270° are not conserved. Those of quartics are replaced by two poles of inversion to similar (but non-coincident) curves in axial inverse loci, and by four poles of inversion to similar (but non-coincident) curves in non-axial inverse loci. Together with one pole of inversion to a similar (but non-coincident) curve possessed by virtue of the initial curve having two lines of symmetry, this makes a total of five poles of inversion to similar curves for the non-axial inverse loci.

8-15-5. Conservation of a focus of self-inversion by the inversion transformation implies conservation of the corresponding angle of self-inversion.

8-15-6. Modularly-inverse pole-pairs about foci of self-inversion at angles of 0°, 180°, 0° and 180°, and θ and $\theta+180°$ ($\theta \neq 90°$ or 270°, except for central curves) are equivalent inversion poles.

8-20-2. The inversion transformation conserves the sum of the number of lines of symmetry and the number of 0°-modes of self-inversion for finite moduli.

8-20-2'. The inversion transformation conserves the sum of the numbers of 0°-modes of self-inversion for finite and infinite moduli.

8-20-3. On each line of symmetry there are two poles about which a biovular central Cartesian inverts to dissimilar biovular central Cartesians; these poles lie at modular distance on each side of center. In these inversions, annular curves invert to two different opposed-oval curves, whereas opposed-oval curves invert to one annular and one dissimilar opposed-oval curve.

8-20-6. On a given line of symmetry, the poles about which a given biovular central Cartesian inverts to similar inverse axial cubics or quartics occur in quartets. The poles in each quartet lie in modularly-inverse pairs on each side of center, i.e., if two poles lie at $\pm nj$, the other two lie at $\pm j/n$, where n can take on any real value except 0 or 1, and j is the modular value (the unit of linear dimension for self-inversion).

8-20-7. All axial inverse loci of biovular central Cartesians are obtained by inverting about poles on three loci: all axial poles from $h = 0$ to $h = j$; all axial poles from $k = 0$ to $k = j$; and all poles incident upon one quadrant of the modular circle, where j is the modular value.

8-20-11. Axial inverse loci of a given biovular central Cartesian possess only two axial poles about which inversions yield curves that are congruent with one another for a common unit of linear dimension. These are the poles at modular distance from the 180°-inter-oval-mode focus of self-inversion.

8-20-12. Axial cubics and quartics inverse to biovular central Cartesians, whose lines of symmetry intersect the curve, possess five axial multi-member sets of exceptional equivalent inversion poles. For any given initial curve, all of the inverse loci about the poles of any given set are similar. Three of these sets consist of two poles each, and the pole-pairs lie at modular distances from the foci of self-inversion; a fourth set consists of the three foci of self-inversion, and the fifth set consists of the three or four vertices. If the line of symmetry of a quartic does not intersect the curve, it has only four such exceptional multi-member sets of equivalent inversion poles, since it possesses no axial vertices.

8-22-1. In the inversion of biovular central Cartesians to dissimilar central Cartesians, the modular distance becomes halved, i.e., if j is modular for self-inversion of the initial curve, $j/2$ is modular for self-inversion of the inverse curves.

8-22-3. A given axial cubic or quartic inverse to a biovular central Cartesian inverts to dissimilar central Cartesians about each of three pairs of poles. One of these central Cartesians is annular, the other two possess opposed ovals.

8-22-4. A monovular central Cartesian inverts to axial cubics and quartics about poles on its lines of symmetry. It inverts to a generally dissimilar central curve about its center.

8-22-5. An axial locus inverse to a monovular central Cartesian inverts to axial monovals about poles on its line of symmetry and on its inverse-axis circle (the circle inverse to the orthogonal line of symmetry of the initial central Cartesian). It inverts to central Cartesians about the two points of intersection of these loci, and self-inverts about the center of the inverse-axis circle. The central Cartesians obtained in this manner invert to one another about their centers and generally are dissimilar.

8-22-6. Non-axial inverse loci of monovular central Cartesians invert to axial monovals about points on the inverse-axis circles (circles inverse to the lines of symmetry of the initial central Cartesians); they invert to the initial central Cartesians about the two points of intersection of these circles. The initial central Cartesians obtained in this manner generally are dissimilar and invert to one another through their centers.

8-22-8. Biovular central Cartesians invert to axial cubics and quartics about points on one or the other line of symmetry or the modular circle (quartics only), excluding the points of intersection of these three loci (they invert to central Cartesians about the points of intersection). They invert to non-axial curves about other points in the plane.

8-22-10. There are three loci in the plane of an axial cubic or quartic inverse to a biovular central Cartesian about which it inverts to other axial curves. These are the two circles of inversion, A and B, centered on the two 0°-foci of self-inversion, and the line of symmetry, L. Inverse curves about points on the third circle of inversion, C, centered on the 180°-focus of self-inversion are non-axial (except for the points where C intersects A, B, and L).

8-22-11. In the inversion of a biovular central Cartesian about any member of a quartet of axial poles consisting of reflected modularly-inverse pole-pairs, the poles inverse to the other three members of the quartet are the foci of self-inversion in the finite plane of the inverse curve (see also MAXIM 8-30-6).

8-22-13. An axial cubic or quartic inverse to a biovular central Cartesian inverts to central Cartesians about the six points of intersection of the line of symmetry, L, and the two circles of inversion, A and B, centered on the two 0°-foci of self-inversion (taking A, B, and L pairwise). The inverse central loci about the two points of intersection of any two of these loci have lines of symmetry inverse to the same two loci, and a modular circle inverse to the third locus.

8-22-15. A non-axial cubic or quartic inverse to a biovular central Cartesian inverts to three pairs of dissimilar central Cartesians about the six points of intersection of three mutually orthogonal 0°-circles of inversion. In a given inversion to a central curve, the two circles providing the point of intersection invert to the lines of symmetry; the third circle inverts to the modular circle.

8-22-16. The line of symmetry of the inverse axial curve obtained by inverting a biovular central Cartesian about a non-axial pole on its modular circle is inverse to the modular circle and passes through the two poles inverse to the reflections of the pole of inversion in the lines of symmetry (these poles become the 0°-foci of self-inversion in the finite plane).

8-22-19. A pair of poles of a given central conic or central quartic that are reflected in a line of symmetry (equivalent inversion poles) are modularly inverse in a focus of self-inversion in the finite or extended plane in all of its inverse loci.

8-22-20. A pair of (equivalent inversion) poles that are modularly-inverse in a focus of self-inversion of a given mode (180°-inter-oval, 0°-inter-oval, or 0°-intra-oval) in the finite or extended plane, in a given curve, are modularly-inverse in a focus of self-inversion of the same mode in all inverse loci of that curve.

8-29-1. Any unexceptional inversion of a locus C to a locus D can be regarded as a self-inversion; conversely, any self-inversion can be regarded as an unexceptional inversion of a locus A to a locus B. In the former case, one regards the union of C and D as the self-inverting curve; in the latter case, one regards the segments A and B of the self-inverting curve as the individual loci that are the inverses of one another.

8-30-2. Poles Q', inverse to equivalent inversion poles Q, of the pole of inversion P, of an initial curve B, are either foci of self-inversion of the inverse locus B' or poles about which B' inverts to a similar (but not coincident) curve. Poles Q' are foci of self-inversion if the corresponding poles Q are:

(a) reflections of P in lines of symmetry,

(b) modularly-inverse to P in 0° or 180°-foci of self-inversion, or

(c) modularly-inverse to P in a 0° and 180°-focus of self-inversion.

8-30-3. If an initial curve that self-inverts about a single pole in multiple modes is inverted about a pole on its modular circle, all of the poles of congruent-inversion of the inverse locus will be collinear.

8-30-5. If an initial curve that self-inverts about a single pole in multiple modes is inverted about a non-modular pole, the poles of congruent-inversion in the finite plane of the inverse locus comprise two ensembles, one concyclic and the other collinear. The concyclic ensemble possesses one more pole than the collinear ensemble.

8-30-6. In a locus B' that is inverse to an intial curve B about pole P, where B self-inverts about its center at 0° and 180°, the pole P*' inverse to pole P*, which is modularly-inverse to P, is the 0°-focus of self-inversion (of B'); the pole $P_R^{*'}$, inverse to P_R^*, the reflection of P* in the center, is the 180°-focus of self-inversion of B' (see also MAXIM 8-22-11).

8-30-9. A focus of self-inversion of an inverse curve that replaces a line of symmetry of an initial curve acquires the mode of self-inversion (0°-inter-oval or 0°-intra-oval) of the focus of self-inversion of the initial curve that lies on the ideal line at infinity in the direction orthogonal to the lost line of symmetry.

8-30-10. In modular axial inversions of a central curve that self-inverts about its center in the 0° and 180°-modes to dissimilar central curves that also self-invert about their centers in the 0° and 180°-modes, the initial 0° center-focus "exchanges positions" (modes of self-inversion) with the initial 0°-focus of self-inversion on the ideal line at infinity in the direction of the axis in question. In non-modular inversions of the same central curve to non-central curves, the inverses of the initial 0°-focus possess the mode of self-inversion of the latter.

8-30-11. Inter-oval mode and intra-oval mode (and inter-segment mode and intra-segment mode) 0°-foci of self-inversion in the extended plane are conserved by the inversion transformation.

8-30-12. The number of foci in the extended plane in each category of self-inversion, that is, 0° and 180°-inter-oval mode (and inter-segment mode) and 0° and 180°-intra-oval mode (and intra-segment mode) is conserved by the inversion transformation; only the locations of the foci change.

8-30-13. If an initial central curve with n lines of symmetry that does not self-invert about the center is inverted about a non-central axial pole, the inverse locus is axial and possesses n-1 foci of self-inversion in the finite plane.

8-36-1. The asymptote(s) of the inverse locus of an initial curve about an incident pole at which the curvature is finite is (are) the inverse(s) of the circle(s) of curvature at that pole.

8-36-2. The inverse locus of a curve about an incident pole at which the curvature is infinite possesses the ideal line at infinity as asymptote.

8-36-3. When an open curve with an asymptote at a finite distance from its center, vertex, DP, etc., is inverted about a non-incident pole, one obtains a closed curve with finite curvature at the reciprocal inversion pole.

8-36-4. When an open curve with the ideal line at infinity as asymptote is inverted about a non-incident pole, one obtains a closed curve with infinite curvature at the reciprocal inversion pole.

8-38-1. Single points of tangency (i.e., excluding bitangents) of radii from foci of self-inversion of non-axial inversions of conics of non-unit eccentricity are poles of inversion to axial-vertex inversion cubics.

8-38-2. In non-axial curves inverse to ellipses, the four points of intersection with the two circles of inversion are poles of inversion to axial-vertex inversion cubics.

8-38-3. In non-axial curves inverse to hyperbolas, the two points of intersection with the inverse transverse-axis circles of inversion (excluding the DP) are poles of inversion to axial-vertex inversion cubics (axial strophoids).

8-40-1. Single points of tangency of radii from foci of self-inversion of cubics and quintics are poles of inversion to axial curves.

8-40-2. If an initial curve B is inverted about a pole P and the circle of curvature of B at P intersects B at points C, then the asymptote of the inverse curve B' intersects B' at points C' inverse to C.

NEW MAXIM

A remarkable, previously unpublished inversion-mapping MAXIM accounts for the finding that the inverse loci of Cassinian monovals about modular incident points possess an ordinary center of symmetry (Kavanau, 1982). Consider a centered Cassinian monoval with its lines of symmetry in coincidence with the rectangular axes. Since Cassinian monovals self-invert about their centers at 90°, the modular incident poles (distance j from center) lie at the four points where the coordinate bisectors intersect the curve. Since all probe-lines, X and Y, of a centered 90°-probe intersect the curve at modularly-inverse distances ($XY = j^2$), the probe-lines of centered α-probes ($0 \leq \alpha \leq 360°$) bisected by a coordinate bisector also must intersect the curve at modularly-inverse distances.
[Assume that one probe-line of a 90°-probe lies along $\theta = 70°$ and the other along $\theta = -20°$. The product of their lengths is j^2. But because Cassinian ovals possess a true center of symmetry, a probe-line at +20° has the same length as the one at -20°. Hence the probe-lines of a 50°-probe lying along $\theta = 20°$ and $\theta = 70°$ (symmetrically-located about the 45°-bisector) also have modularly-inverse lengths, etc., for other probe-angles.]

It is a consequence of the following MAXIM that the cubic obtained by inverting monovular Cassinians about an incident modular pole has an ordinary center of symmetry. More generally, all points in the initial plane divided into ensembles of pairs modularly-inverse (at all angles from 0° to 360°) in a first common center, invert (at 0°) about a second common center to become reflected pairwise in a third common center, where the three centers are collinear and equally spaced.

MAXIM: Given any two rays, A and B, emanating from point Q at an angle α ($0° \leq \alpha \leq 360°$). Given also a point P equidistant from the two rays and at unit distance j from Q. Let all point-pairs, a and b, modularly-inverse ($ab = j^2$) in Q along rays A and B be inverted about point P at 0° (along coincident-radii). Then all the resulting inverse point-pairs, a' and b', are reflected in a common point P* lying midway between P and Q.

APPENDIX III. GLOSSARY OF TERMS

α-Structure rule (point-specified structure rule) see Structure rule, point-specified.
Angle-specified structure rule see Structure rule, angle-specified.
Arc-increment eqs. structure rules employing the derivative ($ds/d\theta$) of arc length along structural curves with respect to the angle of orientation of the X-probe-line to the x-axis in the plane of the initial curve.
  X- expressed in terms of X-intercepts alone, for a specified probe-angle.
  XY- expressed in terms of both X and Y-intercepts for an unspecified probe-angle.
Asymmetries of coordinate systems
  elective assigned by the user, for example, labelling the poles or the use of directed coordinates.
  intrinsic inherent to the system, for example, incongruence of coordinate elements or the use of different types of coordinates (say, distances and angles).
Asymptote the tangent to a curve whose points of contact lie on the ideal line at infinity. If the limit of a tangent line is the ideal line at infinity itself, the latter is said to be the asymptote.
Asymptote-point the point of intersection of an asymptote with a line of symmetry.
Axial pertaining to a line of symmetry.
  curves curves with a single line of symmetry.
  focal locus a line, all points of which have focal rank.
  inverse loci inverse curves with one line of symmetry.
  inversions inversions of curves about poles on a line of symmetry, generally the axis coincident with the x-axis (curves in standard position).
Bipolar coordinate system a coordinate system based upon undirected distances from two fixed points (poles).
Cartesian
  ovals monovular and biovular curves discovered by Descartes and defined by 1st-degree bipolar and polar-circular construction rules.
  condition vanishing of the Cartesian term; the condition on an axial inversion locus of a central Cartesian for which its $\cos^2\theta$-condition foci are in coincidence in the inverse curves, which are Cartesian ovals.
  term radicand of the $\cos^2\theta$-condition of the simple intercept eq. of axial inverse loci of central Cartesians (see Cartesian condition).
Cassinian ovals (Ovals of Cassini) monovular and biovular central curves (members of the central Cartesians) discovered by Cassini and defined by the bipolar eq. $uv = Cd^2$.
Center
  ordinary each line through an ordinary center intersects a non-axial curve at equidistant points on each side of the center (the rectangular eq. of the curve is unchanged if $-x$ replaces x and $-y$ replaces y).
  true the point of intersection of two orthogonal lines of symmetry.
Central Cartesian central curve obtained by inverting Cartesian ovals [rectangular eq. $(x^2+y^2)^2 + B^2x^2 + C^2y^2 + D^4 = 0$].
Central conic conic section with at least two lines of symmetry.
Central curve term most commonly applied to curves with two orthogonal lines of symmetry.

Central modular-focal quartic central Cartesian for which the foci of the $\cos^2\theta$-condition are at modular distance from center.
Central quartic curve obtained by inverting a central conic about its center.
Characteristic distance
  primary   a distance between coordinate elements.
  secondary a distance specifying the size of a coordinate element.
Circle of inversion circle centered on a pole of inversion, of radius equal to the unit of linear dimension for the inversion.
Circumcurvilinear relative to a curve as the single reference element (includes the line).
Circumlinear relative to a line as the single reference element.
Circumpolar relative to a point as the single reference element.
  point-focus for a given curve and a given probe-angle, a point is a circumpolar focus if the rectangular degree of the corresponding structure rule is lower than that of structure rules of the curve about all neighboring points.
  focal locus for a given curve and a given probe-angle, a locus is a circumpolar focal locus if the rectangular degree of the corresponding structure rule is lower than that of structure rules about all neighboring points off the locus.
Closure, immediate see Immediate Closure.
Complete structure rule see Structure rule, complete.
Compound curves curves that consist of joined incomplete arms or segments of rectangular curves.
  mixed   the incomplete arms or segments are dissimilar.
Compound intercept eq. an eq. for a specified probe-angle expressing the relation between the lengths of the X and Y-intercepts of probe-lines from the probe-point to the curve.
Compound intercept format solution of a compound intercept eq. for $\cos^n\theta$ or $\sin^n\theta$, but often for $\cos\theta$ or $\sin\theta$.
Concentric having a common center.
Conchoid   in the classical conchoid construction, equal distances along a line are marked off from its point of intersection with a curve for various positions of the line as it rotates about a point. The conchoid is the locus of the endpoints of the line segments.
Congruent   having the same size and shape.
Congruent-inversion an inversion of an initial curve that yields a similar, congruent, or coincident inverse locus (any similar inverse locus is congruent for the appropriate value of the unit of linear dimension for the inversion).
Conic axial-vertex inversion cubic cubic obtained by inverting a conic about a classical vertex.
Conjugate point or pole the reflection of an initial point or pole in a reference element or a specified point or line.
Conjugate oval a quartic oval which, in combination with a second quartic oval (representable by linear bipolar eqs. that differ only in the sign of a term) constitute biovals that self-invert in three modes, one each about three discrete foci.
Constant-condition a condition obtained by setting the constant term of a simple intercept eq. equal to 0 (defines axial intersections of curves).
Construction rule a coordinate eq. of a curve.
Coordinate curve curve for which the distances from a coordinate element are constant.

$Cos^2\theta$-condition   condition for vanishing of the coefficient of the $cos^2\theta$-
                term of a simple intercept eq.
  focus or foci   focus or foci defined by the $cos^2\theta$-condition.
Covert linear focal locus   a linear focal locus that is not a line of symmetry
                or an asymptote.
Cubic   curve whose rectangular eq. is 3rd-degree in the variables.
Curve of demarcation   typically a member of a group of curves that demarcates
                (with progressive changes in parameter values) between two sub-
                groups of members with different properties, for example, between
                subgroups possessing one and two ovals, while itself possessing
                intersecting loops.
Curve of intersection   a curve of demarcation possessed in common by two dif-
                ferent groups of curves.
Cusp-point   a variety of double-point at which the tangents coincide.
Degree-restriction   a property of bilinear coordinate systems wherein all eqs.
                of a given curve have the same exponential degree.
  partial   a property of, for example, the polar-linear system where the
                presence of the line-pole restricts the degree of axial curves
                to not less than the degree of the corresponding hybrid eq., nor
                greater than twice the degree of that eq.
Dependent bipolar point-foci   the degree of a bipolar eq. of a curve reduces
                from that of the general axial eq. only when it is cast about
                paired dependent point-foci (as compared to being cast about a
                single such focus in combination with an unexceptional second
                axial pole).
Directrix
  circle   a circle-pole for which there exists a point-and-circle (or line-
                and-circle) construction rule of a curve that lacks a constant
                term.
  line   a line-pole for which there exists a point-and-line (or circle-
                and-line) construction rule of a curve that lacks a constant term.
Disparate curves   curves that possess one or more pairs of similar but incon-
                gruent complete arms, ovals, or segments.
  partial   disparate curves for which the segments are incomplete, or com-
                pound curves for which the segments are disparate.
Double-point   a point of single self-intersection of a curve.
Eccentricity   for central conics, $b^2x^2 \mp a^2y^2 = a^2b^2$, and all their inverse loci,
                the eccentricity is $e = \sqrt{(a^2 \pm b^2)}/a$. For the parabola and all
                its inverse loci, $e = 1$.
Eliminant   the eq. obtained by eliminating one or more variables between
                two or more eqs.
Elliptical limacons   see Limacons, elliptical.
Equivalent poles   poles reflected in a line or point of symmetry.
Equivalent inversion poles   points in the plane of a curve or figure about
                which it inverts to similar or congruent curves.
Extremal-distance symmetry   the relation between curves defined by a single
                eq. in undirected-distance coordinate systems, one locus given
                by the greatest-distance construction and the other by the least-
                distance construction.
Focal condition   condition on a construction rule or structure rule that leads
                to a reduction in degree. Also a condition on an intercept eq. or
                format that leads to a reduction in degree thereof or of the de-
                grees of corresponding structure rules.

Focal rank
  circumpolar of a point with respect to a curve or of a curve about a point; the reciprocal of the rectangular degree of the structure rule of the curve about the point (similarly for points of a focal locus).
  point-focal see Circumpolar, point-focus.
Focus a locus about which a curve has greater symmetry (structural simplicity) than about neighboring points not incident upon, or coincident with, the locus.
  angle-independent circumpolar a circumpolar focus defined by a condition on a simple intercept eq. or format.
  axial a line, all points of which have focal rank.
  compound a curvilinear focal locus upon which there lie point-foci.
  eclipsable a point-focus or curvilinear focal locus that loses its focal rank in a certain member of a group of curves.
  loop a point-focus that lies within a loop of a curve.
  of self-inversion a pole about which a curve self-inverts (at any angle).
  multiple two or more point-foci in coincidence.
  simple a curvilinear focal locus upon which there lie no point-foci.
  variable a focus that has different relationships to different members of a group of curves, for example, within a small loop in some, within a large loop in some, and outside of the curve in others.
Harmonic parabolas two-arm parabolas with generally incongruent arms represented by non-degenerate eqs. in collinear multipolar coordinate systems (three or more point poles).
Hyperbolic limacon see Limacons, hyperbolic.
Ideal line at infinity the set of ideal points at infinity, one for each direction in the plane.
Identical coincident with
Immediate closure the property of the inversion transformation whereby in an ensemble of inverse loci of a given initial curve, each member inverts to every other member but to no non-member.
Incomplete-linears linear coordinate eqs. in which not all of the variables participate.
Independent bipolar point-foci the degree of the bipolar eq. of a curve cast about a single independent point-focus and an arbitrary point reduces from the degree of the general bipolar eq.
Individual bipolar focal rank the bipolar focal rank of single poles, as opposed to the bipolar focal rank of pole-pairs.
Inherent properties of curves the structural properties of curves described and defined by their structure rules about points and by their intrinsic eqs.
Initial curve the "starting" curve or beginning curve for an operation.
Inner segment the segment closest to the origin of a structural curve.
Intercept the length of a probe-line from the probe-point to the point of intersection with a curve.
  chord-segment (product) the distances (in the positive sense) from the probe-point to the points of intersection with the curve are measured in opposite directions along a chord through the point.
  coincident-radii (product) the distances (in the positive sense) from the probe-point to the points of intersection with the curve are measured in the same direction along coincident probe-lines.

GLOSSARY OF TERMS

Intercept
  negative   the intercept for the angle $(\theta+180°)$ or $(\theta+\alpha+180°)$.
  positive   the intercept for the angle θ or $(\theta+\alpha)$.
  product   generally the expression for the product of the intercepts from a point to a curve along a line.
  transform   technical term for "structure rule."
Inverse-axis circle   circular locus in the plane of an inverse curve that is inverse to a line of symmetry of the initial curve.
Inverse locus of a curve or figure is the locus obtained relative to a point reference pole by measuring off distances along radius vectors from the pole that are the reciprocals of the distances to the curve along the same radius vectors.
Inversion   the transformation whereby a curve or figure is inverted about a pole (see **Inverse locus**).
  analysis   see Inversion Taxonomy.
  superfamily the ensemble of inverse loci of all members of a group of curves.
  Taxonomy   the classification or systematic organization of curves on the basis of their relatedness through the inversion transformation.
Latus rectum   chord of a conic section at the level of a traditional focus and orthogonal to the axis upon which the traditional focus lies.
Limacons   inverse loci of conic sections about the traditional foci.
  elliptical   inverse loci of ellipses about their traditional foci.
  hyperbolic   inverse loci of hyperbolas about their traditional foci.
  parabolic   inverse locus of the parabola about its traditional focus (the cardioid).
Linear-circular coordinate system a coordinate system based upon undirected distances from a fixed circle and a fixed line.
Line of symmetry a line about which a curve possesses reflective symmetry, or a line of reflective symmetry of a coordinate system.
Loop   a self-intersecting segment of a curve.
MAXIM   either a proven relation or a description of relations based upon current knowledge.
Mimics   see Symmetry mimics.
Modular   for the modular value of distance, i.e., for the unit of linear dimension for self-inversion.
  circle   circle centered upon a focus of self-inversion of radius equal to the unit of linear dimension for self-inversion (the modular value).
  poles   axial poles at modular distance from a focus of self-inversion.
  segment   when a complete structure rule is degenerate and all of the individual factors equated to 0 represent congruent segments, each of the segments is said to be modular.
  value   the unit of linear dimension for self-inversion.
Multiple-focus see Focus, multiple.
Multipolar   relating to more than two poles.
Near-conics   near-circle, near-ellipse, and near-line structural curves about the $-b/2$ focus of limacons and its homologue in Cartesian ovals.
Non-axial curve curve lacking a line of symmetry.
Non-polarized bipolar coordinate system bipolar coordinate system in which the poles are unordered, preserving reflective redundancy in both the bipolar axis and the midline.
Normal pedal the normal pedal P of an initial curve B with respect to a point O is the locus of the feet of the perpendiculars from O on the normals to B.

Oval — a non-self-intersecting closed curve or closed segment of a curve (possessing continuous first derivatives at all points).

Parabolic Cartesian — subgroup of the Cartesian ovals with the bipolar eq. $u^2 = Cdv$.

"Partial translation" — a rectangular translation in which x+h or y+k is not substituted for x or y at every occurrence in the construction rule.

Peripheral vertices — the intersections with the curve of normals to a line of symmetry at the positions of axial foci.

Plane
  extended — the union of the finite Euclidean plane and the ideal line at infinity.
  finite — all points in the Euclidean plane at a finite distance from a reference point.

Polar-circular coordinate system — a coordinate system based upon undirected distances from a fixed point and a fixed circle.

Polar-exchange symmetry — two curves possess polar-exchange symmetry if their construction rules have identical form but with the poles interchanged, for example, $u^2 = Cdv$ and $v^2 = Cdu$.

Polarized bipolar coordinates — a bipolar coordinate system in which the poles bear fixed labels and there is no reflective redundancy in the midline (i.e., the midline is not a line of symmetry).

Polar-linear coordinate system — a coordinate system based upon directed or undirected distances from a fixed point and a fixed line.

Probe — a point (the probe-point) with lines (the probe-lines) emanating from it at a fixed angle (the probe-angle) used to assess the structural simplicities of curves with respect to single fixed reference elements (see Figs. 2, 39, and 40).
  angle — see Probe.
  lines — see Probe.
  point — see Probe.

Quadratic — curve or eq. of rectangular degree 2 in the variables.
Quartic — curve or eq. of rectangular degree 4 in the variables.
Quintic — curve or eq. of rectangular degree 5 in the variables.

Radicand constant-condition — a condition resulting from setting the constant term of the radicand of a simple intercept format equal to 0.

Rank — usually a comparative quantitative measure of symmetry.
  focal — of a focal locus with respect to a curve, or of a curve about a focal locus; the reciprocal of the degree of the rank-determining structure rule.
  subfocal — of a focal or non-focal locus with respect to a curve, or of a curve about a focal or non-focal locus; based upon comparisons of the complexity of the rank-determing structure rules.
  symmetry — non-specific characterization of focal and/or non-focal rank.

Reciprocal inversion pole — the pole in the plane of an inverse curve coincident with the pole of inversion of the initial curve (inversion about this pole restores the initial curve).

Reflective partial curves — incomplete segments of rectangular curves reflected in a line of symmetry of a coordinate system.

RULE — either a proven relation, a description of relations based on current knowledge, or a definition.

Self-inversion — an inversion in which the inverse curve is identical (coincident with) to the initial curve.

GLOSSARY OF TERMS

Self-inversion
   inter-oval two ovals invert to one another.
   intra-oval an oval inverts to itself.
   negative the intercepts are measured in opposite directions (along chord-segments; self-inversion in the 180°-mode).
   omnidirectional circumlinear the circumstance in which the products of the intercepts of two probe-lines with a curve are constant for 0° and 180°-probe-angles and any given probe orientation, for probe-points incident upon a given line.
   positive the intercepts are measured in the same direction from the reference pole (along coincident radii; self-inversion in the 0°-mode).

Simple intercept eq. an eq. that expresses the length of the X-intercept of a probe-line from the probe-point to the curve as a function of θ, the angle of the X-probe-line to the x-axis.
   mixed-transcendental simple intercept eq. involving more than one transcendental function θ.

Singularity of loci refers to the fact that non-degenerate rectangular construction rules of curves represent only a single copy of the curve.

Standard polar construction rule a polar eq. employing only the sines and/or cosines of whole angles θ, with limited ranges (see page 44).

Structural curves the rectangular curves represented by structure rules.

Structural eqs. eqs. describing the structure of a curve with respect to a single reference element (structure rules) or two or more reference elements (construction rules).

Structure probe see Probe.

Structure rules eqs. expressing relations between the structure of an initial curve and the structure of a single reference element (a point, a line, or another curve), as determined with a rotating or translating probe.
   α- see point-specified (below).
   angle-specified a structure rule of a curve about an arbitrary point for a specified angle.
   circumcurvilinear the probe-point travels along a single reference curve.
   circumlinear the probe-point travels along a single line.
   circumpolar the probe-point and probe-lines rotate about a single point.
   complete a structure rule that takes into account all combinations of intercepts--positive, negative, null, real, and imaginary.
   condition a condition on a structure rule which, if satisfied, leads to reduction of its exponential degree.
   formats eqs. for deriving structure rules from simple intercept formats.
   point-specified a structure rule of a curve for an arbitrary probe-angle about a specified point.
   specific a structure rule of a curve for a specified probe-angle and a specified location of the probe-point.
   trivial a structure rule that merely expresses the equality of identical intercepts or for which one intercept is null for all probe orientations.

Subfocal rank a comparative symmetry ranking within common exponential degree categories of structural eqs. based upon measures of the comparative simplicities of the eqs.

Symmetric function a function that is unchanged after exchanging the variables.

Symmetry   a comparative measure of the structural simplicities of curves
               with respect to one or more reference elements based upon the
               simplicities of their structural eqs. with respect to
               the same reference elements.
  circumcurvilinear symmetry with respect to a curve.
  circumlinear symmetry with respect to a line.
  circumpolar   symmetry about a point.
  greater symmetry see more symmetrical.
  more symmetrical having simpler structural eqs. (relative to one or more
               specified reference elements).
  most symmetrical having the simplest structural eqs. (relative to one or
               more specified reference elements).
Symmetry mimics   curves with eqs. of higher rectangular degree than that of an
               initial curve but with identical 0° and 180°-structure rules
               about a homologous pole.
Taxonomy   a paradigm of classification.
Traditional focus   term generally used to refer to the only heretofore recog-
               nized real foci of conic sections in the finite plane, to dis-
               tinguish them from other points about which conics possess ex-
               ceptional structural simplicities, as defined by corresponding
               structure rules.
Transform   see Intercept Transform.
Tripolar   relating to three points.
Trivial structure rule see Structure rules, trivial.
True center   see Center, true.
Uniqueness of identity   property of some coordinate systems wherein all repre-
               sentatives of a curve, for example, a parabola, are similar,
               regardless of location.
Valid locus   a locus for which non-trivial structure rules exist for the
               probe-angle in question.
Variable probe-angle focal conditions   focal conditions for hyperbolas, where-
               in the focal status of a locus, or of a curve about a locus,
               depends upon both the probe-angle and the eccentricity.
Variable focus   see Focus, variable.

REFERENCES

Crofton, M. W., On Certain Properties of the Cartesian Ovals, treated by the
    Method of Vectorial Co-ordinates. *Proc. Lond. Math. Soc.* 1:5-18, 1866.
Kavanau, J. Lee, *Symmetry, An Analytical Treatment*, Los Angeles, Science Soft-
    ware Systems, Inc., 1980.
Kavanau, J. Lee, *Curves and Symmetry*, vol. 1, Los Angeles, Science Software
    Systems, Inc., 1982.
Lockwood, E. H., *A Book of Curves*, Cambridge, Cambridge University Press,1961.
Lockwood, E. H. and R. H. Macmillan, *Geometric Symmetry*, Cambridge, Cambridge
    University Press, 1978.
Salmon, G., *A Treatise on Conic Sections*, London, Longmans, Green and Co.,1879.
Wilker, J. B., Inversive Geometry and the Hexlet. *Geometriae Dedicata* 10:469-
    473, 1981.
Zwikker, C., *Advanced Plane Geometry*. Amsterdam, North-Holland Publ. Co., 1950.

APPENDIX IV.   ANSWERS TO PROBLEMS II.

Chapter II.

1. First, find the general eq. for the slopes of tangents to ellipses, i.e., $dy/dx = f(x,y)$. Set this expression equal to 1, the tangent of the angle that the y-axis makes with the major axis of the "tangent 45°-bisector ellipse." Then eliminate y using eq. 4a for a centered ellipse in standard position. This yields the x-coordinate of the chord through the tangent points. The y-coordinate of the tangent-point then gives the length of the hemi-chord, and these two values enable one to obtain the x-coordinate of the point at which the two tangents intersect. This is the point at which the origin of Fig. 4f is located. The values of h and k are obtained readily from this x-coordinate.

2. transformation to rectangular coordinates: $x_0 = (x - y\cos\theta/\sin\theta)$, $y_0 = y/\sin\theta$

   rectangular equation: $y^2 = 4a\sin^2\theta(x - y\cos\theta/\sin\theta)$

   rectangular eq. in standard form for 45°-oblique system: $(y+a)^2 = 2a(x + a/\sqrt{2})$

3. initial curve: $y_b^2 = 4ax_b$

   transformation to rectangular coordinates: $x_b = y$, $y_b = (x-y)/\sqrt{2}$

   rectangular eq.: $x^2 - 2xy + y^2 = 8ay$

   rotated and in standard form: $(y - a\sqrt{2})^2 = 2\sqrt{2}a(x + a\sqrt{2}/2)$
   (see answer to Problem 5)

   general eq. for y_b: $y_b = x\sin\theta - y\cos\theta$

4. initial curve: $x_b^2 + y_b^2 = 2x_b y_b = j^2/2$

   transformation to rectangular coordinates: $x_b = y$, $y_b = (y-x)/\sqrt{2}$

   rectangular curve: $x^2 + 5y^2 - 4xy = j^2$

   rectangular curve in standard form (axes rotated 22-1/2°; see answer to Problem 5): $(1 - \sqrt{2}/2)^2 x^2 + (1 + \sqrt{2}/2)^2 y^2 = j^2/2$

   eccentricity: $e = \sqrt{(a^2 - b^2)}/a = \sqrt{(2\sqrt{2})/(1 + \sqrt{2}/2)} = 0.98517$

5. initial curve: $x_0 y_0 = j^2$

   transformation to rectangular coordinates: $x_0 = (x - y)$, $y_0 = y\sqrt{2}$

   rectangular eq.: $y(x-y) = j^2/\sqrt{2}$

   To rotate the coordinate axes (CCW) into coincidence with the line(s) of symmetry of a conic section with the general eq., $Ax^2 + Bxy + Cy^2 + D^2x + E^2y + F^3 = 0$, one rotates through an angle, θ, derived from the eq., $\tan 2\theta = B/(A-C) = 1/[0-(-1)] = 1$, whence $\theta = 22-1/2°$. From the rotation eqs., 3a,b, we have, $x = 0.92388x' - 0.38268y'$ and $y = 0.38268x' + 0.92388y'$. Performing these substitutions leads to the rectangular eq. in standard form, $0.20711x^2 - 1.20711y^2 = j^2/\sqrt{2}$.

From the standard form for hyperbolas, $x^2/a^2 - y^2/b^2 = 1$, it is evident that $a^2 = j^2/0.20711\sqrt{2}$ and $b^2 = j^2/1.20711\sqrt{2}$. Since $e^2 = (a^2+b^2)/a^2$, we have $e^2 = 1.1716$, whence $e = 1.0824$.

6. initial curve: $\quad\quad\quad\quad\quad\quad\quad\quad\quad b^2x^2 + a^2y^2 = a^2b^2$
   slope at point (x,y): $\quad\quad\quad\quad\quad dy/dx = -b^2x/a^2y$
   slope at point with abscissa, x: $\quad dy/dx = -bx/a\sqrt{(a^2-x^2)}$
   substitute abscissa of left traditional focus: $\quad\quad\quad\quad\quad dy/dx = \sqrt{(a^2-b^2)}/a = e$

Chapter III.

1. No, the rectangular eq. includes the origin, whereas the polar eq. does not, since $\sec(\theta/3)$ cannot vanish. The rectangular eq. is cast about the loop DP [polar eq., $r\cos\theta = a(4\cos^2\theta-1)$], whereas the polar eq. is cast about the loop focus (see Kavanau, 1980).

2. In converting eq. 6e-left to polar coordinates, two r's are factored out, representing the general double intersections of the curve with lines that include the origin. For the circle, only one such intersection is lost. The corresponding polar eqs. are $r^2 = a^2(2\cos^2\theta-1)$ and $r = \sqrt{2}a(\cos\theta-\sin\theta)$. Solving these eqs. simultaneously for r gives two roots $r = 0$, the intersections for $\theta = 45°$ and $225°$, and two roots $r = \pm 2/\sqrt{5}$, the intersections for $\theta = 161.565°$ ($\cos\theta = -3/\sqrt{10}$) and $\theta = 341.565°$. The $(0, 45°)$ root is the intersection of the first-traced of the coincident bisector circles with the corresponding point of the lemniscate at the polar pole, and the $(0, 225°)$ root is the intersection of the second-traced circle with it at the origin. The $(-2/\sqrt{5}, 161.565°)$ intersection is for the second-traced circle.

3. One has $a^2\cos 2\theta = a^2\cos^2\theta$, whence $-\sin^2\theta = 0$ and $\theta = 0°$ and $180°$. For these values of θ, $r_{0°} = a$ and $\pm a$, $r_{180°} = -a$, $\pm a$.

4. The segment in question is in the 4th-quadrant and it is congruent with the segment in the 1st-quadrant (and its reflection in the 0°-axis).

5. Make the substitutions $e^2a^2 = a^2-b^2$ and $b = a\sqrt{(1-e^2)}$ and simplify.

6. For a line including the origin, one has $y = mx$, where m is the slope. Find the points of intersection of this line with the bisector circle by simultaneous solution of the two eqs. This will reduce to a single point, namely a tangent-point, at the two positions for which the radicand vanishes. Setting the radicand equal to 0 and solving for m yields the slopes of the two tan-

gent-lines, $m = [h^2 \pm R\sqrt{(2h^2-R^2)}]/(h^2-R^2)$.

7. Assume a hyperbolic limacon (i.e., $b > a$). Eq. 17a, for hyperbolic limacons, corresponds to the orientation of Fig. 11a-middle. Since the DP of the curve represented by eq. 17a is at $r = h$ (along the 0°-axis), this corresponds to an x-h translation (assuming that h is positive). Accordingly, for hyperbolic limacons, h must be taken to be $(b^2-a^2)/2b$. [The same expression would be used on the assumption of an elliptical limacon (i.e., $a < b$), since the curve then would have to be translated to the left. Since eq. 17a is for an x-h translation, h would have to be negative for a resulting translation to the left.] Substituting in eq. 17a yields the eq., $r^4 + 2a^2r^3(\cos\theta)/b + r^2[(a^4/b^2)\cos^2\theta - a^2 + (a^4-b^4)/2b^2] + (a^2/2b^3)(a^2-b^2)^2 r\cos\theta + (a^2-b^2)^4/16b^4$. This can be factored into two quadratic eqs. in r, one of which is eq. 17c.

Hint for factoring, not knowing solution 17c: From inspection, it is evident that the two factors must be of the form, $r^2 + a^2 r(\cos\theta)/b \pm Br + (a^2-b^2)^2/4b^2$, or a sign variant thereof. Solve for B by multiplying out and comparing with the above eq. The solution must then be checked for proper values of the signs (as in eq. 17c) by verifying an intercept [Let $\theta = 0$ and check for $r = (b^2-a^2)/2b$].

An analytical solution is obtainable directly using the standard procedures of Structural Equation Geometry (see Chapters VII and VIII). Express the above eq. in powers of $\cos\theta$ and solve for $\cos\theta$ using the quadratic-root formula. The two roots are the required quadratic solutions in r, one of which is eq. 17c.

Chapter IV.

1. The distance eqs. are $u^2 = x^2+y^2$ and $v^2 = (x-d)^2+y^2$, whence $x = (u^2-v^2+d^2)/2d$ and $y = \pm\sqrt{[4d^2u^2-(u^2-v^2+d^2)^2]}/2d$.

2. Add together eq. 22a and the square of eq. 22b (either left or right), whence $x^2+y^2 = (u^2+v^2)/2 - d^2/4$. Since the eq. of a centered circle is $x^2+y^2 = R^2$, we have $(u^2+v^2)/2 = d^2/4 + R^2$. Letting $d = 2R$ yields $u^2+v^2 = d^2$.

3. From the distance eq. for v, $v^2 = (x-x_1)^2+y^2$, we have $x^2+y^2 = v^2+2xx_1-x_1^2$. Eq. 22a' is the general axial distance eq. about poles p_v at $(x_1,0)$ and p_u at $(x_2,0)$. Eliminating x between the two eqs. yields $x^2+y^2 = [(v^2x_2-u^2x_1)/(x_2-x_1)] + x_1x_2$.

4. The initial eqs. are $v'^2 = (x-d/2+h)^2+(y+k)^2$ and $u'^2 = (x+d/2+h)^2+(y+k)^2$, together with eqs. 21 and 22.

5. Relation 26c. This is shown by squaring and adding together the rectangular rotation eqs., $x = x'\cos\theta - y'\sin\theta$ and $y = x'\sin\theta + y'\cos\theta$.

6. Since the v^4/u^4-ratio (see eq. 27a) always is 1 (for general lines), the information sought is obtained from the v^2/u^2-ratio. This is $-(2bm+d)/(2bm-d)$. It is immaterial whether lines of positive or negative slope are taken as examples. Taking lines of negative slope, say $m = -1$, yields the ratio $(d-2b)/(d+2b)$. For this case, increasing positive values of b correspond to lines more and more distant to the right of the midpoint. Accordingly, a v^2/u^2-ratio less than 1 signifies a line closer to pole p_v. For lines normal to the bipolar axis (eq. 23d), a positive coefficient signifies a line closer to p_v. For lines parallel to the bipolar axis (eq. 27c), the larger the constant term, i.e., the larger the absolute value of b, the more distant the line.

7. For $x_1 = x_2 = 2a$, eq. 43a becomes $B^8 - 256a^2(y_2-y_1)^3[y_2v^2-y_1u^2+(y_1y_2-4a^2)\cdot(y_2-y_1)] = 0$. Let the two specific poles be $y_1 = na$ and $y_2 = ma$. Then $B^2 = \pm 4a\sqrt{(ax)}(n-m)$ for a general point on the curve at $[x, \pm 2\sqrt{(ax)}]$. Similarly, we find $[y_2v^2-y_1u^2+(y_1y_2-4a^2)(y_2-y_1)] = ax^2(m-n)$. Substituting these values in the above eq. yields $256a^6x^2(n-m)^4 - 256a^2(m-n)^3a^3\cdot ax^2(m-n) = 0$, which vanishes, verifying the original eq. for points $(2a,na)$ and $(2a,ma)$ and a general point on the curve.

8. Let $y_2 = na$, then $v^2 = 5a^2$ and $u^2 = a^2+a^2(2-n)^2$. Forming $B^2 = (v^2-u^2+y_2^2) = [5a^2-u^2-a^2(2-n)^2+n^2a^2] = 4a^2n$, the left-hand side of eq. 44c yields $B^8 = (4a^2n)^4 = 256n^4a^8$. Similarly, the right-hand side yields $256a^2(v^2-4a^2)y_2^4 = 256a^2\cdot a^2\cdot(an)^4 = 256n^4a^8$. Thus, eq. 44c applies at an LR-vertex $(a,2a)$ for a general value of the position of p_u on the 2a-chord at $(2a,na)$.

9. In one of several possible methods, translate the standard vertex parabola, $y^2 = 4ax$, $d/2 + a = 3d/4$ units to the left to bring the focus to the bipolar pole to the left of the origin, and label this pole p_v, yielding $y^2 = 4a(x+d/2+a) = d(x+3d/4)$. Making the substitutions of eqs. 22, regrouping to give a term $4d^2v^2$ on one side of the eq., and taking roots (only the positive roots give a solution), and regrouping again yields the desired eq. This eq. is not of comparable simplicity to eq. 38f, since when both eqs. are expanded, eq. 38f has only three terms (eq. 38f'), whereas the present eq. has four terms.

10. Pole p_v is taken to be at the focus of the rectangular parabola, $y^2 = 4a(x+a)$, which is translated a units to the left to bring the focus to the origin. Pole p_u is taken to be any point in the plane (x_1,y_1), whereupon $v^2 =$

ANSWERS TO PROBLEMS IV.   467

x^2+y^2 and $u^2 = (x-x_1)^2+(y-y_1)^2$. Substituting for y^2 in the eq. for v^2 leads to the simple linear result, $v = x + 2a$, whence $y = \pm 2\sqrt{[a(v-a)]}$. Substituting in the eq. for u^2 or v^2, or in the difference eq., v^2-u^2, leads to eq. 50.

11. Substitute $a = d/4$, $k = -a = -d/4$. This places the latus rectum in coincidence with the y-axis and the poles at $\pm 2a$, the positions of the LR-vertices.

12. Consider the general rectangular eq. of a curve of degree n in the form, $A_1x^n+B_1y^n+C_1x^{n-1}y+D_1x^{n-2}y + + +A_2^2x^{n-1}+B_2^2y^{n-1}$, etc. The maximum degree of any radical-free term will be 2n. In the cases of terms involving a radical (after making the substitutions of eqs. 22a,b), the term of maximum degree will be the product of a component of degree 2n-2 times the radical, with radicand of degree 4. When all the terms involving radicals are segregated on the right side, the radical factors out. Squaring both sides, the maximum degree will be 4n on the left and $2(2n-2)+4 = 4n$ on the right.

13. Take as the initial curve a standard rectangular hyperbola with its left focus translated to the origin, $b^2[x-\sqrt{(a^2+b^2)}]^2-a^2y^2 = a^2b^2$. Take p_u at $(0,b^2/2)$, the coordinates of the upper left LR-vertex. Letting $\sqrt{(a^2+b^2)} = ae$, the coordinates of a point (x,y) of the curve are x, $\pm(b/a)\sqrt{[(x-ae)^2-a^2]}$. It follows that $u^2 = x^2+[b^2/a\pm(b/a)\sqrt{(x^2-2aex+a^2e^2-a^2)}]^2$ and $v^2 = x^2+(b^2/a^2)[(x-ae)^2-a^2]$. Form the difference, v^2-u^2, and solve for x, giving $x = \pm a\sqrt{[(a^2/4b^6)(v^2-u^2+b^4/a^2)^2+1]}+ae$. Now substitute this value of x in the expression for v^2.

14. The eq. for a minor-axis rectangular ellipse is $a^2x^2+b^2y^2 = a^2b^2$. Making the substitutions of eqs. 22 leads to the sought-after 4th-degree eq. To obtain the eq. for the poles at the b-vertices, make the substitutions $b = d/2$ and $a^2 = d^2/4(1-e^2)$.

15. For the right focus, $x_1 = \sqrt{(a^2-b^2)}$. Substituting in eq. 79, $v^2a^2 = (a^2-b^2)x^2 -2b^2x\sqrt{(a^2-b^2)} + b^4$, and taking the square root yields $\pm va = x\sqrt{(a^2-b^2)} - b^2$. By testing, the upper alternate sign is found to apply, yielding
$v = [x\sqrt{(a^2-b^2)}]/a - b^2/a = ex - a(1-e^2)$

16. Set up the eq. for v^2 for the DP and the eq. for u^2 for the point (x_1,y_1). When this is done, and the radical is segregated on the left, and the entire eq. is squared, one obtains (letting $A^2 = u^2-v^2-x_1^2-y_1^2$), $(A^2+2xx_1)^2 = 4y_1^2[a^2/2 - bx-x^2\pm(a/2)\sqrt{(a^2-4bx)}]$. Segregating terms and squaring again to eliminate the radical, one obtains $x = [1/2(x_1^2+y_1^2)]\{x_1A^2\pm\sqrt{[4v^2y_1^2(x_1^2+y_1^2)-y_1^2A^4]}\}$. The general eq. for v^2 for the DP (for both loops) is $v^2 = a^2/2 - bx\pm(a/2)\sqrt{(a^2-4bx)}$, or $v^4+2v^2bx+b^2x^2 = a^2v^2$, or $v^2 + bx = \pm av$ (upper sign for the large loop, lower

sign for the small loop). Substituting the solution for x in either of the eqs. for both loops gives the 8th-degree eq.

$$[v^4-a^2v^2+bv^2(A^2x_1+by_1^2)/(x_1^2+y_1^2) + b^2A^4(x_1^2-y_1^2)/4(x_1^2+y_1^2)^2]^2$$
$$[b^2y_1^2/4(x_1^2+y_1^2)^4][4v^2(x_1^2+y_1^2)-A^4][2v^2(x_1^2+y_1^2)+A^2bx_1]^2.$$

17. Substituting x of eq. 84 in eq. 85a, squaring the left-hand side, and regrouping and cancelling terms, leads to eq. 92. Making the substitutions called for yields the eq. $v^4-v^2[2v^2-2u^2+3a^2] + (v^2-u^2+a^2)^2 = 0$, which can be tested for points of the e = 2 limacon (b = 2a), for example, v = 0, u = a; v = a, u = 0; v = 3a, u = 2a; and v = a, u = √2a.

18. To avoid confusion, we take the cases of hyperbolic limacons, for which b > a and the distance between the two foci is $d = (b^2-a^2)/2b$. The standard orientation of the initial curve of eq. 81 is with both loop vertices and the focus of self-inversion to the left of the DP (for elliptical limacons, one loop vertex and the focus of self-inversion are to the right of the DP). Since x is taken to be positive to the right and negative to the left, and the focus of self-inversion is to the left of the DP a distance d, we have $x_{f-s.-i.} = x_{DP} + d$. Hence, for a common value of x, we let x = x + d in the eq. for the focus of self-inversion, yielding $u^2+a^2(x+d)/b = bu-d^2$. Solving simultaneously gives $u^2-bu+d^2 = -a^2d/b+a^2v^2/b^2 \mp a^3v/b^2$. For both sides to be perfect squares, we need (after completing the squares) all the constant terms outside the completed squares to vanish from the eq. $(u-b/2)^2+d^2-b^2/4 = a^2(v \mp a/2)^2/b^2 - a^4/4b^2-a^2d/b$. Substituting the value given above for d leads to cancellation of all the constant terms, leaving $(u-b/2)^2 = a^2(v \mp a/2)^2/b^2$, whence $\pm(u-b/2) = a(v \mp a/2)/b$, or $av/b \mp u = \pm a^2/2b \mp b/2$. Testing for a hyperbolic limacon and substituting the value of d, yields the eq. for the correct combination of signs, $\pm av/b + u = d$, whence $\pm av + bu = bd$. For elliptical limacons the treatment is carried out with $d = (a^2-b^2)/2b$ and the focus of self-inversion to the right of the DP.

19. It readily is shown that the pole at -b/2 is to the left of the focus of self-inversion [at $(a^2-b^2)/2b$] in both elliptical (a > b) and hyperbolic (b > a) limacons, and that the distance between the two poles is $a^2/2b$. Hence, the x's of the two eqs. are related as follows, $x_{-b/2} = x_{f-s.-i.} + d = x_{f-s.-i.} + a^2/2b$. Substituting this value of x in eq. 89, eliminating the common value of x from the latter eq. and eq. 88, and rearranging and squaring, yields $(4/a^2)(u^2-a^2/2 - b^2/4)^2 = 2b^2-a^2-(4b^2/a^2)[bv-v^2-(b^2-a^2)^2/4b^2]$, whence it readily is shown that $(u^2-a^2/2 - b^2/4) = \pm b(v-b/2)$ or $u^2+b^2/4 = b(v+d)$. Testing shows that the eq. with the plus sign applies to all portions of all limacons.

20. For this case, $d = a^2/2b$, whence $d = b/4 = \sqrt{2}a/4$. The initial eq. (see Problem 19) is $u^2+b^2/4 = b(v+d)$, or $u^2+4d^2 = 4d(v+d)$, whence $u^2 = 4dv$.

21. This is accomplished by showing that terms in u and v of the same powers pair up either with identical coefficients or with the abscissae of the two poles exchanged in their coefficients. The complete expansion of eq. 92 follows:

$4(x_2v^2-x_1u^2)^2+4bx_2v^2(v^2-u^2)+2b^2(x_2^2-x_1^2)v^2+4a^2x_1(x_2-x_1)v^2+8x_1x_2^2(x_2-x_1)v^2$
$+b^2(v^2-u^2)^2 \quad +4bx_1u^2(u^2-v^2)+2b^2(x_1^2-x_2^2)u^2+4a^2x_2(x_1-x_2)u^2+8x_1^2x_2(x_1-x_2)u^2$
$+4bx_1(x_1^2-2x_2^2+x_1x_2)u^2 \quad +4x_1^2x_2^2(x_2-x_1)^2+b^2(x_2^2-x_1^2)^2-4a^2x_1x_2(x_2^2-x_1^2) +$
$4bx_2(x_2^2-2x_1^2+x_1x_2)v^2 \quad +4x_1x_2b(x_2-x_1)(x_2^2-x_1^2).$

22. This derivation can be carried out in several ways, the easiest of which is to employ eq. 92. Letting $x_1 = 0$ and $x_2 = (a-b)$ and rearranging, yields $b^2[v^2-u^2+(a-b)^2]^2+4(a-b)v^2\{b[v^2-u^2+(a-b)^2]-a^2(a-b)]\}+4(a-b)^2v^4 = 0$. Since the DP is an independent bipolar point-focus, this eq. must reduce. Transferring the term $4(a-b)^2a^2v^2$ to the right leaves both sides perfect squares, whence $b[v^2-u^2+(a-b)^2]+2(a-b)v^2 = \pm 2(a-b)av$ or $b(a-b)^2-bu^2 = \pm 2a(a-b)v+(b-2a)v^2$, where the plus sign is for the large loop and elliptical oval, and the minus sign is for the small loop.

23. The initial eq. is 78b for axial focal ellipses. Since the left directrix is at a distance a/e from center, the directrix-focus distance is $a/e - ae$. Thus, the reflection of the left focus in the left directrix is at a distance $2a(1/e-e)$ from the focus. Accordingly, $x_1 = -2a(1/e-e) = -d$. Substituting for x_1 in eq. 78b (recalling that $b^2 = a^2-a^2e^2$) yields $v^2-u^2+d^2+(2d/ae)(av-a^2-a^2e^2) = 0$ or $v^2+2dv/e = u^2$.

24. For the standard rectangular hyperbola, we have $b^2x^2-a^2y^2 = a^2b^2$. The conventional distance eqs. cast about the axial vertices are $u^2 = (x-a)^2+y^2$ and $v^2 = (x+a)^2+y^2$, whence $x = (v^2-u^2)/4a$. Using the distance eq. for p_v and substituting for x and y yields $v^2 = [(v^2-u^2)/4a+a]^2+(b^2/a^2)[(v^2-u^2)^2/16a^2-a^2]$, whence $(a^2+b^2)(v^2-u^2)^2/16a^4-(u^2+v^2)/2 = (b^2-a^2)$. Making the substitutions, $a = d/2$ and $b^2 = d^2(e^2-1)/4$ yields $e^2(v^2-u^2)^2-2d^2(u^2+v^2) = d^4(e^2-2)$, the same as for ellipses (eq. 75).

25. Letting $x_1 = 0$ in eq. 100b yields $(v^2+a^2/2)^2 = (a^2/4)(a^2+8x^2)$ or $v^4-a^2v^2 = 2a^2x^2$ for the center, while letting $x_1 = a$ yields $(v^2+2ax+3a^2/2)^2 = (a^2/4) \cdot [a^2+8(x+a)^2]$ or $v^4+av^2(3a+4x)+2a^2(x+a)^2 = 0$ for the right vertex. Both eqs. are 4th-degree in v and 2nd-degree in x, but the eq. for the center is much simpler. Hence the center has higher subfocal rank. For comparison, the eq. of the

right-loop focus is 4th-degree in v but only 1st-degree in x. Accordingly, considered independently of the left loop-focus, it has higher subfocal rank than either the center or the right vertex, while considered as a member of a pair with the left-loop focus it has dependent point-focal rank.

26. The initial eq. is 100b. This will not reduce because the right side is not a perfect square. Hence the only possibility for reduction is to rearrange in powers of x and complete the square. This gives
$2(2x_1^2-a^2)[x+(v^2+x_1^2-a^2/2)x_1/(2x_1^2-a^2)]^2 = a^2(v^2+a^2-3x_1^2)^2/2(x_1^2-a^2/2) + a^4/4$. Both sides of this eq. are perfect squares only for the condition $a = 0$, which is not allowed. Hence reduction is not possible.

27. Letting $x_u = x_v + \sqrt{2}a$ in eq. 101b and eliminating x between the resulting eq. and eq. 101a, gives $(u^2+a^2)^2+u^2(v^2+a^2)^2/v^2 = 5a^4(u^2+v^2)/v^2 + 4a^2u^2$. The method of reduction of this eq. is not obvious, but lies in expanding and cancelling terms, yielding $u^2(u^2+v^2) = a^4(u^2+v^2)/4v^2$. Factoring out (u^2+v^2), which does not represent a real solution, yields $u^2 = a^4/4v^2$, whence $uv = a^2/2 = d^2/4$.

28. Let $x_v = x_u - d$ in eq. 113b' and substitute the resulting value of x_u in eq. 113a, giving $A^2(x_u) = u^2(A^2-B^2)/2d - BCu+d(A^2-C^2)/2$. After completing the square on both sides, the remainders cancel each other, yielding $A^2[v-ACd/(A^2-B^2)]^2 = B^2[u-BCd/(A^2-B^2)]^2$. Taking roots yields $Av \mp Bu = Cd(A^2 \mp B^2)/(A^2-B^2)$. On testing, this becomes $Av - Bu = Cd$ for the small oval.

29. Letting $x_u = x_w + (A^2-C^2) d/(A^2-B^2)$, eliminate x_u and x_w from eqs. 113a and 113c', as in Problem 29, yielding $C^2[u-BCd/(A^2-B^2)]^2 = A^2[w+ABd/(A^2-B^2)]^2$. Taking roots yields $Cu \mp Aw = Bd(C^2 \pm A^2)/(A^2-B^2)$. By testing on a specific small oval (see Fig. 21-1 or 23), the correct signs are found to be the lower set, yielding $Cu + Aw = Bd(C^2-A^2)/(A^2-B^2)$. Now $d_{uw} = \pm(C^2-A^2)/(A^2-B^2)$, whence $Aw + Cu = Bd_{uw}$. For the large oval, the eq. is $Av - Cu = -Bd_{uw}$.

30. Eliminate x from eqs. 113b,c, segregate terms, and complete the squares, as in Problems 29 and 30, giving $C^2[v-ACd/(A^2-B^2)]^2 = B^2[w+ABd/(A^2-B^2)]^2$, whence, for the small oval, $Cv + Bw = Ad(C^2-B^2)/(A^2-B^2)$. Now $d_{vw} = \pm[d-(A^2-C^2)d/(A^2-B^2)] = \pm[(C^2-B^2)d/(A^2-B^2)]$, whence $Bw + Cv = Ad_{vw}$. For the large oval, $Bw - Cv = -Ad_{vw}$.

31. We start with the general axial bipolar eq. for poles at x_1 and x_2, expressed in terms of x, the abscissa of a point on the curve, that is, $x = (v^2-z^2-x_1^2+x_2^2)/2(x_2-x_1)$. Let $x_2 = d_{uz}$ and $x_1 = [8d \pm Cd_{uz}\sqrt{(C^2-16)}-C^2d_{uz}]/8$ & equate

ANSWERS TO PROBLEMS IV.   471

the x of this eq. to the x of eq. 119b, the conventional distance eq. from focus p_v. Letting $G^2 = [8+C\sqrt{(C^2-16)}-C^2]/8$, $F = [C-\sqrt{(C^2-16)}]$, and $E = \sqrt{(C^2-16)}$, we have $[v^2-z^2-d_{uz}^2 G^4/64+d_{uz}]/(Cd_{uz}F/4) = [v^2\pm\sqrt{(8C)}d_{uz}v/\sqrt{F}+C^2d_{uz}^2 G^2/32]/(Cd_{uz}F/4)$, where the plus sign is for the small oval. After combining appropriately and simplifying, we obtain the desired eq., $z^2\pm\sqrt{(8C)}d_{uz}v/\sqrt{F} = Cd_{uz}^2 F/4$. Since $d_{vz} = C(C-E)d_{uz}/8$, we also have $z^2\pm 8\sqrt{(8C)}d_{vz}v/C(C-E)^{3/2} = 16d_{vz}^2/C(C-E)$. The Problem also can be solved readily using the two corresponding conventional distance eqs. (eqs. 119b and 120).

32. Set up the distance eq. for a pole at $x = x_1$, $z^2 = (x-x_1)^2+y^2$ and substitute for y^2 from eq. 116b, giving $(z^2-x_1^2-C^2d^2/2+d^2)+2x(x_1-d) = Cd\sqrt{(2dx-d^2-C^2d^2/4)}$. Now let $x_1 = d$, whence $(z^2-C^2d^2/2)^2 = C^2d^2(2dx-d^2+C^2d^2/4)$ and $z^4-C^2d^2z^2+C^2d^4 = 2C^2d^3x$, the sought-after conventional distance eq. Now take the general axial bipolar eq. $x = (z^2-u^2-x_1^2+x_2^2)/2(x_2-x_1)$ and let $x_2 = 0$ for focus p_u, and $x_1 = d$ for pole p_z. Then $x = (z^2-u^2-d^2)/(-2d) = [z^4-C^2d^2z^2+C^2d^4]/2C^2d^3$. Cross-multiplying, cancelling terms, and taking roots, yields the equation $z^2 = Cdu$. The Problem is even more easily solved by using the general axial bipolar eq. for p_u and p_z and the distance eq. $u^2-Cdu+d^2 = 2dx$, for pole p_u.

33. Solve the conventional distance eqs. (119b,c) for these foci for x, and equate the solutions, where $E = \sqrt{(C^2-16)}$. Factor out common terms from the denominators and complete the squares by adding and subtracting the squares of half the linear terms. Equate the remainders to each other and show that they cancel. This leaves the completed squares, $(C+E)[v+\sqrt{(8C)}d_{uz}/2\sqrt{(C-E)}]^2 = (C-E)[w+\sqrt{(8C)}d_{uz}/2\sqrt{(C+E)}]^2$. Take the roots, combine terms, and determine the correct signs by testing with the ovals of Fig. 23, giving $\sqrt{(C+E)}v - \sqrt{(C-E)}w = \pm E\sqrt{(C/2)}d_{uz}$ (plus sign for large oval). Now $d_{vw} = x_1-x_2 = Cd_{uz}E/4$, whence $\sqrt{(C+E)}v - \sqrt{(C-E)}w = 4d_{vw}\sqrt{(1/2C)}$ (eq. 121b').

34. Letting $E = \sqrt{(C^2-16)}$, solve the conventional distance eqs. 119a,b for x and equate the solutions. Complete the squares by adding and subtracting the squares of half the linear terms. Equate the remainders and show that they cancel. You must show that $C(C-E)(1-C^2/4) = C^2(8+CE-C^2)/4 - 16C/(C-E)$. This leaves the completed squares $C(C-E)(u-Cd_{uz}/2)^2 = 8[v\pm\sqrt{(8C)}d_{uz}/2\sqrt{(C-E)}]^2$ (plus sign for the small oval). Taking the roots, simplifying, and determining the appropriate signs by testing (Fig. 23), yields $\sqrt{[C(C-E)}u \pm 2\sqrt{2}v = C\sqrt{[C(C-E)}]d_{uz}/2 - \sqrt{[16C/(C-E)}]d_{uz}$. Now $d_{uv} = d_{uz}(8-CE-C^2)/8$, since $C > 4$ for biovals. Substituting yields $\sqrt{[C(C-e)}]u \pm 2\sqrt{2}v = 4\sqrt{C}d_{uv}/\sqrt{(C-E)}$, the required solution. For the equi-

lateral limacon ($b = \sqrt{2}a$, $e = \sqrt{2}$), which is the curve of demarcation (for $C = 4$), we obtain $4u \pm 2\sqrt{2}v = 4d_{uz}$ (plus sign for small loop).

35. After squaring the left side of eq. 117b and arranging in powers of x on the left, completing the square, and gathering together the remaining terms on the right, one obtains the expression $-4(x_1-d)v^2 - 4(x_1-d)(d^2 - C^2d^2/2 - x_1^2) + d(C^2-4)(x_1^2 - 2dx_1 + d^2) + C^2d^3$. To be able to extract a root on the right side, all but the v^2-term must vanish. Equating the remaining terms to 0 and cancelling yields the desired eq. 118a.

36. Substitute the values for A, B, and C of eq. 121a' in eq. 123d, that is, show that $8 \cdot 16C(C-E) + 8 \cdot C(C-E) = 16C^2$, where $E = \sqrt{(C^2-16)}$.

37. The values of A_L, B_L, and C_L of eqs. 113a,b are those of the linear Cartesian eq., $B_L u + A_L v = C_L d_{uv}$. These parameters are expressed in terms of C_p by eqs. 124a-c. Making these substitutions in eqs. 113a,b gives for p_u the distance eq. $8x - (8-C^2+CE)u^2/2d_{uv} - [8-16C/(C+E)]d_{uv}/2 = \pm 4C_p u$, whereas for p_v it gives $C(C-E)x - (8-C^2+CE)v^2/2d_{uv} - [C(C-E)+16C/(C-E)]d_{uv}/2 = \pm 8\sqrt{[2C/(C-E)]}v$. From eq. 118c, focus p_v lies at $x_1 = (8+CE-C^2)d_{uz}/8$ relative to focus p_u, whence $d_{uv} = -(8+CE-C^2)d_{uz}/8$. The minus sign is employed because focus p_v lies to the left of focus p_u in biovular parabolic Cartesians ($C > 4$; it is imaginary in the monovular members, since $E = \sqrt{[C^2-16]}$ is imaginary). Using this value of d_{uv}, replace d_{uz} in the above eqs. and substitute C^2-16 for E^2, leading to eqs. 119a,b.

38. Deriving eq. 137 from the hybrid distance eqs., let $x = x-a$ in eq. 136c, yielding $x = a(a^2-v^2)/(a^2+v^2)$. Now substitute this value of x in eq. 136b, leading to the eq. $u^2(a^2+v^2) = (a^2-v^2)^2$. To derive eq. 137 from the general axial distance eq., let $x_1 = 0$ for p_v at the loop vertex and $x_2 = -a$ for p_u at the DP (since eq. 136c is cast about the loop vertex), whereupon the general axial distance eq. becomes $v^2 - u^2 + a^2 = -2ax$. Now eliminate x between this eq. and eq. 136b, yielding $u^2(a^2+v^2) = (a^2-v^2)^2$.

39. Eliminate x and y between eqs. 136a, $v^2 = (x-x_1)^2 + y^2$, and $(v^2-u^2-x_1^2+x_2^2) = 2x(x_2-x_1)$, where p_v is at $(x_1,0)$ and p_u is at $(x_2,0)$. Letting $A^2 = (v^2-u^2-x_1^2+x_2^2)$, yields $v^2 = \{[A^2-2x_1(x_2-x_1)]/2(x_2-x_1)\}^2 + [A^4/4(x_2-x_1)^2][a-A^2/2(x_2-x_1)]/[a+A^2/2(x_2-x_1)]$. Eliminating fractions and combining terms, yields $2av^2(x_2-x_1)^2 + A^2v^2(x_2-x_1) = A^4(a-x_1) + A^2x_1(x_2-x_1)(x_1-2a) + 2ax_1^2(x_2-x_1)^2$. Substituting for A^2 and A^4 and cancelling and combining terms leads to eq. 138.

40. The initial eq. is the rectangular eq. of central conics, $b^2(x-a)^2 \mp a^2y^2 = a^2b^2$ (upper sign for hyperbolas), cast about the left vertex (x-a translation). First convert to polar coordinates, $r^2(r\cos\theta-a)^2 \mp a^2r^2\sin^2\theta = a^2b^2$. Now invert by replacing r by a^2/r (one does not replace r merely by 1/r because that does not maintain dimensional balance), $b^2(a^2\cos\theta/r-a)^2 \mp a^6\sin^2\theta/r^2 = a^2b^2$. Now clear of fractions and restore to rectangular coordinates by multiplying through by r^4, yielding $b^2(a^2x-ax^2-ay^2)^2 \mp a^6y^2 = a^2b^2(x^2+y^2)^2$. Expand, cancel terms, and regroup to $a^2b^2x^2 \mp a^4y^2 = 2ab^2x(x^2+y^2)$. Dividing by $2ab^2$ and regrouping, yields eq. 135a.

41. The distance eq. 119a for p_u of parabolic Cartesians is taken as the simplest example but the results apply in general. Let $x_2 = 0$ for p_u and eliminate x between eqs. 119a and the general axial bipolar distance eq., $(v^2-u^2-x_1^2+x_2^2) = 2x(x_2-x_1)$, yielding the general axial focal eq. for focus p_u, $dv^2+u^2(x_1-d) - Cdx_1u = dx_1(x_1-d)$. For $x_1 = d$, this eq. is parabolic, for $x_1 > d$, it is elliptical, and for $x_1 < d$, it is hyperbolic. For $x_1 = 2d$, it is "circular." The equilateral hyperbolic case, $x_1 = 0$, is not allowed, since this places pole p_v in coincidence with p_u. The corresponding eq. for pole p_u for the general eq. of linear Cartesians (derived from eq. 113a) is $v^2+[x_1(A^2-B^2)-A^2d]u^2/A^2d \pm 2x_1BCu/A^2 = x_1[x_1A^2-d(A^2-C^2)]/A^2$, from which similar conclusions can be drawn in relation to x_1 and distance $dA^2/(A^2-B^2)$. For this eq. to simplify to the standard eq. of a vertex parabola, $v^2 = Cdu$, the two bracketed expressions must vanish. Equating them to 0 and eliminating x_1 leads to the symmetry condition, $A^2B^2+A^2C^2 = B^2C^2$, for parabolic Cartesians.

42. This is a consequence of the facts that limacons are the curves inverse to conics about their traditional foci, and that the slopes of conics at their LR-vertices are equal in magnitude to their eccentricities (RULE 5). Recall that the limacon LR-vertices are the points inverse to the conic LR-vertices. Consider also that the inversion transformation is negatively conformal, that is, it preserves the magnitudes of angles but reverses their senses. Accordingly, the angle that the radius vector for the inversion makes with the tangent at a conic LR-vertex is the same as the angle that it makes with the tangent at a limacon LR-vertex, but of reverse sense. Since the radius vector with which we are concerned is normal to the x-axis for the conic to limacon inversion, a reversal of the angle results directly in a reversing of the slope; the same phenomenon occurs for radius vectors parallel to the x-axis (but not for radius vectors to the contralateral LR-vertices). For radius vectors at other angles, adjustments must be applied (see Problem 44). On the other hand, if all angles are specified with respect to the radius vectors (rather than the x-axis), no adjustments need be made.

43. For radius vectors between inverse curves (from the pole of inversion) directed at ±90° or ±180° to the x-axis, the slope angles are either equal (at 0°

or 90°) or equal in magnitude but of opposite sign. At all other angles, the following alternative relationships apply: (1) the angles of the two tangents become equal in magnitude and opposite in sign if the angle of the radius vector is added to one of them, (2) same as rule (1) but if twice the angle of the radius vector is subtracted from one of them; and (3) same as rule (1) if ±(180° - twice the radius-vector angle) is added to one of them.

44. Segregating the terms in u^4 and v^4 from eq. 128, yields $[x_2^2-(C^2-B^2)/4]v^4$ and $[x_1^2-(C^2-B^2)/4]u^4$, where x_2 is the abscissa of p_u and x_1 is the abscissa of p_v. The condition, $4h^2 = (C^2-B^2)$, defines the *$cos^2\theta$-condition circumpolar foci*, which lie within the ovals (see page 192). Accordingly, since the coefficient of the near exterior pole is the square of the abscissa of the far exterior pole, less the square of the abscissa of an interior pole, the coefficient of the near pole is larger than that of the far pole. This gives the near pole greater weight in the eq. (based on quartic terms). The argument for the quadratic terms is more complex.

45. Applying the results of Problem 3, according to which $x^2+y^2 = (v^2x_2-u^2x_1)/(x_2-x_1) + x_1x_2$, and using eqs. 22a' and 6k, yields the general axial bipolar eq. $[v^2(x_2-a/2)-u^2(x_1-a/2)+x_1x_2(x_2-x_1)-(a/2)(x_2-x_1)^2]^3 = (27/4)a^2(x_2-x_1)[v^2x_2-u^2x_1 + (x_2-x_1)x_1x_2]^2$. From examining this eq., it is evident that the pole at x_1 or $x_2 = 0$ (the small-loop vertex) and the pole at x_1 or $x_2 = a/2$ (a pole within the large loop) have the highest subfocal rank (each condition leads to greater simplification of the eq. than any third condition). Letting $x_2 = 0$ and $x_1 = a/2$, yields $(a^2/4-v^2)^3 = -(27/4)a^2u^4$. The pole at $a/2$ is highest ranking because its variable is weighted heaviest and occurs to the lowest degree in the eq. There is no focal condition for the general axial eq. The only obvious possibility--the vanishing of the 6th-degree terms--requires $x_1 = x_2$, which is non-productive, since it requires coincident poles.

46. The polar eq. is $r^6 = (ar^2+br^2\cos^2\theta)^2$, whence, taking roots, $r^3 = ar^2+br^2\cos^2\theta$ and $v^3 = av^2+bx^2$. Since this hybrid eq. is of lower degree than the rectangular eq., the pole about which the eq. is cast (a true center of the curve) is a bipolar focus (RULE 26). Employing the general axial bipolar eq. in terms of x, $(v^2-u^2-x_1^2+x_2^2)/2(x_2-x_1) = x$, letting $x_1 = 0$ for the focus at the origin, and eliminating x between the two eqs., yields the 4th-degree general focal axial bipolar eq. $4x_2^2v^3 = 4av^2x_2^2+b(v^2-u^2+x_2^2)^2$. The corresponding general axial bipolar eq. is 8th-degree (see Problem 47). It can be shown, both by expanding and by solving for x_2^2, that there are no conditions for which the general focal axial bipolar eq. reduces, nor are there axial poles of high subfocal rank for which

the eq. simplifies (say, by the vanishing of a term or the simplification of a coefficient).

47. The expressions for x^2 and y^2 are obtained from eqs. 22a,a',b and the general conventional distance eq. for v, $v^2 = (x-x_1)^2+y^2$. Making these substitutions in the sextic eq., yields $[(u^2+v^2)/2 - d^2/4]^3 = \{a[(u^2+v^2)/2 - d^2/4] + b(u^2-v^2)^2/4d^2\}^2$, for equivalent axial poles. The corresponding eq. for arbitrary axial poles is $[(v^2x_2-u^2x_1)/(x_2-x_1) + x_1x_2]^3 = \{a[(v^2x_2-u^2x_1)/(x_2-x_1) + x_1x_2] + b(v^2-u^2-x_1^2+x_2^2)/4(x_2-x_1)^2\}^2$. The general axial distance eq., $(v^2-x_1^2+2xx_1)^3 = [a(v^2-x_1^2+2xx_1)+bx^2]^2$, is obtained by solving the conventional distance eq. for v (given above) for y^2 and substituting in the initial sextic eq.

48. The bipolar construction rules in u and v are $\sqrt{10}u + \sqrt{8}v = \sqrt{40}d$ (small oval) and $\sqrt{10}u - \sqrt{8}v = \sqrt{40}d$ (large oval), where $d = j/4$ (eqs. 104 and 124, and Fig. 23). Using the relation, $\pm Eu \pm Fv = d = 3.75j$, together with the values of the vertices of the ovals (Fig. 23), one finds the corresponding construction rules in v and w to be $\sqrt{(5/4)}w - \sqrt{5}v = 3.75j$ (small oval) and $\sqrt{5}v - \sqrt{(5/4)}u = 3.75j$. Eliminating the constant term between the two eqs. for the small oval and the two eqs. for the large oval, yields the tripolar construction rules, $\sqrt{10}u + [\sqrt{8} + \sqrt{(8/9)}]v = \sqrt{(2/9)}w$, and $\sqrt{10}u + \sqrt{(2/9)}w = [\sqrt{8} + \sqrt{(8/9)}]v$, for the small and large ovals, respectively.

To re-express these eqs. in the form of bipolar eqs. in u and w, set up the distance eqs. of a point in the plane (x_1,y_1) from pole p_u, p_v, and p_w, eliminate x and y, and solve for v in terms of u and w, yielding $v^2 = (15u^2+w^2-15j^2)/16$. Substituting this value of v in the tripolar eq. for the small oval and combining terms, yields $u^2+2\sqrt{(0.2)}uw+0.2w^2 = 4j^2$, whence $u + \sqrt{(0.2)}w = d_{uw}/2 = 2j$, or $2u + 2\sqrt{(0.2)}w = d_{uw}$. Since the corresponding eq. for the large oval differs only in signs, it is readily shown to be $2u - 2\sqrt{(0.2)}w = d_{uw}$.

[Note that if the original bipolar system is unpolarized, each bipolar construction rule represents two congruent ovals (reflected in the midline). After transforming to the tripolar system, only one oval is represented by each construction rule (since the three foci are not equally spaced, the tripolar system has no orthogonal line of symmetry). On transforming back to an unpolarized bipolar system, two congruent ovals again are represented by each construction rule.]

Chapter V.

1. The initial locus is the parabola, $y^2 = 4ax$, cast about its focus and a normal line-pole at $x = a/2$. The polar-linear eq. for the left half-plane segment of the initial locus is $u+v = 3d = 3a/2$. To find the complementary locus in the right half-plane, let $v^2 = (x-a/2)^2 + y^2$ and $u = x$, where the line-pole is the y-axis. Substituting leads to the eq. $y^2 = -2a(x-a)$. This is a confocal parabola of 1/2 the size of the initial parabola, opening to the left, with its vertex at a distance $a/2$ units to the right of its focus, and intersecting (coming into contact with) the initial parabola at the line-pole. The chord of intersection is midway between the vertex and the focus of the initial parabola, but at twice the distance from the vertex to the focus in the complementary parabola (the reflection of the vertex in the focus).

2. Either use eqs. 156a,b' and substitute in the eq. $x^2 = 4a(y+h)$, or displace the standard vertex parabola, yielding $y^2 = 4a(x+h)$, and make the substitutions $u = \pm y$ and $v^2 = x^2 + y^2$. Either approach yields the general eq. $16a^2(v^2-u^2) = (u^2-4ah)^2$. Expanding yields $16a^2v^2 = u^4 + 8au^2(2a-h) + 16a^2h^2$. For both sides to be perfect squares, one needs $h = a$, whence $4av = u^2 + 4a^2$. The most-symmetrical 4th-degree parabola in this group is the one for which $h = 0$, yielding the simplest 4th-degree eq. $u^4 = 16a^2(v^2-u^2)$. Next-most symmetrical is the parabola with the point-pole at the reflection of its vertex in the focus, with the eq. $u^4 = 16a^2(v^2-4a^2)$.

3. With the line-pole assumed to be to the right of the point-pole, and the point-pole at the origin, and assuming a locus to exist to the right of the vertex-tangent, the transformation eqs. are not 155a,b', but $v^2 = x^2+y^2$ and $u = \pm(d-x)$. Substituting in eq. 173a, squaring once, and putting in standard form, leads to the eq. $[x+de(e\pm1)/(1-e^2)]^2 - y^2/(e^2-1) = d^2(1\pm e)^2/(1-e^2)^2$. Letting $d = a(e-1)$ and simplifying for the case of the upper alternate sign, yields eq. 174a, while for the lower alternate sign it yields the complementary locus of eq. 174b. For an ellipse, the procedures are the same, except for the initial substitution, $u = (d-x)$, and the fact that $d = a(1-e)$, yielding $(x+ae)^2+y^2(1-e^2) = a^2$. There can be no complementary locus in the right half-plane of the ellipse (see Problem 5).

4. Eq. 170e requires that $v+eu = 2a$. The distance from the left focus to the right directrix is $ae+a/e = a(e^2+1)/e$. For e less than unity (the condition for ellipses), this fraction exceeds $2a$. Thus, since v alone exceeds $2a$ and u cannot be negative, there can be no locus to the right of the right directrix.

5. For the case of eq. 173a, v-eu = d, let x be the distance of an axial point of a presumed complementary locus beyond the left vertex-tangent. Then $u = x$, $v = a-ae+x$, and $d = a-ae$. Eq. 173a then requires that $x = ex$ for this point. Since this is not allowed for ellipses (but applies to the parabola), there can be no such point. For a non-axial point with the ordinate y, the corresponding eq. is $y^2+(1-e^2)x^2+2ax(1-e)^2 = 0$, for which there can be no real solution since all terms are positive. The demonstrations for eq. 173c follow the same course.

6. Let eq. 173b represent a right arm and be cast about the right focus and left vertex-tangent. This arrangement satisfies the conditions of eqs. 155a,b, b', so they are employed to eliminate u and v from eq. 173b. Squaring once, arranging in standard form, and letting $d = a(1+e)$, yields $[x-a(1\mp e)/(1-e)]^2 - y^2/(e^2-1) = a^2(1\mp e)^2/(1-e)^2$. Employing the upper signs for the locus in the right half-plane, yields $(x-a)^2 - y^2/(e^2-1) = a^2$. This is the initial hyperbola cast about the left vertex-tangent as the y-axis and p_v, the focus of the right arm at $x = a(1+e)$, for which eq. 173b represents the right arm. Now taking the lower signs of the above eq., for the complementary locus in the left half-plane, yields $[x+a(1+e)/(e-1)]^2 - y^2(e^2-1) = a^2(1+e)^2/(e-1)^2$. This hyperbola is magnified in size by a factor $(1+e)/(1-e)$, since the term on the right is the square of the length of the complementary hyperbola's semi-transverse axis and a is the length of the initial hyperbola's semi-transverse axis. Accordingly, from examining the left-most term, it is evident that the right vertex is at the origin and that the right vertex-tangent is coincident with the y-axis. The left arm is in the left half-plane. Testing points of this arm in eq. 173b confirms that eq. 173b applies to it. For example, the left vertex is at $x = -2a(1+e)/(e-1)$, whence $u = 2a(1+e)/(e-1)$ for the left vertex. Since p_v is at $x = a(1+e)$, $v = 2a(1+e)/(e-1) + a(1+e)$ and $d = a(1+e)$; these values satisfy eq. 173b.

7. Assume the arm of the initial hyperbola to be the right arm, as in Fig. $30b_1$. The transformation eqs. 155a,b then apply. Letting $d = ae$ and eliminating u and v yields the degenerate eq. $x^2(1-e^2)-2aex(1\mp 1)+y^2 = a^2(1-e^2)$. Employing the upper sign and putting the resulting eq. in standard form, yields the eq. of the initial hyperbola, $x^2+y^2/(e^2-1) = a^2$. For the lower sign, one obtains the eq. $[x+2ae/(e^2-1)]^2 - y^2/(e^2-1) = a^2(e^2+1)^2/(e^2-1)^2$. The linear size magnification thus is $(e^2+1)/(e^2-1)$. As a fraction of the semi-transverse axis, the displacement of the line-pole to the left is by the amount $2e/(e^2+1)$, or $2a\sqrt{(a^2+b^2)}/(2a^2+b^2)$, which is not a significant location (i.e., it is neither the location of the conjugate axis, a directrix, an LR, nor a vertex-tangent).

8. The rectangular eq. of axial circles is $(x-h)^2+y^2 = R^2$. Since $x = \pm u$, this becomes $u^2 \mp 2hu+h^2+y^2 = R^2$. The distance eq. for v^2 is $v^2 = (x-d)^2+y^2 = (\pm u-d)^2 + y^2$, whence $y^2 = v^2-(\pm u-d)^2$. When y is eliminated, a resulting term in u^2 is subtracted, leaving only linear terms in u.

9. The initial eqs. are $y^2 = 4a(x+x_1)$, $v^2 = x^2+(y-d)^2$, and $u = \pm y$. Eliminating x and y yields $16a^2v^2 = u^4+8au^2(2a-x_1)+16a^2(d^2+x_1^2)\pm32a^2du$. The negative alternate sign is for the portions of the initial parabolas in the upper half-plane (reflected in the polar-linear axis), and the positive alternate sign is for the portions in the lower half-plane. Letting $x_1 = 2a$ and $a = d/\sqrt{8}$ (for incidence), yields $2d^2v^2 = u^4-4d^3u+3d^4$ as the construction rule for the portions of the initial parabolas in the upper half-plane. Since this construction rule is simpler than that of the parabolas of Fig. 29d$_2$, the parabolas in question are more symmetrical (more symmetrically-positioned with respect to the reference elements).

10. The eq. of the dual hemi-parabolas of Problem 9 in the upper half-plane (eq. of Fig. 29d$_1$ for $x_1 = 2a$ and an incident point) is $2d^2v^2 = u^4-4d^3u+3d^4$. Let y be the intercept of the 2a-chord with the complementary locus represented by this eq. in the lower half-plane (i.e., the distance from p_u), and let $d = \sqrt{8a}$, whereupon $u = y$ and $v = y+\sqrt{8a}$. The eq. then becomes $16a^2(\sqrt{8a}+y)^2 = y^4-32\sqrt{8a^3}y + 192a^4$. Any positive roots of this eq. represent intersections of the 2a-chord with the complementary locus in the lower half-plane. The two positive roots are $y = 6.505a$ and $0.3432a$ (obtained by a zeros of functions calculation).

11. Letting $d = \sqrt{8a}$, since p_v is incident upon both the 2a-chord and the curve, yields construction rule $16a^2v^2 = u^4-32\sqrt{8a^3}u +192a^4$. To derive the rectangular eq. of the complementary locus in the lower half-plane, one lets $u = y$ and $v^2 = (y+\sqrt{8a})^2+x^2$ (or replaces y by $-y$ in both eqs., if desired, reflecting the curve in the x-axis). This leads to the eq. $x^2+4\sqrt{8}ay = (y^2/4a - 2a)^2$, for the complementary locus. This quartic eq. is not degenerate and does not represent parabolas.

12. Beginning with the construction rule, $v-eu = d/e$, and assuming that it is for the left arm cast about the focus of the right arm, let $v = \sqrt{[(d-x)^2+y^2]}$ and $u = +x$ (since we seek the locus in the right half-plane). Substituting, and putting the resulting eq. in standard form, yields eq. 171b. Had we started with the eq. $eu-v = d/e$, for the right arm cast about its own focus (eq. 171a-right), we would have let $u = -x$ and achieved the same result. To answer the second part of the Problem, we find the fractional displacement of the comple-

mentary hyperbola from its center in terms of its semi-transverse axis. Letting $d = ae$, the length of its semi-transverse axis is $a(1+e^2)/(e^2-1)$, while its displacement from center is by the amount $2ae/(e^2-1)$. Accordingly, the fractional displacement is $2e/(1+e^2)$, which is unexceptional (the exceptional fractional displacements are 0, 1, e, and 1/e).

13. For this situation, the relation between u and x is $u = x+d$ (eq. 157b), whence $x = u-a(1+e)$. Substituting in eq. 184 yields $v^2+aeu-a^2e(1+e) = \pm av$, where the lower sign is for the small loop.

14. Set the displacement of the directrix-line from the DP equal to the corresponding displacement of the vertex-tangent, whence $a(1-e^2)/2e - a(1-e^2)^2/4e = -a(1+e)$. This simplifies to $2(1-e) - (1-e)^2(1+e) = -4e$. The three roots of this eq. are -1, $(1-\sqrt{2})$, and $(1+\sqrt{2})$, of which only $(1+\sqrt{2})$ is positive and is the sought-after eccentricity.

15. For the left half-plane, $v = \sqrt{[(x+d)^2+y^2]}$ and $u = x$. Making these substitutions in eq. 191a yields the sought-after loci, $[(x+d)^2+y^2+ax/e]^2 = a^2e^2 \cdot [(x+d)^2+y^2]$, where $e > (\sqrt{2}+1)$ and $d = a(1+e)-(e^2-1)a/2e$.

16. For the coefficient of the du-term to be unity, set $(1-e^2)^2 = 4$, whence $e = \sqrt{3}$ and the eq. becomes $v^2 - 3dv \pm du = 0$. This is the eq. of the most-symmetrical limacon with a 2nd-degree eq. cast about an intersecting directrix and the focus of self-inversion (in the non-incident system).

17. For the directrix to lie to the left of the -b/2 pole, e must be greater than 2 (see eq. 192b). Accordingly, we must have $ae/2 + ae(e^2/4 - 1) = a(1+e)$, where the first term is the distance from the DP to the -b/2 (-ae/2) pole, the second term is the positive distance from the -b/2 pole to the directrix, and the third term is the distance from the DP to the -(a+b)-vertex. The roots of this eq. are 4, -2, and -2, whence $e = 4$.

18. The locus in question is $v^4-d^2v^2+4d^3u/27 = 0$. The transformation eqs. are $v^2 = (x+d)^2+ y^2$ and $u = x$, whence $y^4+2y^2[(x+d)^2-d^2/2]+(x+d)^4-d^2(x+d)^2+4d^3x/27 = 0$. Solving for y^2 yields $y^2 = -(x+d)^2+d^2/2 \pm d\sqrt{(d^2-16dx/27)}/2$, for which there is no real solution for y for the case in which both x and d are positive.

19. To find the rectangular eqs. of the complementary locus in the right half-plane, convert the eq. of the locus in the left half-plane to its rectangular equivalent in the right half-plane. The initial eq. is $v^2+2du = Cdv$, and the transformation eqs. are $v^2 = (x+d/2)^2+y^2$ and $u = x$, whence the locus in question

is represented by the portions of the rectangular curve, $[(x+d/2)^2+y^2+2dx]^2 = C^2d^2[(x+d/2)^2+y^2]$, in the 1st and 4th-quadrants. The x-intercepts of this locus are given by the eq. $x^2+(3\pm C)dx+(1/4 \pm C/2)d^2 = 0$. For $C = 5$, the intercepts are $x = (2\pm\sqrt{13})d/2$ and $(-8\pm\sqrt{53})/2$. Only one of these intercepts is positive, namely $x = 2.8028d$. The values $v = 3.3028d$ and $u = 2.8028d$ satisfy the initial polar-linear eq.

20. The conventional $C = 5$ distance eq. is $v^2/d \pm \sqrt{20}v - 25d/16 = 5x/2$. Since focus p_v is $d/4$ units to the left of the monoval-bioval focus, $x_v = x_{M-B} + d/4$, whence the hybrid eq. is $v^2/d \pm \sqrt{20}v - 25d/16 = 5(x-d/4)/2$ or $v^2/d \pm \sqrt{20}v - 15d/16 = 5x/2$ (plus sign for small oval). From this eq. it is evident that the directrix must be to the left of p_v (because a negative constant term is needed on the right to cancel the negative constant term on the left). Hence we employ the transformation eq. 157a, $u = \pm(x+d_{F-D})$ or $x = \pm u - d_{F-D}$, where d_{F-D} is the focus-directrix distance. Accordingly, the eq. becomes $v^2/d \pm \sqrt{20}v - 15d/16 = 5(\pm u - d_{F-D})$. Letting $d_{F-D} = 3d/8$ yields the construction rule $v^2/d \pm \sqrt{20}v = \pm 5u/2$ (plus signs for small oval and locus to right of p_u; left minus sign for large oval, right minus sign for locus to left of p_u). Since the small $C = 5$ oval is entirely to the right of p_u (see Fig. 23a$_2$), its construction rule is the 2nd-degree eq. $v^2/d + \sqrt{20}v = 5u/2$. For the large oval, the construction rule is $(v^2/d-\sqrt{20}v - 5u/2)(v^2/d-\sqrt{20}v + 5u/2) = 0$, while for both ovals it is $(v^2/d-\sqrt{20}v - 5u/2)(v^2/d-\sqrt{20}v + 5u/2)(v^2/d + \sqrt{20}v - 5u/2) = 0$.

21. The conventional distance eq. is $w^2/d \pm \sqrt{5}w - 25d = 10x$. Since focus p_w is $4d$ units to the left of the monoval-bioval focus, $x_w = x_{M-B} + 4d$, and the hybrid eq. is $w^2/d \pm \sqrt{5}w - 25d = 10(x-4d)$ or $w^2/d \pm \sqrt{5}w + 15d = 10x$ (plus sign for small oval). From this eq. it is evident that the directrix must lie to the right of p_w (since a positive term is needed on the right to cancel the positive constant term on the left). Accordingly, we use the transformation eq. 158a, $x = \pm u + d_{F-D}$, giving $w^2/d \pm \sqrt{5}w + 15d = 10(\pm u + d_{F-D})$. Letting $d_{F-D} = 1.5d$ (see Fig. 23a$_2$), we obtain $w^2/d \pm \sqrt{5}w = 10u$ (plus sign for small oval and locus to right of p_u). Since the small $C = 5$ oval is entirely to the right of p_u (Fig. 23a$_2$), its eq. is $(w^2/d + \sqrt{5}w - 10u) = 0$. The eq. of the large oval is $(w^2/d - \sqrt{5}w - 10u) \cdot (w^2/d + \sqrt{5}w - 10u) = 0$, while the eq. for both ovals is the product of all three parenthetical expressions equated to 0.

22. The pole at $-b/2$ is at $h = -\sqrt{2}a/2$ (with the DP as the origin) in the equilateral limacon ($e = \sqrt{2}$, $b = \sqrt{2}a$), while the focus of self-inversion is at $h = (a^2-b^2)/2b = -\sqrt{2}a/4$. From eq. 190a, the directrix for the focus of self-inver-

sion is $\sqrt{2}a/8$ units to the left of the focus, while from eq. 192b the directrix for the pole at $-b/2$ is $\sqrt{2}a/8$ units to the right of the pole. These values bring the directrices into coincidence midway between the two poles.

23. For the right vertex, the eq. is $z^4 - d^2z^2 = 2d^3u$. With the directrix for p_z at $x = d/2$ ($d/2$ units to the right of p_u and to the left of p_z), substitution yields $6.853d^4 - 2.617d^4 = 4.236d^4$, which confirms the rule for the right vertex. For the left vertex, the eq. is $z^4 - d^2z^2 = -2d^3u$. Substitution yields $0.1459d^4 - 0.3820d^4 = -0.2361d^4$, which confirms the construction rule for the left vertex.

24. For the initial circle in the p_v-containing half-plane to possess line-pole p_u as a secant (i.e., to overlap line-pole p_u), we need $R > h$. From eq. 180a, an existing complementary circle in the opposite half-plane is centered at $2d-h$ and has a radius of $\sqrt{[4d(d-h)+R^2]}$. This circle will overlap line-pole p_u if $\sqrt{[4d(d-h)+R^2]} > (h-2d)$. Squaring and cancelling terms yields the condition $R^2 > h^2$. But this also is the condition for overlap of the line-pole by the initial circle. Hence, if the initial circle overlaps the line-pole, the complementary circle in the opposite half-plane also will overlap it. By the Pythagorean theorem, an initial circle with $R > h$ intersects the line-pole p_u at $y = \pm\sqrt{(R^2-h^2)}$. By the same theorem, the complementary circle in the opposite half-plane also intersects the line-pole p_u at $\sqrt{\{[4d(d-h)+R^2] - (h-2d)^2\}} = \sqrt{(R^2-h^2)}$. Hence these two circle possess segments that join at the line-pole.

25. To derive the hybrid distance eq., substitute v for r and x for $r\cos\theta$ in the polar eq., yielding $v^3 = av^2 + bx^2$. The polar-linear distance u from line-pole p_u is $\pm x$ of the hybrid eq. Since x occurs only as the square in the hybrid eq., we have $x^2 = u^2$ and the polar-linear cubic construction rule $v^3 = av^2 + bu^2$. For the 0°-intercept, $r = a+b = v = u$, which satisfies the polar-linear eq. For the 90°-intercept, $r = v = a$ and $u = 0$, which also satisfies the eq.

Chapter VI.

1. Rearranging and squaring eqs. 199a,a' yields $(u_L \pm R)^2 = x^2 + y^2$ (upper sign for distances to exterior points) and $(u_G - R)^2 = x^2 + y^2$. Expanding, forming the differences $u_L^2 - v^2$ and $u_G^2 - v^2$ (employing eq. 199b for v^2), and rearranging yields $2dx = (u_L \pm R)^2 - v^2 + d^2$ (upper sign for exterior points) and $2dx = (u_G - R)^2 - v^2 + d^2$. These are combined to yield eq. 200a.

2. For ellipses, $a = R/2$ and $b = \sqrt{(R^2 - d^2)}/2$, while $e = d/R$. Accordingly, the focus-center distance is $ae = d/2$ and the focus-focus distance is d. Since the center of the curve is displaced to the right a distance $d/2$ from the origin (x_c, the center of the circle-pole), this brings the two foci to the positions in question. For hyperbolas, the demonstration follows the same course, except that $b = \sqrt{(d^2 - R^2)}/2$.

3. Let w be the distance from the origin to the left arm. For a greatest-distance construction, $u = R + w$. Accordingly, the hybrid distance eq. 143, $w = a(e^2 - 1) - ex$, which is for the left arm cast about the left focus, becomes $(u - R) = a(e^2 - 1) - ex$. In terms of the abscissa x, the distance to a point of the locus from x_c is $x = (w^2 - v^2 + d^2)/2d$. In terms of u and v this becomes $x = [(u-R)^2 - v^2 + d^2]/2d$. Eliminating x between the hybrid eq. and the latter eq. yields $u - R = a(e^2 - 1) - e[(u-R)^2 - v^2 + d^2]/2d$. Letting $a = d/2e$, cancelling terms in $\pm ed/2$, and taking roots, yields $(u - R + d/e)^2 = \pm v^2$. For $R = d/e$, this leads to the sought-after greatest-distance construction rule for the left arm, $u = v$, which applies to the right arm for least distances.

4. In terms of the parameters a and b of the polar eq. of limacons, $r = a - b\cos\theta$, $a = \pm 2R/(C^2 - 1) = \pm 2Cd/(C^2 - 1)$, and $b = -2Cd/(C^2 - 1)$, whence $r = \pm 2R/(C^2 - 1) + [2C^2 d/(C^2 - 1)]\cos\theta$, or $r = [2Cd/(C^2 - 1)](\pm 1 + C\cos\theta)$. The sign of the constant term only influences the phase of the plot, whereas the sign of the $\cos\theta$-term influences its orientation. Since $C = e$, C^2 is greater than 1 for hyperbolic limacons; the hyperbolic limacon of this eq. has the reverse orientation to that of the standard limacon (Figs. 7a, 12d_1,d_2,f_2, and 19). Accordingly, the focus of self-inversion is at an axial point $x = (a^2 - b^2)/2b$. Substituting for a and b, with $R = Cd$, yields $(b^2 - a^2)/2b = d$. This places the focus of self-inversion d units to the right of the DP in coincidence with p_v, with the DP at the center of the circle-pole. For elliptical limacons, and $C = e < 1$, the orientation is standard, and $(a^2 - b^2)/2b = [2Cd/(C^2 - 1)](1 - C^2)/(-2C) = d$, which is the same result.

5. We can set up the following three equalities between the coefficients of these eqs., $A^2/(A^2-C^2) = C_{pc}^2/(C_{pc}^2-1)$, $C_{pc}^2 d^2 - R^2 = (C_{pc}^2-1)d^2$, and $R^2(C_{pc}^2-1)^2 = C^4 d^2/(A^2-C^2)^2$. Simultaneous solution of these eqs. yields $R = d$ and $C_{pc} = A/B$. But for $B^2 = C^2$, we also have $A/B = 1/e$ (see eqs. 94a',b').

6. Solve the symmetry condition for A^2 and substitute this value in the expression for the location of the focus in question. This yields $x = C^4 d/B^4$, which is positive. Hence this focus always is to the right of the origin.

7. Let x be the radius of the circle-pole centered on the monoval-bioval focus at x_c. Assume first that the circle-pole intersects the line of symmetry between the two ovals (as in Fig. 32e) and set up the least-distance eqs. for the two vertices of the large oval, $2.6180j - x = 1.3820jC$ and $6.8541j - x = 10.8541jC$, where j is the unit of linear dimension ($d = j/4$ for the given vertex positions). Solving simultaneously yields $x = 2j$ and $C = 1/\sqrt{5}$. Thus, with $r = 2j$, the construction rule for the large oval is $u_L = w/\sqrt{5}$. This same least-distance construction rule with $R = 2j$ applies to the small oval. For example, the eq. for the right vertex is $2j - 0.3820j = (4j - 0.3820j)/\sqrt{5}$.

8. As an example, let the circle-pole be centered on p_w and let the monoval-bioval focus be at the origin, x_c. Testing for the vertex locations of the large oval in the eq. $w = Cv$ for an assumed radius, x, of the circle-pole, reveals that a greatest-distance construction is required. For example, $4j + x - 2.6180j = 2.6180jC$ is the eq. for the right vertex, and $4j + 6.8541j + x = 6.8541jC$ is the eq. for the left vertex for a greatest-distance construction. Simultaneous solution yields $x = 10j/\sqrt{5}$ and $C = \sqrt{5}$. Employing these same values in tests using the locations of the vertices of the small oval reveals that a least-distance construction rule applies for them. For example, for the right vertex, the eq. for a least-distance construction is $0.4721j + 0.3820j = 0.3820j\sqrt{5}$. Testing other combinations of poles confirms this situation; unless the circle-pole is centered on the monoval-bioval focus, both least and greatest-distance constructions must be used with the construction rule $u = Cv$ to produce both ovals.

9. Since u equals either $\pm[\sqrt{(x^2+y^2)}-R]$ or $\sqrt{(x^2+y^2)} + R$, and using eq. 199b for v, eliminate u and v from eqs. 213a',a" for both alternatives. The first squaring gives eqs. of the form $(x-d)^2 + y^2 = (Cd\pm R)^2 \pm 2(Cd\pm R)\sqrt{(x^2+y^2)} + x^2 + y^2$. Eq. 214 represents all the various eqs. obtained after cancelling the x^2 and y^2-terms, segregating the radical on one side, squaring again, and putting the

resulting four eqs. in standard form for central conics (by completing the squares).

10. Substituting from eqs. 199a,a',b in $u^2 = Cdv$ and eliminating radicals yields the non-degenerate 8th-degree eq. $16C^2d^2R^2(x^2+y^2)[(x-d)^2 + y^2] = \{(x^2+y^2-R^2)^2-C^2d^2[(x+d)^2+y^2]\}^2$. Following the same procedure for $v^2 = Cdu$ yields the degenerate 8th-degree eq. $[(x-d)^2+y^2\pm CdR]^2 = C^2d^2(x^2+y^2)$. The latter represents two sets of Cartesian ovals. These include parabolic Cartesians as special cases. They become exclusively parabolic Cartesians for $R = 0$ (the bipolar system). The non-degenerate 8th-degree eq. represents parabolic Cartesians only for $R = 0$, but the latter are cast with pole p_z (the limacon $-b/2$ homologue) at the origin (since the eq. for $R = 0$ has undergone an $x+d$ translation from the standard eq. 116, which shifts the curve a distance d to the left, bringing pole p_z to the origin).

11. Combine eqs. 217a,a' to give $u = \pm[\sqrt{(x^2+y^2)}\pm R]$--which allows for all valid combinations of signs (the combination $u = -\{\sqrt{[x^2+y^2]}+R\}$ is disallowed)--leading to the eq. $\pm[\sqrt{(x^2+y^2)}\pm R] = \pm(x-d)$, whence $\pm\sqrt{(x^2+y^2)} = \pm(x-d) \pm R$. Squaring both sides, cancelling the x^2-terms, and putting in standard form, leads to eq. 218a. Because of the presence of alternate signs of the term $\pm(x-d)$ for v in the above eq., it is immaterial whether the least-distance, greatest-distance, or disallowed combination of signs is employed in the expressions for u. In all three cases, eq. 218 results.

12. Let $d = nR$ (n, positive), then the vertex of arm 1 lies at $x = (n+1)R/2$, which is to the right of x_c. Since p_v lies at $x = nR$, and the nearest point on the circle-pole lies at $x = R$, the location of the vertex is seen to be midway between these positions.

13. The initial eq. is $w = -ex_w + a(1-e^2)$. Letting $w = u \pm R$ (for greatest distances and least distances to exterior points), this becomes $u \pm R = -ex_w + a(1-e^2)$. Now the directrix distance v, from line-pole p_v at $x = 5R/2$, is given by the eq. $v = 5R/2 - x_w$, assuming, for the moment, that the locus is entirely to the left of the line-pole, as in the polar-linear case. Thus, $u \pm R = a(1-e^2) - e(5R/2 - v)$. In order for this eq. to simplify to $u = ev$, it is necessary that $\pm R = a - ae^2 - 5eR/2$. Letting $e = 1/2$ yields $R(5/4 \pm 1) = 3a/4$, whence $a = R/3$ and $3R$ (see Fig. 35a). Both ellipses are, indeed, to the left of line-pole p_v. The relation for least distances to interior points, $w = R-u$, does not apply to this case.

14. We begin with the hybrid distance eq. 147, $w^2 - Cdw + d^2 = 2dx_W$ and carry through the solution for least distances to exterior points and greatest distances. Letting $w = u \pm R$ (upper sign for least distances to exterior points), yields $(u \pm R)^2 - Cd(u \pm R) + d^2 = 2dx_C$. In order for the linear terms in u to cancel, we must have $Cd = 2R$ and a least-distance construction (and this also applies when $w = R - u$ for least distances to interior points). Accordingly, we obtain $u^2 = 2d_{uz}x_c + R^2 - d_{uz}^2$. In order for this eq. to take the form $u^2 = Cdv$ we must have any one of the three conditions, $x_c = v \pm d_{uv}$ or $x_c = v$ (eqs. 157b, 158b, and 159b). In all three cases the curve must be to the right of the line-pole. For $x_c = v + d_{uv}$, the line-pole is to the right of x_c (allowable only for monovals), for $x_c = v - d_{uv}$ it is to the left of x_c, while for $x_c = v$ it includes x_c (the diametric system). The latter case is ruled out, since then $d_{uv} = 0$. Allowing for the other two cases yields the eq. $u^2 = 2d_{uz}v \pm 2d_{uz}d_{uv} + R^2 - d_{uz}^2$. For p_v to be a directrix, we require $R^2 \pm 2d_{uz}d_{uv} - d_{uz}^2 = 0$ (upper sign for the line-pole to the right of x_c). Since d_{uz} must be positive, this yields the root $d_{uz} = \pm d_{uv} + \sqrt{(d_{uv}^2 + R^2)}$.

Accordingly, the required construction rule is $u^2 = 2[\sqrt{(d^2+R^2)} \pm d]v$, whence $C = 2\sqrt{[1+(R/d)^2] \pm 2}$. In other words, the linear-circular eq. $u^2 = Cdv$ represents parabolic Cartesians cast about the monoval-bioval focus and a directrix when the parameter C of this eq. is equal to $2\sqrt{[1+(R/d)^2] \pm 2}$ (upper sign for the directrix to the right of x_c). Since the curve must be to the right of the line-pole, these curves are represented by the upper signs in eq. 223 (the loci of which are to the right of the directrix in Fig. 38).

Applying these results to Fig. 21-1a, since d_{uz} is the unit of distance, all distances are expressed in units of d_{uz}. Since the bipolar C is equal to 1 for Fig. 21-1a, and $C = 2R/d_{uz}$, we have $R = 1/2$. Also, since $d_{uz} = \pm d_{uv} + \sqrt{(d^2+R^2)}$, we have $d_{uv} = 3/8$ and the directrix is to the right of x_c. Accordingly, the constant C of the polar-linear eq. is 16/3 and the construction rule is $u^2 = (16/3)d_{uv}v = (16/3)(3/8)d_{uz}v = 2d_{uz}v$ (in units of d_{uz}), as also is given directly by the eq. above that defines the condition for the line-pole to be a directrix. With the circle-pole of radius 1/2 at the origin of Fig. 21-1a, and the directrix at $x = 3/8$, a few simple tests (for example, at the vertices) show that the linear-circular eq. $u^2 = Cdv$ applies for least distances. Since the circle-pole intersects the monoval of Fig. 21-1a, the curve also must be bitangent to the directrix at the points of intersection (easily confirmed from the hybrid eq., which has real roots only for $x \geq 3/8$ and ≤ 2.6180).

15. Substituting from the transformation eqs. 217a,a'b and squaring to eliminate the radical leads to eq. 223. Both eqs. 217a,a' lead to the same result, since alternate signs are eliminated on the first squaring and the different signs of the cross-product involving the radical are lost when it is segregated and squared.

16. Comparing eqs. 223 and 224 we find that $b = 2R$, $a^2 = 2CdR$, and $R^2 + Cd^2 = (a^2-b^2)^2/4b^2$. Eliminating a^2 and b^2 yields $C = 4(R+d)/d$.

17. The common solution applies when $Cd^2 = R^2$ and $C = 4(R+d)/d$ (see text and Problem 16). Solving these eqs. simultaneously yields $R = (2 + 2\sqrt{2})d$, whence $C = (2 + 2\sqrt{2})^2$. Employing this value for the limacon cast about its DP (Fig. 38e-left; compare eqs. 81 and 223) yields $b = Cd = (2 + 2\sqrt{2})^2 d$ and $a = 2R = 2d\sqrt{C} = 2(2 + 2\sqrt{2})d$, whence $e = b/a = (1 + \sqrt{2})$. Employing the same value of C for the limacon cast about its focus of self-inversion (Fig. 38e-right), yields $b^2 = 4R^2 = 4(2 + 2\sqrt{2})^2 d^2$ and $a^2/b = (2 + 2\sqrt{2})^2 d$, whence $a = \sqrt{2}(2 + 2\sqrt{2})^{3/2} d$ and $e = b/a = 1/\sqrt{[1 + \sqrt{2}]} = 0.64359$.

18. From comparing eqs. 223 and 224, the conditions to be fulfilled are $Cd = a^2/b$, $b^2 = 4R^2$, and $(a^2-b^2)^2/4b^2 = R^2 - Cd^2$, whence $b/a = e = \sqrt{(5/2)}$.

19. Putting eq. 223 in polar form and taking roots (one of the advantages of polar eqs.) yields $r^2 - 2r[R \pm (Cd\cos\theta)/2] + (R^2 \pm Cd^2) = 0$. The products of the roots of this eq. for the upper and lower signs, respectively, are $R^2 + Cd^2$ and $R^2 - Cd^2$. These sum to $2R^2$.

20. The curves in question have the linear-circular construction rule $v^2 = Cdu$. Substituting for u and v from eqs. 217a,a'b yields the degenerate 8th-degree eq. $[(x-d)^2 \pm R^2]^2 = C^2 d^2 (x^2+y^2)$. This represents two conchoid-like quartics (one each, for the plus and minus signs). For $C = 1$ and $d = R$ (the values of Fig. 38e), the positive sign leads to a two-arm locus with a least-distance construction rule. One arm is tangent to both p_u and p_v, and lies to the left of p_v. The other lies to its right. The negative sign also leads to a curve with two arms. One arm is looped; its vertex is tangent to p_u and p_v, and is the product of a least-distance construction. The remainder of this arm (i.e., all but the loop vertex) and the other arm (which lies to the right of p_v) are the products of greatest-distance constructions.

Chapter VII.

1. The polar eq. of the line is $r(\sin\theta+\cos\theta) = \sqrt{2}d$, whence the lengths of the X and Y (90°) probe-lines are defined by the eqs., $X(\sin\theta+\cos\theta) = \sqrt{2}d$ and $Y(\cos\theta-\sin\theta) = \sqrt{2}d$. Simultaneous solution gives $2XY\cos\theta = \sqrt{2}d(X+Y)$ and $2XY\sin\theta = \sqrt{2}d(Y-X)$. Equating the sum of the squares of the sine and cosine to unity and clearing of fractions yields eq. 226b, $d^2(X^2+Y^2) = X^2Y^2$.

2. The initial eq. is $b^2[x - \sqrt{(a^2-b^2)}]^2 + a^2y^2 = a^2b^2$, while the polar eq. in powers of $\cos\theta$ is $(b^2-a^2)r^2\cos^2\theta - 2b^2r\sqrt{(a^2-b^2)}\cos\theta + (a^2r^2-b^4) = 0$. Solving for $\cos\theta$ with the quadratic-root formula and cancelling terms in $\sqrt{(a^2-b^2)}$ from the numerator and denominator of the solution yields $\cos\theta = (\pm ar-b^2)/r\sqrt{(a^2-b^2)}$.

3. The initial eq. is $b^2(x-h)^2 - a^2y^2 = a^2b^2$. Transforming to polar coordinates and arranging in powers of $\cos\theta$ yields the simple intercept eq. $(a^2+b^2)r^2\cos^2\theta - 2b^2hr\cos\theta + (b^2h^2-a^2r^2-a^2b^2) = 0$. Solving for $\cos\theta$ yields the simple intercept format $\cos\theta = \{b^2h \pm a\sqrt{[(a^2+b^2)r^2+b^2(a^2+b^2-h^2)]}\}/(a^2+b^2)r$.

4. Letting $h = \sqrt{(a^2+b^2)}$ in the simple intercept format of Problem 3 yields the corresponding simple intercept format for the left focus, $\cos\theta = (b^2 \pm ar)/r\sqrt{(a^2+b^2)}$, where the upper alternate sign is for the intercept with the right arm (ascertained by testing) and the lower alternate sign is for the intercept with the left arm. To obtain the 0°-structure rule, since $\cos\theta = \cos(\theta+0°)$, one simply equates the $\cos\theta$-expression for the X-probe-line to the identical expression with X replaced by Y, for the Y-probe-line, yielding $b^2(X-Y) = \mp aXY \pm aXY$, where all combinations of the alternate signs must be taken into account. If we take only the two upper signs or the two lower signs, we obtain the trivial 0°-structure rule $X = Y$ (for both intercepts with the same arm; the only valid combination is for both intercepts positive or both intercepts negative). On the other hand, if we take the upper sign of one term and the lower sign of the other term, we obtain the 0°-structure rules $b^2(X-Y) = \pm 2aXY$, where the upper alternate sign applies when the X-intercept is for the right arm (the far arm) and the lower alternate sign when it is for the left arm (the near arm). Again, there is no valid 0°-structure rule involving both arms for the combination of one positive and one negative intercept. The complete 0°-structure rule, allowing for all valid combinations of intercepts, is $(X-Y)^2[b^2(X-Y)+2aXY][b^2(X-Y)-2aXY] = 0$, or, since a single $(X-Y)$ factor applies for both the left arm alone and the right arm alone, this factor need not be squared.

5. In the process of obtaining the eliminant for the case of the line, eqs. 225b,c are squared to obtain the squares of $\cos\theta$ and $\sin\theta$. Since $\cos^2\theta = \cos^2(\theta+180°)$ and $(-\sin\theta)^2=[-\sin(\theta+180°)]^2$, the squares of the two trigonometric functions automatically apply for both positive and negative intercepts (because positive and negative intercepts are out of phase by 180°). If alternate signs were present in the linear expressions involving $\sin\theta$ and $\cos\theta$, and these persisited after squaring, all combinations of them would have to be taken into account in order for the resulting structure rule to apply for all combinations of intercepts (see Problem 6). In this case, since no alternate sign is present, and the resulting structure rule contains only even powers of both X and Y, the structure rule applies to all combinations of intercepts at 90° to one another.

For ellipses, $\cos\theta$ already occurs as the square in the simple intercept format (eq. 230b), so the eqs. defining the X and Y intercepts (eqs. 230c,d) again apply for angles of both θ and $(\theta+180°)$, that is, for both positive and negative intercepts. Again, no alternate sign occurs in these expressions, so no additional squaring is necessary (nor a need to express the structure rule in degenerate form), and both X and Y occur in the 90°-structure rule only in even powers. In the analogous case for a focus (Problem 6), alternate signs are present and two additional squarings are required to eliminate them, giving a degenerate 16th-degree eq. that represents all combinations of intercepts (or the structure rule must be expressed as the product of four factors, each of 4th-degree).

6. The initial eq. is $b^2[x - \sqrt{(a^2-b^2)}]^2 + a^2y^2 = a^2b^2$, leading to the polar eq. $(b^2-a^2)r^2\cos^2\theta - 2b^2r\sqrt{(a^2-b^2)}\cos\theta + (a^2r^2-b^4) = 0$, which leads to the simple intercept format $\cos\theta = (\pm ar-b^2)/r\sqrt{(a^2-b^2)}$. Testing reveals that the upper sign applies for positive intercepts and the negative sign for negative intercepts. The intercept for the X-probe-line is given by the eq. $\cos\theta = (\pm aX-b^2)/X\sqrt{(a^2-b^2)}$, while that for the Y-probe-line is $\cos(\theta+90°) = -\sin\theta = (\pm aY-b^2)/Y\sqrt{(a^2-b^2)}$. Squaring both expressions, equating the sum to unity, and clearing of fractions yields $(a^2+b^2)X^2Y^2 + b^4(X^2+Y^2) \mp 2ab^2XY^2 \mp 2ab^2X^2Y = 0$. The upper and lower signs of the XY^2-term are for positive and negative X-intercepts, respectively, and those of the X^2Y-term for positive and negative Y-intercepts, respectively. Expressing the 90°-structure rule as a degenerate 16th-degree eq. consisting of the product of four 4th-degree factors, and identifying each factor, yields

ANSWERS TO PROBLEMS VII.   489

$[(a^2+b^2)X^2Y^2 + b^4(X^2+Y^2) - 2ab^2XY(X+Y)] \cdot [(a^2+b^2)X^2Y^2 + b^4(X^2+Y^2) + 2ab^2XY(X+Y)] \cdot$
   two positive intercepts   two negative intercepts

$[(a^2+b^2)X^2Y^2 + b^4(X^2+Y^2) - 2ab^2XY(X-Y)] \cdot [(a^2+b^2)X^2Y^2 + b^4(X^2+Y^2) + 2ab^2XY(X-Y)] = 0$
  positive X, negative Y-intercepts   positive Y, negative X-intercepts

7. Letting $h = 3R/2$ and $\theta = 0°$ in eq. 232c or 232d yields $X = R/2$ and $5R/2$, while letting $\theta = 0°$ in eq. 232e yields $Y = \pm\sqrt{5}Ri/2$ ($i^2 = -1$), representing two imaginary intercepts. Substituting $h = 3R/2$ in eq. 233a' yields $(X^2+Y^2) \cdot (X^2Y^2+25R^4/16) = 4X^2Y^2R^2$. Any of the four combinations of paired X and Y-intercepts satisfies this eq.

8. Rearranging eq. 233a' to $X^2Y^2(X^2+Y^2-4R^2) + (X^2+Y^2)(h^2-R^2)^2 = 0$, it readily is seen that for real solutions to exist we must have $X^2 + Y^2 \leq 4R^2$. To obtain the maximum value that the two intercepts can take on simultaneously, let $X = Y$, whence the maximum value is given by the eq. $2X^2 = 4R^2$, whence $X = \sqrt{2}R$, in which case $h = R$ (deduced from the right term of the eq.), which is the condition for incidence. The maximum of an intercept for a real solution is obtained by setting one intercept equal to 0, whence $X^2 = 4R^2$, or $X = 2R$ (or $Y = 2R$), for $\theta = 0°$ (or 270°).

9. Since the focus-directrix distance for ellipses is $p = a/e - ae = a(1-e^2)/e$, we have $ep = a(1-e^2)$. Accordingly, for $e = 0$ (and $p = \infty$), $ep = a$. Substituting in eq. 237a yields 237b.

10. The $\cos^2\theta$-simple intercept format of ellipses about their centers is given by eq. 230b, and the eq. for the X-intercepts is 230c. For an α-probe, the corresponding eq. for the Y-intercepts is $\cos^2(\theta+\alpha) = a^2(Y^2-b^2)/Y^2(a^2-b^2)$. Expanding, we have $\cos^2(\theta+\alpha) = \cos^2\theta\cos^2\alpha - 2\sin\theta\sin\alpha\cos\theta\cos\alpha + \sin^2\theta\sin^2\alpha = a^2(Y^2-b^2)/Y^2(a^2-b^2)$. Eliminating functions of θ from this eq. and eq. 230c yields eq. 244. Alternatively, one could substitute for $f(X)$ and $f(Y)$ in the $\cos^2\theta$-structure rule format of eq. 245b.

11. For the spiral of Archimedes, $X = a\theta$ and $Y = a(\theta+\alpha)$, whence $Y = a(X/a+\alpha) = X + a\alpha$. For the logarithmic spiral, $X = be^{\theta/a}$, $Y = be^{(\theta+\alpha)/a} = Xe^{\alpha/a}$.

12. The X and Y-intercept eqs. are $X = a + b\cos^4\theta$ and $Y = a + b\sin^4\theta$. Eliminating θ gives the 90°-structure rule $(X-Y)^2 - 2ab(X+Y) + b(4a+b)$. Rotating 45° to eliminate the XY-term reveals this to be the eq. of the parabola $Y^2 = \sqrt{2}bX - b(2a+b/2)$.

13. Letting $X^2 = a^2 - b^2\cos^2\theta$ and $Y^2 = a^2 - b^2\sin^2\theta$, and differentiating, yields $(dX/d\theta)^2 = [b^2(a^2-X^2)-(a^2-X^2)^2]/X^2$ and $(dY/d\theta)^2 = [b^2(a^2-Y^2)-(a^2-Y^2)^2]/Y^2$. Since $(ds/d\theta)^2 = \dot{s}^2 = (dX/d\theta)^2 + (dY/d\theta)^2$, after substituting and clearing of fractions, we have $X^2Y^2\dot{s}^2 = (X^2+Y^2)(-a^4+a^2b^2-X^2Y^2) + 2X^2Y^2(-b^2+2a^2)$, which is 4th-degree in X and Y.

14. We have $\cos\theta = f(X)$ and $\cos(\theta+\alpha) = f(Y)$, whence $dX/d\theta = -(\sin\theta)/f'(X)$ and $dY/d\theta = -[\sin(\theta+\alpha)]/f'(Y)$. Substituting for $\sin\theta$ and $\sin(\theta+\alpha)$, we obtain $(dX/d\theta)^2 = [1-f^2(X)]/[f'(X)]^2$ and $(dY/d\theta)^2 = [1-f^2(Y)]/[f'(Y)]^2$, whence it follows that $(ds/d\theta)^2 = (dX/d\theta)^2 + (dY/d\theta)^2$ is independent of α.

15. We know from the text and Problem 14 that XY-arc-increment eqs. are independent of α, so the 90°-XY-arc-increment eq. of Problem 13 actually applies for any probe-angle. To obtain the 90°-X-arc-increment eq., one eliminates Y from the XY-arc-increment eq. by means of the 90°-structure rule. Using the values of X^2 and Y^2 of Problem 13 for $\alpha = 90°$, and eliminating θ, yields the 90°-structure rule $X^2 + Y^2 = 2a^2 + b^2$. Substituting the value of Y^2 from this structure rule into the XY-arc-increment eq. of Problem 13 yields the 4th-degree X-arc-increment eq. $X^2(2a^2+b^2-X^2)\dot{s}^2 = X^4(3b^2-2a^2) + X^2(4a^4-4a^2b^2-3b^4) + a^2(b^4+a^2b^2-2a^4)$.

16. The simple intercept format for the parabola about its focus is $\cos\theta = (-2a\pm r)/r$ (eq. 227d). For ellipses, the simple intercept format about the left focus is obtained by casting the polar eq. about the left focus, yielding $\cos\theta = (-b^2\pm ar)/r\sqrt{(a^2-b^2)}$ (Problem 2). The upper alternate signs of both formats are for positive intercepts. Substituting X for r (for the X-probe-line) and differentiating with respect to θ yields $(dX/d\theta)^2 = X^2(X-a)/a$ for the parabola and $(dX/d\theta)^2 = X^2(2aX-X^2-b^2)/b^2$ for ellipses. Since these expressions are independent of the probe-angle (see text and Problem 14), the same forms apply to the Y-intercepts. Accordingly, for the parabola, $\dot{s}^2 = (dX/d\theta)^2 + (dY/d\theta)^2 = [(X^3+Y^3) - a(X^2+Y^2)]/a$, while for ellipses, $\dot{s}^2 = [2a(X^3+Y^3) - (X^4+Y^4) - b^2(X^2+Y^2)]/b^2$. The eq. for the parabola is both simpler and of lower degree. Letting X = a and 2a, respectively, for the parabola, yields $(ds/d\theta)^2 = 0$ and $4a^2$ for the axial vertex and the LR-vertices. For the ellipse, letting X = a + ae and a - ae at the a-vertices, yields $(ds/d\theta)^2 = 0$, while at the near LR-vertices, letting $X = b^2/a$ yields $(ds/d\theta)^2 = a^2e^2(1-e^2)^2$.

17 To obtain the 180°-X-arc-increment eqs. from the XY-arc-increment eqs. of Problem 16, one eliminates Y from these eqs. using the corresponding 180°-

structure rules (eqs. 228a and 229a). For the parabola, with $Y = aX/(X-a)$, substituting, clearing of fractions, and combining terms, yields $a(X-a)^3 \overset{\circ}{s}{}^2 = X^2[(X-a)^4 + a^4]$. For the ellipse, with $Y = b^2X/(2aX-b^2)$, the same operations lead to $b^2(2aX-b^2)^4 \overset{\circ}{s}{}^2 = X^2(2aX-b^2-X^2)[(2aX-b^2)^4+b^8]$. Accordingly, the focal 180°-X-degrees for the parabola and ellipses are 6 and 8, respectively--twice the degrees of the XY-arc-increment eqs.

18. Since the manner in which h and k enter the simple intercept eq. about a point in the plane is through association with x and y, respectively, i.e., $x \pm h$ and $y \pm k$, no simple intercept eq. with a term in r^n, with $n \geq 3$, can arise in the absence of association with terms in r^{n-1}, r^{n-2},.....r. Since obtaining the roots, r_1, r_2, r_3, ...r_n, of these eqs. will involve general solutions of eqs. of 3rd-degree or higher, a simple expression for the product of any pair of roots, $XY = f(\theta,h,k)$ that applies for all points in the plane will not exist.

Chapter VIII.

1. Since the coefficients of the X^2 and Y^2-terms are equal, a 45°-rotation is required to eliminate the XY-term. Letting $X = \sqrt{2}(X'-Y')/2$ and $Y = \sqrt{2}(X'+Y')/2$, in the bracketed expression of eq. 258a, cancelling and combining terms, and equating to 0, yields the eq. $X^2(1-\cos\alpha) + Y^2(1+\cos\alpha) = 4h^2 \sin^2\alpha$. This equation represents ellipses of eccentricities $\sqrt{[(\pm 2\cos\alpha)/(1 \pm \cos\alpha)]}$ for all probe-angles except 0° and 180° (upper signs for $\cos\theta > 0$). For 90°-probes, it represents circles.

2. From eq. 257b, $(dX/d\theta)^2 = h^2 X^2 \sin^2\theta/(h\cos\theta - X)^2$, where X replaces r for θ unchanged. Eliminating the trigonometric functions gives $(dX/d\theta)^2 = X^2[4h^2X^2 - (X^2+h^2-R^2)^2]/(h^2-R^2-X^2)^2$. $(dY/d\theta)^2$ is obtained in the same manner by replacing r by Y and θ by $(\theta+\alpha)$. This operation, however, gives the identical function (since the X and Y-intercepts simply are out of phase, the general expressions for $dX/d\theta$ and $dY/d\theta$, with θ eliminated, must be the same). Accordingly, the XY-arc-increment eq. $\overset{\circ}{s}{}^2 = (dX/d\theta)^2 + (dY/d\theta)^2$ is obtained by summing these expressions, yielding eq. 259b.

3. Solving eq. 258a (bracketed expression equated to 0) for Y and forming Y^2 yields $Y^2 = X^2\cos^2\alpha \pm 2\sqrt{(4R^2-X^2)}X\sin\alpha\cos\alpha + (4R^2-X^2)\sin^2\alpha$. Substituting in the incident XY-arc-increment eq. $\overset{\circ}{s}{}^2 = 8R^2 - (X^2+Y^2)$, yields $\overset{\circ}{s}{}^2 = 8R^2 - 2X^2\cos^2\alpha - 4R^2\sin^2\alpha \pm 2\sqrt{(4R^2-X^2)}X\sin\alpha\cos\alpha$. After squaring to eliminate the radical, this becomes 4th-degree in X. For 0° and 180°, it reduces to 2nd-degree, while for 90° and 270°, it reduces to 0-degree.

4. Eq. 263 is the eliminant between eqs. 261c and 262c. Segregate the linear terms in $\sin\theta$ and $\cos\theta$ of eq. 261c on the right and square, yielding $[X^2\sin^2\theta + (k^2-4ah)]^2 = 16a^2X^2\cos^2\theta + 16akX^2\sin\theta\cos\theta + 4k^2X^2\sin^2\theta$. Segregating the $\sin\theta\cos\theta$-term on the right and squaring again, yields $\{[X^2\sin^2\theta + (k^2-4ah)]^2 - 16a^2X^2\cos^2\theta - 4k^2X^2\sin^2\theta\}^2 = 256a^2k^2X^4\sin^2\theta\cos^2\theta$. Substituting for $\sin^2\theta$ and $\cos^2\theta$ using the $\sin^2\theta$-compound intercept format, $\sin^2\theta = \pm(k^2-4ah)/XY$, expanding, combining terms, and segregating in powers of XY, yields $256a^4X^4Y^4 \mp 128a^2A^2(k^2+4a^2)X^3Y^3 + 16A^4(k^2+4a^2)^2X^2Y^2 - 32a^2A^4(X\pm Y)^2X^2Y^2 \pm 8A^6(4a^2-k^2)(X\pm Y)^2XY + A^8(X\pm Y)^4 = 0$, where $A^2 = (k^2-4ah)$. This rearranges to eq. 263.

5. Let $k = 0$ in the X-intercept eq. 265 for the parabola about incident points and solve for $\cos\theta$, yielding the $\cos\theta$-simple intercept format $\cos\theta = [-2a + \sqrt{(4a^2+X^2)}]/X$. For 90°-probes, the corresponding simple intercept format for Y is $\sin\theta = -[-2a + \sqrt{(4a^2+Y^2)}]/Y$. Using the identity $\sin^2\theta + \cos^2\theta = 1$, substituting, clearing of fractions, and combining terms yields the eq. $8a^2(X^2+Y^2) + X^2Y^2 = 4a[X^2\sqrt{(4a^2+Y^2)} + Y^2\sqrt{(4a^2+X^2)}]$. Squaring and cancelling terms, yields $[128a^4X^2Y^2+X^4Y^4] = 32a^2X^2Y^2\sqrt{(4a^2+X^2)}\sqrt{(4a^2+Y^2)}$. X^2Y^2 now is factored out, representing the two trivial incident structure rules $XY = 0$, and both sides are squared again. Combining and cancelling terms, yields eq. 268 for the 90°-structure rule about the axial vertex.

6. Let $k = 2a$ in the X-intercept eq. 265 for the parabola about incident points, yielding $X\sin^2\theta - 4a\sin\theta = 4a\cos\theta$. The corresponding Y-intercept eq. for a 90°-probe is $Y\cos^2\theta + 4a\sin\theta = 4a\cos\theta$. Taking the quotient to eliminate the linear term in $\cos\theta$, and solving for $\sin\theta$, yields the 90°-compound intercept format $\sin\theta = [4a \pm \sqrt{(16a^2+XY+Y^2)}]/(X+Y)$. Substituting in either of the above intercept eqs. and squaring, as necessary, to eliminate radicals, leads, after much cancelling, combining, segregating of terms, etc., to the sought-after 90°-structure rule 269.

7. The 180°-structure rules about points on the transverse and major axes are $4a^2h^2X^2Y^2 = (a^2\pm b^2)(h^2-a^2)^2(X-Y)^2 \pm 4b^2h^2(h^2-a^2)XY$ (upper signs for hyperbolas). Letting $h^2 = (a^2\pm b^2)$ and factoring out $(a^2\pm b^2)$, leaves $4a^2X^2Y^2 = b^4[(X-Y)^2+4XY]$. Taking the positive root for two positive intercepts, yields $2aXY = b^2(X+Y)$ for both ellipses and hyperbolas. The similar treatment for the 0°-structure rule yields $4a^2X^2Y^2 = b^4[(X+Y)^2-4XY]$, whence $2aXY = b^2(X-Y)$, which applies to hyperbolas for two positive intercepts, where X is the intercept with the far arm, and to ellipses for a positive Y-intercept and negative X-intercept. [With the X-intercept negative, this eq. describes the same relationship as the 180°-

structure rule for ellipses; thus, $2a(-X)Y = b^2(-X-Y)$ becomes $2aXY = b^2(X+Y)$.]

8. The general 0° and 180°-structure rules about points on the minor axis are (from eq. 276) $4b^2k^2X^2Y^2 = (b^2-a^2)(k^2-b^2)^2(X\pm Y)^2 \pm 4a^2k^2(k^2-b^2)XY$, with the upper alternate signs for 0°. Letting $k^2 = (b^2-a^2)$, yields $4b^2X^2Y^2 = a^4(X\pm Y)^2 \mp 4a^4XY$, whence the structure rules for the imaginary foci are $2bXY = a^2(X\mp Y)$. The two imaginary roots of the corresponding simple intercept eq. are $\sqrt{(b^2-a^2)} - b$ and $\sqrt{(b^2-a^2)} + b$. Setting the former equal to X and the latter equal to Y satisfies the eq. [The alternative assignment of the roots satisfies the negative root-equation for the structure rules $-2bXY = a^2(X\mp Y)$.]

9. By substitution and simplification, these are found to be $\cos\theta = (\pm aX-b^2)/X\sqrt{(a^2-b^2)}$, $\pm a\sqrt{(X^2-b^2)}/X\sqrt{(a^2-b^2)}$, and $\{-ab^2 \pm a\sqrt{[b^4+X^2(a^2-b^2)]}\}/X(a^2-b^2)$, respectively.

10. Employing the $\cos\theta$-structure-rule format 245a, which for 90°-structure rules is simply $\sin^2\theta + \cos^2\theta = 1$, yields $(aX+b^2)^2/X^2(a^2-b^2) + (aY-b^2)^2/Y^2(a^2-b^2) = 1$. Clearing of fractions, expanding, and regrouping, yields eq. 293a (lower alternate signs).

11. From eqs. 294, the $\cos^2\theta$-simple intercept formats are $\cos^2\theta = a^2(X^2\pm b^2)/X^2(a^2\pm b^2)$, where the upper alternate signs are for hyperbolas. Substituting in the $\cos^2\theta$-structure-rule format 245b, yields $a^4(X^2\pm b^2)^2/X^4(a^2\pm b^2)^2 + a^4(Y^2\pm b^2)^2/Y^4(a^2\pm b^2)^2 - 2[a^2(X^2\pm b^2)/X^2(a^2\pm b^2) + a^2(Y^2\pm b^2)/Y^2(a^2\pm b^2)]\sin^2\alpha - 2a^4(X^2\pm b^2)(Y^2\pm b^2)(\cos^2\alpha - \sin^2\alpha)/X^2Y^2(a^2\pm b^2)^2 = (\cos^2\alpha - 1)\sin^2\alpha$. Multiplying through by $X^4Y^4(a^2\pm b^2)^2$ to clear of fractions, expanding, and gathering terms in powers of X^nY^m, yields eq. 296. If one wishes merely to confirm the variable-angle reduction condition, one need gather terms only for the X^4Y^4-term and show that this term is $[(b^2\mp a^2)^2 - (b^2\pm a^2)^2\cos^2\alpha]X^4Y^4\sin^2\alpha$.

12. We have $\tan(\alpha/2) = b/a = \sqrt{(e^2-1)}$, whence, using the formula for $\tan 2\alpha$, $\tan\alpha = 2\sqrt{(e^2-1)}/(2-e^2)$. Converting to the cosine function yields $\cos\alpha = \pm(2-e^2)/e^2$.

13. The coefficient of the 4th-degree term is $[2a^2(1\pm\cos\alpha) - (a^2+b^2)(1-\cos^2\alpha)]$. Equating to 0 and solving for $\cos\alpha$, yields $\cos\alpha = (\pm a^2\pm b^2)/(a^2+b^2)$. For the solutions, $\cos\alpha = \pm 1$, the reductions are to 2nd-degree or trivial 1st-degree structure rules. The condition for other angles is $\cos\alpha = \pm(a^2-b^2)/(a^2+b^2)$, which is identical with the variable probe-angle condition 297b for probe-angles equal to the asymptote angles.

14. If the simple intercept format 290b is for a 90°-probe-angle, letting $\theta = 0$

for the X-intercept should correspond to ($\theta+90°$) for the Y-intercept, in which case the length of the Y-intercept should be 2b. Letting $\theta = 0°$, yields $[Y(b^2-a^2) + a^2b]^2 = b^2[a^4 + Y^2(b^2-a^2)]$, whence $-Y^2a^2(b^2-a^2) + 2a^2b(b^2-a^2)Y = 0$, or $Y = 2b$.

15. Performing an x+h translation on the initial eq., converting to polar coordinates, expanding terms, and grouping in powers of $\cos\theta$, yields the simple intercept eq. $2(2h^2-a^2)X^2\cos^2\theta + 2hX(2X^2+2h^2-a^2)\cos\theta + (h^2+X^2)^2 - a^2(h^2-X^2) = 0$. Forming the radicand, expanding, simplifying, and grouping terms, yields $[2a^2X^2(X^2+a^2-2h^2) + a^2h^2(2h^2-a^2)]$, which satisfies the requirement of the Problem.

16. Examination of eq. 320, $36b^8 - 76b^6a^2 + 53b^4a^4 - 14b^2a^6 + a^8 = 0$, reveals that the roots must occur in oppositely signed pairs $\pm n_1$, $\pm n_2$, etc., since all exponents are even. Also, the product of the roots, $n_1(-n_1)(n_2)(-n_2)\cdots$ must equal 36 (or 1/36th). Since $n_1 = 1$ (b = a) is seen to be a root on mere inspection, dividing by $(b-a)(b+a) = (b^2-a^2)$ yields the 6th-degree eq., $36b^6 - 40b^4a^2 + 13b^2a^4 - a^6 = 0$. The product of the roots again must be 36. Since 36 factors to $2\cdot2\cdot3\cdot3$, the most likely six roots are $\pm\sqrt{2}$, $\pm\sqrt{2}$, ± 3. Dividing through by $(b^2-a^2)/9$, yields $36b^4 - 36b^2a^2 + 9a^4 = 0$. Factoring out a 9 leaves the perfect square $(2b^2-a^2)^2 = 0$, representing the other four roots.

17. Letting $h = -b/2$ in eq. 315 yields the simple intercept eq. $4k^2X^2\sin^2\theta - 2kX(a^2-2B^2-2X^2)\sin\theta + X^2[X^2+(2B^2-a^2)] + (B^4-a^2C^2) = -a^2bX\cos\theta$, where $B^2 = (k^2-b^2/4)$ and $C^2 = (k^2+b^2/4)$. Substituting Y for X yields the Y-intercept eq. for 0°. Using the latter eq. together with the above eq., to eliminate $\cos\theta$, yields the compound 0°-intercept eq. $4k^2XY\sin^2\theta + 4kXY(X+Y)\sin\theta + XY(X^2+XY+Y^2) + (2B^2-a^2)XY - (B^4-a^2C^2) = 0$. Solving for $\sin\theta$, yields the compound 0°-intercept format $2k\sin\theta = -(X+Y) \pm \sqrt{[(a^2-2B^2)+XY+(B^4-a^2C^2)/XY]}$.

18. In the derivation of the polar eq. from eq. 314, one obtains $(r^2-brcos\theta)^2 = a^2r^2$. In taking the root (and factoring out an r) to obtain $r - b\cos\theta = a$, from which the simple intercept format 317a is derived, the negative root has been discarded. Since the curve is traced only once for θ ranging from 0° to 360°, there is only one value of r for each value of θ, and only the trivial 0°-structure rule exists. The eq. of the limacon that includes the negative root is $r = \pm a - b\cos\theta$. For this eq., two roots exist for each value of θ, and one also obtains the 0°-structure rule $X - Y = 2a$ (and $Y - X = 2a$) for one positive and one negative intercept.

19. Using the simple intercept format $\cos\theta = (a-X)/b$ yields $(dX/d\theta)^2 = b^2-a^2 + 2aX - X^2$, whence the XY-arc-increment eq. $\dot{s}^2 = 2(b^2-a^2) + 2a(X+Y) - (X^2+Y^2)$ is 2nd-degree, accounting for the linear and circular XY-arc-increment eq. entries in Table 10. To obtain the 180°-X-arc-increment eq., one substitutes $Y = 2a - X$ from the 180°-structure rule 321c, yielding $\dot{s}^2 = 2b^2 - a(a-X)^2$, which also is 2nd-degree, and accounts for the other linear X-arc-increment eq. entry in Table 10. Similarly, substituting for Y from eq. 321d yields the 90°-X-arc-increment eq. $\dot{s}^2 = b^2$, accounting for the other circular X-arc-increment eq. entry in Table 10.

20. Rotating the axes CCW 45° yields the eq. $(X-\sqrt{2}a)^2/b^2(1+\cos\alpha) + Y^2/b^2(1-\cos\alpha) = 1$. This represents ellipses with $e^2 = (2\cos\alpha)/(1+\cos\alpha)$ for 0°-90° and 270°-360°, and $e^2 = (-2\cos\alpha)/(1-\cos\alpha)$ for 90°-270°.

21. From eq. 336a, the vertices lie at $h^2 = [-B^2 \pm \sqrt{(B^4-4D^4)}]/2$. Using these values of h^2 in the Cartesian condition (expression 338c equated to 0) leads to the requirements $B^4 = 4D^4$ or $D^4 = 0$, neither one of which is allowed (see restrictions on eq. 335a).

22. The poles of the Cartesian condition are defined by the roots of the eq. formed by setting the Cartesian term 338c equal to 0, namely, $h^4(C^2-B^2) + h^2(B^2C^2-4D^4) + D^4(C^2-B^2) = 0$. The $\cos^2\theta$-condition foci are defined by setting the coefficient of the $\cos^2\theta$-term of eq. 338a equal to 0, namely, $4h^2 + B^2 - C^2 = 0$. Eliminating h from these eqs. by substituting from the latter into the former, yields $B^2 = -C^2$. Referring to eq. 127, this is seen to be the condition for Cassinian ovals. Letting $B^2 = -C^2$ in the $\cos^2\theta$-condition, places the $\cos^2\theta$-condition foci at $\pm\sqrt{(C^2/2)} = \pm d/2$. Since the Cartesian-condition poles are modularly-inverse to one another, the product of the distances of the two poles from the center must be $\pm\sqrt{D^4}$ (see eqs. 337 and 338d). Hence, the other pair of poles must be at distances $\pm\sqrt{(2D^4/C^2)}$. The same result can be obtained by letting $B^2 = -C^2$ in the Cartesian condition.

23. The initial eq. is 325. An x+h translation and conversion to polar coordinates yields $[r^2+2(h-d)r\cos\theta + (h-d)^2]^2 = C^2d^2(r^2+2hr\cos\theta + h^2)$, where C is the parameter of the bipolar eq. $u^2 = Cdv$. Replacing r by X, expanding, and grouping terms, yields the simple intercept eq. $4(h-d)^2X^2\cos^2\theta + 2X[2X^2(h-d) + 2(h-d)^3 + C^2d^2h]\cos\theta + [X^2+(h-d)^2]^2 - C^2d^2(X^2+h^2) = 0$. Solving for $\cos\theta$ and factoring an X from the radicand leaves the radicand constant-term $[2(h-d)^3 - C^2d^2h]^2 - 4(h-d)^2[(h-d)^4-C^2d^2h^2]$. Equating to 0 and solving for h, yields the three

roots (foci of self-inversion), $h = 0$, $[8-C^2+C\sqrt{(C^2-16)}]/8$, and $[8 - C^2 - C\sqrt{(C^2-16)}]/8$. All three roots are real for $C > 4$ (the biovular members) but only the $h = 0$ root is real for $C < 4$ (the monovals). For $C = 5$, the three roots are $h = 0, -1/4, -4$, which checks against the locations of the foci of self-inversion of Fig. 23 (recall that $d = d_{uz}$).

24. Converting eq. 338 to polar coordinates (by substituting r for X), inverting (by substituting j^2/r for r at each occurrence), clearing of fractions, converting to rectangular coordinates, and performing an $x+H$ translation, yields $[j^4+2hj^2(x+H)+h^2K^2]^2 + B^2[j^4(x+H)^2+2hj^2(x+H)K^2+h^2K^4] + C^2j^4y^2 + D^4K^4 = 0$, where $K^2 = [(x+H)^2+y^2]$. To obtain the simple intercept eq., expand, convert to polar coordinates, replace r by X, and group in powers of $\cos\theta$, yielding eq. 8-20-10 of Kavanau (1982).

25. Letting $H = -j^2/2h$ in the solution to Problem 24, the simple intercept eq. of the translated inverse locus of the translated initial central Cartesian becomes $16h^2j^4(h^4-C^2h^2+D^4)X^2\cos^2\theta + 8hj^2(4h^2X^2+j^4)(h^4-D^4)X\cos\theta + 16h^4(h^4+h^2B^2+D^4)X^4 + 8h^2j^4X^2(h^4+D^4-h^2B^2+2h^2C^2) + j^8(h^4+h^2B^2+D^4) = 0$. Solving for $\cos\theta$ and examining the coefficients of the terms in the radicand reveals that both the constant term and the coefficient of the X^4-term involve the Cartesian term 338c as a multiplicative factor. Accordingly, if the original pole of inversion is one of the poles of the Cartesian condition, the X^4-term and constant term of the radicand vanish and X^2 can be factored from the radical. The simple intercept format then becomes $\cos\theta = \{4h^2(h^4-D^4)X^2 + j^4(h^4-D^4) \pm 4h^3j^2X\sqrt{(C^4-4D^4)}\}/4hj^2(h^4-C^2h^2+D^4)X$, where h is one of the roots of the Cartesian condition. In other words, this simple intercept format defines the specific curve obtained by inverting a specific central Cartesian about a Cartesian-condition pole h (defined by eq. 338d). [If $h^4 = D^4$, the inversion locus is another central Cartesian.]

26. The eq. of the initial Cassinian ovals is $(x^2+y^2)^2 - d^2x^2/2 + d^2y^2/2 + d^4(1/16 - C_{bp}^2) = 0$, where C_{bp} is the parameter of the bipolar eq. $uv = C_{bp}d^2$. It follows that $B^2 = -d^2/2$, $C^2 = d^2/2$, and $D^4 = d^4(1/16 - C_{bp}^2)$. Evaluating h for poles of the Cartesian condition yields $h = \pm d/2$ and $h = \pm 2d\sqrt{(1/16 - C_{bp}^2)}$. We select the pole at $h = d/2$ (which is coincident with the $\cos^2\theta$-condition focus of the right oval). Substituting in the solution to Problem 25 yields $\cos\theta = [d^2X^2+j^4 \pm dj^2X/C_{bp}]/(-2dj^2X)$. Now assume that $j = d$ (the distance between the $\cos^2\theta$-condition foci--the bipolar poles), whence $\cos\theta = (X^2+d^2 \pm dX/C_{bp})/(-2dX)$. This will be recognized as the simple intercept format of parabolic Cartesians

(eq. 325b), but with the bipolar Cassinian "C_{bp}," of $uv = C_{bp}d^2$, replacing the bipolar parabolic Cartesian "C_{bp}," of $u^2 = C_{bp}dv$ (these C's are reciprocals of one another). [The difference in signs influences only the phase of the plots.] Letting $C_{bp} = 1/4$ yields $(X^2+d^2 \pm 4dX)/(-2dX)$, the same as eq. 325b for $C_{bp} = 4$. For the equilateral limacon, the simple intercept format about the focus of self-inversion is derived from eq. 317d by letting $b = \sqrt{2}a$, yielding $\cos\theta = (8\sqrt{2}aX-8X^2-a^2)/4\sqrt{2}aX$. The reciprocal inversion pole in the plane of the limacon is the $-b/2$ focus, and the focus of self-inversion of the limacon is the inverse of the opposite $\cos^2\theta$-condition focus of the equilateral lemniscate, at a distance d from the pole of inversion. Accordingly, the distance d in the plane of the equilateral lemniscate corresponds to the distance between the $-b/2$ focus and the focus of self-inversion in the plane of the equilateral limacon, whereupon $d = b/2 - (b^2-a^2)/2b = a^2/2b = a\sqrt{2}/4$, for which the formats become identical.

27. The solution is eqs. 8-20-14a,b of Kavanau (1980).

28. Simple long division of polynomials yields the quadratic eq. in H, $2hH^2[2h^2(C^2-B^2)+(B^2C^2-4D^4)] + 4h^2Hj^2(C^2-B^2) + 2hj^4(C^2-B^2) = 0$.

29. Substituting $C_{bp} = 1/4$, and letting $j = d$, yields $H = -d$ and a double-root $H = -2d$, since the radical vanishes for the latter solution. The initial equilateral lemniscate is inverted about the Cartesian-condition pole at $h = d/2$, the $\cos^2\theta$-condition focus. The reciprocal inversion pole in the plane of the limacon is the $-b/2$ focus; the focus of self-inversion of the limacon (the inverse of the reflection of the pole of inversion in the line of symmetry) is d units to the left of this focus, at $-d$, and the DP (at distance $d/2$ from the pole of inversion in the plane of the equilateral lemniscate) is 2d units to its left at $-2d = -d^2/(d/2)$. [In the equilateral limacon, the focus of self-inversion lies midway between the DP and the $-b/2$ focus.]

30. Letting $C_{bp} = 1/5$ and $j = d$, yields $H = -d$, $-5d/4$, and $-5d$. Again, as in Problem 29, the reciprocal inversion pole is at the positions of the $-b/2$ double-focus and, for $j = d$, the focus of self-inversion (the monoval-bioval focus) is d units to its left at $-d$. Consulting Fig. 23, the locations of the other two foci are confirmed at $-5d/4$ and $-5d$ (actually, at $-d/4$ and $-4d$ relative to the focus of self-inversion).

31. The poles of the Cartesian condition are defined by the eq. $h^4(C^2-B^2) + h^2(B^2C^2-4D^4) + D^4(C^2-B^2) = 0$. To find the condition for coincidence of these poles with poles at $h^2 = C^2/2$, substitute the latter values of h in the former eq., yielding the degenerate eq. $(C^2+B^2)(C^4/4-D^4) = 0$. The first factor equated to 0 yields $B^2 = -C^2$, the condition defining Cassinian ovals for eq. 335a. The second factor equated to 0 yields $C^4 = 4D^4$, one of the prohibited conditions of eq. 335a. For this condition to be valid for real values of the coefficients of eq. 335a, D^4 must be positive (otherwise C^2 would be imaginary), whereupon the curve must consist of annular or opposed biovals (see eqs. 336). For this condition, eq. 335a becomes $(x^2+y^2)^2 + B^2x^2 \pm 2D^2y^2 + D^4 = 0$. This is the degenerate eq. of two congruent, real or imaginary, concentric or non-concentric circles. [The manner of factoring this eq. is not evident, but one can easily confirm the circumstances by deriving the degenerate quartic eq. of two congruent, real or imaginary, non-concentric circles from the initial eq. $\{(x+a)^2 + y^2 + R^2\} \cdot \{(x-a)^2 + y^2 + R^2\} = 0$, by multiplying out and segregating terms (R^2, positive, negative, or 0), whereupon one finds that $D^4 = (a^2+R^2)^2$ and $B^2 = 2(R^2-a^2)$. If one employs the initial eqs. of concentric, real or imaginary, non-congruent circles, $\{x^2 + y^2 + R^2\}\{x^2 + y^2 + r^2\} = 0$, one finds that r^2 must equal R^2 if eq. 335a with $C^4 = 4D^4$ is to be satisfied, that is, the circles must be congruent.]

INDEX

Apollonius, 13
Arc-increment eqs., 325,326,337,338, 347,349-351,368,369,455
Asymmetries of coordinate systems, 65-70
Asymptote-point and asymptote-point focus, 195,404,408,455
Asymptotes, xii,195,455
  as focal loci, 324,388
  imaginary, 388
  structure rules about,388
Bernoulli, J., 41 (1645-1705)
Bicircular coordinates, 260
Bolyai, J., vi (1802-1860)
Bipolar coordinate system, xv,xvi, 90-212,444-446,455
  foci, focal loci, and point-foci, 105-107,121,444,457,458
  history, 93-95
  polarized versus unpolarized, 91, 95-98,217,218,459,460
  rotations, 103-105
  translations, 102-104
  weighting of poles, 91,112,124, 126,127,131,144,147,148,153, 159,168,190,196
Brianchon, J., v (1785-1864)
Bullet-Nose Curve, 322
Carathéodory, C., xii (1873-1950)
Cardioid, xviii,161,249,298,357,418, 420
Cartesian condition, 191,429-432, 436,455
  poles of, 191,429-433,436,439
Cartesian near-conics, see Near-conics
Cartesian ovals, 6,14,16,93-95,97, 102,296,455
  bitangent monovals, 297-300
  confocal monovals, 297-300
  linear, 97,100,176,263,423-427
    bipolar, 170-179
      deriving the foci, 174,175
      "master construction rules," 172,173
      parabolic construction rules, 185,186
      rectangular eq., 173,269
    distance eqs., 174,177,198
    limacon near-conic homologues, 424,427
    polar-circular, 271
    structure rules, 424-427

Cartesian ovals
  linear-circular, 294-301
  monoval-bioval focus, 180,183,252, 253,273-275,357
  parabolic, x,6,97,100,176,178,188, 189,191,263,421,423,436,460
    bipolar, 179-191
      construction rules, 184-186
      deriving the foci, 182,183
      limacon near-conic homologues, 423
      symmetry condition for, 185,186, 272
    distance eqs., 182,183,198,252,254
    polar-circular, 272-275
    polar-linear, 251-254
    rectangular eq., 182,422
    structure rules, 347,422,423
  self-inversion, 183,189,253,298-301, 345,358,424-426,435,436
Cartesian term, 429-432,455
Casey, J., 93 (1820-1891)
Cassini, G.D., vii,164 (1625-1712)
Cassinian ovals, vii,x,47,93,94,100, 164-170,187-192,416,419,427,431, 436-440,455
  bipolar, 187-192
  polar, 47,77,80-82
  rectangular eq., 47,187
  self-inversion, vii,167,331,345,454
  simple intercept eq., 85
  single-oval, 80-82,331,454
  structure rules, 332,347
  two-oval, 77,80,82
Cayley, A., 93 (1821-1895)
Cayley's sextic, 47,48,102
Center of symmetry, 58,446
  ordinary, 341,342,454,455
  true, x,342,438,446,455
Central Cartesians, x,100,427-440,455
  and Cartesian ovals, x,171,187,430
  axial vertices, 428,429
  bipolar, 187-193
  distance eqs., 190
  homologies with central conics, x, 432,433
  inverse curves, 436-439,454
  parabolic, 187,188
  rectangular eq., 187,428
  vertex cubics, 434,435,454
  vertices, 429

Central conics, 455
  as curves of demarcation, 180
  axial inversions, 415
  axial-vertex inversion cubics, 193-197
  bipolar, 143-154
  hybrid distance eqs., 198,207, 234,235
  linear-circular, 287-290
  polar-circular, 264-267,275-280
  polar-linear, 234-240
    focus-line-of-symmetry, 236,237
    focus-vertex-tangent, 239,240
    incident, 242
    vertex-line-of-symmetry, 238,240
  simple intercept eqs., 85,321,384, 391,392
  simple intercept formats, 321
  structure rules, 320-322,326,332-338,344,347,383-403
Central curves with 4th-degree maximum inverse loci, 427
Central modular-focal quartics, 439, 456
Central quartics, 75-78,427,456
Cesáro, E., 4 (1859-1906)
Characteristic distance, 57,66-69, 92,98,129,130,163,178,456
Chasles, M., 93 (1793-1880)
Circles,
  arc-increment eqs., 369
  bipolar, 104,111-121,134
    axial, 113-118,134
    general, 120,121
    midline, 119,120,134
  circumcurvilinear symmetry, 309,310
  circumpolar symmetry, 312,322,323, 364-369
    structure rules, 312,322
  compound, 219,243,246
  disparate, 219,243
  "4th-degree," 330,332,360
  homologies with equilateral hyperola and/or line, 334,336,338, 340,360,361
  hybrid distance eqs., 198
  monopolar, 106,111,114,222,243,245
  of Apollonius, 94-96,111-118,121
  of inversion, 456
  of Pythagoras, 112-117,162,163
  polar-linear, 219,242-246
    disparate, 219,244-246
  rectangular, xiii,18-23
  simple intercept eq., 322

Circles
  simple intercept format, 322,364,365
  structural curves, 317
  structure rules, 312,313,322,333,347, 348,364-369
  self-inversion, 313
Circumcurvilinear symmetry, 308-310
Circumlinear symmetry, 309,310
Circumpolar symmetry, 305-442
Cissoid of Diocles, xviii,xix,16,47, 193,194,207,410
  simple intercept eq., 85
Classification of curves, xi,xii,326-329,352,359-361
  by circumpolar symmetry analysis, 328,329,352,361
  by Inversion Taxonomy, 327,328
Cochleoid, 47
Complementary loci, 224,237-246,446
  shapes of, 240-242,446
Compound curves, 216-220,278-280,456
  mixed conics, 278-280,456
Compound intercept eqs., see Intercept eqs.
Conchoids, 7,213,244,456
  of Nicomedes, 47
Congruent-inversion, 166,345,359,437, 456
Conic axial-inversion quartics, 415
Conjugate oval, 97,100,456
Conjugate point or pole, 456
Constant condition, see Focal conditions
Construction rules,
  and structural simplicity, xiii,3, 11-13,18,339,443,444
  as relationships between curves, 301
  versus structure rules, 311,312,443
Coordinate curves, 33,62,222,456
Coordinate eqs. versus structure rules, 311,312
Coordinate system comparisons, 91, 202-210,255-257,260-262,280-282,301,302
Coordinate transformations,
  bipolar-rectangular, 48,71-82
  linear-circular-rectangular, 283,284
  polar-circular-bipolar, 267-269
  polar-circular-rectangular, 262-264
  polar-hybrid, 137,138,199-201
  polar-linear-hybrid, 221
  polar-linear-rectangular, 220-222
$Cos\theta$-condition, see Focal conditions
$Cos^2\theta$-condition, see Focal conditions
$Cos^2\theta$-group, 53-55
Covert focal loci, see Focal loci

INDEX

Cramer, G., x,340 (1704-1752)
Crofton, M.W., 93,95,107,171,462 (1826-1915)
Cubics, 188,193-197,403-412,433-435,446
Curves of demarcation, 80,82,100, 165,167,178-181,273,274,296-302,332,357,416,423,427,435. 436,447,457
   elliptical limacon, 296,299,300
Curves of intersection, 167,457
Cusp-point, 357,410,418,420,457
Degree-restriction, xvi,15-17,35, 36,57,66-69,142,457
   partial, 17,69,255,256,457
Degrees of structure rules, see Structure rules
Demarcating locus, 288-290
Descartes, R., v,13,14 (1596-1650)
Design and synthesis of highly-symmetrical curves, 329,331
Devil's Curve, x,330,340,360
Dimensional uniformity of eqs.,xiv, 41,42
Directed versus undirected distances, see Undirected distances
Directrices,
   common, 176,191,254,257
   consisting of a circle, 1,457
   consisting of a line, xi,1,28, 156,447,457
      of Cartesian ovals, 97,176,191, 252,253,295-300
      of conics, 219,234-236,243,244, 285-290
      of limacons, 156,247-250,295-300
Disparate curves, 457
   hyperbolas, 232,238-240,265,275, 276
   parabolas, 217,284-287
Distance eqs.,
   conventional, 3,137-143,445
   hybrid, 137-143,153,154,197-200, 234,235,266-277,294,295
Double-point, 457
Eccentricity
   and symmetry, 161-164,277,287, 288,298,339
   and the exceptional lines of conics, xviii
   as a visible property, vii,xvii, 443
   of conics, 28,92
   of limacons, 161,339
   of vertex cubics, 193

Eclipsed and eclipsable foci, 334-338, 362
Eliminant, 457
Ellipses,
   bipolar, 92,143-154,209
      axial, 146-154
      bisector, 152
      comparisons with limacons, 162-164
      focal, 148-151,153
         focus-center, 149-151
         focus-vertex, 148-151
   circumpolar foci, 323,324
   compound, 237-239
   disparate, 265, 288-290
   distance eqs., 154
   linear-circular, 287-290
   polar, 62-65,68,73-75
   polar-circular, 264-267,275-280
   polar-linear, 232-240
   rectangular, 23-31
   simple intercept eq., 85,391,392
   simple intercept format, 321,389, 394,396,398,400
   structural curves, 317,385,387,392-396
   structure rules, 315,320-322,332-338,347,383-403
Equilateral curves, 405
   circle, see Circle
   hyperbola, vii,163-165,333-338, 357-362,366-368,398-400,404
   lemniscate, vii,xviii,80,330,332, 360,416
      bipolar, 164-170
         affinities, 164-167
         distance eqs., 167-170
         point-foci (dependent), 169, 170,190,192
      polar, 5-78,81,82
      simple intercept eq., 85
   limacon, xviii,78-80,97,163,164,254, 255,263,404,405,427,436
   strophoid, xviii,327,404-410,412,431
      bipolar, 193-197
      eqs., 197
      distance eqs., 194,198
      self-inversion, 194-196,327
      structure rules, 347,408-412
Equivalent inversion poles, 457
Equivalent poles, 457
Exponential degree-restriction, see Degree-restriction
Extremal-distance constructions, 1, 262-304
   greatest versus least-distance, 266

502 INDEX

Extremal-distance symmetry, 261,263,
  265,270,271,276,285-290,296,457
Fermat, P. de, v,13,14 (1601-1665)
Focal conditions, 133,372,373,379,
  388,406,457
  angle independent, 389,425,434,
    435
  asymptote, 388
  constant, 85,86,370,371,384,406,
    414,416,429,435,447,456,460
  $\cos\theta$, 85,86
  $\cos^2\theta$ and $\sin^2\theta$, 86,169,406,415,
    420-423,425,429-433,439,457
  radicand, 389,394,447
  radicand constant, 86,389,415,
    416,425,426,430,435,447
  structure-rule, 416,417,425,435,
    461
  variable probe-angle, 402,403,462

Focal rank, 458,460
  bipolar, 131,132,445,446
  circumpolar, 324,458
Foci and focal loci, 458
  bipolar, 105-107,121,148-150,
    158,444-446
    definition, 105-107,121,141,142,
      160,444-446
  circumpolar, 134,323-326
    and positional symmetry (struc-
      tural simplicity), 323,324,447
    angle-dependent versus angle-
      independent, 325,416,425,433-
      435,458
    asymptote-point, 195,404,407,
      408,410
    at infinity, 345,391,415,432
    compound, 338,358,458
    $\cos^2\theta$-condition, 192,406,415,
      420-424,429-433,439,457
    covert, 324,358,410,414,418,457
    definitions, 323,326,356-358,
      447,455,456
    detection, 343
    eclipsable, 334-338,358,362,458
    imaginary or complex, 324,389,
      415
    incident variable, 357
    loop-pole at $h = 2b/3$, 410,412
    loop vertex of cubics, 194-197,
      409,410
    loss or gain in inverse locus,
      345,447
    "monoval-bioval," 180,183,252,
      253,422

Foci and focal loci,
  circumpolar,
    multiple, 329,357,358,432,447,458
    number of, 324
    of self-inversion, vii,86,162,167,
      176,189,195,298-301,331,345,359,
      406,408,419-424,427,429,433-435,
      447,454,458
      trivial, 358
    ranking of, 324,325,360
    simple, 338,358,458
    triple, 188,196,329
    types, 324
    quadruple, 358,416
    quintuple, 416
    variable, 329,356,357,406,431,432,
      447,458
      conditionally penetrating, 357
      penetrating, 357,407
  dependent, 169,170,190,192,196,446
Folium of Descartes, 327,410
4-Leaf rose, see Rose
Galois, E., vi (1811-1832)
Geometric correlates, 181,182,194,254,
  255
Gergonne, J.D., xix (1771-1859)
Green's function, xii
Half-planes, non-equivalence of, 218
Harmonic mean eqs., 320,373,388
Harmonic parabolas, 303,458
Hierarchical ordering of curve-pole combinations by circumpolar symmetry,
  324,347
Hilbert, D., vi (1862-1943)
Hybrid polar-rectangular coordinate system, xi,3,17,136-143,213-216
Hyperbolas,
  asymptotes, 167,330,355,356
  bipolar, 96,122,130,162-164
  circumlinear self-inversion, 330,355,
    356
  circumpolar foci and focal loci, 324,
    386,388-391
  disparate, 238
    two-arm, 232,238-240,265,275,276
  equilateral, see Equilateral curves
  one-arm, 1,218,232
  partial, 232
  polar-circular, 264-267,275-280
  polar-linear, 230-240
  reflective two-arm, 238,242
  structural curves, 396,399
  structure rules, 320,321,332-338,347,
    383-403

INDEX 503

Hyperbolas with incongruent arms,
see Hyperbolas, disparate
Hyperbola axial-vertex inversion
  cubics, 403-412
  foci, 406-412
    asymptote-point, 407,410-412
    DP, 406,408-410
    loop vertex, 406-410
    variable, 406,408-412
  loop pole at $h = 2b/3$, 407-410,412
  structure rules, 407-412
Identification of curves, xviii,xix
Immediate closure, xi,xii,327,328,
  360,456,458
Incomplete linears, 222,458
Infinity,
  line at, 345,437,458
  point or points at, 164,167,345,
  437
Inherent structure of curves, viii,
  2,4-7,52,84,193,310,314,358,
  359,419,458
Inner segment, 458
Intercept eqs.,
  compound, 375,377,456
  simple, 84-87,319,320,377,381,406,
    424,429,447,461
    derivation, 319,353,354
    mixed-transcendental, 370,371,
      384,391,392,413,447,461
Intercept formats,
  compound, 371,372,377,392,394,398,
    456
  derivation, 371,372
  simple, 320,321,381,389,394,396,
    400,406,409,411,414,416,422,
    424,429,447
Intercept products,
  simple, 353-355,359,372,375,377,
    384,458,459
Intercept transforms, xxi,365,459
Intercept transform curves, xxi
Intersections at the polar pole, 50,
  51
Intrinsic eqs., viii,4,5
Inverse modular circle, 188
Inverses of central Cartesians, versus inverses of central conics,
  436-439
Inverse x-axis circle, 188,459
Inversion analysis, 327,328,459
Inversion superfamily, 328,459
Inversion transformation, vii,x-xii,
  xviii,xix,42,301,360,430,436-
  440,459

Inversion transformation,
  reciprocal inversion pole, 357,460
  self, see Self-inversion
Inversive plane, 345
Kappa curve, 47
Klein, F., vi (1849-1925)
Laplacian, xii
Latus rectum, 459
  and eccentricity, vii,xviii,28,443
Lemniscate of Bernoulli (see also Equilateral lemniscate), vii,xviii,47,
  50,75-81,164-170
Lemniscate of Gerone, 47
Lie, S., vi (1842-1899)
Limacons, 6,102,413-421,459
  bipolar, 96-99,154-165
    axial, 159,160
    axial foci, 158
    construction rules versus those of
      conics, 161-164
    general, 155-158
  circumpolar symmetry, 412-421
  comparisons with conics, 161
  construction from circles, 54-56,313,
    314
  covert linear focal locus, 324,358,
    410,414,418
  distance eqs., 157-160,198
  eccentricity, 161
    $e = 0.5$; 96,156,248,263,297,298
    $e = 0.75$; 156
    $e = 1$; 161,249,258,298
    $e = \sqrt{2}$; xviii,78-80,97,163,164,254,
      255,263,404,405,427,436
    $e = 2$; xviii,45,77,96,156,163,193,
      254,255,263,420,431
    $e = 4$; 250,251
    $e = 8$; 330
  eqs. about different foci, 414
  foci,
    $(a^2-b^2)/2b$, 86,158,357,358,410,415
    $-b/2$, 338,339,356-358,410,415
    DP, 158,357,358,410,413,415
    vertices, 410
  hybrid distance eqs., 198,247,249,
    251,294,295
  inversion correspondences with conics,
    356,357
  linear-circular, 294-301
  polar, 45,47,55,56,77-80
  polar-circular, 269-271,274
  polar-linear, 246-251
    focal-axial, 247-250
    non-focal-axial, 250,251
  rectangular eq., 47,155,413

Limacons,
  self-inversion, 156,248,249
  simple intercept eqs., 85
  structure rules, 313,314,347,410,
    417-423
    -b/2 focus, 338,339,410,420,421
    DP, 313,314,410,418,419
    focus of self-inversion, 410,
      419,420
    vertices, 410,421
  vertices, 417,418,421
Limacon term, 430
Linear-circular coordinate system,
  282-304,459
  degrees of eqs., 283
  diametric, 286-290
  tangent, 285,286,289,290,296,298,
    302
Line at infinity, 458
Lines,
  bipolar, xiii,102,103,107-110,
    121,130,133,134
  hybrid distance eq., 198
  intersecting, 103,108,231,232
  itersections with curves, 36
  linear-circular, 285,286
  of symmetry, 459
    as focal loci,324,373
    coordinate, 69,90,95,98,204,205,
      225,257,280,287,444,445,455
    equivalence with 0°-foci of
      self-inversion, 345,359,437
    intrinsic versus elective, 95,455
    loss and gain in inversions,
      345,359,438,447
  parallel line-pair, 103,130,219,
    223,233,234,321,333,347
  polar-circular, 278-280
  polar-linear, 229-234
  segments and rays,
    bipolar, 109,110
    polar-circular, 279,280
    polar-linear, 229-234
  structural curves, 317
  structure rules, 313,316-318,321,
    333,348
Lituus, 47
Lobachevsky, N.I., vi (1793-1856)
Lockwood, E.H., 9,93,462
Menaechmus, 13
Macmillan, R.H., 9,462
MAXIMS, inversion-mapping, 448-454,
  459

Modular,
  circle, 188,438,439,459
  distance, 428
  poles, 187-191,428,437,439,440,459
  segment, 381,382,459
Modularly-inverse poles,431,433,437,
  447
Multipolar coordinate systems, 90,303
Near-conics, 459
  Cartesian, 423,424,427
  near-circles, 421
  near-ellipses, 330,421
  near-lines, 420,427
Nephroid, 47
Nestling curves, 368,391
Newton, I., 41,92,93 (1642-1727)
Normal pedals, 322,459
Norwich spiral, 8
Oblique coordinates, 14,37,39
Omnidirectional circumlinear self-
  inversion, 355,356,461
Oresme, N., 13 (∼1323-1382)
Panton, A.W., 93
Parabola,
  135°-bilinear, 217
  bipolar, 122-139,217
    axial, 127-132
    focal, 129, 134-136
      comparative structural sim-
        plicities, 135,136
    focal axial, 129
    general, 122-127
    midline, 132-134
  compound, 215,225,227,292,293
  disparate,
    partial, 215,225-228
    two-arm, 217,284-287
  harmonic, 303
  hybrid distance eq., 135,198
  linear-circular, 217,284-287,290-294
    confocal, 285
  multipolar, 217
  partial, 215,225
  polar-circular, 217
  polar-linear, 217,223-229,231,233
  reflective dual and dual-hemi, 227,
    228,242
  reflective partial, 225-228
  simple intercept eq., 319,369,370
  simple intercept format, 320,381
  structural curves, 317,374-381
  structure rules, 319,320,333,336,
    343,347,369-383
  two-arm, 32-34,226,228

INDEX

Parallel line-pair, see Lines
Plücker, J., vi,xix,1 (1801-1868)
Polar coordinate system, ix,xvi,
    xvii,41-89,142,143
  asymmetries, 66-70
  history, 41
  plotting curves, 43,53-57,71-82
Polar-circular coordinate system,
    260-281, 460
  centered, 266
  degrees of eqs., 264
  incident, 266
  "master eqs." for conics, 269
Polar curves as resultants, 53-57
Polar eqs.,
  and structure rules, 51-53,312
  basic, 49,50,140,141,355,444
  degrees of, 43-46,49,50
  inventory of, 47
  standard, ix,44,46,68,71-82,312,
    443,461
Polar-exchange symmetry, 216,218,
    225,229,234,257,271,287,288,460
Polar-linear coordinate system, 6,
    17,88,213-259,446,447,460
  incident, 225,229,230,233,234,
    245,247,250,257
Polynomial distance relations and
    structural simplicity, 87,88
Poncelet, J.-V., v (1788-1867)
Probe, 460
"Punctuated" curves, 51
Quetelet, A., 93 (1796-1874)
Radicand, conditions on, see Focal
    conditions
Reciprocal inversion pole, 357,460
Rectangular coordinate system, xvi,
    9-40,142,143
  advantages, 34-39
  history, 13,14
  with undirected distances, 31-34
Roberts, Samuel, 93
Roberval, G.P. de, 14 (1602-1675)
Rose, 4-leaf, 330,337,347,350-353
  inverse, 334-338,347
Rotation eqs.,
  bipolar, 104,105
  rectangular, 16,17,23
RULE, 460
Russell, B., vi (1872-1970)
Salmon, G., 356,462 (1819-1904)
Schwartz, H.A., xii (1843-1921)

Self-inversion, 460,461
  $0°$, 79,80,167,176,189,195,298-301,
    345,359,406-408,419-424,427,429,
    433,434,437
  $90°$, vii,167,331,345,440,454
  $180°$, 162,176,189,300,301,345,424,
    427,429,433,434,437
  as special cases of congruent-
    inversion, 359
  circumlinear, 330,355,356
  foci of, see Foci and focal loci
  generalized, 331
  positive, 461
  negative, 461
  trivial foci of, 358
Simple intercept eqs., see Intercept eqs.
Simple intercept products, see Inter-
    cept products
Singularity of loci, xvi,15,17,35,461
"Single-point" coordinate systems, 82
Sommerfeld, A., xiv
Spirals, 41
  Archimedes, 47,345-348
  Fermat, 47
  logarithmic, 47,345-348
  Norwich, 8, 346
  reciprocal, 347
Steiner, J., vi (1796-1863)
Structural curves, 461
  and lines of symmetry, 341,342,379
  circle, 317,365-367
  complete, 378,380,399,
  ellipses, 317,385,387,392-396,401
  equilateral hyperbola, 334-340
  equilateral lemniscate, 330,332
  hyperbolas, 334-340,396-399
  limacons, 330,338,339
  linear Cartesians, 426,427
  lines, 316-318,338
  modular segments, 381,382,459
  parabolas, 317,374-381
  parabolic Cartesians, 423
  phase relations, 342,367
  symmetry properties of, 318,340-342,
    366,375,379
Structural Equation Geometry defined, 1-8
Structural eqs. defined, 2,4,461
Structure rules, 305-442,461
  and constructions, 313,314
  and exceptional loci, ix
  and polynomials, 379
  circumcurvilinear, 330,461
  circumpolar, 84,305-442,461
    about vertices, 334-337,362,377,
      398,409,434

Structure rules,
  circumpolar,
    α-, see point-specified (below)
    and inherent structure, viii,4-
      7,52,84,310,314,358-361
    and structural simplicity, ix,
      314,315,325,339,343,359,360,
      379,419,425,431
    angle-specified, 8,346,370-373,
      461
    Cartesian ovals, 347,422-427
    Cassinian ovals, 332,347
    central conics, 332-338,347,
      383-403
    complete, 34-342,371,382,400,461
    complex or imaginary solutions,
      367,402
    circle, 312,313,322,323,347,348,
      364-369
    cubics, 347,407-412
    degrees of, 332,336,343,344,390,
      396,410,414,443,447
    derivation of, 51-53,316-322
    general, viii,8,364-366,369,370
    hierarchical trees, vii,ix
    inventory of low-degree, 347
    limacons, 313,314,347
    lines, 313,316-318,321,333,348
    LR-vertices, 334-337
    most common, 331,332,345,346
    parabola, 319,320,343,347,369-
      383
    parabolic Cartesians, 347,422,423
    point-specified, 8,314,343-348
      379,390,402,418,461
    specific, viii,8,346,369,461
    trivial, 312,320,323,333,334,
      369,371,418,422,426,461
    variable probe-angle, 402,403
  comparisons of, 314,315
  in the complex domain, 319
  properties of, 318,319
  versus construction rules, 311,312
Structure-rule formats, 344,461
Subfocal ranking and conditions,129,
  133,134,142,144,150-154,158,177,
  183-185,192,196,205,207,336,337,
  362,379,408,409,412,430,431,445-
  447,460
Sylvester, J.J., 93 (1814-1897)
Symmetric functions, 341,342,419,461
Symmetry (of position and form), 9-
  31,462
  and structural simplicity, xiii,
    9-17,87,88,205-207,312,314,323,
    351,359,443

Symmetry (of position and form)
  and structural simplicity
    parabola versus central conics,
      351,352,362
    circumcurvilinear, 308-310,312,330,
      340,462
      identical reciprocal, 340
    circumlinear, 308-310,312,462
    circumpolar, 87,305-442,462
    classical, vii,9,10,58,91,306-308,
      358,359
    condition for parabolic Cartesians,
      185,186,272
    condition for vertex coincidence, 426
      431
    conics, comparative, 33,34
    generalized, 11,34,37,87,306,308,443
    mimics, 53-55,329-331,347,350,351,
      361,462
    mirror image, see lines of symmetry
    positional, 9-31,57-68,204,205,443
    subfocal conditions, 129,133,154
Synthesis of highly symmetrical curves,
  329,331
Tangent circles, 329-331,347,361
Trisectrix of Maclaurin, xviii,16,47,
  207,407-412,431
Tschirnhausen's cubic, 47,48,410
Undirected distances, 31-34,37-40,66-70
Uniqueness of identity, xv,xvi,35,462
Unicursal Cross Curve, 316-318,360
Valid locus defined, 462
Vectorial coordinates, see Bipolar Co-
  ordinate system
Vertex-coincidence symmetry condition,
  426,431
Vertex cubics, 188,403-412,433-435
Vertices, structure rules about, 334-
  337,362,377,398,409,434
Whewell, W., 4 (1794-1866)
Wilker, J.B., xii,462
Williamson, B., 93
Wolstenholme, J. 93
X-degrees, 347,350,352
XY-degrees, 347,350-352
Zwikker, C., 1,164,413,462